Energy, Environment, and Sustainability

Series Editor

Avinash Kumar Agarwal, Department of Mechanical Engineering, Indian Institute of Technology Kanpur, Kanpur, India

AIMS AND SCOPE

This books series publishes cutting edge monographs and professional books focused on all aspects of energy and environmental sustainability, especially as it relates to energy concerns. The Series is published in partnership with the International Society for Energy, Environment, and Sustainability. The books in these series are edited or authored by top researchers and professional across the globe. The series aims at publishing state-of-the-art research and development in areas including, but not limited to:

- Renewable Energy
- Alternative Fuels
- Engines and Locomotives
- Combustion and Propulsion
- Fossil Fuels
- Carbon Capture
- Control and Automation for Energy
- Environmental Pollution
- Waste Management
- Transportation Sustainability

Review Process

The proposal for each volume is reviewed by the main editor and/or the advisory board. The chapters in each volume are individually reviewed single blind by expert reviewers (at least four reviews per chapter) and the main editor.

Ethics Statement for this series can be found in the Springer standard guidelines here https://www.springer.com/us/authors-editors/journal-author/journal-author-helpdesk/before-you-start/before-you-start/1330#c14214

Andreas Bartl · Franz Winter · Stefan Pflügl ·
Avinash Kumar Agarwal
Editors

Alcohol Production Processes and Their Utilisation in the Transport Sector

Springer

Editors
Andreas Bartl
Institute of Chemical, Environmental and Bioscience Engineering
TU Wien
Vienna, Wien, Austria

Franz Winter
Institute of Chemical, Environmental and Bioscience Engineering
TU Wien
Vienna, Wien, Austria

Stefan Pflügl
Institute of Chemical, Environmental and Bioscience Engineering
TU Wien
Vienna, Wien, Austria

Avinash Kumar Agarwal
Engine Research Laboratory, Department of Mechanical Engineering
Indian Institute of Technology Kanpur
Kanpur, Uttar Pradesh, India

ISSN 2522-8366 ISSN 2522-8374 (electronic)
Energy, Environment, and Sustainability
ISBN 978-981-96-7383-4 ISBN 978-981-96-7384-1 (eBook)
https://doi.org/10.1007/978-981-96-7384-1

This Springer imprint is published by the registered company Springer Nature Singapore Pte Ltd.
The registered company address is: 152 Beach Road, #21-01/04 Gateway East, Singapore 189721, Singapore

Preface

Policymakers in various sectors are advocating the use of alcohol fuels. Methanol, ethanol, and butanol are the main candidates receiving attention. Innovative alcohol production technologies are being explored. Simultaneously, their use in internal combustion engines requires also minor hardware modifications. In parallel, challenges related to the supply and demand of alcohol fuels need to be addressed.

The International Society for Energy, Environment and Sustainability (ISEES) was founded at the Indian Institute of Technology Kanpur (IIT Kanpur), India, in January 2014 to spread knowledge and awareness and to catalyze research activities in the fields of energy, environment, sustainability, and combustion. Its goal is to contribute to clean, affordable, and secure energy resources and a sustainable environment for society, spread knowledge in the areas mentioned above, and create awareness about the environmental challenges the world is facing today. ISEES has taken a unique approach by breaking conventional silos of specializations (engineering, science, environment, agriculture, biotechnology, materials, fuels, etc.) to tackle the problems related to energy, environment, and sustainability holistically. This is evident from the participation of experts from all fields to resolve these issues. The ISEES is involved in various activities, such as conducting workshops, seminars, and conferences, in its areas of interests. The society also recognizes the outstanding works of young scientists, professionals, and engineers for their contributions by conferring awards in various categories.

The Eighth International Conference on "Sustainable Energy and Environmental Challenges" (VIII-SEEC) was organised under the auspices of ISEES from December 04 to 06, 2023, at the Malaviya National Institute of Technology Jaipur, India. This conference provided a platform for discussions between eminent scientists and engineers from several countries, including India, the USA, Brazil, Canada, France, Greece, and Africa. At this conference, eminent international speakers presented their views on energy, combustion, emissions, and alternative energy resources for sustainable development and a cleaner environment. The conference featured two high-profile plenary talks by Prof. G. D. Yadav, Institute of Chemical Technology, Mumbai, and Prof. Norma A. Alcantar, University of South Florida, USA.

The conference included 21 technical sessions on energy and environmental sustainability topics, including two plenary talks, 23 keynote talks, 35 invited talks from prominent scientists, and 95 contributed talks by students and researchers. The conference included technical sessions on alternative fuels for sustainable transport, advanced engines and fuels, sustainable technologies, new energy, biomass and gasification, advanced diagnostics and optical techniques, policies for environmental sustainability, solid waste remediation, wastewater treatment, air pollution and control, natural treatment system, pollution mitigation, environmental sustainability: challenges and priorities, energy and environment, sustainable processing of biomass, biofuels and biorefineries, environmental bioengineering, microbial processes, bioremediation, human health and environmental sustainability, and several assorted topics.

More than 250 participants and speakers attended the conference, where six ISEES books published by Springer (Singapore) in the dedicated series, "Energy, Environment and Sustainability", were released. This conference laid the roadmap for technology development, opportunities, and challenges in the energy, environment, and sustainability domain. These topics are relevant for the country and the world in the present context. We acknowledge the support from various funding agencies and organisations for the successful conduct of the VIII-SEEC, where the idea of these books originated. We would therefore like to express our gratitude to acknowledge Malaviya National Institute of Technology Jaipur (Special thanks to Dr. Nikhil Sharma); SERB, Government of India; Department of Scientific and Industrial Research (DSIR) (Special thanks to Dr. Vipin Shukla), Sponsors (Dynomerk Controls India Pvt. Ltd., Rebound Envirotech Pvt. Ltd., BlueDrop Enviro Pvt. Ltd.) and our publishing partner Springer (Special thanks to Ms. Swati Mehershi).

The editors would like to express their sincere gratitude to many authors worldwide for submitting their high-quality work on time and revising it appropriately at short notice. We want to express our special gratitude to our prolific reviewers : Dr. Bugra Arda Zincir, Dr. Dhruv Raj Karana, Dr. Sridhar Sahoo, and several potential researchers who reviewed chapters of this monograph and provided their valuable suggestions to improve them.

This book discusses the various aspects of alcohol production processes and their utilisation in the transport sector. This book explores the feedstocks for methanol, ethanol, and butanol production. This book includes a few chapters on alcohol production technologies, describing the various pathways for alcohol production. This book also contains a few chapters on alcohol-fuelled engines. A few chapters include recent results and are focused on current trends in alcohol production and engine technologies. This book is a valuable reference for those seeking to stay at the forefront of these evolving fields. We hope the book greatly interests the professionals, researchers, and students working on alcohol production and their utilisation.

Vienna, Austria — Andreas Bartl
Vienna, Austria — Franz Winter
Vienna, Austria — Stefan Pflügl
Kanpur, India — Avinash Kumar Agarwal

Contents

Part III Alcohol Fueled Engines

Editors and Contributors

About the Editors

Dr. Andreas Bartl studied Technical Chemistry and Chemical Process Engineering at TU Wien, where he also received his Ph.D. After four years working in the pyrotechnics research and development and production industry, he returned to TU Wien. For almost 20 years, he has been at the Institute of Chemical, Environmental, and Bioscience Engineering within the Research Unit of "Mechanical Process Engineering and Clean Air Technology". In 2012, he received the conferment of the right to lecture and the academic title "Privatdozent." Since returning to TU Wien, Andreas has worked in the broader field of fiber technology. In his research group, the dry grinding of (short) fibers and fiber characterization have been the focus of research activities. Andreas is a member of national and international organizations in the fiber processing chain (e.g. Austrian Fibre Institute, Global Fibre Congress) and waste management (e.g., IWWG, CEC4Europe).

Dr. Franz Winter has published more than 100 peer-reviewed journal publications, more than 200 reviewed international conference papers, more than 90 invited lectures, and 40 book chapters and patents. Co-editor of the Handbook of Combustion (5 volumes, ca. 3000 pages), author of the book Combustion. Chairman of the 19th International Conference on Fluidized Bed Combustion 2006, chairman of the 4th European Combustion Meeting 2009, chairman of the International Conference on Fuel Quality 2012. Chairman 2000–2002 and 2011–2013 and Vice-chairman of the International Energy Agency (IEA)—Fluidized Bed Conversion (FBC) group. Co-editor of *Journal of Energy and Environmental Sustainability* (JEES). Since 2017, Scientific Chair of the IRRC International Waste-to-Energy Conference takes place yearly in Vienna.

Dr. Stefan Pflügl has 10 years of experience in the area of developing sustainable bioprocess solutions towards a biobased industry. Among others, he was part of a team that developed a bioprocess for valorizing crude glycerol from biodiesel production, which will be commercialized as a pilot plant by the involved industrial partner in 2020. The core competence of Stefan Pflügl is developing and optimizing bioprocesses to utilize industrial residual streams (e.g., gas streams containing CO_2, residual streams from the pulp and paper industry, molasses, etc.) as a substrate for producing value-added chemicals. Furthermore, Dr. Pflügl has expertise in the field of microbial strain development and improvement. Since 2016, he has been a team leader for “Sustainable Bioprocess Solutions” in the “Biochemical Engineering” research area at TU Wien, responsible for 6 Ph.D. students, 2 M.Sc. students, and 1 technician.

Prof. Avinash Kumar Agarwal is currently Director at IIT Jodhpur. He obtained his Undergraduate Degree in Mechanical Engineering (1994) from Malaviya Regional Engineering College, Jaipur, and his M.Tech. (Energy, 1996) and Ph.D. (Energy, 1999) from the Indian Institute of Technology (IIT) Delhi. After his Post-Doctoral Fellowship (1999–2001) at the ERC, UW, Madison, USA, he returned to India in 2001 and joined IIT Kanpur. Prof. Agarwal has established a state-of-the-art "Engine Research Laboratory" at IIT Kanpur. Prof. Agarwal is interested in researching IC engines, combustion, conventional and alternative fuels, optical diagnostics, laser ignition, HCCI, particulates and emission control, and large bore engines. Prof. Agarwal has published over 550+ peer-reviewed international journal and conference papers, 65 edited books, and 129 book chapters, attracting 18,300+ Scopus and 27,400+ Google Scholar citations. For his outstanding contributions, Prof. Agarwal is conferred upon Sir J. C. Bose National Fellowship (2019) by SERB, Prestigious Shanti Swarup Bhatnagar Prize (2016) in Engineering Sciences, Rajib Goyal Prize in Physical Sciences (2015); NASI-Reliance Industries Platinum Jubilee Award (2012); INAE Silver Jubilee Young Engineer Award (2012); INSA Young Scientist Award (2007); UICT Young Scientist Award (2007); INAE Young Engineer Award (2005); Devendra Shukla Research Fellowship (2009–2012), Poonam and Prabhu Goyal Endowed Chair Professorship (2013–2016), SBI Endowed Chair Professorship (2018–2021, 2022–2025) at IIT Kanpur. He is an elected Fellow of TWAS, SAE, ASME, INAE, ISEES, RSC, NASI, WSSET, AAAS and Combustion Institute USA.

Contributors

Avinash Kumar Agarwal Engine Research Laboratory, Department of Mechanical Engineering, Indian Institute of Technology Kanpur, Kanpur, India

Samareh Ahmadi School of Mechanical Engineering, Iran University of Science and Technology, Tehran, Iran

Madhava Anil Kumar Centre for Rural and Entrepreneurship Development, National Institute of Technical Teachers Training and Research, Chennai, Tamilnadu, India

S. Anil Lal Department of Mechanical Engineering, National Institute of Technology Sikkim, Ravangla, Sikkim, India

Anbarasan Baluchamy Department of Mechanical Engineering, PSNA College of Engineering and Technology, Dindigul, Tamil Nadu, India

Andreas Bartl Institute of Chemical, Environmental and Bioscience Engineering, TU Wien, Vienna, Austria

P. Baskaralingam Department of Chemistry, College of Engineering Guindy, Anna University, Chennai, Tamilnadu, India

Jeny Rachel Biju Environmental Management Laboratory, Department of Applied Science and Technology, Alagappa College of Technology, Anna University, Chennai, Tamilnadu, India

Rohit Das Amity Institute of Biotechnology, Amity University Kolkata, Kadampukur Village, Rajarhat, Newtown, Kolkata, West Bengal, India

Ayat Gharehghani School of Mechanical Engineering, Iran University of Science and Technology, Tehran, Iran

Shiladitya Ghosh Amity Institute of Biotechnology, Amity University Kolkata, Kadampukur Village, Rajarhat, Newtown, Kolkata, West Bengal, India

Krishnangshu Guha Engine Research Laboratory, Department of Mechanical Engineering, Indian Institute of Technology, Kanpur, India

Jai Gopal Gupta Department of Mechanical Engineering, National Institute of Technology Sikkim, Ravangla, Sikkim, India

T. J. Jacklien Emema Rose Environmental Management Laboratory, Department of Applied Science and Technology, Alagappa College of Technology, Anna University, Chennai, Tamilnadu, India

Aditya Kashyap Department of Chemical Engineering and Biochemical Engineering, Rajiv Gandhi Institute of Petroleum Technology, Jais, Amethi, India

Nikolaos C. Kokkinos Hephaestus Laboratory, Department of Chemistry, School of Sciences, Democritus University of Thrace, Kavala, Greece;
Petroleum Institute, Democritus University of Thrace, Kavala, Greece

Rakesh Kumar Department of Chemical Engineering and Biochemical Engineering, Rajiv Gandhi Institute of Petroleum Technology, Jais, Amethi, India

Neeraj Kumbhakarna Department of Mechanical Engineering, Indian Institute of Technology Bombay, Maharashtra, India

Sudhir Kumar Pathak Research and Development Division, PLUSS Advanced Technologies Limited, Bawal, Rewari, Haryana, India

Stefan Pflügl Institute of Chemical, Environmental and Bioscience Engineering, TU Wien, Vienna, Austria

Antti Pitkäoja Lappeenranta-Lahti University of Technology LUT, LUT School of Energy Systems, Lappeenranta, Finland

S. Prabakaran Department of Mechanical Engineering, Indian Institute of Technology Bombay, Maharashtra, India

Ajeet Kumar Prajapati Department of Chemical Engineering and Biochemical Engineering, Rajiv Gandhi Institute of Petroleum Technology, Jais, Amethi, India

Alexandros Psalidas Department of Chemistry, School of Sciences, Democritus University of Thrace, Kavala, Greece

Reetu Raj Department of Mechanical Engineering, Indian Institute of Technology Bombay, Maharashtra, India

Md Shadab Reza Department of Mechanical Engineering, National Institute of Technology Sikkim, Ravangla, Sikkim, India

Jabraeil Ahbabi Saray School of Mechanical Engineering, Iran University of Science and Technology, Tehran, Iran

Pravin Omprakash Sharma Mechanical Engineering Department, Dr. D. Y. Patil Institute of Technology, Pimpri, Pune, Maharashtra, India

Amirali Shirzad School of Mechanical Engineering, Iran University of Science and Technology, Tehran, Iran

Har Mohan Singh Department of Physics and Environmental Sciences, Sharda School of Engineering & Science, Sharda University, Uttar Pradesh, Greater Noida, India

Jason Thamizhakaran Stanley Environmental Management Laboratory, Department of Applied Science and Technology, Alagappa College of Technology, Anna University, Chennai, Tamilnadu, India

Sivanesan Subramanian Environmental Management Laboratory, Department of Applied Science and Technology, Alagappa College of Technology, Anna University, Chennai, Tamilnadu, India

Tan Thakur Amity Institute of Biotechnology, Amity University Kolkata, Kadampukur Village, Rajarhat, Newtown, Kolkata, West Bengal, India

Amutha Thanarasu St. Joseph College of Engineering, Chennai, Tamilnadu, India

Franz Winter Institute of Chemical, Environmental and Bioscience Engineering, TU Wien, Vienna, Austria

Part I
General

Chapter 1
Introduction to Alcohol Production Processes and Their Utilization in the Transport Sector

Andreas Bartl, Franz Winter, Stefan Pflügl, and Avinash Kumar Agarwal

Abstract Reducing greenhouse gas emissions can be achieved by replacing fossil fuels in the transport sector. Fossil energy resources are finite and rapidly depleting. In addition, fossil energy use has negative impacts on the atmosphere. The use of alternative fuels has become a necessity. Representative members of the alcohol family include methanol, ethanol, butanol, and pentanol. Among various options, alcohols represents a promising solution, as their physicochemical properties are in several aspects superior to conventional diesel and gasoline. From a sustainability viewpoint, alcohols are advantageous because their inherent oxygen reduces exhaust emissions. The book reviews mechanical, chemical, and biochemical processes for alcohol production and examines current developments, barriers, prospects, and utilization pathways.

Keywords Methanol · Ethanol · Butanol · Production technologies · Engine

Climate change mitigation requires a reduction in global carbon dioxide emissions, which requires moving away from fossil fuels. Because fossil fuel resources are finite, they are rapidly depleating, with serious impacts on ecosystems and the environment. This underscores the urgency of developing alternative fuels, such as alcohol, biodiesel, vegetable oils, and gaseous fuels like hydrogen, natural gas, and biogas, that can be produced from locally accessible resources. Alcohol fuels stand out from the rest because they have performance, ignition, and combustion characteristics that are similar to those of conventional petroleum-derived fuels. Alcohols can be used directly in internal combustion engines as oxygenated fuels or biofuels. Alcohol fuels are a good substitute for traditional fuels and help create a more sustainable

A. Bartl · F. Winter · S. Pflügl
Institute of Chemical, Environmental and Bioscience Engineering, TU Wien, Vienna, Austria

A. K. Agarwal (✉)
Engine Research Laboratory, Department of Mechanical Engineering, Indian Institute of Technology Kanpur, Kanpur, India
e-mail: akag@iitk.ac.in

A. Bartl et al. (eds.), *Alcohol Production Processes and Their Utilisation in the Transport Sector*, Energy, Environment, and Sustainability,
https://doi.org/10.1007/978-981-96-7384-1_1

energy future because their inherent oxygen content improves engine performance and lowers exhaust emissions.

The first section of this book is General, which contains one chapter entitled Introduction to Alcohol Production Processes and their Utilization in the Transport Sector, which introduces the monograph.

The second section of this book is about Alcohol Production Technologies, which contains six chapters. The first chapter explores feedstocks for methanol, ethanol, and butanol production. This chapter examines the potential feedstocks for producing alcohols such as methanol, ethanol, and butanol, highlighting their significance in the context of sustainable energy solutions. The chapter then delves into the importance of sustainable feedstock for alcohol production, examining the environmental impact, economic benefits, and long-term sustainability. This comprehensive overview serves as a crucial resource for researchers and practitioners seeking to enhance the sustainability and efficiency of alcohol production. The second chapter is Towards Carbon–Neutral Methanol: Emerging Technologies, Key Challenges, and the Pathway to Net-Zero Emissions. Recent advancements in membrane-based methanol synthesis and purifying technologies are highlighted in the review, covering innovative membrane components such as hybrid frameworks, ceramics, and polymers. The present obstacles and potential outcomes for achieving environmentally friendly methanol production on a large scale are examined. The third chapter is about Processes Used in the Production of Methanol. This chapter delves into the various processes employed for methanol synthesis and production. The importance of process parameters such as temperature, pressure, and gas composition, has been investigated in methanol synthesis. The fourth chapter is Methanol Production Processes in the Transport Sector. The biochemical processes of methane to methanol and carbon dioxide to methanol are also remarkable, as they utilize microorganisms and enzymes as catalysts under mild conditions. The fifth chapter is about biomass gasification and the design of a gasification island for a methanol synthesis plant. This chapter describes syngas production for methanol synthesis by biomass thermochemical conversion. The chapter focuses on the biomass gasification process and gasification island design for methanol production. The overall process design of a methanol synthesis plant depends on the selected gasification technology. This chapter describes the biomass gasification process and reviews biomass gasification processes suitable for methanol synthesis. The sixth chapter is about bioethanol production from second-generation feedstocks for the transportation sector. Biochemical production processes are gaining popularity due to their potential waste-to-energy and eco-friendly attributes. This chapter presents the current state of the biochemical ethanol production processes and strategies for improvement to facilitate its implementation as a dependable transportation fuel.

The book's third section is alcohol-fueled engines, which contains six chapters. The book's first chapter is about alcohol fuels for sustainable energy solutions. Alcohol fuels stand out from the rest because they have performance, ignition, and combustion characteristics that are quite like conventional fuels derived from petroleum. The second chapter of the book is about alcohol for fueling engines. Ethanol and methanol are the primary alcohol fuels studied for this purpose. Ethanol is valued for its higher octane number, allowing for increased compression ratios and improved engine thermal efficiency. Its oxygen content promotes complete combustion, reducing carbon monoxide emissions and unburned hydrocarbons. On the other hand, methanol is noted for its high-octane rating and potential for use in various engine types. Both ethanol and methanol are produced from renewable sources, supporting their role as sustainable alternatives to traditional fossil fuels. The combustion properties of alcohol fuels, such as their energy density and viscosity compared to gasoline and diesel, are also examined. The book's third chapter concerns current challenges and state-of-the-art reviews of methanol-fueled engines. This review synthesizes recent research and technological advancements in methanol-fueled ICEs, addressing critical challenges and highlighting state-of-the-art developments. Significant issues such as fuel properties, combustion characteristics, and emissions control are analyzed. This review explores engine modification advancements to enhance efficiency and emission reduction using methanol as a fuel. The book's fourth chapter comprehensively reviews regulated, unregulated, and particulate emissions in methanol-powered engines. Methanol is among the most widely used fuels in the alcohol family due to its several advantageous properties, including a high-octane number, high heat of vaporization, high flame speed, a low stoichiometric air–fuel ratio, etc., making it a suitable alternative fuel for engines. With a high oxygen-to-carbon and hydrogen-to-carbon ratio, methanol generally produces less soot and smoke emissions. The main challenge in applying methanol as a fuel is the unregulated emissions such as unburned methanol, formaldehyde, benzene, toluene, and Soluble Organic Fraction (SOF). This chapter includes an in-depth review of methanol-fueled engines, regulated, unregulated, and particulate characteristics. The book's fifth chapter is about innovative fuel injection strategies for enhancing methanol utilization in internal combustion engines. The chapter provides a comprehensive examination of the different fuel injection methodologies employed to optimize the use of methanol in internal combustion engines. This chapter delves into multiple strategies, including blending, port fuel injection, high-pressure direct injection, and emulsions, each tailored to maximize the advantages of methanol while addressing their inherent challenges. The chapter also explores emulsions, where methanol is combined with water or other fluids to form a stable mixture, resulting in a more complete combustion process and significant emission reductions. This chapter also aims to thoroughly review different fuel injection strategies, such as adjusting injection pressure, altering injection timing, and implementing multiple injections to enhance engine performance and control emissions. The sixth chapter of the book is Bioethanol: Production Processes and Strategies for Utilization in Diesel Engines. The chapter starts with a discussion on the physicochemical properties of ethanol relevant to engine application. It is followed by a description of various processes to produce ethanol and its classification based on the feedstock used. There are four main strategies for using ethanol in CI engines: fumigation, blending by adding cosolvents, blending by emulsification, and dual fuel mode. The chapter addresses the technical challenges of blending ethanol with diesel and discusses the potential solutions. The safety aspects of using ethanol as a fuel have also been brought into the scope of discussion.

This monograph is divided into three different sections: (I) General, (II) Alcohol Production Technologies, and (III) Alcohol Fueled Engines. Specific chapters covered in the manuscript include:

- Introduction to Alcohol Production Processes and Their Utilization in the Transport Sector.
- Exploring Feedstocks for Methanol, Ethanol, and Butanol Production.
- Towards Carbon–Neutral Methanol: Emerging Technologies, Key Challenges, and the Pathway to Net-Zero Emissions.
- Processes Used in the Production of Methanol.
- Methanol Production Processes in the Transport Sector.
- Biomass Gasification and Design of Gasification Island for Methanol Synthesis Plant.
- Production of Bioethanol from Second-Generation Feedstocks for the Transportation Sector.
- Alcohol Fuels for Sustainable Energy Solutions.
- Alcohol for Fuelling IC Engines.
- Current Challenges and State-of-the-Art Reviews of Methanol-Fueled ICEs.
- Comprehensive Review of Regulated, Unregulated and Particulate Emissions in Methanol-Powered Engines.
- Innovative Fuel Injection Strategies for Enhancing Methanol Utilization in Internal Combustion Engines.
- Bioethanol: Production Processes and Strategies for Utilization in Diesel Engines.

Part II
Alcohol Production Technologies

Chapter 2
Exploring Feedstocks for Methanol, Ethanol, and Butanol Production

Ayat Gharehghani, Jabraeil Ahbabi Saray, and Amirali Shirzad

Abstract This chapter explores the potential feedstocks for the production of alcohols—methanol, ethanol, and butanol—highlighting their significance in the context of sustainable energy solutions. It begins with a brief introduction to these alcohols, elucidating their roles and applications. The chapter then delves into the importance of sustainable feedstock for alcohol production, examining the environmental impact, economic benefits, and long-term sustainability. A review of conventional feedstocks and procedures provides a foundation for understanding current practices. The chapter categorizes various renewable feedstocks, discussing their viability and potential advantages. Key considerations for feedstock selection are outlined to guide optimal choices for alcohol production. Finally, future prospects and challenges in the field are addressed, with an emphasis on the technological and regulatory developments required to advance the industry. This comprehensive overview serves as a crucial resource for researchers and practitioners seeking to enhance the sustainability and efficiency of alcohol production.

Keywords Methanol feedstock · Ethanol feedstock · Butanol feedstock · Renewable energy · Biofuel synthesis

2.1 Brief Introduction to Alcohols (Methanol, Ethanol, and Butanol)

Alcohols are a diverse and essential class of organic compounds characterized by the presence of one or more hydroxyl (–OH) groups attached to carbon atoms, and they play a critical role in both industry and daily life. Among these, methanol, ethanol, and butanol stand out for their widespread applications and unique properties (Cardona et al. 2005).

A. Gharehghani (✉) · J. A. Saray · A. Shirzad
School of Mechanical Engineering, Iran University of Science and Technology, Tehran, Iran
e-mail: ayat_gharehghani@iust.ac.ir

A. Bartl et al. (eds.), *Alcohol Production Processes and Their Utilisation in the Transport Sector*, Energy, Environment, and Sustainability,
https://doi.org/10.1007/978-981-96-7384-1_2

Alcohols, including methanol, ethanol, and butanol, are highly versatile compounds extensively utilized as biofuels and in various industrial applications. Each of these alcohols possesses distinct properties and serves unique functions across different industries. Together, methanol, ethanol, and butanol illustrate the diversity within the alcohol family. This chapter will explore their chemical structures, properties, production methods, and applications, while highlighting their importance in both everyday life and advanced industrial processes (Saray et al. 2024).

These alcohols not only offer renewable energy solutions but also have crucial roles in the chemical industry, aiding in the creation of many vital products. As global demand for sustainable, eco-friendly alternatives rises, the significance of methanol, ethanol, and butanol as both biofuels and industrial chemicals becomes increasingly important. Their contribution to reducing environmental impact and fostering sustainability highlights their growing importance (Praveena et al. 2024; Shukla et al. 2021). The development and use of these alcohols as biofuels are vital in promoting a more sustainable and environmentally friendly energy landscape, minimizing our reliance on non-renewable resources and lowering greenhouse gas emissions. The unique characteristics of methanol, ethanol, and butanol make them invaluable in both industrial and energy sectors.

2.2 Significance of Sustainable Feedstock for Alcohol Production

Environmental Impact

Sustainable feedstocks, including biomass and agricultural residues, compared to fossil fuels significantly reduce greenhouse gas emissions. The carbon dioxide absorbed by these plants during growth compensates for the emissions generated during alcohol combustion, resulting in a reduced net carbon footprint. Additionally, using waste materials such as crop residues, forest residues, and municipal solid waste for alcohol production helps manage waste and decreases landfill usage (Ge et al. 2018). This approach not only lessens the environmental impact of waste disposal but also generates additional revenue for waste management. Furthermore, sustainable feedstock production often incorporates practices that enhance biodiversity and land use management (Singh et al. 2021). The environmental effects of biofuel production, such as changes in land use and water usage, need careful management to ensure sustainability. Effective policies and incentives will be crucial in advancing biofuel adoption (Segaran and Guruswami 2023).

Techniques like crop rotation, intercropping, and cover cropping improve soil health, reduce erosion, and increase biodiversity. These methods contribute to a more sustainable and environmentally friendly approach to biofuel production, promoting a healthier ecosystem and better resource management (Anto et al. 2020).

Economic Benefits

By utilizing renewable feedstocks sourced locally, nations can lessen their reliance on imported fossil fuels, thereby bolstering energy security and economic autonomy. The cultivation and processing of sustainable feedstocks stimulate rural economies by creating jobs and increasing farmer income. These activities also help enhance livelihoods and promote economic stability in rural regions (Vasco-Correa and Shah 2019). The future of biofuels looks promising, but it faces several challenges. Major concerns include the economic feasibility of production methods, scalability, and the establishment of distribution and usage infrastructure. Technological advancements and economies of scale are reducing the cost of producing alcohols from sustainable feedstocks. This makes biofuels more competitively priced compared to traditional fossil fuels, thereby promoting their wider adoption (Kendir Çakmak and Ugurlu 2020). Measures like subsidies, tax benefits, and blending mandates can stimulate investment and innovation. Moreover, collaborations between public and private sectors can expedite the development and commercialization of advanced biofuel technologies.

Long-Term Sustainability

Sustainable feedstocks typically optimize resource utilization, encompassing water and land efficiency. Techniques such as precision agriculture, agroforestry, and cultivating drought-resistant crops play crucial roles in fostering enduring sustainability (Ferdeş et al. 2020). Embracing sustainable feedstocks further advances circular economies by promoting material recycling and reuse in production. This approach helps curb waste, preserve resources, and establish a self-sustaining cycle for long-term viability (Li et al. 2022b).

In essence, transitioning alcohol production from fossil-based to sustainable feedstocks—like those used for methanol, ethanol, and butanol—is critical. This shift not only reduces environmental impact but also boosts economic benefits, drives technological progress, and gains regulatory backing. Long-term sustainability in these practices promotes a greener future, paving the way for a more sustainable, circular economy (Delrue et al. 2012). Alcohol production, traditionally reliant on fossil resources, is now shifting toward sustainable, renewable feedstocks. This shift is vital for addressing environmental concerns, reducing carbon footprints, and ensuring energy security. The importance of sustainable feedstocks in alcohol production is underscored by several crucial factors (Pereira et al. 2012).

This chapter delves into diverse feedstocks suitable for methanol, ethanol, and butanol production, detailing their origins, accessibility, and conversion processes into valuable alcohols. Insight into these potential feedstocks is pivotal for advancing alcohol production and fostering their integration, thereby facilitating the shift toward a more sustainable, circular economy.

2.3 Conventional Feedstock and Procedures for Alcohols Production

The historical production of alcohols like methanol, ethanol, and butanol has traditionally depended on conventional feedstocks sourced from fossil fuels and specific agricultural materials. Familiarity with these traditional feedstocks forms a basis for recognizing the progress and advantages associated with transitioning to more environmentally sustainable alternatives. This segment examines the principal conventional feedstocks utilized in alcohol production and their respective manufacturing procedures.

2.3.1 Conventional Feedstock and Procedures for Methanol Production

Methanol (CH_3OH), the simplest of these alcohols, is predominantly synthesized from natural gas, although it can also be produced from biomass, municipal waste, and carbon dioxide. It is an essential feedstock in the manufacture of formaldehyde, acetic acid, and numerous other chemicals. Additionally, methanol is becoming increasingly popular as an alternative fuel because of its high-octane rating and ability to reduce emissions when used in internal combustion engines (Kendir Çakmak and Ugurlu 2020; Ferdeş et al. 2020).

Traditionally, methanol has relied on natural gas for its production, chosen for its widespread availability and economic feasibility. The production process starts with steam reforming natural gas to create synthesis gas, comprising hydrogen and carbon monoxide. This gas mixture undergoes catalytic conversion into methanol, forming the basis of its manufacturing process (Wang et al. 2010). As shown in Fig. 2.1, worldwide methanol production is considerably high, and its demand was anticipated to grow.

The current methods for producing methanol rely on two main catalytic steps: creating synthesis gas and then converting this syngas into methanol. The key raw material for producing synthesis gas include natural gas, methane from petroleum, shale gas, coal, and biomass (Galadima and Muraza 2015). Reforming technologies, such as dry and steam reforming, effectively produce synthesis gas from fossil fuels. On the other hand, biomass can be processed through pyrolysis and gasification to generate synthesis gas (Bruijnincx and Weckhuysen 2013).

The growing global emphasis on the production and utilization of shale gas indicates a future shift in syngas production toward greater reliance on shale gas as a primary feedstock (King 2010). Currently, the extraction of shale gas is predominantly carried out in Canada, the United States and other North American nations. Despite significant technically recoverable reserves found in the Asia Pacific region, commercialization efforts there remain relatively limited. There are clear signs that

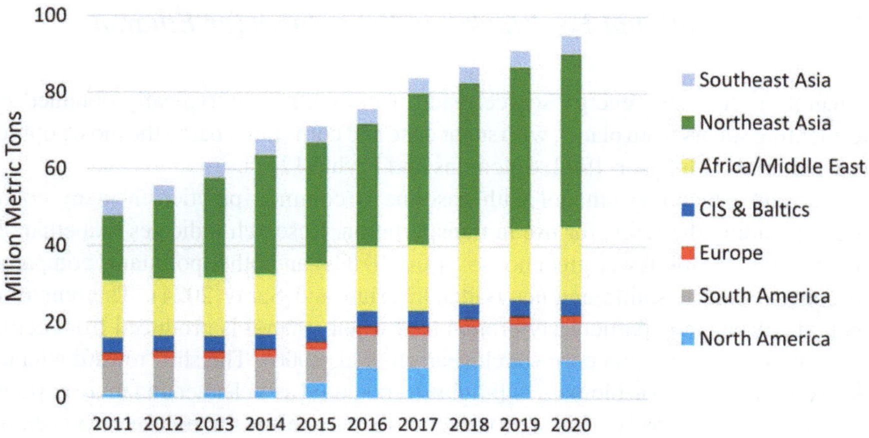

Fig. 2.1 Worldwide methanol output by 2020 (Tabibian and Sharifzadeh 2023)

shale gas is increasingly replacing conventional natural gas in these North American regions. Production is expected to continue rising in the coming years (Ehlinger et al. 2014). Among the different market sectors, the transportation industry is recognized as the fastest-growing market for shale gas in the near future. Additionally, the overall global market for shale gas is projected to experience a growth rate of 5.3%, reaching an estimated value of at least 9.19 billion US dollars by the year 2020. This anticipated growth underscores the expanding role of shale gas in the global energy landscape and its potential impact on future energy markets (Galadima and Muraza 2015).

Methanol, initially produced in the nineteenth century through wood alcohol extraction during pyrolysis, became commercially viable in the early twentieth century (Spath and Dayton 2003). This development resulted from extensive research and development efforts, leading to the modern processes used today for its large-scale production (Sikarwar et al. 2017).

Natural gas reforming is traditionally the primary method for methanol synthesis. However, advancements in biomass gasification (BG) have broadened the scope of biomass feedstocks that can be used to produce syngas, which is subsequently converted into methanol. The syngas derived from BG contains various chemical constituents, such as water, alcohols, and ketones. It undergoes methanol synthesis in the presence of a catalyst within a dedicated reactor (Spath and Dayton 2003). The resulting crude methanol is a complex blend of chemical species and undergoes separation in a distillation unit to isolate pure methanol. These developments in BG offer a promising alternative pathway for methanol production. They utilize renewable biomass resources effectively, alongside conventional natural gas feedstocks (Holmgren et al. 2014).

2.3.2 Conventional Feedstock and Procedures for Ethanol

Ethanol, a renewable energy source made from biomass, is typically obtained by fermenting sugars from plants, with sugar cane and corn starch being the most popular raw materials (Morrison 1992; Solomons and Fryhle 1277).

Currently, blending ethanol with gasoline is common practice in many countries, including the USA, for use in transportation. Research indicates that ethanol-fueled engines emit fewer greenhouse gases (GHGs) and other pollutants compared to conventional gasoline engines (Ghandehariun and Saray 2024). This environmental advantage is particularly pronounced when ethanol is produced from cellulosic biomass rather than corn starch (Farrell et al. 2006). The shift toward ethanol derived from cellulosic biomass highlights its potential as a cleaner and more sustainable fuel option globally. This approach contributes to efforts to reduce carbon emissions and improve air quality (Ng et al. 1983). A major environmental benefit of ethanol is its ability to support a closed carbon cycle, as some of the CO_2 emitted in its production and use, reabsorbed by plants (Gerbrandt et al. 2016; Ghandehariun et al. 2024). Ethanol serves as a precursor for various fuels, chemicals, and polymers (Palsson et al. 1981). Additionally, it holds promise as a source for producing green hydrogen to fuel cells (Subramani and Song 2007). Nonetheless, the social, environmental, and economic implications of ethanol use remain subjects of ongoing scrutiny and evaluation (Deluga et al. 2004).

Bioethanol is industrially synthesized through two main routes. The first method involves extracting sugars from energy crops or food such as corn or sugarcane, followed by fermentation to produce ethanol (Kadam and McMillan 2003). The second method utilizes ethylene derived from petroleum, which is hydrated over a solid acid catalyst to yield ethanol. Biochemical ethanol production has become a well-established conventional process. Table 2.1 provides a global overview of countries producing substantial amounts of ethanol using energy crops as feedstock via traditional methods. This highlights the widespread adoption and significance of bioethanol in the global energy sector (Sikarwar et al. 2017). The United States (US) leads in fuel ethanol production, contributing 54% to the global output, followed by Brazil, Europe, China, Canada, Thailand, India, and Argentina (Fig. 2.2). While Brazil primarily uses sugarcane for ethanol production, other countries mainly rely on starch-based crops (Li et al. 2022a).

Selecting the most suitable feedstocks for ethanol production is largely determined by local conditions. North America and Europe have primarily based their ethanol industry on starchy materials. This is due to their specific agricultural conditions, which are not suitable for cultivating sugar cane, the highest-yielding feedstock (Shapouri et al. 2003). The utilization of starchy crops, especially corn, for bioethanol production has led to intense debate over the suitability of these raw material. This debate focuses on the significant energy input required for their production. Critics argue that the energy required to produce ethanol from corn may offset its benefits, raising concerns about its sustainability and efficiency (Wang et al. 1999). Therefore,

Table 2.1 Nations with significant bioethanol output from various energy crops (Speight 2011)

Country	Bioethanol feedstocks	Bioethanol yields (liters/hectare)
USA	Corn, 98%	3770
	Sorghum, 2%	1365
Brazil	Sugarcane, 100%	6641
EU-27	Wheat, 48%	1702
	Sugar beet, 29%	5145
China	Corn, 70%	2011
	Wheat, 30%	1730
Canada	Corn, 70%	3460
	Wheat, 30%	1075

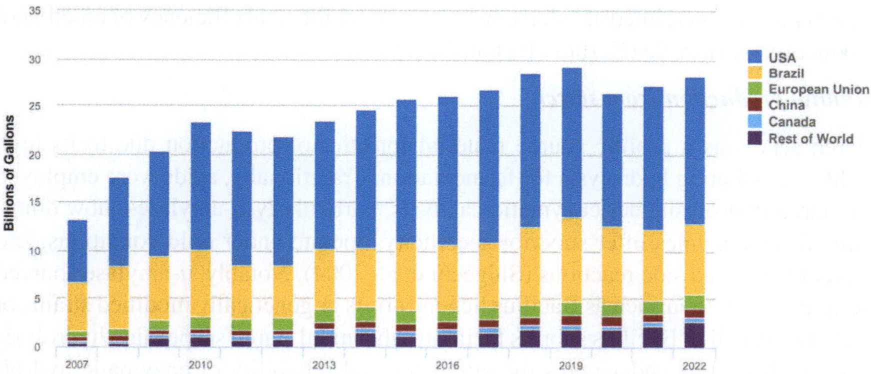

Fig. 2.2 Global fuel ethanol production by country in 2020 (Li et al. 2022a)

finding the most energy-efficient and sustainable feedstock that matches regional agricultural capabilities remains a critical discussion.

A comprehensive environmental and economic assessment of various feedstocks is crucial for selecting the most suitable raw material for ethanol production in each scenario. Utilizing simulation tools based on accurate data from current ethanol production facilities, mathematical models or pilot plants is a valuable method for such evaluations.

Ethanol production from sugar

Ethanol production relies heavily on sugar cane as a primary feedstock, typically in the form of either cane juice or molasses, a by-product from sugar mills. In Brazil, a significant portion of ethanol, around 79%, is produced directly from fresh sugar cane juice, with the remainder coming from cane molasses (Wilkie et al. 2000). In contrast, in India, ethanol production primarily utilizes sugar cane molasses, as cane juice is not commonly employed for this purpose. Beet molasses also serve as a source of fermentable sugars for ethanologenic fermentation (Ghosh and Ghose 2003).

The yeast *Saccharomyces cerevisiae* is widely favored for ethanol production because it can hydrolyze cane sucrose into glucose and fructose. Both of these are easily assimilated hexoses. Aeration is crucial for the growth and ethanol production of *S. cerevisiae.* While the yeast can function anaerobically, small amounts of oxygen are necessary for synthesizing essential substances such as fatty acids and sterols (Narendranath et al. 2000). Oxygen levels can be increased by adding chemicals such as urea hydrogen peroxide (carbamide peroxide). This addition improves ethanol production efficiency and helps reduce bacterial contamination. These insights underscore the multifaceted approaches and technological advancements employed to optimize ethanol production from diverse feedstocks (Goyes and Bolaños 2005).They also highlight regional variations and preferences in ethanol production methods. These differences reflect local agricultural practices and technological innovations aimed at enhancing the sustainability and efficiency of bioethanol production worldwide (Gerbrandt et al. 2016).

Ethanol production from starch

Starch serves as a prolific source material for ethanol production due to its high yield, necessitating hydrolysis for fermentation. Traditionally, acids were employed for starch hydrolysis, but enzymatic catalysts, particularly α-amylases, now dominate. These enzymes offer superior specificity, operate under mild conditions, and prevent unwanted side reactions (Shigechi et al. 2004). Notably, α-amylase sourced from robust bacteria such as Bacillus licheniformis or genetically modified strains of Escherichia coli or Bacillus subtilis facilitates the initial starch suspension hydrolysis process. This shift underscores the efficiency and reliability of enzymatic hydrolysis in producing ethanol from starch. It reflects advancements in biotechnological approaches to sustainable fuel production (Apar and Özbek 2004).

To enable amylases to degrade starch, the suspensions must be heated to high temperatures (90–110 °C) to effectively break down starch granules (Apar and Özbek 2004). Recent studies detail how operational factors impact the enzymatic hydrolysis of corn starch with commercial α-amylase. Recently, researchers have explored the potential of hydrolyzing starch at lower temperatures to conserve energy (Robertson et al. 2006).

Ethanol production from corn

Ethanol production in the United States primarily relies on corn as its feedstock. The process begins with milling the corn to extract starch, which undergoes enzymatic treatment to produce glucose syrup. This syrup serves as the basis for ethanol fermentation. There are two main milling techniques employed: wet milling and dry milling (Sánchez and Cardona 2008). In wet milling, the corn kernel is separated into its constituents, with starch converted into ethanol and the remaining components sold as co-products. Dry milling, on the other hand, involves processing whole grains without fractionation. All nutrients are utilized, and the resulting co-product, Dried Distiller's Grains with Solubles (DDGS), is utilized as animal feed. Both milling methods follow similar steps of liquefaction, saccharification, and fermentation. *Saccharomyces cerevisiae* is the preferred yeast for fermentation, conducted at temperatures between 30 and 32 °C. Nitrogen sources, such as ammonium sulfate or urea, are added to support yeast growth. Proteases may also be added to hydrolyze corn proteins, providing additional nitrogen for yeast nutrition (Bothast and Schlicher 2005). Burmaster in 2007 (Burmaster 2006) patented a method to enhance fermentation efficiency by monitoring and controlling the oxidation reduction potential. This innovation aims to increase ethanol yields, shorten fermentation times, and reduce by-product formation from corn mashes and other raw materials. In summary, ethanol production from corn in the USA relies on sophisticated milling techniques, enzymatic processes, and yeast fermentation. Ongoing innovations aim to optimize efficiency and yield while minimizing environmental impact (Cardona and Sánchez 2007).

Ethanol production from wheat

In France, ethanol production primarily relies on beet molasses, although wheat serves as an alternative feedstock using methods akin to those for corn. Various studies have sought to optimize fermentation conditions. For instance, Wang et al. in 1999 (1999a) determined optimal temperatures and specific gravity for fermenting wheat mash. Soni et al. (2003) focused on optimizing starch hydrolysis conditions using enzymes derived from wheat bran via solid-state fermentation. To improve fermentation efficiency, researchers have proposed high gravity fermentations, particularly for wheat mash. These processes involve initial dissolved solids concentrations exceeding 200 g/L, allowing for higher substrate loads. Consequently, they yield higher ethanol concentrations, require less process water, and reduce energy costs. Challenges include extended fermentation times and occasional incomplete fermentations. These issues arise from factors such as product inhibition, inadequate nutrition, and high osmotic pressures (Barber et al. 2002). To expedite high gravity fermentations, controlled additions of small amounts of acetaldehyde during fermentation have been employed. This technique has successfully reduced fermentation times from 790 to 585 h without compromising ethanol yields. The beneficial effect is attributed to acetaldehyde's ability to replenish intracellular pools and restore cellular redox balance (Barber et al. 2002).

In summary, while France predominantly uses beet molasses for ethanol production, wheat is also utilized, employing methodologies similar to those for corn. Ongoing research aims to refine fermentation processes and optimize conditions for starch hydrolysis. It also seeks to enhance fermentation efficiency through innovations such as high gravity fermentations and controlled acetaldehyde additions. These advancements seek to improve ethanol yields while mitigating operational challenges associated with alternative feedstocks in ethanol production (Sánchez and Cardona 2008).

Very high gravity (VHG) fermentation of wheat mashes has been proposed as an advanced technique. These mashes consist of wheat starch hydrolyzates with 200 g or more of dissolved solids per liter. VHG technology allows for the production of high ethanol concentrations from highly concentrated sugar solutions (Sánchez and Cardona 2008). Thomas et al. (1996) highlight the significant water savings achievable through this method in ethanol production. Additionally, implementing VHG fermentation can increase the throughput of an ethanol plant without requiring capacity expansion. Bayrock and Ingledew (2001) developed and tested a system that integrates multistage continuous culture fermentation with VHG cultivation. This system is suitable for a feed stream containing 150–320 g per liter of glucose and utilizes *S. cerevisiae.*

Ethanol production from cassava

Cassava serves as a significant alternative source of starch for both ethanol and glucose syrup production. This tuber has garnered considerable attention, especially in tropical regions, due to its widespread availability and its status as one of the ten most crucial tropical crops. Ethanol can be produced from cassava by using either the entire tuber or the starch derived from it. The extraction of starch from cassava can be performed using a high-efficiency, large-scale industrial method, such as the Alfa Laval extraction technique (Dai et al. 2006). It can also be done through traditional methods, which are suitable for small and medium-sized operations. This extraction process is analogous to the wet-milling process used for converting corn into ethanol. Producing cassava with a high starch content (85–90% dry matter) and lower protein and mineral levels is relatively straightforward. In addition to its role in ethanol production, cassava starch is a key ingredient in the production of glucose syrups, which are widely used in various food and beverage industries. The adaptability of cassava to different climatic conditions, particularly in tropical countries, makes it an invaluable crop for both local consumption and industrial applications (López-Ulibarri and Hall 1997). Its ability to thrive in poor soil conditions further underscores its importance as a reliable and sustainable source of starch. The growing interest in cassava is driven by its potential to meet the increasing global demand for renewable resources and its versatility in various industrial processes (Sánchez and Cardona 2008).

Cassava starch provides greater solubility for amylases and has a lower gelatinization temperature than corn starch. For glucose production, the hydrolysis of cassava flour has been proposed in an enzymatic hollow-fiber reactor, achieving a conversion rate of 97.3% (López-Ulibarri and Hall 1997). This is due to the fact that cassava flour production is simple and cost-effective than cassava starch production. Nonetheless, using the whole cassava tuber for ethanol production might yield better economic outcomes, especially for small-scale producers. Ethanol production from whole cassava mirrors the dry-milling technology used for producing ethanol from corn. Because cassava has a high moisture content (approximately 70%), it must be transported and processed rapidly to avoid spoilage, ideally within 3–4 days after harvesting.

To address the rapid deterioration of cassava, one effective method involves using sun-dried cassava chips (Situation and Needs 2000). Farmers transport cassava roots to small chipping facilities where they are chopped and peeled into smaller pieces. These chips are then dried in the sun over 2–3 days, reducing the moisture content to about 14% while the starch content reaches 65%. This drying process preserves the cassava, making it more suitable for ethanol production and extending its usability. By converting cassava into sun-dried chips, the logistics of transporting and processing the crop become more manageable. This provides a practical solution for small producers and enhances the overall efficiency of ethanol production from cassava (Dai et al. 2006).

Ethanol production from other starchy materials

In addition to wheat and corn, ethanol production can also utilize rye, triticale, barley (Wang et al. 1997), and sorghum (Zhan et al. 2003). Various pretreatment methods have proven effective for these grains. For instance, Wang et al. (1997) employed pearling on barley, wheat, triticale, and rye to increase the starch content of the feedstock by an average of 12%, resulting in a 6.5–22.5% improvement in ethanol yield during fermentation. Additionally, VHG (Very High Gravity) technology has been successfully applied to oats, barley, rye, and triticale (Wang et al. 1999b). Ethanol production has also been explored from other high-starch plant sources. For example, Abd-Aziz in 2002 suggested the use of sago palm for ethanol production in Malaysia. These advancements in pretreatment and technology demonstrate the potential for a diverse range of grains and plants to contribute to ethanol production. They offer flexibility and sustainability in biofuel resources.

Hammond et al. (1996) investigated ethanol production from banana waste and bananas utilizing commercial glucoamylase and α-amylase, achieving an ethanol yield of 0.5 L Ethanol/kg dry matter of ripe bananas. Additionally, Chung and Nam (2002) patented a method for processing starch-containing food wastes by adding malt to the pulverized feedstock. One of the most promising crops for fuel ethanol production is sweet sorghum, which yields grains with high starch content, stalks rich in sucrose, and leaves and bagasse abundant in lignocellulosic material. This versatile crop can be grown in both temperate and tropical regions, requiring only one-third of the water needed for sugarcane and half the water required for corn. Furthermore, sweet sorghum is highly tolerant to drought, saline alkalinity, and flooding (Network 2002), making it an exceptionally resilient and resource-efficient option for ethanol production.

Ethanol is likely the most recognized bio alcohol on the market. Since the 1970s, Brazil has utilized sugarcane for large-scale ethanol production. The predominant and extensively implemented production model relies on first-generation plants. These facilities convert sugars from sugarcane juice into ethanol and sugar, while the sugarcane bagasse, the largest byproduct, is burned to produce steam and electricity (Jones and Woods 1986).

Ethanol production is highly developed, and within this framework, biobutanol would be generated through ethanol conversion. In Brazil, approximately half of the yearly demand of around 60,000 tonnes of butanol is met by domestic production (Rodrigues 2011). The automotive market is not the main focus of refineries due to the lack of n-butanol usage as a transportation fuel. Instead, refineries are targeting the chemical market, which offers higher profitability. Concurrently, efforts are underway to lower production costs to more competitive levels, potentially enabling entry into the biofuels market in the near future (Naik et al. 2010).

Nevertheless, the competitiveness of Brazilian cane ethanol has been extensively proven, particularly when evaluating its energy output/input ratio (refer to Table 2.2). In addition to energy factors, crop yield should be assessed. Table 2.2 shows that ethanol yield from corn surpasses that from sugarcane due to the higher quantity of fermentable sugars (glucose) derived from the original starchy material.

2.3.3 Conventional Feedstock and Procedures for Butanol

Another promising alcohol, butanol (C_4H_9OH) is though less well-known, renewable fuel with many benefits for use in an internal combustion engine. Like ethanol, butanol is producing from biomass through fermentation (Bankar et al. 2013). As a biofuel, butanol stands out due to its higher energy content and compatibility with existing gasoline engines and infrastructure (Ezeji et al. 2007). It also serves as a solvent in the production of resins, coatings, and plasticizers, demonstrating its wide-ranging utility (Osat et al. 2023).

There are essentially two main methods for producing butanol: from biomass (known as "biobutanol") and from fossil fuels (referred to as "petrobutanol"). Despite their different sources, biobutanol and petrobutanol possess identical chemical properties (Trindade and dos Santos 2017). Presently, the majority of butanol produced is synthetic and derived from petrochemical reactions. Petrobutanol is typically obtained through the hydroformylation of propylene, a process commonly referred to as the oxo route (Nejame 2010; Heaton 1991). While this method is prevalent, it is not the most advantageous because the production costs of synthetic butanol are closely tied to the propylene market, which in turn is heavily influenced by crude oil prices (Green 2011). Furthermore, as society increasingly seeks solutions that are independent of petroleum derivatives, this approach is neither the most sustainable nor environmentally friendly. Consequently, there is a growing interest in shifting toward biobutanol production, which offers a more eco-friendly and sustainable alternative for future butanol needs (Trindade and dos Santos 2017) (Fig. 2.3).

Butanol, known as biobutanol when derived from biological processes, is produced through fermentation driven by microorganisms, commonly Clostridium bacteria. *Clostridium acetobutylicum* (Raganati et al. 2016; Lee et al. 2016) is a prevalent species utilized for converting sugars into butanol, alongside acetone and ethanol as secondary outputs. Another effective genus, *Clostridium cellulolyticum*, utilizes cellulosic biomass (Gaida et al. 2016) for biobutanol production. Compared to chemical methods, biological production offers distinct advantages: it leverages renewable resources, yields a reduced number of byproducts that are easier to separate. Beyond conventional feedstocks like sugar and starch, alternative biomasses such as barley, straw, bagasse, and corn cobs can also serve as viable sources for biobutanol production. This approach aligns with sustainable practices by minimizing reliance on finite fossil fuel sources and reducing environmental impact. As research progresses, optimizing biological pathways for butanol production promises

Table 2.2 Comparative metrics for the three primary feedstock types used in fuel ethanol production

Item	Sucrose-containing materials	Starchy materials	Lignocellulosic biomass	Remarks	References
Feedstock yield, ton/ha	70–122.9			Sugar cane; Brazil, Colombia	Cardona and Sánchez (2007)
		20		Cassava	Cardona and Sánchez (2007)
		35		Sweet sorghum	Cardona and Sánchez (2007)
		1.5–3.0		Wheat	Cardona and Sánchez (2007)
		6–10.07		Corna; USA	Cardona and Sánchez (2007)
			19.6–34.40	Cane bagasse	Moreira (2000)
			6.59–11.06	Corn stover	Kim and Dale (2004)
			1.93–3.86	Wheat straw	Kim and Dale (2004)
Feedstock cost, US$/kg	0.0100			Sugar cane; Brazil	Macedo and Nogueira (2005)
		0.0760		Corn; USA	McAloon et al. (2000)
	0.0124	0.1300		Cane/ corn; Colombia	Quintero et al. (2008)
			0.0295	Dry corn juice	Aden et al. (2002)
EtOH yield, L/ton	70			Sugar cane juice	Moreira and Goldemberg (1999)
	100			Sugar beet	Berg (2024)
		180		Cassava	Cardona and Sánchez (2007)

(continued)

Table 2.2 (continued)

Item	Sucrose-containing materials	Starchy materials	Lignocellulosic biomass	Remarks	References
		86		Sweet sorghum	Cardona and Sánchez (2007)
		340–350		Wheat	Cardona and Sánchez (2007)
		370		Corn	Cardona and Sánchez (2007)
		403.1		Corn, wet milling	Gulati et al. (1996)
		419.4–460.6		Corn, dry milling	Gulati et al. (1996)
			140	Cane bagasse	Moreira (2000)
			261.3	Wheat straw	Kim and Dale (2004)
			227.7	Corn stover	Kim and Dale (2004)
			330	Dry corn stover; USA	Kadam and McMillan (2003)
Annual EtOH yield, L/(ha year)	5345–9381			Sugar cane; Brazil, Colombia	Cardona and Sánchez (2007)
	6600			Sugar beet; France	Poitrat (2000)
		3600		Cassava	Cardona and Sánchez (2007)
		9030		Sweet sorghum	Cardona and Sánchez (2007)
		1020–3214		Wheat	Cardona and Sánchez (2007)
		6600		Corn	Cardona and Sánchez (2007)

(continued)

Table 2.2 (continued)

Item	Sucrose-containing materials	Starchy materials	Lignocellulosic biomass	Remarks	References
Production costs, US$/L	0.1980			Sugar cane; Brazil	Xavier (2007)
Anhydrous EtOH	0.2153	0.3381		Sugar cane/corn; Colombia	Quintero et al. (2008)
		0.2325	0.3963	Corn/corn stover; USA	McAloon et al. (2000)
Output/input energy ratio	0.8			Sugar cane	Berg (2024)
	1.9			Sugar beet	Berg (2024)
		1.34–1.53		Corn; USA	Berg (2024)
			6.0	Biomass	Berg (2024)
Possibility of co-generation	Yes	No	Yes		
Co products	Concentrated stillage for fertilization	DDGS (dry milling)	Lignin as feedstock for chemicals		

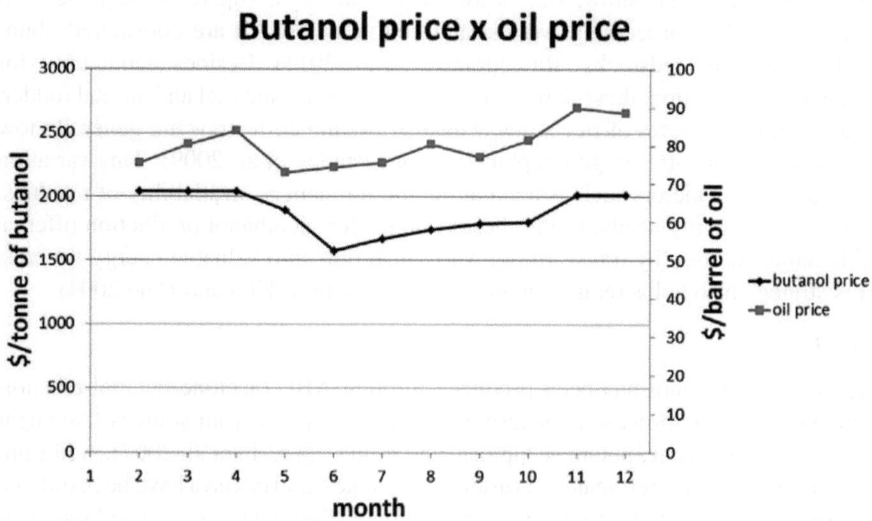

Fig. 2.3 The relationship between crude oil prices and synthetic butanol prices in China (Green Jun. 2011)

continued advancements in biofuel technology, offering a promising avenue for renewable energy solutions.

2.4 Categories of Renewable Feedstock

Biofuels have become a tangible reality, capturing growing interest from society, industry, and academia. This surge is driven by uncertainties in fossil fuel prices, the imperative to reduce emissions, and the exploration of diverse energy options. In the automotive sector, extensive research is focused on different biofuels derived exclusively from renewable feedstocks, distinct from traditional gasoline or diesel sources (Trindade and dos Santos 2017). This transition underscores a pivotal shift toward sustainable energy solutions, aligning with global efforts to mitigate environmental impacts and enhance energy security through renewable resources (Kim and Dale 2004).

2.4.1 Sugars and Starches from Non-food Crops

Ethanol

Globally, a wide range of agroresidues have been employed to produce bioethanol. These include wheat straw, rice straw, sugarcane tops, sugarcane bagasse, soft bamboo, bamboo processing wastes, and cotton stalks. All are considered abundantly available feedstocks (Ravindranath et al. 2011). Besides being used for bioethanol production, these agro-residues serve as domestic fuel and animal fodder. The consumption rates of rice straw, wheat straw, and corn straw are generally low and vary significantly by geographical region (Pandey et al. 2009). This variation is influenced by factors such as local agricultural practices, availability of residues, and regional energy needs. Using these residues for bioethanol production offers a sustainable solution by transforming waste materials into valuable energy sources. This strategy also helps reduce environmental pollution (Kim and Dale 2004).

Butanol

Traditionally, butanol has been produced through ABE (acetone-butanol-ethanol) fermentation. This process uses carbon substrates derived from sources like sugar beets, corn, sugarcane, potatoes, tapioca, and millet (Qureshi et al. 2007). Additionally, molasses, whey permeates, Jerusalem artichokes, and cassava have been utilized as fermentation substrates for butanol production. In South Africa, cane molasses was effectively used for industrial-scale butanol production until the 1980s (Jones and Woods 1986). While sucrose and starch are effective substrates for butanol fermentation, the increasing demand for these materials for biofuel production has driven up their prices. This price increase has led to a significant food versus fuel debate, as the same crops used for biofuel are also essential for food supply (Jang et al. 2012b). The competition for these resources underscores the need for alternative feedstocks that do not directly compete with food crops.

Innovative approaches are being explored to address these challenges. Researchers are investigating the use of non-food biomass, agricultural residues, and other renewable sources to produce butanol (Jang et al. 2012a). These alternatives not only alleviate the competition with food supplies but also contribute to a more sustainable and environmentally friendly biofuel production process. Diversifying the feedstocks for butanol production makes it possible to develop more resilient and flexible biofuel systems. This method allows adaptation to varying agricultural and economic conditions, ensuring a steady supply of renewable energy without compromising food security (Osat et al. 2023).

2.4.2 *Lignocellulosic Biomass*

Ethanol

Converting abundant lignocellulosic biomass into ethanol as a transportation fuel presents a significant opportunity. This conversion can help reduce the trade deficit, enhance energy security, lower greenhouse gas emissions, and stabilize prices (Wyman 1999). It is important to clarify why lignocellulosic biomass should be converted into ethanol rather than other biofuels like biodiesel, pyrolytic oil, biogas, and others. The primary reasons are its abundance and low cost, making the delignification of lignocellulosic biomass highly attractive for ethanol production (Singh et al. 2014).

Numerous lignocellulosic materials have been tested for bioethanol production. Potential lignocellulosic materials for fuel ethanol can be categorized into six groups: crop residues (such as cane bagasse, corn stover, wheat straw, rice straw, rice hulls, barley straw, pulp, sweet sorghum bagasse, and olive stones), hardwoods (like aspen and poplar), softwoods (such as pine and spruce), cellulose wastes (including newsprint, recycled paper sludge, waste office paper), herbaceous biomass (such as switchgrass, alfalfa hay, coastal Bermudagrass, reed canary grass, and Timothy grass), and municipal solid wastes (MSW). Detailed compositions of these materials can be found in various sources (Sun and Cheng 2002). The lignocellulosic feedstock selection is driven by each country's goal to add value to agricultural wastes, particularly those without food value. In the United States, corn stover stands out as one of the promising feedstocks because of its extensive availability. Estimates suggest that the sustainable collection and use of corn stover in the U.S. amount to about 80–100 million dry tons per year. This makes corn stover the most abundant agricultural residue in the country (Li et al. 2022a). Kadam and McMillan (Kadam and McMillan 2003), referencing unpublished data from the NREL, report that the theoretical ethanol yield from corn stover is 480 L per dry ton, provided that both pentoses and hexoses can be fermented into ethanol. This potential yield highlights corn stover as a highly attractive and sustainable option for bioethanol production, leveraging its abundance and converting agricultural waste into valuable biofuel (Gerbrandt et al. 2016). This approach not only supports energy production but also enhances waste management and sustainability efforts.

The focus on lignocellulosic biomass is due to its unique nature and composition. The biomass consists primarily of cellulose, hemicellulose, and lignin. To convert lignocellulosic biomass into ethanol or other valuable products, these main polymers must be broken down into fermentable sugars (Sánchez and Cardona 2008). However, this degradation process is complex, energy-intensive, and not yet fully developed. Despite these challenges, the abundance and low cost of lignocellulosic biomass make it an essential and promising resource for future ethanol production (Afedzi et al. 2022).

One of the main hurdles in using lignocellulosic biomass is its recalcitrant structure, which makes it resistant to breakdown. This resistance necessitates pretreatment processes to make the cellulose and hemicellulose more accessible to enzymatic hydrolysis (Khan and Dwivedi 2013). Common pretreatment methods include physical, chemical, and biological processes that disrupt the lignin structure and increase the surface area of the cellulose fibers (Bridgwater 2012). Following pretreatment, enzymes are used to hydrolyze the cellulose and hemicellulose into simple sugars, which can then be fermented by microorganisms to produce ethanol.

By overcoming these technical challenges, it is possible to unlock a sustainable and economically viable pathway for large-scale bioethanol production. The utilization of lignocellulosic biomass provides a renewable source of energy. It also contributes to waste management and environmental sustainability by converting agricultural residues and other waste materials into valuable fuels (Sukumaran et al. 2010). This approach aligns with global efforts to reduce reliance on fossil fuels and mitigate the impact of climate change. It makes lignocellulosic ethanol a key component of the future renewable energy landscape.

Butanol

A promising approach to obtaining sufficient fermentable carbon substrates involves efficiently utilizing the abundant lignocellulosic biomass found on Earth. This method does not compete with food supplies. The main polysaccharides in this biomass—cellulose, hemicellulose, and pectin—make up the bulk of industrial, agricultural, wood and forest waste residues (Buschke et al. 2011). Biofuels produced from lignocellulosic biomass offer the additional benefit of reducing greenhouse gas emissions compared to those made from corn (Cai et al. 2012). However, despite its abundance, extracting fermentable sugars from lignocellulosic materials is quite difficult. Consequently, significant research efforts have been focused on using lignocellulosic hydrolysate for butanol fermentation (Lu et al. 2012). This research aims to develop effective methods for breaking down complex polysaccharides into simple sugars. These sugars can then be fermented into butanol, ultimately providing a sustainable and environmentally friendly alternative to conventional biofuels.

2.4.3 Algae

Butanol

Biofuel production from terrestrial biomass is only advantageous if it is produced sustainably without harming the environment and biodiversity; otherwise, it reignites the food versus fuel debate. Fuels derived from land-based plants are contentious because they require agricultural land that could be used to grow food crops, which are essential for feeding an increasing global population. A viable alternative is the use of abundant algal biomass for sustainable biofuel production (Qureshi et al. 2007). Algal biomass offers several benefits over terrestrial biomass: it doesn't require agricultural land, can grow in seawater and wastewater, has a higher growth rate, and yields more per area than land-based plants. This approach mitigates the competition for arable land and freshwater resources, making algal biofuels a more sustainable and efficient option. Additionally, the higher productivity of algae ensures that biofuel production can meet energy demands without compromising food security (Tran et al. 2010). This makes algae an attractive and environmentally friendly alternative for biofuel production, addressing both energy needs and ecological concerns.

Recently, the "Photanol" concept has been introduced for the direct conversion of solar energy into the reducing power needed for producing fermentation end products such as biofuels, using photosynthetic algae (Hellingwerf and Teixeira de Mattos 2009). This innovative concept involves constructing a metabolic network that merges phototrophic and fermentative metabolisms utilizing synthetic biology techniques. In this process, H_2O, CO_2, and solar energy serve as inputs to generate fermentation products, an approach termed "photofermentation." The aim is to enhance the economic appeal of biofuel production by increasing photosynthetic efficiency (Angermayr et al. 2009). This approach has been validated in ethanol production by introducing genes for pyruvate decarboxylase and alcohol dehydrogenase II into the cyanobacteria Synechocystis sp. PCC 6803 (Hellingwerf and Teixeira de Mattos 2009). The Photanol concept, however, is not limited to ethanol; it holds potential for producing other biofuels, such as butanol, theoretically achievable through this technology. By leveraging the principles of synthetic biology and enhancing photosynthetic processes, the Photonol concept aims to provide a sustainable and efficient method for biofuel production, addressing both energy and environmental challenges (Atsumi et al. 2009).

2.4.4 Waste Streams

Methanol: Sustainable feedstocks for bio-methanol include biogenic waste streams such as manure, agricultural residues, and food waste, which can be transformed into bio-methane and subsequently upgraded to bio-methanol (Ali et al. 2015). Current estimates indicate that there is a substantial potential for bio-methane feedstocks, yet this potential is hindered by high demand across various industries. Additionally, the existing infrastructure, established practices, and current regulatory framework do not facilitate the large-scale collection of appropriate waste streams (Ostadi et al. 2023). Another source for bio-methanol production is woody biomass, which can be transformed into syngas through gasification.

Butanol: Municipal Solid Waste (MSW) is made up of various materials discarded by communities, including organic components like paper and food waste, which have substantial potential for generating biofuels such as butanol (Beig et al. 2021).

Food waste, rich in biodegradable materials like proteins, carbohydrates, and fats, serves as an ideal feedstock for fermentation processes where microorganisms convert these components into butanol. Similarly, paper waste, mainly consisting of cellulose, can be pre-treated to release fermentable sugars, which are then utilized in microbial fermentation to produce butanol (Yukesh Kannah et al. 2020). The conversion of organic MSW fractions helps reduce the amount of waste sent to landfills and contributes to renewable energy generation, providing a sustainable alternative to fossil fuels. However, this process faces challenges such as potential contamination with hazardous materials or non-organic, the variability in MSW composition, and the necessity of ensuring economic viability for large-scale production. Overcoming these obstacles is essential for enhancing the efficiency and cost-effectiveness of producing biofuels from MSW (Guo et al. 2022).

2.5 Conclusion

The transition to biofuels like ethanol, butanol, and methanol represents a pivotal shift in global energy strategies. It aims at increasing sustainability and reducing the environmental impact of energy consumption. Derived from renewable feedstocks, these biofuels are emerging as viable alternatives to traditional fossil fuels, offering both environmental and economic benefits. Ethanol has long played a central role in biofuel production, particularly as a gasoline additive to help reduce emissions. Producing ethanol from lignocellulosic biomass, such as corn stover, sugarcane bagasse, and other agricultural by-products, provides a sustainable option. It capitalizes on the availability and low cost of these materials. This approach mitigates the food-versus-fuel debate by relying on waste biomass rather than food crops, thus turning potential waste into a valuable energy source. However, converting lignocellulosic biomass into fermentable sugars remains challenging due to the complex composition of these materials. Advanced pretreatment is necessary to break down lignin and hemicellulose. This is followed by enzymatic hydrolysis to release sugars for fermentation. Breakthroughs in biotechnology and process engineering are essential. They help improve enzyme efficiency, cut production costs, and boost yields. As these technological advancements continue, large-scale lignocellulosic ethanol production will become increasingly viable, providing a sustainable energy solution.

Butanol offers additional advantages over ethanol, such as greater energy density, lower hygroscopicity, and compatibility with existing gasoline infrastructure. These qualities make it an attractive option for blending or even as a standalone fuel. While traditional ABE (acetone-butanol-ethanol) fermentation processes are well-established, they are now being refined through biotechnological innovations that enhance both yield and efficiency. Using non-food biomass sources, including agricultural waste and industrial by-products, for butanol production further alleviates food security concerns while supplying a sustainable raw material. Moreover, butanol's flexibility in using diverse feedstocks—ranging from sugars and starches to cellulosic materials—enhances its potential as a renewable fuel. Advances in genetically engineered microorganisms and optimized fermentation processes are critical to increasing butanol yields and reducing production costs. As the biofuel industry progresses, butanol is expected to play an increasingly important role due to its favorable fuel properties and the shift toward reducing carbon emissions.

Producing methanol from renewable resources, such as biogenic waste and woody biomass, also presents a promising route to sustainable fuel production. Beyond serving as a versatile fuel, methanol is an essential industrial feedstock for manufacturing chemicals, plastics, and synthetic materials. Methods such as anaerobic digestion and gasification allow waste materials to be converted into methanol, supporting a circular economy by transforming waste into valuable products. However, scaling up bio-methanol production faces challenges, including the need for robust infrastructure to efficiently gather and process biomass feedstocks. Additionally, supportive policies and regulatory frameworks are crucial to expanding bio-methanol production and integrating it into established fuel supply chains. Addressing these challenges will require coordinated efforts among governments, industry, and research institutions to create an environment conducive to bio-methanol's growth.

Transitioning to biofuels is a crucial element of the global effort to achieve a more sustainable and resilient energy system. With ongoing technological advancements enhancing the efficiency and cost-effectiveness of biofuel production, these renewable fuels are poised to significantly address the world's increasing energy needs. Utilizing a variety of feedstocks and cutting-edge technologies, biofuels have the potential to foster a cleaner, more sustainable future by reducing reliance on fossil fuels and alleviating climate change impacts.

References

Abd-Aziz S (2002) Sago starch and its utilisation. J Biosci Bioeng 94(6):526–529. https://doi.org/10.1016/S1389-1723(02)80190-6

Aden A et al (2002) Lignocellulosic biomass to ethanol process design and economics utilizing co-current dilute acid prehydrolysis and enzymatic hydrolysis for corn stover. National Renewable Energy Lab. (NREL), Golden, CO

Afedzi AEK, Rattanaporn K, Parakulsuksatid P (2022) Impeller selection for mixing high-solids lignocellulosic biomass in stirred tank bioreactor for ethanol production. Bioresour Technol Rep 17:100935. https://doi.org/10.1016/J.BITEB.2021.100935

Ali KA, Abdullah AZ, Mohamed AR (2015) Recent development in catalytic technologies for methanol synthesis from renewable sources: a critical review. Renew Sustain Energy Rev 44:508–518. https://doi.org/10.1016/J.RSER.2015.01.010

Angermayr SA, Hellingwerf KJ, Lindblad P, Teixeira de Mattos MJ (2009) Energy biotechnology with cyanobacteria. Curr Opin Biotechnol 20(3):257–263. https://doi.org/10.1016/J.COPBIO.2009.05.011

Anto S et al (2020) Algae as green energy reserve: technological outlook on biofuel production. Chemosphere 242:125079. https://doi.org/10.1016/J.CHEMOSPHERE.2019.125079

Apar DK, Özbek B (2004) σ-Amylase inactivation during corn starch hydrolysis process. Process Biochem 39(12):1877–1892. https://doi.org/10.1016/J.PROCBIO.2003.09.014

Atsumi S, Higashide W, Liao JC (2009) Direct photosynthetic recycling of carbon dioxide to isobutyraldehyde. Nat Biotechnol 27(12):1177–1180

Bankar SB, Survase SA, Ojamo H, Granström T (2013) Biobutanol: the outlook of an academic and industrialist. Rsc Adv 3(47):24734–24757

Barber AR, Henningsson M, Pamment NB (2002) Acceleration of high gravity yeast fermentations by acetaldehyde addition. Biotechnol Lett 24(11):891–895. https://doi.org/10.1023/A:1015541331460

Bayrock DP, Michael Ingledew W (2001) Application of multistage continuous fermentation for production of fuel alcohol by very-high-gravity fermentation technology. J Ind Microbiol Biotechnol 27(2):87–93. https://doi.org/10.1038/sj.jim.7000167

Beig B et al (2021) Current challenges and innovative developments in pretreatment of lignocellulosic residues for biofuel production: a review. Fuel 287:119670. https://doi.org/10.1016/J.FUEL.2020.119670

Berg C (2024) World fuel ethanol—analysis and outlook. Sugar trading manual. https://www.distill.com/World-Fuel-Ethanol-A%26O-2004.html. Accessed 25 July 2024

Bothast RJ, Schlicher MA (2005) Biotechnological processes for conversion of corn into ethanol. Appl Microbiol Biotechnol 67(1):19–25. https://doi.org/10.1007/s00253-004-1819-8

Bridgwater AV (2012) Review of fast pyrolysis of biomass and product upgrading. Biomass Bioenerg 38:68–94. https://doi.org/10.1016/J.BIOMBIOE.2011.01.048

Bruijnincx PCA, Weckhuysen BM (2013) Shale gas revolution: an opportunity for the production of biobased chemicals? Angew Chemie Int Ed 52(46)

Burmaster BM (2006) Ethanol fermentation using oxidation reduction potential

Buschke N, Schröder H, Wittmann C (2011) Metabolic engineering of *Corynebacterium glutamicum* for production of 1, 5-diaminopentane from hemicellulose. Biotechnol J 6(3):306–317

Cai Z, Zhang B, Li Y (2012) Engineering *Saccharomyces cerevisiae* for efficient anaerobic xylose fermentation: reflections and perspectives. Biotechnol J 7(1):34–46

Cardona CA, Sánchez OJ, Montoya MI, Quintero JA (2005) Analysis of fuel ethanol production processes using lignocellulosic biomass and starch as feedstocks. In: Seventh world congress of chemical engineering, Glasgow, Scotland, UK

Cardona CA, Sánchez ÓJ (2007) Fuel ethanol production: process design trends and integration opportunities. Bioresour Technol 98(12):2415–2457. https://doi.org/10.1016/J.BIORTECH.2007.01.002

Chung BH, Nam JG (2002) Process for producing high concentration of ethanol using food wastes by fermentation. Pat. KR20020072326

Dai D, Hu Z, Pu G, Li H, Wang C (2006) Energy efficiency and potentials of cassava fuel ethanol in Guangxi region of China. Energy Convers Manag 47(13–14):1686–1699. https://doi.org/10.1016/J.ENCONMAN.2005.10.019

Delrue F et al (2012) An economic, sustainability, and energetic model of biodiesel production from microalgae. Bioresour Technol 111:191–200. https://doi.org/10.1016/J.BIORTECH.2012.02.020

Deluga GA, Salge JR, Schmidt LD, Verykios XE (2004) Renewable hydrogen from ethanol by autothermal reforming. Science 303(5660):993–997

Ehlinger VM, Gabriel KJ, Noureldin MMB, El-Halwagi MM (2014) Process design and integration of shale gas to methanol. ACS Sustain Chem Eng 2(1):30–37

Ezeji TC, Qureshi N, Blaschek HP (2007) Bioproduction of butanol from biomass: from genes to bioreactors. Curr Opin Biotechnol 18(3):220–227. https://doi.org/10.1016/j.copbio.2007.04.002

Farrell AE, Plevin RJ, Turner BT, Jones AD, O'hare M, Kammen DM (2006) Ethanol can contribute to energy and environmental goals. Science 311(5760):506–508

Ferdeş M, Dincă MN, Moiceanu G, Ştefania Zăbavă B, Paraschiv G (2020) Microorganisms and enzymes used in the biological pretreatment of the substrate to enhance biogas production: a review. Sustainability 12(17):7205

Gaida SM, Liedtke A, Jentges AHW, Engels B, Jennewein S (2016) Metabolic engineering of *Clostridium cellulolyticum* for the production of n-butanol from crystalline cellulose. Microb Cell Fact 15(1):6. https://doi.org/10.1186/s12934-015-0406-2

Galadima A, Muraza O (2015) From synthesis gas production to methanol synthesis and potential upgrade to gasoline range hydrocarbons: a review. J Nat Gas Sci Eng 25:303–316. https://doi.org/10.1016/J.JNGSE.2015.05.012

Ge JC, Kim HY, Yoon SK, Choi NJ (2018) Reducing volatile organic compound emissions from diesel engines using canola oil biodiesel fuel and blends. Fuel 218:266–274. https://doi.org/10.1016/J.FUEL.2018.01.045

Gerbrandt K et al (2016) Life cycle assessment of lignocellulosic ethanol: a review of key factors and methods affecting calculated GHG emissions and energy use. Curr Opin Biotechnol 38:63–70. https://doi.org/10.1016/J.COPBIO.2015.12.021

Ghandehariun S, Saray JA (2024) Economic assessment and environmental challenges of carbon capture using membrane techniques. In: Advances and technology development in greenhouse gases: emission, capture and conversion. Elsevier, pp 515–532

Ghandehariun S, Gharehghani A, Saray JA (2021) Ethylene and ethanol production from CO_2. In: Advances and technology development in greenhouse gases: emission, capture and conversion. Elsevier, pp 427–442

Ghosh P, Ghose TK (2003) Bioethanol in India: recent past and emerging future. Adv Biochem Eng Biotechnol 85:1–27. https://doi.org/10.1007/3-540-36466-8_1

Goyes A, Bolaños G (2005) Un estudio preliminar sobre el tratamiento de vinazas en agua supercrítica. XXIII Congr Colomb Ing Química 1:13

Green EM (2011) Fermentative production of butanol—the industrial perspective. Curr Opin Biotechnol 22(3):337–343. https://doi.org/10.1016/J.COPBIO.2011.02.004

Gulati M, Kohlmann K, Ladisch MR, Hespell R, Bothast RJ (1996) Assessment of ethanol production options for corn products. Bioresour Technol 58(3):253–264. https://doi.org/10.1016/S0960-8524(96)00108-3

Guo Y et al (2022) Production of butanol from lignocellulosic biomass: recent advances, challenges, and prospects. RSC Adv 12(29):18848–18863. https://doi.org/10.1039/d1ra09396g

Hammond JB, Egg R, Diggins D, Coble CG (1996) Alcohol from bananas. Bioresour Technol 56(1):125–130. https://doi.org/10.1016/0960-8524(95)00177-8

Heaton CA (1991) The chemical industry. https://books.google.com/books/about/The_Chemical_Industry.html?id=YvnpBwAAQBAJ. Accessed 22 July 2024

Hellingwerf KJ, Teixeira de Mattos MJ (2009) Alternative routes to biofuels: light-driven biofuel formation from CO_2 and water based on the 'photanol' approach. J Biotechnol 142(1):87–90. https://doi.org/10.1016/J.JBIOTEC.2009.02.002

Holmgren KM, Andersson E, Berntsson T, Rydberg T (2014) Gasification-based methanol production from biomass in industrial clusters: characterisation of energy balances and greenhouse gas emissions. Energy 69:622–637. https://doi.org/10.1016/J.ENERGY.2014.03.058

Jang Y, Lee J, Malaviya A, Seung DY, Cho JH, Lee SY (2012a) Butanol production from renewable biomass: rediscovery of metabolic pathways and metabolic engineering. Biotechnol J 7(2):186–198

Jang YS, Malaviya A, Cho C, Lee J, Lee SY (2012b) Butanol production from renewable biomass by clostridia. Bioresour Technol 123:653–663. https://doi.org/10.1016/J.BIORTECH.2012.07.104

Jones DT, Woods D (1986) Acetone-butanol fermentation revisited. Microbiol Rev 50(4):484–524

Kadam KL, McMillan JD (2003) Availability of corn stover as a sustainable feedstock for bioethanol production. Bioresour Technol 88(1):17–25. https://doi.org/10.1016/S0960-8524(02)00269-9

Kendir Çakmak E, Ugurlu A (2020) Enhanced biogas production of red microalgae via enzymatic pretreatment and preliminary economic assessment. Algal Res 50:101979. https://doi.org/10.1016/J.ALGAL.2020.101979

Khan Z, Dwivedi AK (2013) Fermentation of biomass for production of ethanol: a review. Univ J Environ Res Technol 3(1)

Kim S, Dale BE (2004) Global potential bioethanol production from wasted crops and crop residues. Biomass Bioenerg 26(4):361–375. https://doi.org/10.1016/J.BIOMBIOE.2003.08.002

King GE (2010) Thirty years of gas shale fracturing: what have we learned? In: SPE annual technical conference and exhibition? SPE, p. SPE-133456

Lee SH, Kim S, Kim JY, Cheong NY, Kim KH (2016) Enhanced butanol fermentation using metabolically engineered *Clostridium acetobutylicum* with ex situ recovery of butanol. Bioresour Technol 218:909–917. https://doi.org/10.1016/J.BIORTECH.2016.07.060

Li J, Zhao R, Xu Y, Wu X, Bean SR, Wang D (2022a) Fuel ethanol production from starchy grain and other crops: an overview on feedstocks, affecting factors, and technical advances. Renew Energy 188:223–239. https://doi.org/10.1016/J.RENENE.2022.02.038

Li S, Li F, Zhu X, Liao Q, Chang JS, Ho SH (2022b) Biohydrogen production from microalgae for environmental sustainability. Chemosphere 291:132717. https://doi.org/10.1016/J.CHEMOSPHERE.2021.132717

López-Ulibarri R, Hall GM (1997) Saccharification of cassava flour starch in a hollow-fiber membrane reactor. Enzyme Microb Technol 21(6):398–404. https://doi.org/10.1016/S0141-0229(97)82260-0

Lu C, Zhao J, Yang ST, Wei D (2012) Fed-batch fermentation for n-butanol production from cassava bagasse hydrolysate in a fibrous bed bioreactor with continuous gas stripping. Bioresour Technol 104:380–387. https://doi.org/10.1016/J.BIORTECH.2011.10.089

Macedo IC, Nogueira LAH (2005) Evaluation of ethanol production expansion in Brazil. Cad. NAE/Núcleo Assuntos Estrategicos da Presidência da Repú blica. Biofuels, Sect., vol 2, pp 113–220

McAloon A, Taylor F, Yee W, Ibsen K, Wooley R (2000) Determining the cost of producing ethanol from corn starch and lignocellulosic feedstocks. National Renewable Energy Lab. (NREL), Golden, CO

Moreira JR (2000) Sugarcane for energy—recent results and progress in Brazil. Energy Sustain Dev 4(3):43–54. https://doi.org/10.1016/S0973-0826(08)60252-5

Moreira JR, Goldemberg J (1999) The alcohol program. Energy Policy 27(4):229–245. https://doi.org/10.1016/S0301-4215(99)00005-1

Morrison BRK (1992) Organic chemistry, 6th edn. Benjamin Cummings

Naik SN, Goud VV, Rout PK, Dalai AK (2010) Production of first and second generation biofuels: a comprehensive review. Renew Sustain Energy Rev 14(2):578–597. https://doi.org/10.1016/J.RSER.2009.10.003

Narendranath NV, Thomas KC, Ingledew WM (2000) Urea hydrogen peroxide reduces the numbers of lactobacilli, nourishes yeast, and leaves no residues in the ethanol fermentation. Appl Environ Microbiol 66(10):4187–4192. https://doi.org/10.1128/AEM.66.10.4187-4192.2000

Nejame S (2010) Butanol as a fuel—view from the field. In: Promotion present to NREL, vol 453

Network W (2002) Village level bioenergy system based on sweet sorghum

Ng TK, Busche RM, McDonald CC, Hardy RWF (1983) Production of feedstock chemicals. Science 219(4585):733–740

Osat M, Shojaati F, Osat M (2023) Techno-economic assessment of butanol and pentanol productions from sorption enhanced chemical looping gasification of a lignocellulosic biomass. Renew Energy 217:119176. https://doi.org/10.1016/J.RENENE.2023.119176

Ostadi M, Bromberg L, Cohn DR, Gençer E (2023) Flexible methanol production process using biomass/municipal solid waste and hydrogen produced by electrolysis and natural gas pyrolysis. Fuel 334:126697. https://doi.org/10.1016/J.FUEL.2022.126697

Palsson BO, Fathi-Afshar S, Rudd DF, Lightfoot EN (1981) Biomass as a source of chemical feedstocks: an economic evaluation. Science 213(4507):513–517

Pandey A, Biswas S, Sukumaran RK, Kaushik N (2009) Study on availability of Indian biomass resources for exploitation: a report based on a nation-wise survey, vol 105. TIFAC, New Delhi

Pereira RG et al (2012) Sustainability and mitigation of greenhouse gases using ethyl beef tallow biodiesel in energy generation. J Clean Prod 29–30:269–276. https://doi.org/10.1016/J.JCLEPRO.2012.01.007

Poitrat E (2000)"The potential of liquid biofuels in France. In: World renewable energy congress VI. Elsevier, pp 1294–1299

Praveena V, Martin LJ, Matijošius J, Aloui F, Pugazhendhi A, Varuvel EG (2024) A systematic review on biofuel production and utilization from algae and waste feedstocks—a circular economy approach. Renew Sustain Energy Rev 192:114178. https://doi.org/10.1016/J.RSER.2023.114178

Quintero JA, Montoya MI, Sánchez OJ, Giraldo OH, Cardona CA (2008) Fuel ethanol production from sugarcane and corn: comparative analysis for a Colombian case. Energy 33(3):385–399. https://doi.org/10.1016/J.ENERGY.2007.10.001

Qureshi N, Saha BC, Cotta MA (2007) Butanol production from wheat straw hydrolysate using *Clostridium beijerinckii*. Bioprocess Biosyst Eng 30(6):419–427

Raganati F et al (2016) Butanol production by *Clostridium acetobutylicum* in a series of packed bed biofilm reactors. Chem Eng Sci 152:678–688. https://doi.org/10.1016/j.ces.2016.06.059

Ravindranath NH, Sita Lakshmi C, Manuvie R, Balachandra P (2011) Biofuel production and implications for land use, food production and environment in India. Energy Policy 39(10):5737–5745. https://doi.org/10.1016/J.ENPOL.2010.07.044

Robertson GH, Wong DWS, Lee CC, Wagschal K, Smith MR, Orts WJ (2006) Native or raw starch digestion: a key step in energy efficient biorefining of grain. J Agric Food Chem 54(2):353–365. https://doi.org/10.1021/jf051883m

Rodrigues JAR (2011) Do engenho à biorrefinaria: a usina de açúcar como empreendimento industrial para a geração de produtos bioquímicos e biocombustíveis. Quim Nova 34:1242–1254

Sánchez ÓJ, Cardona CA (2008) Trends in biotechnological production of fuel ethanol from different feedstocks. Bioresour Technol 99(13):5270–5295. https://doi.org/10.1016/J.BIORTECH.2007.11.013

Saray JA, Gharehghani A, Hosseinzadeh D (2024) Towards sustainable energy Carriers: a solar and wind-based systems for green liquid hydrogen and ammonia production. Energy Convers Manag 304:118215

Segaran B, Guruswami C (2023) The impact of bioenergy resources for sustainable environment BT—bioenergy: impacts on environment and economy. In: Ramanujam PK, Parameswaran B, Bharathiraja B, Magesh A (eds) Springer Nature Singapore, Singapore, pp 67–78. https://doi.org/10.1007/978-981-99-3002-9_5

Shapouri H, Duffield JA, Wang M (2003) The energy balance of corn ethanol revisited. Trans ASAE 46(4):959

Shigechi H, Fujita Y, Koh J, Ueda M, Fukuda H, Kondo A (2004) Energy-saving direct ethanol production from low-temperature-cooked corn starch using a cell-surface engineered yeast strain co-displaying glucoamylase and α-amylase. Biochem Eng J 18(2):149–153. https://doi.org/10.1016/J.BEJ.2003.08.003

Shukla PC, Belgiorno G, Di Blasio G, Agarwal AK (2021) Introduction to alcohol as an alternative fuel for internal combustion engines BT—alcohol as an alternative fuel for internal combustion engines. In: Shukla PC, Belgiorno G, Di Blasio G, Agarwal AK (eds) Springer Singapore, Singapore, pp 1–4. https://doi.org/10.1007/978-981-16-0931-2_1

Sikarwar VS, Zhao M, Fennell PS, Shah N, Anthony EJ (2017) Progress in biofuel production from gasification. Prog Energy Combust Sci 61:189–248. https://doi.org/10.1016/J.PECS.2017.04.001

Singh R, Shukla A, Tiwari S, Srivastava M (2014) A review on delignification of lignocellulosic biomass for enhancement of ethanol production potential. Renew Sustain Energy Rev 32:713–728. https://doi.org/10.1016/J.RSER.2014.01.051

Singh D et al (2021) A comprehensive review of physicochemical properties, production process, performance and emissions characteristics of 2nd generation biodiesel feedstock: *Jatropha curcas*. Fuel 285:119110. https://doi.org/10.1016/J.FUEL.2020.119110

Situation P, Needs D (2000) Cassava's potential in Asia in the 21. Cassava's Potential Asia 21:4–632

Solomons G, Fryhle C (2009) Organic chemistry, 10th edn. Wiley, p 1277

Soni SK, Kaur A, Gupta JK (2003) A solid state fermentation based bacterial α-amylase and fungal glucoamylase system and its suitability for the hydrolysis of wheat starch. Process Biochem 39(2):185–192. https://doi.org/10.1016/S0032-9592(03)00058-X

Spath PL, Dayton DC (2003) Preliminary screening--technical and economic assessment of synthesis gas to fuels and chemicals with emphasis on the potential for biomass-derived syngas. National Renewable Energy Lab (NREL), Golden, CO

Speight JG (ed) (2011) The biofuels handbook. The Royal Society of Chemistry. https://doi.org/10.1039/9781849731027

Subramani V, Song C (2007) Advances in catalysis and processes for hydrogen production from ethanol reforming

Sukumaran RK et al (2010) Lignocellulosic ethanol in India: prospects, challenges and feedstock availability. Bioresour Technol 101(13):4826–4833

Sun Y, Cheng J (2002) Hydrolysis of lignocellulosic materials for ethanol production: a review. Bioresour Technol 83(1):1–11. https://doi.org/10.1016/S0960-8524(01)00212-7

Tabibian SS, Sharifzadeh M (2023) Statistical and analytical investigation of methanol applications, production technologies, value-chain and economy with a special focus on renewable methanol. Renew Sustain Energy Rev 179:113281

Thomas KC, Hynes SH, Ingledew WM (1996) Practical and theoretical considerations in the production of high concentrations of alcohol by fermentation. Process Biochem 31(4):321–331. https://doi.org/10.1016/0032-9592(95)00073-9

Tran HTM, Cheirsilp B, Hodgson B, Umsakul K (2010) Potential use of *Bacillus subtilis* in a co-culture with *Clostridium butylicum* for acetone–butanol–ethanol production from cassava starch. Biochem Eng J 48(2):260–267. https://doi.org/10.1016/J.BEJ.2009.11.001

Trindade WRS, dos Santos RG (2017) Review on the characteristics of butanol, its production and use as fuel in internal combustion engines. Renew Sustain Energy Rev 69:642–651. https://doi.org/10.1016/J.RSER.2016.11.213

Vasco-Correa J, Shah A (2019) Techno-economic bottlenecks of the fungal pretreatment of lignocellulosic biomass. Fermentation 5(2):30

Wang S, Sosulski K, Sosulski F, Ingledew M (1997) Effect of sequential abrasion on starch composition of five cereals for ethanol fermentation. Food Res Int 30(8):603–609. https://doi.org/10.1016/S0963-9969(98)00020-9

Wang S, Ingledew WM, Thomas KC, Sosulski K, Sosulski FW (1999a) Optimization of fermentation temperature and mash specific gravity for fuel alcohol production. Cereal Chem 76(1):82–86. https://doi.org/10.1094/CCHEM.1999.76.1.82

Wang S, Thomas KC, Sosulski K, Ingledew WM, Sosulski FW (1999b) Grain pearling and very high gravity (VHG) fermentation technologies for fuel alcohol production from rye and triticale. Process Biochem 34(5):421–428. https://doi.org/10.1016/S0032-9592(98)00097-1

Wang L et al (2010) Promoting effect of an aluminum emulsion on catalytic performance of Cu-based catalysts for methanol synthesis from syngas. Fuel Process Technol 91(7):723–728. https://doi.org/10.1016/J.FUPROC.2010.02.003

Wang M, Saricks C, Santini D (1999) Effects of fuel ethanol use on fuel-cycle energy and greenhouse gas emissions

Wilkie AC, Riedesel KJ, Owens JM (2000) Stillage characterization and anaerobic treatment of ethanol stillage from conventional and cellulosic feedstocks. Biomass Bioenerg 19(2):63–102. https://doi.org/10.1016/S0961-9534(00)00017-9

Wyman CE (1999) Biomass ethanol: technical progress, opportunities, and commercial challenges. Annu Rev Energy Environ 24(1):189–226
Xavier MR (2007) The Brazilian sugarcane ethanol experience
Yukesh Kannah R et al (2020) Food waste valorization: biofuels and value added product recovery. Bioresour Technol Rep 11:100524. https://doi.org/10.1016/j.biteb.2020.100524
Zhan X, Wang D, Tuinstra MR, Bean S, Seib PA, Sun XS (2003) Ethanol and lactic acid production as affected by sorghum genotype and location. Ind Crops Prod 18(3):245–255. https://doi.org/10.1016/S0926-6690(03)00075-X

Chapter 3
Toward Carbon–Neutral Methanol: Emerging Technologies, Key Challenges, and the Pathway to Net-Zero Emissions

T. J. Jacklien Emema Rose, Amutha Thanarasu, P. Baskaralingam, Madhava Anil Kumar, Jeny Rachel Biju, Jason Thamizhakaran Stanley, and Sivanesan Subramanian

Abstract The ecologically friendly methanol can be produced using a variety of ways from earth-abundant raw materials and remains simple to handle, store, and logistics, it is regarded as a productive, affordable, and safe substitute for fossil fuels. For reducing carbon emissions in the chemical and energy industries, clean methanol can be a significant contributor to reaching net-zero emission goals. Green methanol is generated from e-methanol, which is made from green hydrogen electrolysis and CO_2 collected and biomethanol, which is made by distilling biomass and crop residues. Recent advancements in membrane-based methanol synthesis and purifying technologies are highlighted in the review, covering the use of innovative membrane components such as hybrid framework, ceramics, and polymers. Water can be separated both during and after production using water-selective zeolite membranes. Zeolite membranes within the methanol synthesizing reactor can increase the yield of methanol generation. Furthermore, the present obstacles and potential outcomes for achieving environmentally friendly methanol production on a big scale are examined. In summary, the review highlights topics for further investigation and offers perspectives on the present status of membrane-based techniques for the synthesis and purifying of methanol. The advancement of renewable energies and the enhancement

---

T. J. Jacklien Emema Rose · J. R. Biju · J. T. Stanley · S. Subramanian (✉)
Environmental Management Laboratory, Department of Applied Science and Technology, Alagappa College of Technology, Anna University, Chennai, Tamilnadu, India
e-mail: sivanesh1963@gmail.com

A. Thanarasu
St. Joseph College of Engineering, Chennai, Tamilnadu, India

P. Baskaralingam
Department of Chemistry, College of Engineering Guindy, Anna University, Chennai, Tamilnadu, India

M. Anil Kumar
Centre for Rural and Entrepreneurship Development, National Institute of Technical Teachers Training and Research, Chennai, Tamilnadu, India

A. Bartl et al. (eds.), *Alcohol Production Processes and Their Utilisation in the Transport Sector*, Energy, Environment, and Sustainability,
https://doi.org/10.1007/978-981-96-7384-1_3

of the technological affordability and productivity depend heavily on the creation of efficient membranes and the optimal configuration of membrane-based activities.

Keywords Green methanol · Zeolite · Membrane technology · Carbon dioxide emission · Sustainable environment

3.1 Introduction

Global energy consumption has increased exponentially as a result of evolving lifestyles brought about by technology advancements and growing population. This need has been positively correlated with gross domestic product (GDP) development, particularly in developing nations. According to current data, 34% of the world's energy comes from petroleum, 27% from coal, and 24% from natural gas. This indicates that fossil fuels account for 85% of the total energy supply worldwide (Gautam et al. 2020; Kumar et al. 2022). The combustion of fossil fuels generates anthropogenic greenhouse gases (GHGs), notably carbon dioxide (CO_2), which have led to significant ecological concerns including climate change, acid rain, and depletion of ozone layer (Lee et al. 2020). The ambient concentration of CO_2 climbed from 280 parts per million (ppm) prior to the industrial revolution to 420 ppm in 2022 as an outcome of the increased consumption of carbon-based fossil fuels. By the end of the decade, predictions indicate that CO_2 levels might face raised to 570 ppm (Patil et al. 2024a). The goal of California's Low Carbon fuel Standard is to reduce the carbon footprint of automobile fuel by 20% from 2010 levels by the year 2030. States must cumulatively attain an average of 32% renewable energy by 2030 in accordance with the Renewable Energy Directive (RED II) of the European Union. The worldwide goal of reaching net-zero emissions by 2050 is furthered by both measures (Harris et al. 2021). Attempting to keep global warming significantly under 2°C exceeding pre-industrial rates by 2100, the Paris Agreement, which was signed by 196 nations in 2015, will consolidate national and regional initiatives to reduce greenhouse gas emissions and mitigate against the effects of climate change (Garcia et al. 2021; Battaglia et al. 2021). Therefore, the development of renewable and environmentally friendly energy carriers is imperative.

Methanol (MeOH) is produced at about 90 places worldwide, extensively assisting in a manufacturing output of nearly 110 million metric tons yearly and volunteering as an essential chemical in the global economy. It is recognized for its exceptional energy density (22.7 MJ/kg, 18 MJ/L HHV) and adaptability as a fuel, solvent, and precursor for numerous chemicals, which includes olefins, gasoline, kerosene, and dimethyl ether (DME) (Adnan and Kibria 2020; Deka et al. 2022). The methanol economy provides a contemporary approach to progressively lessening our dependency on coal and oil, which, because of their widespread use, are significant causes of global warming. Methanol is an essential chemical in many different sectors in addition to being used as fuel for internal combustion engines and domestic energy requirements (Sonthalia et al. 2023). By using methanol as a cleaner fuel substitute, India's methanol economy seeks to improve air quality, lessen reliance on fossil fuels, and increase energy security. The nation wants to develop methanol in industrial, transportation, and cooking applications by utilizing its coal reserves and agricultural waste through NITI Aayog program. Methanol is produced through an exothermic reaction at temperatures between 220 and 300 °C and pressures between 50 and 100 bar utilizing syngas (CO and H_2), which usually originates from steam-reforming natural gas. The extraction of methanol from natural gas frequently ends up in carbon levels that vary from 250 to 540 kg of CO_2 equivalent (CO_2e) per ton, emphasizing the critical demand of finding greener manufacturing strategies (Scomazzon et al. 2024). Global green methanol output is presently insignificant, at a mere 0.2 Mt/yr or roughly 0.2% of overall production. It is predominantly generated by biomass gasification, municipal solid waste and sludge, but it has great potential for the energy sector and may be essential in decarbonizing the chemical and energy sectors, as suggested by Nobel laureate G.A. Olah (Sollai et al. 2022; Bazaluk et al. 2020; Iribarren et al. 2022). Green methanol, which is made through sustainable and environmentally friendly sources and methods, serves numerous applications in a variety of sectors because of its clean and adaptable nature. It can hasten the defossilization process in industries like chemical manufacture and logistics (Reddy et al. 2023).

Unrefined methanol contains impurities, requiring purification for optimal recovery. Membrane distillation (MD) is a separation technique that maximizes operational efficiency by using a hydrophobic or superhydrophobic membrane. It is a cutting-edge, new technique renowned for its extraordinarily high separation efficiency. MD is a cost-effective and energy-efficient method for methanol purification, outperforming traditional techniques like liquid–liquid extraction, adsorption, crystallization, and azeotropic distillation. The employing of mild chemicals, environmental friendliness, and modular design provide intrinsic benefits to membrane systems that improve process safety and financial feasibility (Chakrabortty et al. 2020; AbdulKadir et al. 2020). Zeolite membranes offer some unique characteristics among the various types of membranes that can be used in membrane reactors: (a) they have good chemical and physical stability; (b) they allow for pore size customization; and (c) it is possible to alter the zeolite's chemical composition, allowing for an ideal collaboration of the zeolite surface with the molecules of chemicals (Gorbe et al. 2018). Zeolites are crystalline aluminosilicates that are porous and have found beneficial uses in adsorption, catalysis, and ion exchange. Because zeolite membranes are hydrophilic by nature, they are frequently utilized to remove specific amounts of water, particularly from aqueous-organic combinations. The most often utilized zeolite materials in the construction of membranes include mordenite, zeolite-A, chabazite, sodalite, zeolite-T, and Zeolite Socony Mobil (ZSM)-5. Zeolite compounds were discovered to have higher thermal and chemical stability, making them more advantageous for this sort of usage (Ullah et al. 2023).

This analysis offers a succinct evaluation of the state-of-the-art methanol manufacturing technology, emphasizing recent developments and tackling industry issues.

The article examines renewable possibilities that are essential for improving production efficiency using membrane-based technologies, with a focus on clean methanol as a possible energy source. It also looks at the way zeolite materials might enhance manufacturing procedures. Together with a brief lifecycle study of resource efficiency, the review provides insights into the sustainability and economic feasibility of large-scale renewable green methanol production. While pointing out knowledge gaps, the discussion constructively assesses the material that is presently accessible.

3.2 Tracing the Evolution of Methanol Production

Methanol often referred to as wood alcohol, methyl alcohol, or carbinol, is a colorless, polar liquid that has a freezing point of 97.6 °C and a boiling point of 64.6 °C at room temperature. Robert Boyle is credited with its discovery in 1661 when he created pure methanol by purification of crude wood vinegar over lime milk (Araya et al. 2020). Methanol is therefore regarded as the most prospective fuels because of its liquid state at atmospheric temperatures and extensively used in the chemical and energy fields. As input for the production of acetic acid, MTBE, DME, and formaldehyde, all of which are utilized in goods like adhesives, foams, plywood subfloors, solvents, windshield washer fluid, etc. (Meunier et al. 2020). In order to improve the fuel grade and increase oil economic benefits, methanol is also utilized to generate biodiesel and mixed to regular fuel (85% methanol + 15% fuel), like gasoline. Olefin synthesis using methanol is currently at a commercial level (Biswal et al. 2022). Methanol is a fuel that is highly efficient, safe, environmentally friendly, dependable, and applicable. It is a great alternative to conventional internal combustion machines (Tian et al. 2022a). There are several reported synthesis pathways for methanol production. Traditionally, low-pressure methanol synthesis is used to generate it from carbon monoxide and hydrogen. There are several ways to obtain this synthesis gas. Steam reforming, partial oxidation, and autothermal reforming are the common methods to derive natural gas from fossil fuels. Other fossil fuel resources comprise coal or coke oven gas in addition to natural gas (Schorn et al. 2021). The traditional methanol production pathway was illustrated in Fig. 3.1. Methanol has been proposed to serve as a practical hydrogen transporter that is simple to store and move. Methanol is useful in automobiles too because it may be mixed with conventional gasoline without requiring any technological changes to the portfolio of vehicles. With its numerous uses in manufacturing goods and as an additive ingredient for transportation, methanol is currently produced in excess of 70 million tons annually globally (Wolday and Ramteke 2024). A vast range of synthetic goods, including adhesives, LCD screens, silicones, paints and medications, use methanol as a base material. China is a major manufacturer of methanol, accounting for 65% of the methanol generated globally. It uses this fuel to transport blends, ranging from M5 to M100 (Srivastava et al. 2022).

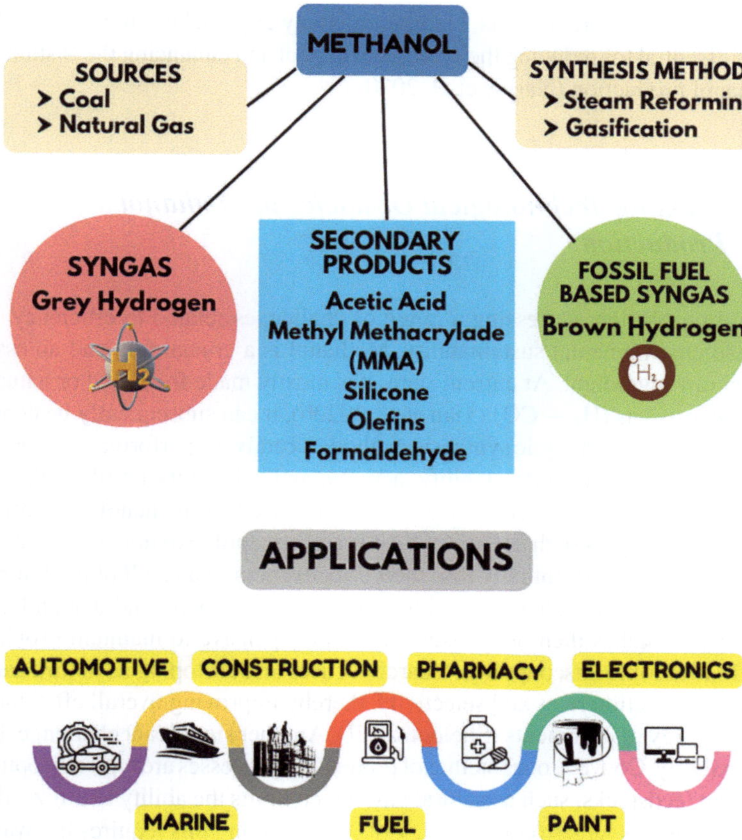

Fig. 3.1 Conventional method of producing methanol

3.3 Ecological Hurdles in Production of Methanol

Methanol production primarily stem from its reliance on non-renewable resources and associated environmental impacts. Traditional methods using natural gas or coal emit greater amount of carbon dioxide, contributing to climate change, and consume large amounts of water, straining local ecosystems (Narine et al. 2020). The energy-intensive nature of the process, often fueled by fossil sources, adds to the carbon footprint, while air and water pollution from pollutants like sulfur oxides and nitrogen oxides pose public health risks (Jocz et al. 2019). Additionally, land use for raw material extraction can lead to habitat destruction and biodiversity loss, and the transportation of methanol and its feedstocks carries the risk of spills and environmental contamination (Gutzmer et al. 2019). To address these challenges, sustainable alternatives such as renewable feedstocks, green hydrogen, carbon capture and storage, energy efficiency improvements, circular economy approaches, and advanced catalysts are essential for reducing the ecological impact and enhancing the sustainability of methanol production (Narine et al. 2020a).

3.3.1 Tackling Technological Obstacles in Methanol Production

Methanol production addressing a range of challenges related to efficiency, scalability, and environmental sustainability. Methanol is a crucial fuel and an essential chemical raw ingredient. At current state, it is mostly made from coal or natural gas using synthesis gas ($H_2 + CO$) (Tian et al. 2022b), it can subsequently be converted into methanol by the catalytic synthesis method. A catalyst's performance is primarily determined by three factors: stability, activity, and selectivity (Patil et al. 2024b). One major obstacle is the efficiency of catalysts used in methanol synthesis. The conventional catalysts such as $Cu/ZnO/AlO_3$ are susceptible to impurities, have poor thermal stability, and eventually lose their effectiveness as a result of agglomeration and toxicity. They also challenge with direct CO_2 conversion and demand a lot of energy, which makes them less sustainable and expensive to maintain (Toyir et al. 2009). To overcome this, ongoing research focuses on developing advanced catalysts that enhance reaction rates and selectivity, thereby improving overall efficiency and reducing energy requirements (Sheldon 2017). Another significant challenge is feedstock flexibility. Conventional methanol production processes are typically optimized for specific feedstocks, such as natural gas, which limits the ability to utilize alternative or renewable sources (Gupta et al. 1998). Addressing this requires innovation in process design to create systems capable of efficiently converting diverse feedstocks, including biomass, captured CO_2, and various waste products, into methanol. These flexible processes can enhance sustainability and reduce reliance on non-renewable resources (Sarp et al. 2021). Some cumulative technological obstacles and feasible solutions in methanol production are mentioned in Table 3.1.

Integration of renewable energy into methanol production is another critical technological hurdle. Current production methods often rely on fossil fuels, making it difficult to incorporate renewable energy sources without significant modifications to existing infrastructure (Andersen et al. 2006). Developing processes that utilize renewable energy for direct production and the generation of green hydrogen, which can then be used in methanol synthesis, is essential for reducing the carbon footprint of methanol production. Additionally, efficient carbon capture and utilization (CCU) remains a technical challenge. Capturing CO_2 and converting it into methanol is complex and can be cost-prohibitive (Garay et al. 2019). Advancements in CCU technologies, coupled with supportive economic and policy measures, are necessary to make these processes viable at scale. By addressing these technological obstacles through innovation and investment, the methanol industry can move toward more sustainable and environmentally friendly production methods (Harmsen et al. 2019).

Table 3.1 Technological obstacles and feasible solutions in methanol production

Technological obstacle	Description	Feasible solution	References
High energy consumption	Demands high pressures (50–100 bar) and temperatures (200–300 °C)	Produce catalysts with low pressure and low temperature	Tabibian and Sharifzadeh (2023)
CO_2 utilization challenges	Inadequate conversion rate of direct CO_2 to methanol using conventional catalysts	Use cutting-edge catalysts such as MOFs or compounds based on zirconium	Patil et al. (2024b)
Feedstock purity requirements	Conventional catalysts are poisoned by impurities such as sulfur	Make use of self-regenerating or impurity-tolerant catalysts	Bozzano and Manenti (2016)
Catalyst deactivation	The lifespan and efficiency of catalysts are decreased by hardening and contamination	Create stronger catalysts with greater durability and reliability	Bram et al. (2024)
Limited use of non-traditional feedstocks	The composition of syngas created from biomass and recyclables varies	Create adaptable catalysts and reactors to deal with varying raw materials	Hoang et al. (2023)
Carbon footprint	Mostly depends on hydrogen produced from fossil fuels	Utilize water electrolysis to produce sustainable hydrogen	Peter (2018)

3.3.2 *Social and Intellectual Constraints in Methanol Production*

Addressing social and intellectual constraints in methanol production involves overcoming a variety of challenges related to public perception, regulatory frameworks, and the dissemination of knowledge and innovation (Arora et al. 2018). Intellectually, the field faces the challenge of integrating multidisciplinary knowledge from chemistry, engineering, environmental science, and economics to develop innovative and sustainable production processes. Collaboration between academia, industry, and government is essential to foster the exchange of ideas and promote research and development (Collodi et al. 2017). Furthermore, regulatory frameworks need to evolve to support the adoption of cleaner technologies while ensuring safety and sustainability standards are met (Naik et al. 2010). One of the primary intellectual constraints is the need for extensive research and development (R&D) to improve production methods, which requires substantial funding and long-term investment. The complexity of methanol synthesis, especially when integrating renewable energy sources and carbon capture technologies, demands a deep understanding of multiple scientific and engineering disciplines (Schwarz et al. 1995). Moreover, the pace of innovation is often slowed by the fragmented nature of research efforts, with various institutions and companies working in isolation rather than collaboratively (Whittet et al. 2011). To overcome these barriers, fostering a collaborative environment where knowledge and resources are shared is crucial. This could involve public–private partnerships, open-access research initiatives, and international cooperation. Encouraging interdisciplinary education and training programs can also help build a workforce capable of tackling the complex challenges associated with sustainable methanol production (Jocz et al. 2019a).

3.3.3 Addressing Concerns Along Legislative, Public Recognition, Environmental, and Scientific Fronts

Methanol production requires a multifaceted approach that spans legislative, public recognition, environmental, and scientific fronts (Narine et al. 2020). Governments can provide incentives such as tax breaks, grants, and subsidies for companies investing in renewable feedstocks, carbon capture and utilization (CCU), and energy-efficient processes (Collodi et al. 2017). Additionally, establishing stringent environmental standards and monitoring systems can ensure compliance and drive innovation toward greener solutions (Schorn et al. 2021). Public awareness campaigns can educate communities about the environmental and economic advantages of sustainable methanol production, such as reduced greenhouse gas emissions and the potential for job creation in green industries (Svitnič et al. 2023). Current situation of CO_2 emission is illustrated in Fig. 3.2. Engaging with local communities and stakeholders through transparent communication and participatory decision-making processes can build trust and mitigate opposition to new technologies. From an environmental perspective, it is vital to adopt practices that minimize the ecological footprint of methanol production. This includes utilizing renewable energy sources, optimizing water usage and implementing waste reduction strategies (Calixto 2023). Environmental impact assessments (EIAs) and continuous monitoring can help identify and mitigate potential ecological risks. On the scientific front, fostering interdisciplinary research is key to overcoming technical barriers in methanol production. Funding for R&D in areas such as catalyst development, renewable feedstock conversion, and CCU technologies is crucial (Hoang et al. 2023). By addressing these concerns holistically across legislative, public, environmental, and scientific domains, the methanol industry can progress toward more sustainable and socially accepted production methods (Franklin et al. 2000).

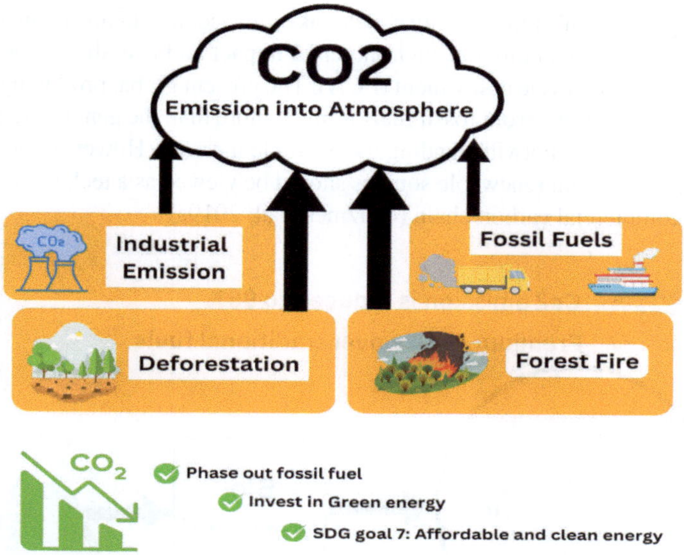

Fig. 3.2 Current situation of CO_2 emission

3.4 Sustainable Technologies and Processes of Methanol Production

Methanol is a conventional one-carbon organic molecule that has a wide range of notable uses, including fuel for direct methanol fuel cells, automobiles, ships and energy generators, as well as the synthesis of pure H_2 (Tian et al. 2022b). The energy quantity in methanol is 56,000–64,000 BTU/gal, roughly 48.69–52.45% that of gasoline and 39.8–41.2% that of methane (Varela et al. 2024). Green methanol is characterized as being synthesized without emitting any pollutants and utilizing renewable resources. The various approaches for producing green methanol are explained in Fig. 3.3. Green methanol exhibits potential as a workable substitute for fossil fuels, considering the growing demand for sustainable energy sources and the intensifying environmental problems. According to Bazaluk et al. (2020a), the following kinds of green methanol are distinguished: renewable, biomethanol, and low-carbon methanol (LCM). Organic substrate is used to manufacture biomethanol. Carbon dioxide and renewable electricity are used to generate renewable methanol. By combining carbon dioxide from processing plants with natural gas or another fossil fuel, LCM is created. Bowker (2019) looks upon methanol synthesis, which is commonly referred to as "green." Because it releases no additional net CO_2 emissions entering the environment, it is referred to as "green." Furthermore, since about half of the heat of combustion of methanol comes from the renewable energy found in its hydrogen components, it may also lower CO_2 emissions. An evaluation of a methanol production's renewability should take into account all upstream and downstream processes, including the environmental impact and overall carbon footprint, such as through life cycle assessment (LCA). The present global producing potential for methanol generated from fossil fuels is much more than the generating ability for renewable methanol, notwithstanding its favorable increase. However, the manufacture of methanol from renewable sources should be viewed as a technique that helps close the commercial carbon chain (Gutzmer et al. 2019a).

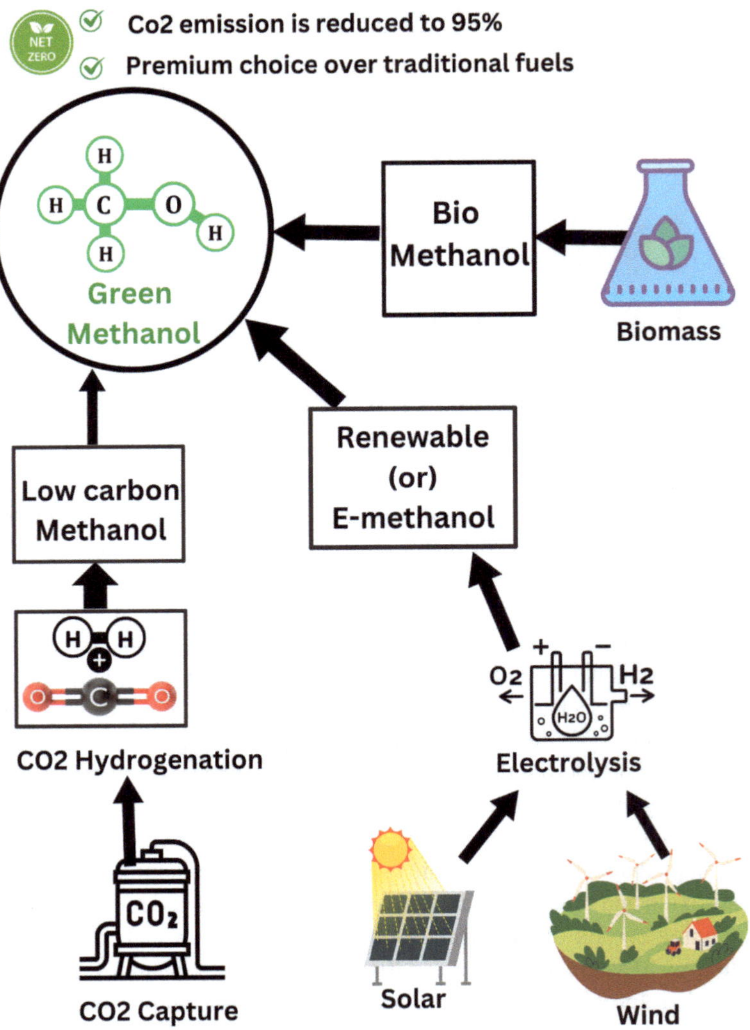

Fig. 3.3 Various approaches for producing green methanol

3.4.1 Biomass to Methanol Production

The advancement of technologies for effective utilization of biomass feedstock in methanol production has been accelerated recently. The U.S. Department of Energy (DOE) has released an analysis indicating that, over one billion tons of domestic biomass might be obtainable by 2040 (Harris et al. 2021). The regenerative characteristics of virgin and waste biomasses, along with their minimal environmental impact, make them a promising sustainable substitute for traditional energy resources. Global efforts are underway to mitigate greenhouse gas emissions in the fossil fuel-based economy by substituting or augmenting it with biofuels (Gautam et al. 2020). Among many intriguing techniques over converting biomass is biomass gasification, which can produce significant worth syngas or H_2. Investigators have focused a lot of emphasis on sorption-enhanced gasification method above other biomass gasification techniques. Some of the biomass that are extensively used for the synthesis of methanol is listed in Table 3.2. Liu et al. (2024a) conquered the barrier stating that syngas produced through biomass gasification might not be directly employed for methanol synthesis within a combined framework via biomass sorption-enhanced gasification in conjunction with a carbon utilization model which was established through Aspen Plus. This resulted in high syngas efficiency, integrated CO_2 capture, and utilization, and adjustable H_2/CO ratio. The primary disadvantage of producing biomethanol from biomass is that it is more expensive than fossil resources, due to differences in raw material, pretreatment, and refining. Yet, rising CO_2 emission from fossil fuels will significantly narrow this gap. Additionally, the shift toward sustainable development objectives emphasizes techno-economic aspects, favoring biomethanol process feasibility (Bobadilla et al. 2024).

A "cradle to gate" life cycle assessment, element utilization analysis, and energy utilization analysis of the "coal and biomass co-gasification to methanol cogeneration process" was carried out by Liu et al. (2024b). In comparison with syngas obtained from coal, the addition of biomass modifies the H/C in the syngas. This modification additionally impacts on later process variables, like the syngas split ratio, the water gas shift (WGS) functionality and the energy usage for the components engaged. The biomass ratio causes the consumption performance of C and H to grow, while O utilization efficiency achieves its optimum level at biomass-to-fuel ratio (B/F) = 0.5 along with the best energy efficiency. In conclusion, the findings demonstrated that adding biomass to methanol manufacturing systems greatly improves element usage, energy efficiency, increases methanol yield while also lowering CO_2 emissions and the well-being of humans and the environment. Hernández-Camacho et al. (Camacho et al. 2024) provide a mathematical model that represents the supply network for the production of methanol using renewable resources from 32 Mexican states, identifying the optimum course of action in terms of balancing profit, social impact, and CO_2 emissions. Agricultural wastes, urban solid waste, vegetable oil and beer industry debris, and manure are among the several possible source materials that are analyzed. With a marginal cost index of 2592.536, CO_2 emissions of – 0.021 Mt/y and a revenue of 7,334,100 USD/y, the best-compromising option is identified.

Table 3.2 Biomass used for the synthesis of methanol

Biomass	Method	Temperature °C	Other process details	References
Oil palm residues	Gasification process	750	Equivalent ratio = 0.25 Energy efficiency = 38.57% Exergy efficiency = 25.44%	Im-orb et al. (2020)
Municipal solid waste	Refuse-derived fuel gasification	900	Methanol yield = 0.629 kg_{MEOH} / kg_{RDF}	Šuhaj et al. (2017)
Rice straw	Integrated biomass pyrolysis, gasification and methanol synthesis	800	Methanol yield = 0.23 Kmol h^{-1} Bio-oil yield = 0.09 Kmol h^{-1} Energy efficiency = 60.07%	Im-orb and Arpornwichanop (2020)
Corncob	Thermochemical conversion	560	Methanol yield = 53% Energy recovery = 72.7%	Brigagão et al. (2019)
Forest residues	Catalytic gasification	225	Methanol purity = 99.85%	Carvalho et al. (2017)
Pine biomass	Gasification	900	Steam to biomass mass ratio = 0.9 Reduced CO_2 emission = 80%	Gamero et al. (2018)
Residues from agriculture and forest maintenance	Gasification	900	Methanol to biomass ratio = 0.43 steam to biomass ratio = 0.89 oxygen to biomass ratio = 0.57	Haydary et al. (2021)

The research by Rajaee et al. (2023) offers a thorough examination of three different biomass to methanol plant configurations, all of which include a high temperature solid oxide electrolysis cell (SOEC) and a circulating fluidized bed blown biomass gasifier (CFBG). Whether the SOEC was fed with steam (StE), syngas (SyE), or a combination of CO_2 and steam (StCE) allowed for the differentiation of the setups. The size of the SOEC units was designed to attain high carbon efficiency, or to hold onto as much biogenic carbon as possible in the methanol that is generated. Wu et al. (2024) suggested a novel approach to methanol synthesis that combines a biomass integrated gasification combined cycle (MS-BIGCC) with methanol synthesis. To evaluate the unique principle, in-depth thermodynamic analysis, economic appraisal, and sensitivity analyses were carried out based on a 41.8 MS BIGCC power plant and a 100 kt/yr methanol production plant shows a 3.46% gain in energy conservation and a 61.9% performance in energy efficiency. Furthermore, over the course of its lifetime, this design generates 72.6 million dollars more in profit than individuals do. Vulnerability evaluation additionally shows that about 250 °C is the ideal temperature for the polygeneration system and 60 bar, having a 0.90 MS recycling ratio being preferred.

3.4.2 *Carbon Capture and Effective Utilization*

In power stations that rely on coal for energy production, it is possible to mitigate CO_2 emissions by trapping the gas and utilizing it for alternative purposes that yield advantages. CO_2 capture and utilization (CCU) are the term for this combination technology (Yang et al. 2023). Based on the precise step in the procedure when CO_2 is separated, there exist three primary types of capture technologies. The three such systems are the oxyfuel combustion carbon capture system, the post-combustion carbon capture system, and the pre-combustion system (Ibigbami et al. 2024). After igniting the fossil fuel, CO_2 is extracted through the exhaust gas via post-combustion capturing. Currently, the principal coals used in all conventional coal-fired power plants are burned instantly to produce electricity. Therefore, as an "end-of-pipe" technology, post-combustion capture makes more sense for coal-fired power plants that are currently in operation (Kajolinna et al. 2024). CO_2 is captured via a variety of post-combustion techniques, such as adsorption, membrane separation, and absorption (Akeeb et al. 2022).

Consequently, efficient methods for producing methanol from captured CO_2 have drawn a lot of interest as a long-term way to address four problems: lowering CO_2 emissions, increasing CO_2 worth, getting deal with the requirement for CO_2 storage, and satisfying the requirement of methanol (Zhang et al. 2024). Itemized cost estimation, sensitivity analysis, uncertainty analysis, and carbon footprint analysis were carried out by Kim et al. (2022) for four cases of methanol production (steam methane reforming, coal gasification, water electrolysis, and direct methanol production by carbon dioxide electrolysis) with methanol production capacities of 1, 10, 20, and 50 ton d^{-1} in order to determine the optimal process with regard to both financial and environmental viewpoints. With a methanol output of 50 tons per day, the corresponding unit methanol production expenses of 3.66, 2.99, 3.69, and 0.55 \$ kg^{-1} were determined by itemized cost estimation that took into account the results of the carbon footprint study. The direct methanol production by carbon dioxide electrolysis has become more practical than other scenarios, in comparison with unit methanol production cost (0.396 \$ kg^{-1}) from fossil fuel-based synthesis, in order to generate electricity, artificial methane, and methanol with the least amount of CO_2 emissions, Eisavi et al. (2024) propose and examine a holistic approach from an economic perspective. A protone exchange memberane electrolyzer (PEME), fuel synthesis units, molten carbonate fuel cell (MCFC), Rankine cycle (RC), organic Rankine cycle (ORC) and externally fired gas turbines (EFGT) make up the intended system. MCFC is used to raise the effluent's CO_2 content, which facilitates easier capture. Methane and methanol are produced when the CO_2 from the EFGT unit exhaust gas is collected and hydrogenated by electrolyte H_2. Alharthi et al. (2022) synthesized nanocrystalline palladium metal (Pd) based on mesoporous silica for CO_2 hydrogenation-assisted methanol production. The results of the structure–activity tests indicated that the degree of Pd crystallinity and higher dispersions were among the primary elements influencing the Pd/SiO_2 catalysts' ability to hydrogenate pure CO_2 in a slurry reactor.

3.4.3 Renewable Energy Integration

One important technological advancement for the move to fossil fuel-free energy facilities is renewable power-to-fuel (PtF). For a system to be deemed completely sustainable, all of its raw materials must come from solar, wind, hydro, or geothermal energy. Green hydrogen, which is generated from clean energy, and collected CO_2 are combined to form e-methanol. The international renewable energy agency (IRENA) states that pilot and experimental units for the production of e-methanol have been built globally (Nemmour et al. 2023). The independent transformation, utilizing air capture to produce CO_2 and H_2O, is evaluated by Bos et al. (2020) along with the economics of employing wind and air energy to produce renewable methanol. Pendse and Prajapati (2024) investigate the combination of electrochemistry and solar electricity to operate effective methanol manufacturing systems, that emphasize upon the methane oxidation reaction (MOR) and electrochemical reduction (ECR) of CO_2 as routes for methanol synthesis. Employing a hot reservoir rock (HRR) system for methanol synthesis with high temperature and pressure within an underground massive natural reservoir reactor, the Wojnarowski and Janiga (2024) provide a novel method to utilize in situ thermal energy. Involving the immediate application of renewable resources, this strategy seeks to incorporate renewable energy technology into a single energy generating and fuel manufacturing facility. Depending upon an hourly power supply through hybrid PV-wind platforms, Fasihi and Breyer (2024) explore the ideal system design for the least expensive green e-methanol synthesis using electrolytic hydrogen and atmospheric carbon dioxide at a $0.45° \times 0.45°$ geographical accuracy. The findings indicate that, at the optimal sites, e-methanol could be generated for a price that varies between 1200–1500, 600–680, 390–430, and 315–350 € per t_{MeOH} in 2020, 2030, 2040, and 2050, respectively, with a mean weighted price of capital of 7%.

3.4.4 Advanced Catalysis Used in Methanol Production

Currently, the development of novel and extremely productive catalysts is the primary goal of methanol synthesis studies. To improve activity and selectivity, multiple experiments were carried out to create catalysts with a large surface area, high active site dispersion, and reduced particle size (Ali et al. 2015). Within the past few decades, the technological and scientific study of catalysis have caused advancements in a number of industries. Many heterogeneous and homogeneous catalytic systems have been documented currently enabling the capture of CO_2, and transforming it into beneficial substances might be additionally exploited as important sources of energy in the manufacturing sector (Darji et al. 2023). One of the key techniques in CCU is the catalytic hydrogenation of CO_2, which is a remarkable green and ecologically friendly technique. CO_2 is converted into energy products like methanol, olefins, and paraffin by the process of catalytic hydrogenation (Atsbha et al. 2021). Homogeneous catalysts must be capable to break down H_2 to allow insertion into CO_2 to generate formatting species, along with sufficient hydricity to reduce formate to methanol, in order to facilitate the production of methanol from CO_2. The process might prove thermodynamically advantageous at lower temperatures because CO_2 hydrogenation to methanol is exothermic in general (Sen et al. 2022). Le Chatelier's theory states that greater pressure will encourage the methanol production. Various metal catalysts, including as Cu, Zn, Ag, Cr, and Pd, were used to catalyze the hydrogenation of CO_2 into methanol (Zhang et al. 2021). The catalysts $Cu/ZnO/Al_2O_3$ are employed to produce methanol experience significant deactivation when in operation, mostly as a result of compaction, Barrow et al. (2024) describe compositions where the deliberate employment of a modest amount of a Si-based activator has significantly decreased deactivation, leading to accumulated active improvements that can surpass a factor of 1.8 in comparison with unpromoted catalysts. Extended lifetimes up to twice as long as those of earlier era catalysts, are another benefit of this improved stability. These results have significant ramifications for the sustainable manufacture of methanol from H_2 and CO_2.

3.4.5 Membrane-Based Methanol Production Toward Net Zero

Sustaining the product grade at a satisfactory level requires downstream filtration of the methanol created. Hydrogenating CO_2 produces three products: water, methyl alcohol, and the intermediate product. Since the water in the process mixture functions as a catalyst toxin and stops the synthesis from happening, it is imperative to separate the water out of the solution (Qadeer et al. 2023). Methanol can be refined using a variety of methods. Distillation is frequently the main technique used in solvent collection. Distillation is a simple procedure that effectively isolates solvents from water (Araya et al. 2020). Incondensable components like CO_2, CO, hydrogen gas, and water are extracted from methanol by the distillation phase. Energy-intensive filtration processes need a lot of energy. In regard to lowering expenses, the distillation procedure is not very effective due to the modest volatility found in the composition. Membrane-based techniques are now widely acknowledged as essential components of many commercial methods of segregation. Gas discrimination, pervaporation, reverse osmosis, nanofiltration, dialysis, and membrane contactors are examples of popular membrane operations (Mohammed et al. 2024). Membrane panels for the gas segregation system are able to achieve this kind of precise isolation. Selective membranes allow for the achievement of purification in just two steps. Due to the high pressure demands, membrane-based gas separation technology suffers from limited selectivity, leakage, difficulty handling gases, and a high energy consumption (Chakrabortty et al. 2020). Organic materials can be dewatered, organic pollutants can be removed from water-based solutions in negligible levels, and organic matter can be separated with the use of modifications to the pervaporation procedure. Depending on the elements that follow it in the manufacturing network, this method can be classified as either hydrophilic or organophilic pervaporation. Methanol was extracted from solutions containing water using organophilic membranes (Rostovtseva et al. 2022; Golubev et al. 2021). In pervaporation procedures, membranes composed of inorganic zeolite and composite polydimethylsiloxane (PDMS) were employed. Pervaporation membranes come in three different kinds: symmetric, asymmetric, and composite. The liquid cannot pass through the membrane because of obstacles that are created during the pervaporation process, which begins at the surface layer (Liu and Jin 2021). Membrane distillation (MD) has many advantages over other membrane separation techniques, such as the following: Its advantages over traditional pressure-driven membrane separation processes include: (1) nearly 100% removal of stable materials, large molecules, and ionized elements; (2) lower operating temperatures; (3) lower operating pressures; and (4) reduced membrane mechanical property requirements. Owing to its unique advantages, the MD approach is extensively employed in numerous applications, such as wastewater treatment and saltwater desalination. Two reasons contribute to the drawbacks of the MD process: (1) the scarcity of substantial-performance membranes and (2) the high energy required (Karim et al. 2021). Numerous experiments were conducted employing zeolite, cation exchange, and polyimide membranes for the production of methanol. The main features that have been found consist of thermal and chemical stability, temperature-dependent water selectivity over methanol and the reagents, and membrane penetration (Prašnikar et al. 2024). Membranes used for synthesis of methanol are discussed in Table 3.3.

3.4.5.1 Membrane Reactors

The 1950s marked the initial development of the membrane reactor (MR) concept as a solution to issues with CO_2 release, which demonstrated poor conversion, harsh operating circumstances, and high budget (Jazani et al. 2023). According to the standard description provided by IUPAC, a membrane reactor (MR) is a device

Table 3.3 Membranes used for synthesis of methanol-related products

Membrane type	Materials	Use	Properties	References
Poly(vinyl alcohol)-based composite membrane	Titanium dioxide (TiO_2) and carbon nanoparticles	Alkaline direct methanol fuel cells	Tensile strength = 163 MPa Permeability = 45,000 Ss/cm^3 Ionic conductivity = 10^{-2} S/cm	Khalid et al. (2024)
Organosilica membranes	Ultrathin and chemically stable Bis(triethoxysilyl)acetylene	Organic/organic pervaporation	High separation factor = 120 Permeation flux = 2–4 kg m^{-2} h^{-1}	Dong et al. (2020)
MFI zeolite membrane	Silica substrate in a parent gel of Si/Al = 100	Methanol-to-olefins (MTO) reactions	Propylene selectivity = 48.5% C_3H_6/C_3H_8 ratio = 236	Tanizume et al. (2021)
Zeolite membrane	Zeolite A, mordenite, zeolite T, chabazite and Ti-Chabazite	Hydrogenation of CO_2 to methanol	Zeolite A provided the best H_2O/H_2 separation factor	Raso et al. (2021)
Ion transport membrane	ITM oxygen integrated into the autothermal reformer (ATR) process	Natural gas to methanol	Power production = 68% Overall plant exergy losses decreased by up to 21%	Fankomo and Greeff (2024)

that may conduct both endothermic and exothermic reactions and membrane-based separations simultaneously within an identical physical device (Palma et al. 2020). A membrane reactor combines the processes of membrane separation with chemical conversion. The membrane reactor serves as a contactor, dividing a second sector that contains a catalyst from the resultant stream in the first sector (Kancherla et al. 2021). In addition to boosting methanol production, membrane reactors also enable greater level of quality and a reduction in catalyst breakdown rate, which lowers process expenses. Four fundamental concepts are used to categorize membrane reactors: the type of reactor (extractor, distributor, or contactor); the type of membrane utilized during the process (organic, inorganic, porous, or dense); if the membrane reactor is catalytic or inert; and the type of reaction that takes place in the membrane reactor (e.g., dehydrogenation, esterification, water dissociation, or wastewater treatment) (Shuit et al. 2012). The intrinsic advantages of membrane systems, which improve process economies and safety, include their straightforward design, lack of harmful substances, sustainability, and great flexibility in operation owing to its modular structure (Qadeer et al. 2023). Additionally, a number of innovative reactor designs were put up for the instantaneous conversion of CO_2 into methanol. Recently, Bazmi et al. (2024) presented a novel design for MeOH production (MeS-MCR) using waste CO_2 by integrating a reverse water gas shift reactor (RWGSR) and a membrane contactor reactor. By overcoming the difficulties that direct CO_2 conversion into MeOH encounters, the incorporated RWGSR/MeS-MCR process yields carbon conversions that are higher than 80%, which is much higher than the optimal conversion of about 20–30% for the methanol synthesis (MeS) reaction. In order to generate methanol, Hamedi and Brinkmann (2021) designed a modified MR model and then used it for CO_2 hydrogenation. They assessed if employing MR affected sensitivity and reaction efficiency. According to the findings, the use of MR can increase reaction rate for greater operating pressures by up to 70.0% in the adiabatic mode and 96.2% in the non-adiabatic mode. In an adiabatic system, this results in a drop of 16.3% in reaction specificity, whereas in a non-adiabatic system, it is just 4.9%. This indicates that MRs have a greater impact on the non-adiabatic configuration's functionality. Two distinct approaches are used Samimi et al. (2019) to improve the indirect CO_2 hydrogenation technique for the generation of green methanol. In the first scenario, water toxicity of the catalyst in the methanol synthesis reactor occurs when the syngas formed in a RWGS reactor travels through a separator. In scenario 2, water was separated throughout the process and the proposed divider was eliminated by applying a water permselective membrane to the RWGS reactor. In scenario 2, methanol synthesis climbed by 109 tons per day (a percentage of 50.23) compared to CO_2 reduction.

3.4.5.2 Membranes Used in Membrane Reactor

Non-volatile components are basically removed from the blend using membrane separation. When purifying methanol using a traditional distillation method, the energy required to separate the methanol out of its aqueous solution can account for up to 40% of the net energy used. As a result, in terms of usage of energy, the membrane distillation technique is economical with the conventional distillation column (Chakrabortty et al. 2023). Le Chatelier's principle states that the membrane isolates the end results from the process mixture when they develop, shifting the equilibrium in favor of end-product production (Elvira et al. 2023). A membrane initial categorization in a membrane reactor is based on its properties (Fig. 3.4). The utilization of biological membranes has a number of disadvantages, despite their ease of manufacturing. These include a restricted pH range, a working temperature below 100 °C, difficulties with cleaning, and a susceptibility to pathogenic infiltration. Many inorganic membranes show promising results in methanol synthesis. A great option for the production of methanol is a microporous carbon membrane. They are very resistant to corrosion, have good thermodynamic strength, and excellent permeability accuracy. The capacity to alter the porosity structure and hydrophobicity/hydrophilicity of these membranes by choosing the proper polymer precursors and operating at the proper temperature is what gives them their exceptional durability (Gallucci 2018).

In order to improve the each single-pass transformation, an alternative high-pressure MR method using a liquid sweep was used by Li and Tsotsis (2019) to study the methanol synthesis process. A silylating chemical was used to alter an asymmetric ceramic membrane, making its pore surface hydrophobic. Polotskaya et al. (2018) focused on comparing two kinds of membranes derived from poly (amidoimide acid) that had 2,2′-biquinoline-6,6′ units in the framework. The components were made from the polymer-metal complex (containing Cu(I)) and from the original polymer. The methanol-hexane azeotropic combination could be separated using the polymer-Cu(I) complex membrane having an isolation coefficient of 980 and a pervaporation separation index of 66.6; as a result, this approach is far more efficient compared to utilizing conventional membranes. To find ways to improve the efficiency of the autothermal methanol synthesis (AMS) process, Rahmani et al. (2012) investigated upon the viability of in situ H_2O removal and H_2 addition techniques using two distinct membranes (H-SOD and Pd/Ag), which were examined using a one-dimensional model.

Membranes are used to improve the methanol production process. In addition, the yields of methanol and hydrogen were increased by 10.55 and 9.28% and 17.22 and 18.5%, respectively. These findings imply that this arrangement may be a strong means of increasing the synthesis of hydrogen and methanol. Zeolite-based membranes are highly intriguing because of their ability to be employed as molecular sieves, which are made possible by the zeolites distinct and constrained pore size variation. Other polymeric membranes are typically not suitable for the synthesis of methanol due to their limited thermal stability. Since only inorganic zeolite membranes can withstand the high pressure and temperature required for

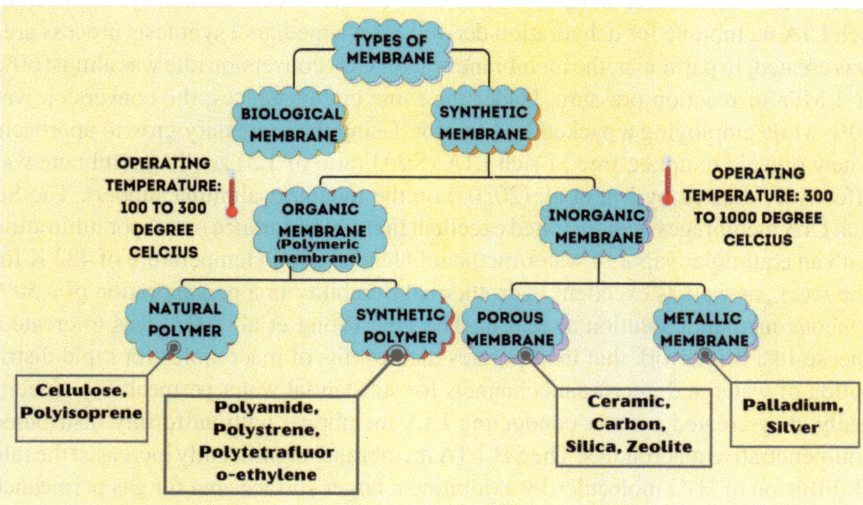

Fig. 3.4 Different types of membrane

the methanol production process, only these membranes appear to be appropriate. Hydrothermal production is used to create zeolite microporous membranes by Gautier et al. (2023), which are then placed on tubular alumina support structures, which give the membrane rigidity. Because of its high preference for water, Linde Type A (LTA), and surface-modified sodalite (SOD) zeolites were chosen for the membrane manufacturing process for methanol synthesis.

3.4.5.3 Zeolite Membranes for Methanol Production

A crystalline hydrated aluminosilicate substance with a standard pore arrangement is called a zeolite. It has high adsorption, catalysis, shape selectivity, and ion exchange feature due to its distinct chemical and physical qualities. Zeolites are more commonly utilized as adsorbents, ion exchangers, and catalysts than other inorganic materials due to their superior hydrothermal rigidity, configurable pore shapes, and customizable chemical characteristics (Li et al. 2021). The combination of fundamental substances like Si and Al determines the physical characteristics of various types of zeolite compounds. The most often utilized zeolite materials in the construction of membranes include mordenite, zeolite-A, chabazite, sodalite, zeolite-T, and Zeolite Socony Mobil (ZSM)-5 (Ullah et al. 2023). Various methods were developed to fabricate zeolite membranes that are both self-supporting and reinforced. They consist of subsequent stages: secondary growth, embedding, seeding, and in situ hydrothermal synthesis (Mohammed et al. 2024). Zeolite membranes are excellent choices for the construction of catalytic membrane reactors due to their consistent pore arrangement and strong temperature stability. At various temperatures and H_2/CO_2 feed ratios, an A-type zeolite membrane reactor converts CO_2 more efficiently and yields more methanol than a traditional catalytic fixed bed reactor (CFBR) (Yue et al. 2021). Seshimo et al. (2021) using seed-assisted hydrothermal synthesis, a Si-rich LTA membrane for dehydration derived from a methanol synthesis process area was created. In particular, the membrane reactor CO_2 conversion rate was almost 60% at 4 MPa of reaction pressure. Under the same circumstances, the conversion was 20% while employing a packed-bed reactor. Using the secondary growth approach, a new organic template-free Si-rich LTA (Si/Al ratio of 1.5) zeolite membrane was effectively created by Liu et al. (2020a) on the tubular α-alumina support. The Si-rich LTA membranes demonstrated excellent thermal endurance in a vapor infiltration with an equimolar vapor of water/methanol blend at a high temperature of 423 K for one week, as well as excellent hydrothermal durability in a pervaporation of a 50% aqueous methanol solution at 333 K. The goal (Song et al. 2023) was to create a cheese-like framework that incorporates the benefits of macroholes for rapid distribution of water and micro-nanochannels for substantial water permeability. Specifically, they created a water-conducting LTA membrane with uniformly distributed non-penetrative macroholes. The SH-LTA membranes significantly increased the rate of diffusion of H_2O molecules by exhibiting a larger surface area for gas permeance and an increased variation in pressure surrounding dispersed macroholes owing to an acute rise in close molecular concentration, as contrasted with intact-LTA and low-defect LTA membranes.

In conclusion, by boosting separation and purification, zeolite membranes offer a great deal of possibilities to enhance the processes involved in the manufacturing of methanol compared to all other methanol production methods. Zeolite membranes aid in the separation of methanol from additional gases (CO, CO_2, and H_2) during the manufacture of methanol. Zeolites are uniformly sized porous substances that have the ability to precisely penetrate methanol while holding on to larger compounds or contaminants. Zeolite membranes have the ability to remove methanol directly from the synthesis gas flow, hence minimizing the requirement for energy-consuming separation procedures. Zeolites are valuable for manufacturing activities because they retain their structure and functionality at high temperatures. They may also provide high permeability rates, which translate to more effective separation. The development of sophisticated zeolite materials with enhanced characteristics for particular separation obligations, upgraded membrane fabrication to lower costs and increase zeolite membrane strength and flexibility, and the combination of zeolite membranes with traditional separation methods to maximize overall process productivity are some of the ongoing study efforts aimed at strengthening the reliability of zeolite membranes.

3.5 Economic Viability and Market Trends of Sustainable Methanol Synthesis

There are about 650 coal gasifier units in operation or undergoing development as of 2014; with 1300 units scheduled for completion by 2019, coal gasification production would have roughly doubled. Nevertheless, the process of converting coal into methanol requires more greenhouse gas emissions that present worldwide carbon reduction initiatives, with output fluctuating between 81.8 and 143.8 g CO_2 e/MJ methanol (Harris et al. 2021). As we move toward more environmentally friendly energy resources and lower greenhouse gas emissions, the subject of sustainable methanol synthesis is becoming progressively more essential. Biomass, CO_2, and renewable electricity are among the substrates that can be used to manufacture sustainable methanol. When weighed against conventional fossil fuels, feedstocks especially renewable ones may be more expensive. Much funding is typically required for sustainable methanol synthesis processes, such as CO_2 collection or electrolysis for hydrogen production. Yet, as technology develops and becomes more widespread, costs are coming down. These can include labor, energy usage, and management.

The three novel "Power-to-Methanol" routes; one-step CO_2-to-methanol electrolysis, two-step synthesis using H_2O electrolysis, and three-step synthesis involving H_2O electrolysis and CO_2-to-CO electrolysis are compared by Adnan and Kibria

(2020) for their techno-economic and climate benefits. The primary economic variables and scientific objectives for "Power-to-Methanol" pathways that are viable are identified. As technological developments occur, sustainable procedures might start with higher operating costs but eventually grow affordable. Because of economics of scale, the price per unit of methanol can drop as output increases. Eventually, industrial plants and technological developments might contribute to cost reduction. Hank et al. (2018) demonstrate how heavily the cost of producing energy and H_2, as well as CO_2, affects the economic viability of a certain power to liquid (PtL) process, especially for eMeOH. In addition, the market's capacity to pay for eMeOH will be impacted by indirect factors like carbon taxes. Germany is presently not financially viable in producing eMeOH under the assessed procedure parameters in this research. Prior to implementing any PtL process, it's critical to consider any possible benefits: Carbon Recycling International's methanol production is supported by inexpensive geothermal electricity and CO_2 from a nearby geothermal power station. Supportive laws and regulations have a big impact on the sustainability of the economy. The greater upfront costs of producing methanol sustainably can be somewhat mitigated by subsidies, tax breaks, and carbon pricing schemes. The shifting balance of supply and demand, feedstock costs, and geopolitical variables can all have an impact on the price of methanol on the international market. Competitiveness with traditional methanol, which is frequently produced from less expensive fossil fuels, is a requirement for sustainable methanol. Growing consumer awareness and stronger environmental restrictions have led to an increase in demand for sustainable chemicals and fuels. Sustainable solutions are being investigated by industries like chemicals, transportation, and automobiles. The Carbon Capture, Utilization and Storage (CCUS) study done by Szima and Cormos (2018) discussed (i) offers a successful technique for lowering an energy-intensive plant's carbon footprint, (ii) offers a technology for CO_2 utilization, and (iii) offers an alternate method of meeting the increasing need for MeOH in Europe, a sector that is predicted to increase in the near future. The plant's high degree of unity, as demonstrated by the mass and energy balances, keeps operating costs below zero. The plant uses 17,650 kg of CO_2 per hour with a carbon conversion rate of 97%, whereas its net CO_2 usage is 17150 kg. In this instance, the value exceeds the annual sales estimate of 50 million euros. Innovation, prototype, and commercial-scale projects, alongside advancements in research from the corporate and public sectors, are all being funded by higher expenditures in sustainable methanol projects.

Reducing carbon footprints is one of the many lofty environmental goals that many organizations are adopting. Methanol derived from sustainable sources offers businesses a practical way to achieve these goals. Based on local assets and regulations, various areas are implementing distinct approaches. For example, a combination of robust environmental rules and policy support is driving significant investment in sustainable methanol production in North America and Europe. Bellotti et al. (2019) examine the financial viability of a 5 MW electrolyzer base-methanol production plant combined with a photovoltaic (PV) power plant, at the pre-feasibility stage. The most favorable economic circumstances whereby the facility can be regarded as an effective investment in terms of average rate of return (ARR) were determined by performing an economic sensitivity analysis relying on the response surface methodology (RSM) methodology. The electrolyzer initial expenditure and the expense of energy were found to be the characteristics that had the biggest impact on the system's economic viability through a sensitivity study. Considering an environmental perspective, it is important to note that the methanol manufacturing plant suggested here is able to avoid consuming around 5.2 million metric tons of natural gas annually and recycle approximately 5800 tons of CO_2 annually in comparison with regular operation. The result could lessen the total carbon impact of methanol production and assist in reducing energy use. Sustainable methanol is expected to play an increasingly important role in the international chemical and energy sectors as long as these developments persist.

3.6 Environmental Impact on Reduction of Carbon Footprint Using Membranes

The conventional way of producing methanol uses a lot of fossil fuels, which increases carbon emissions. Green hydrogen, CO_2 from industrial operations, biomass, and sustainable methanol synthesis are all important. The use of membranes for methanol synthesis usually entails efficient gas separation, which can be used to separate methanol from other byproducts or carbon dioxide (CO_2) from hydrogen (H_2) in order to enhance the chemical process.

Reduction in Carbon Emissions—Membranes can increase the effectiveness of hydrogen separation and CO_2 removal, which can result in improved resource utilization and a decrease in total carbon emissions per unit of methanol generated. By capturing CO_2 from atmospheres or during manufacturing activities, membranes can lessen the quantity of CO_2 that would otherwise be emitted into the environment.

Integration with Renewable Feedstocks—When compared to conventional methanol production processes, the methanol generated has a significantly lower carbon footprint because green hydrogen and collected CO_2 are used as feedstocks. The procedure supports a circular carbon economy, in which carbon emissions are recycled rather than released, by employing captured CO_2.

Energy Efficiency—Membranes have the potential to reduce the amount of energy needed to produce methanol by increasing efficiency of separation. Lowering consumption of energy immediately lowers the carbon footprint because it is a major contributor to carbon emissions. Reduced waste and operating expenses are two benefits of improved techniques, both of which can substantially lower carbon emissions.

Waste Minimization—Lower total emissions and a cleaner production process are the results of less byproduct generation and enhanced separation. Improved material separation and recycling limit waste and cut down on the need for extra resources, all of which help to lessen carbon emissions.

Scalability and Integration—Sustainable membrane-based systems have a wide range of applications in industry and can significantly lower carbon emissions. It may be possible to adapt membrane systems in current methanol factories to lower their carbon footprint without having to completely redesign them.

Comparing zeolite membranes with all alternatives, sustainable methanol production has significant advantages for the environment. Because of its characteristics as molecular sieves, zeolite membranes are well-known for having great selectivity and permeability for specific gases. They are useful in effectively separating gases of different sizes and shapes, like CO_2 from H_2 or methanol from other wastes. Zeolite membranes improve the purity of the methanol generated by better separating it from other byproducts, which can result in fewer processing stages and lower overall emissions. Zeolite membranes' contribution to lowering carbon emissions is projected to increase as technology advances and becomes more affordable. Consequently, an innovative investigation path is required to decarbonize the methanol sector. This approach introduces climate smart systems, such as resource circular CCU and CCS projects and renewable energy pathways, enabling more sustainable operations. A multi-criteria system architecture is supported by a techno-economic and ecological assessment of existing and upcoming methanol technologies. This approach, along with the implications aggregated via legislative priorities and adverse effects, helps to inform decisions among global methanol suppliers. Through improved process efficiency, financial viability, and environmental sustainability using workable decarbonization methodologies, net-zero emission can be attained.

3.7 Lifecycle Analysis of Resource Efficiency

The process of assessing total environmental effects of a product or facility "from cradle to grave" is known as life cycle assessment (LCA) (Liu et al. 2020b). Evaluation of life cycle usage of energy and evaluation of effects on the environment and public health are done through LCA. Moreover, it is used to vehicle assessment, encompassing both vehicle and fuel cycles. The fuel cycle is comprised of five stages: feedstock extraction and transportation, fuel generation and transformation, supply and shipping, and fuel consumption and operation (Li et al. 2024). The majority of environmental evaluations of methanol production concentrate on a small number of particular methods or pathways. Nonetheless, there is a lack of comprehensive studies on both the more conventional older energy-to-methanol processes and the most recent carbon dioxide hydrogenation to methanol systems. Particularly, there has been plenty to explore in the financial and ecological aspects of the Carbon Capture and Utilization (CCU) pathway (Liu et al. 2024c). Next, in order to determine the actual environmental impact of producing green methanol from

CO_2, Lanzac et al. (2022) have done a life cycle assessment involving the simulated system.

Studies show that if renewable energy is used to power the operation, up to 1.75 tons of CO_2 can be avoided for every ton of methanol generated. However, sensitivity analysis suggests that in order for a methanol plant to be financially viable, either extremely low prices for H_2 (1.5 \$ kg 1) or high taxes on CO_2 (300 \$ per ton) are required. Three primary processes were analyzed for the synthesis of CO_2 to methanol by Salkuyeh et al. (2021): tri-reforming of methane, dry reforming, and direct CO_2 hydrogenation. Direct CO_2 hydrogenation is an ecologically friendly option, according to the LCA calculations, but only if the electricity's GHG intensity is less than 0.17 kg CO_2 equivalent per kWh of electricity. Galusnyak et al. (2022) assess the generation of methanol (MeOH) by direct CO_2 hydrogenation. Employing GaBi software and ReCiPe as the impact assessment method, the environmental analysis is completed in accordance with the LCA methodology. The manufacturing of MeOH and its conversion into alkali biodiesel method has the least impact score for all energy scenarios, according to the ecological assessment. Using a range of metrics, such as capital costs, operational expenses, economic viability, the minimal market rate for methanol, and the CO_2 reduction score needed for every approach to be economically feasible, thus LCA models assess the financial viability of the suggested pathways.

3.8 Challenges and Future Research Direction

Due to our ever-rising global temperature and quickly fluctuating climate, numerous nations are attempting to switch from fossil fuels to clean energy sources and green hydrogen. Despite its volatile nature makes it challenging to transport and store, methanol has been shown to be suitable substitute for vehicular fuel. Although methanol is generated using environmentally friendly and sustainable practices, it still has the same inherent toxicity. Methanol's toxicity poses possible risks to the health and safety of those who produce, store, and transport it. To reduce these dangers, strict safety procedures and guidelines are required (Shi et al. 2023). On the production chain, green methanol remains significantly more expensive than conventional methanol. In addition to giving access to less expensive raw materials, improving the performance of the overall process can result in significant production cost savings. The overall complexity of producing methanol may rise due to the requirement for CO_2 capture and conversion technology. Some environmental gains may be countered by increased production costs and energy consumption brought on by this complexity (Yadav et al. 2020). Resources like water and renewable energy are needed to produce green methanol. The viability of large-scale production may be impacted by regional variations in these resources' availability. For green methanol production to have positive environmental effects, these resources must be sourced sustainably. Government incentives, consumer demand, and the capacity to gradually reduce production costs are all critical to the economic feasibility of green methanol production (Ahmed 2020).

Thermodynamic constraints hinder the methanol manufacturing process performance from CO_2 hydrogenation. By using zeolite membranes within the methanol production unit to remove byproduct water in situ, this obstacle can be eliminated (Brunetti et al. 2020). This allowed for higher CO_2 conversions and methanol selectivity than could be obtained with conventional reactors. In order to determine whether these zeolite membrane reactors can be used for massive methanol production, more research is required. Currently, these reactors are only being used in laboratories. Developing a membrane with outstanding durability and large surface area, effective process control methods, and experimental research are some crucial topics that still require investigation when it comes to industrial use. Furthermore, in order to entirely eradicate the downstream distillation process, greater methanol purity from the synthesis reactor requires more effective membrane components. The invention of efficient membranes with enhanced selectivity, permeability, and durability is a key component of forthcoming objectives for membrane-based technologies in methanol production and purification. By regulating parameters like feed flow rate, temperature, and pressure, membrane-based systems can function more efficiently generally and use less energy. These future paths can support the sustainable generation of renewable energy and further encourage the growth of membrane-based methanol production and purification technology.

3.9 Conclusion

Improved clean energy innovation and the rapid and widespread adoption of all affordable, dependable, and sustainable energy technologies are essential to achieving net zero. This review thoroughly examines the prospect benefits of green methanol to limit the use of fossil fuels in the process of decarbonizing the earth's atmosphere. The article starts off by outlining the core ideas of green methanol synthesis. Green methanol production process must have essentially no environmental impact because it is made from biomass, CO_2 that has been absorbed, and renewable energy. The only byproducts of methanol usage are CO_2 and H_2O; however, the same quantity of CO_2 released during methanol synthesis is subsequently captured, closing the CO_2 loop and achieving net-zero emissions. There is also discussion of the disadvantages of methanol produced from fossil sources. The formulation of efficient membranes with enhanced discrimination, accessibility, and durability is one of potential approaches for the synthesis and processing of methanol using membrane-based technologies. Membrane-based technologies can operate more efficiently in general and use less resources if operating parameters like pressure, temperature, and feed flow rate are optimized. Controlling fouling and scaling problems is essential to ensuring the effectiveness and longevity of membranes in the production and purification of methanol. Comparing membrane-based systems with conventional techniques requires a comprehensive cost-effectiveness analysis, which sheds light on their economic feasibility. To improve these membrane systems' scalability, dependability and cost, more study and development are also essential. Developments in these fields will be essential to the commercialization and expansion of membrane-based technologies, which will help to improve the synthesis and purification of methanol and provide renewable energy in a sustainable manner. Following is a brief summary of fundamental membrane-based technology, which has promise as an affordable, green, and economically feasible approach to carbon collection and use in methanol synthesis processes. Since circular and sustainable economies offer a way to turn waste materials into useful resources, they may also have wider ramifications for the development of these economies. In general, sustainable growth will be ensured by this kind of effort to promote technological innovation for the manufacture of green methanol. Making the switch to green methanol production could also aid in accomplishing the Sustainable Development Goals (SDGs) of the UN, including SDG 7 on affordable and clean energy and SDG 13 on climate action.

Declaration of Competing Interest The authors declare no competing interests.

References

AbdulKadir WAFW, Ahmad AL, Seng OB, Lah NFC (2020) Biomimetic hydrophobic membrane: a review of anti-wetting properties as a potential factor in membrane development for membrane distillation (MD). J Ind Eng Chem 91. https://doi.org/10.1016/j.jiec.2020.08.005

Adnan MA, Kibria MG (2020) Comparative techno-economic and life-cycle assessment of power-to-methanol synthesis pathways. Appl Energy 278. https://doi.org/10.1016/j.apenergy.2020.115614

Ahmed U (2020) Techno-economic feasibility of methanol synthesis using dual fuel system in a parallel process design configuration with control on green house gas emissions. Int J Hydrogen Energy 45. https://doi.org/10.1016/j.ijhydene.2019.12.169

Akeeb O, Wang L, Xie W, Davis R, Alkasrawi M, Toan S (2022) Post-combustion CO_2 capture via a variety of temperature ranges and material adsorption process: a review. J Environ Manag 313. https://doi.org/10.1016/j.jenvman.2022.115026

Alharthi AI, Din IU, Alotaibi MA, Bagabas A, Naeem A, Alkhalifa A (2022) Low temperature green methanol synthesis by CO_2 hydrogenation over Pd/SiO_2 catalysts in slurry reactor. Inorg Chem Commun 142. https://doi.org/10.1016/j.inoche.2022.109688

Ali KA, Abdullah AZ, Mohamed AR (2015) Recent development in catalytic technologies for methanol synthesis from renewable sources: a critical review. Renew Sustain Energy Rev 44. https://doi.org/10.1016/j.rser.2015.01.010

Andersen TM, Thuestad L, Thorstensen TA (2006) RAPID: a new approach for improved regularity and decreased maintenance costs. https://doi.org/10.4043/17827-ms

Araya SS, Liso V, Cui X, Li N, Zhu J, Sahlin SL, Jensen SH, Nielsen MP, Kær SK (2020) A review of the methanol economy: the fuel cell route. Energies 13(3). https://doi.org/10.3390/en13030596

Arora A, Iyer SS, Bajaj I, Hasan MMF (2018) Optimal methanol production via sorption-enhanced reaction process. Ind Eng Chem Res 57(42). https://doi.org/10.1021/acs.iecr.8b02543

Atsbha TA, Yoon T, Seongho P, Lee CJ (2021) A review on the catalytic conversion of CO_2 using H_2 for synthesis of CO, methanol, and hydrocarbons. J CO_2 Util 44. https://doi.org/10.1016/j.jcou.2020.101413

Barrow N, Bradley J, Corrie B, Cui Y, Tran TG, Erden TE, Fish A, Garcia M, Glen P, Mistry N, Nicholson M, Standring SR, Sheldon D, Smith T, Summer A, Din KU, Macleod N (2024) Doubling the life of Cu/ZnO methanol synthesis catalysts via use of Si as a structural promoter to inhibit sintering. Sci Adv 10. https://doi.org/10.1126/sciadv.adk2081

Battaglia P, Buffo G, Ferrero D, Santarelli M, Lanzini A (2021) Methanol synthesis through CO_2 capture and hydrogenation: Thermal integration, energy performance and techno-economic assessment. J CO_2 Util 44. https://doi.org/10.1016/j.jcou.2020.101407

Bazaluk O, Havrysh V, Nitsenko V, Baležentis T, Streimikiene D, Tarkhanova EA (2020) Assessment of green methanol production potential and related economic and environmental benefits: the case of china. Energies 13(12). https://doi.org/10.3390/en13123113

Bazmi M, Gong J, Jessen K, Tsotsis T (2024) Waste CO_2 capture and utilization for methanol production via a novel membrane contactor reactor process: techno-economic analysis (TEA), and comparison with other existing and emerging technologies. Chem Eng Process Process Intensif 201. https://doi.org/10.1016/j.cep.2024.109825

Bellotti D, Cassettari L, Mosca M, Magistri L (2019) RSM approach for stochastic sensitivity analysis of the economic sustainability of a methanol production plant using renewable energy sources. J Clean Prod 240. https://doi.org/10.1016/j.jclepro.2019.117947

Biswal T, Shadangi KP, Sarangi PK, Srivastava RK (2022) Conversion of carbon dioxide to methanol: a comprehensive review. Chemosphere 298. https://doi.org/10.1016/j.chemosphere.2022.134299

Bobadilla LF, Azancot L. Castaño MG, López ER, Pérez LP, Olivencia FJD, Ye R, Chong K, Sánchez PHB, Wu Z, Reina TR, Odriozola JA (2024) Biomass gasification, catalytic technologies and energy integration for production of circular methanol: new horizons for industry decarbonization. J Environ Sci 140. https://doi.org/10.1016/j.jes.2023.09.020

Bos MJ, Kersten SRA, Brilman DWF (2020) Wind power to methanol: renewable methanol production using electricity, electrolysis of water and CO_2 air capture. Appl Energy 264. https://doi.org/10.1016/j.apenergy.2020.114672

Bowker M (2019) Methanol synthesis from CO_2 hydrogenation. Chem Cat Chem 11. https://doi.org/10.1002/cctc.201900401

Bozzano G, Manenti F (2016) Efficient methanol synthesis: perspectives, technologies and optimization strategies. Prog Energy Combust Sci 56. https://doi.org/10.1016/j.pecs.2016.06.001

Bram MV, Liniger J, Majidabad SS, Shabani HR, Teles MPR, Cui X (2024) Challenges in power-to-X: a perspective of the configuration and control process for e-methanol production. Int J Hydrog Energy 76. https://doi.org/10.1016/j.ijhydene.2024.05.273

Brigagão GV, Araújo ODQF, Medeiros JLD, Mikulcic H, Duic N (2019) A techno-economic analysis of thermochemical pathways for corncob-to-energy: fast pyrolysis to bio-oil, gasification to methanol and combustion to electricity. Fuel Process Technol 193. https://doi.org/10.1016/j.fuproc.2019.05.011

Brunetti A, Migliori M, Cozza D, Catizzone E, Giordano G, Barbieri G (2020) Methanol conversion to dimethyl ether in catalytic zeolite membrane reactors. ACS Sustain Chem Eng 8. https://doi.org/10.1021/acssuschemeng.0c02557

Calixto I (2023) Master thesis technology pathways for the supply chain of green ammonia, methanol, and hydrogen for maritime applications

Camacho NVH, Castro FIG, Ortega JMP, Martín M (2024) Production of methanol from renewable sources in Mexico: supply chain optimization. Comput Chem Eng 188. https://doi.org/10.1016/j.compchemeng.2024.108780

Carvalho L, Furusjö E, Kirtania K, Wetterlund E, Lundgren J, Anheden M, Wolf J (2017) Techno-economic assessment of catalytic gasification of biomass powders for methanol production. Bioresour Technol 237. https://doi.org/10.1016/j.biortech.2017.02.019

Chakrabortty S, Nayak J, Ruj B, Pal P, Kumar R, Banerjee S, Sardar M, Chakraborty P (2020) Photocatalytic conversion of CO_2 to methanol using membrane-integrated Green approach: a review on capture, conversion and purification. J Environ Chem Eng 8. https://doi.org/10.1016/j.jece.2020.103935

Chakrabortty S, Kumar R, Nayak J, Jeonc BH, Dargar SK, Tripathy SK, Pal P, Ha GS, Kim KH, Jasi´nski M (2023) Green synthesis of MeOH derivatives through in situ catalytic transformations of captured CO_2 in a membrane integrated photo-microreactor system: a state-of-art review for carbon capture and utilization. Renew Sustain Energy Rev 182. https://doi.org/10.1016/j.rser.2023.113417

Collodi G, Azzaro G, Ferrari N, Santos S (2017) Demonstrating large scale industrial CCS through CCU—a case study for methanol production. Energy Proc 114:122. https://doi.org/10.1016/j.egypro.2017.03.1155

Darji HR, Kale HB, Shaikh FF, Gawande MB (2023) Advancement and state-of-art of heterogeneous catalysis for selective CO_2 hydrogenation to methanol. Coord Chem Rev 497. https://doi.org/10.1016/j.ccr.2023.215409

Deka TJ, Osman AI, Baruah DC, Rooney DW (2022) Methanol fuel production, utilization, and techno-economy: a review. Environ Chem Lett 20. https://doi.org/10.1007/s10311-022-01485-y

Dong G, Nagasawa H, Yu L, Wang Q, Yamamoto K, Ohshita J, Kanezashi M, Tsuru T (2020) Pervaporation removal of methanol from methanol/organic azeotropes using organosilica membranes: experimental and modeling. J Memb Sci 610. https://doi.org/10.1016/j.memsci.2020.118284

Eisavi B, Nami H, Ranjbar F, Sharifi A (2024) Economic assessment and optimization of low-carbon biomass-based power, methane, and methanol production. Int J Hydrogen Energy 52. https://doi.org/10.1016/j.ijhydene.2023.11.025

Elvira I, Lasobras J, Soler J, Herguido J, Menéndez M (2023) Preparation of polymeric-ceramic composite membranes for use in the methanol synthesis reaction. Front Membr Sci Technol 2. https://doi.org/10.3389/frmst.2023.1267374

Fankomo PP, Greeff IL (2024) Ion transport membrane heat engine integration with autothermal reforming-based methanol production. Chem Eng Res Des 201. https://doi.org/10.1016/j.cherd.2023.11.026

Fasihi M, Breyer C (2024) Global production potential of green methanol based on variable renewable electricity. Energy Environ Sci 17(10). https://doi.org/10.1039/d3ee02951d

Franklin PM, Koshland CP, Lucas D, Sawyer RF (2000) Clearing the air: using scientific information to regulate reformulated fuels. Environ Sci Technol 34(18). https://doi.org/10.1021/es0010103

Gallucci F (2018) Inorganic membrane reactors for methanol synthesis. Methanol Sci Eng. https://doi.org/10.1016/B978-0-444-63903-5.00018-2

Galusnyak SC, Petrescu L, Chisalita DA, Cormos CC (2022) Life cycle assessment of methanol production and conversion into various chemical intermediates and products. Energy 259. https://doi.org/10.1016/j.energy.2022.124784

Gamero MP, Santamaria JA, Valverde JL, Sánchez P, Silva LS (2018) Three integrated process simulation using aspen plus®: pine gasification, syngas cleaning and methanol synthesis. Energy Convers Manag 177. https://doi.org/10.1016/j.enconman.2018.09.088

Garay AG, Frei MS, Qahtani AA, Mondelli C, Gosálbez GG, Ramírez JP (2019) Plant-to-planet analysis of CO_2-based methanol processes. Energy Environ Sci 12(12). https://doi.org/10.1039/c9ee01673b

Garcia G, Arriola E, Chen WH, Luna MDD (2021) A comprehensive review of hydrogen production from methanol thermochemical conversion for sustainability. Energy 217. https://doi.org/10.1016/j.energy.2020.119384

Gautam P, Neha, Upadhyay SN, Dubey SK (2020) Bio-methanol as a renewable fuel from waste biomass: current trends and future perspective. Fuel 273. https://doi.org/10.1016/j.fuel.2020.117783

Gautier V, Hertz A, Chaise A, Roger AC (2023) Membrane reactors for methanol synthesis: from material science to process simulation. Int Congr Catal. cea-04314272

Golubev GS, Volkov VV, Borisov IL, Volkov AV (2021) High free volume polymers for pervaporation. Curr Opin Chem Eng 36. https://doi.org/10.1016/j.coche.2021.100788

Gorbe J, Lasobras J, Francés E, Herguido J, Menéndez M, Kumakiri I, Kita H (2018) Preliminary study on the feasibility of using a zeolite A membrane in a membrane reactor for methanol production. Sep Purif Technol 200(164–168). https://doi.org/10.1016/j.seppur.2018.02.036

Gupta S, Mohan K, Prasad R, Gupta S (1998) Solid waste management in India: options and opportunities. Resour Conserv Recycl 24

Gutzmer QIR, Kaiser D, Bertau M (2019) Renewable methanol synthesis. Chem Bio Eng Rev 6(6). https://doi.org/10.1002/cben.201900012

Hamedi H, Brinkmann T (2021) Rigorous and customizable 1D simulation framework for membrane reactors to, in principle, enhance synthetic methanol production. ACS Sustain Chem Eng 9(22). https://doi.org/10.1021/acssuschemeng.1c01677

Hank C, Gelpke S, Schnabl A, White RJ, Full J, Wiebe N, Smolinka T, Schaadt A, Henning HM, Hebling C (2018) Economics & carbon dioxide avoidance cost of methanol production based on renewable hydrogen and recycled carbon dioxide-power-to-methanol. Sustain Energy Fuels 2(6). https://doi.org/10.1039/c8se00032h

Harmsen J, Nesterova N, Bekdemir C, Kranenburg KV (2019) Green maritime methanol WP2 initiation and benchmark analysis

Harris K, Grim RG, Huang Z, Tao L (2021) A comparative techno-economic analysis of renewable methanol synthesis from biomass and CO_2: opportunities and barriers to commercialization. Appl Energy 303. https://doi.org/10.1016/j.apenergy.2021.117637

Haydary J, Šuhaj P, Husár J (2021) Waste biomass to methanol–optimisation of gasification agent to feed ratio. Biomass Convers Biorefin 11(2). https://doi.org/10.1007/s13399-020-00692-3

Hoang AT, Pandey A, Osés FJMD, Chen WH, Said Z, Ng KH, Ağbulut U, Tarełko W, Olçer AI, Nguyen XP (2023) Technological solutions for boosting hydrogen role in decarbonization strategies and net-zero goals of world shipping: challenges and perspectives. Renew Sustain Energy Rev 188. https://doi.org/10.1016/j.rser.2023.113790

Ibigbami OA, Onilearo OD, Akinyeye RO (2024) Post-combustion capture and other carbon capture and sequestration (CCS) technologies: a review. Environ Qual Manag https://doi.org/10.1002/tqem.22180

Im-orb K, Arpornwichanop A (2020) Process and sustainability analyses of the integrated biomass pyrolysis, gasification, and methanol synthesis process for methanol production. Energy 193. https://doi.org/10.1016/j.energy.2019.116788

Im-orb K, Phan AN, Arpornwichanop A (2020) Bio-methanol production from oil palm residues: a thermodynamic analysis. Energy Convers Manag 226. https://doi.org/10.1016/j.enconman.2020.113493

Iribarren D, Calvo-Serrano R, Martín-Gamboa M, Galán-Martín A, Guillén-Gosálbez G (2022) Social life cycle assessment of green methanol and benchmarking against conventional fossil methanol. Sci Total Environ 824. https://doi.org/10.1016/j.scitotenv.2022.153840

Jazani O, Bennett J, Liguori S (2023) Carbon-low, renewable hydrogen production from methanol steam reforming in membrane reactors—a review. Chem Eng Process Process Intensif 189. https://doi.org/10.1016/j.cep.2023.109382

Jocz JN, Medford AJ, Sievers C (2019) Thermodynamic limitations of the catalyst design space for methanol production from methane. Chem Cat Chem 11. https://doi.org/10.1002/cctc.201801438

Kajolinna T, Melin K, Linjala O, Roine JN (2024) Performance of novel carbonate-based CO_2 post-combustion capture process. Int J Greenh Gas Control 136. https://doi.org/10.1016/j.ijggc.2024.104173

Kancherla R, Nazia S, Kalyani S, Sridhar S (2021) Modeling and simulation for design and analysis of membrane-based separation processes Comput Aided Chem Eng 148. https://doi.org/10.1016/j.compchemeng.2021.107258

Karim AA, Leaper S, Skuse C, Zaragoza G, Gryta M, Gorgojo P (2021) Membrane cleaning and pretreatments in membrane distillation—a review. Chem Eng J 422. https://doi.org/10.1016/j.cej.2021.129696

Khalid F, Roy AS, Parveen A, Muñoz RC (2024) Fabrication of the cross-linked PVA/TiO_2/C nanocomposite membrane for alkaline direct methanol fuel cells. Mater Sci Eng B Solid-State Mater Adv Technol 299. https://doi.org/10.1016/j.mseb.2023.116929

Kim H, Byun M, Lee B, Lim H (2022) Carbon-neutral methanol synthesis as carbon dioxide utilization at different scales: economic and environmental perspectives. Energy Convers Manag 252. https://doi.org/10.1016/j.enconman.2021.115119

Kumar A, Daw P, Milstein D (2022) Homogeneous catalysis for sustainable energy: hydrogen and methanol economies, fuels from biomass, and related topics. Chem Rev 122(385–441). https://doi.org/10.1021/acs.chemrev.1c00412

Lanzac TC, Ramirez A, Navajas A, Gevers L, Brunialti S, Gandía LM, Aguayo AT, Sarathy SM, Gascon J (2022) A techno-economic and life cycle assessment for the production of green methanol from CO_2: catalyst and process bottlenecks. J Energy Chem 68. https://doi.org/10.1016/j.jechem.2021.09.045

Lee B, Lee H, Lim D, Brigljeví B, Cho W, Cho HS, Kim CH, Lim H (2020) Renewable methanol synthesis from renewable H_2 and captured CO_2: how can power-to-liquid technology be economically feasible?. Appl Energy 279. https://doi.org/10.1016/j.apenergy.2020.115827

Li Y, Zhu G, Wang Y, Chai Y, Liu C (2021) Preparation, mechanism and applications of oriented MFI zeolite membranes: a review. Micropor Mesopor Mater 312. https://doi.org/10.1016/j.micromeso.2020.110790

Li C, Hao Q, Wang H, Hu Y, Xu G, Qin Q, Wang X, Negnevitsky M (2024) Assessing green methanol vehicles' deployment with life cycle assessment-system dynamics model. Appl Energy 363. https://doi.org/10.1016/j.apenergy.2024.123055

Li Z, Tsotsis TT (2019) Methanol synthesis in a high-pressure membrane reactor with liquid sweep. J Memb Sci. https://doi.org/10.1016/j.memsci.2018.09.071

Liu G, Jin W (2021) Pervaporation membrane materials: recent trends and perspectives. J Membr Sci 636. https://doi.org/10.1016/j.memsci.2021.119557

Liu B, Kita H, Yogo K (2020a) Preparation of Si-rich LTA zeolite membrane using organic template-free solution for methanol dehydration. Sep Purif Technol 239. https://doi.org/10.1016/j.seppur.2020.116533

Liu Y, Li G, Chen Z, Shen Y, Zhang H, Wang S, Qi J, Zhu Z, Wang Y, Gao J (2020b) Comprehensive analysis of environmental impacts and energy consumption of biomass-to-methanol and coal-to-methanol via life cycle assessment. Energy 204. https://doi.org/10.1016/j.energy.2020.117961

Liu H, Tang Y, Ma X, Tang J, Yue W (2024a) Biomass gasification based on sorption-enhanced hydrogen production coupled with carbon utilization to produce tunable syngas for methanol synthesis. Energy Convers Manag 309. https://doi.org/10.1016/j.enconman.2024.118428

Liu L, Li S, Zhuang Y, Du J (2024b) Exploring biomass addition in methanol production: a life cycle analysis of efficiency and environmental impacts. J Clean Prod 468. https://doi.org/10.1016/j.jclepro.2024.143072

Liu J, Zhao J, Wei H, Zhu Q, Li Y (2024c) Comparative environmental assessment of methanol production technologies: a cradle-to-gate life cycle analysis. Energy Convers Manag 302. https://doi.org/10.1016/j.enconman.2024.118128

Meunier N, Chauvy R, Mouhoubi S, Thomas D, Weireld GD (2020) Alternative production of methanol from industrial CO_2. Renew Energy 146(1192–1203). https://doi.org/10.1016/j.renene.2019.07.010

Mohammed MG, Hashim NA, Daud WMAW, Hartley UW, Aroua MK, Wohlrab S (2024) Overview of the latest progress and prospects in the catalytic hydrogenation of carbon dioxide (CO_2) to methanol in membrane reactors. Int J Hydrogen Energy 77. https://doi.org/10.1016/j.ijhydene.2024.06.138

Naik SN, Goud VV, Rout PK, Dalai AK (2010) Production of first and second generation biofuels: a comprehensive review. Renew Sustain Energy Rev 14(2). https://doi.org/10.1016/j.rser.2009.10.003

Narine K, Mahabir J, Koylass N, Samaroo N, Gryzbon SS, Baboolal A, Guo M, Ward K (2020). Climate smart process design for current and future methanol production. J CO_2 Util 44(101399). https://doi.org/10.1016/j.jcou.2020.101399

Nemmour A, Inayat A, Janajreh I, Ghenai C (2023) Green hydrogen-based e-fuels (e-methane, e-methanol, e-ammonia) to support clean energy transition: a literature review. Int J Hydrogen Energy 48. https://doi.org/10.1016/j.ijhydene.2023.03.240

Palma V, Barba D, Meloni E, Ruocco C, Martino M (2020) Membrane reactors for H_2 and energy production. Curr Trends Future Dev Bio-Membr Membr Syst Hydrogen Prod. https://doi.org/10.1016/B978-0-12-817110-3.00002-3

Patil T, Naji A, Mondal U, Pandey I, Unnarkat A, Dharaskar S (2024a) Sustainable methanol production from carbon dioxide: advances, challenges, and future prospects. Environ Sci Pollut R 31. https://doi.org/10.1007/s11356-024-34139-3

Patil T, Naji A, Mondal U, Pandey I, Unnarkat A, Dharaskar S (2024b) Sustainable methanol production from carbon dioxide: advances, challenges, and future prospects. Environ Sci Pollut R 31. https://doi.org/10.1007/s11356-024-34139-3

Pendse A, Prajapati A (2024) A perspective on solar-driven electrochemical routes for sustainable methanol production. Sustain Chem 5(1). https://doi.org/10.3390/suschem5010002

Peter SC (2018) Reduction of CO_2 to chemicals and fuels: a solution to global warming and energy crisis. ACS Energy Lett 3:1557. https://doi.org/10.1021/acsenergylett.8b00878

Polotskaya G, Pulyalina A, Goikhman M, Podeshvo I, Rostovtseva V, Shugurov S, Gofman I, Saprykina N, Gulii N, Loretsyan N, Toikka A (2018) Novel polyheteroarylene membranes for separation of methanol–hexane mixture by pervaporation. Sci Rep 8(1). https://doi.org/10.1038/s41598-018-36118-4

Prašnikar A, Linec M, Jurković DL, Bajec D, Sarić M, Likozar B (2024) Understanding membrane-intensified catalytic CO_2 reduction reactions to methanol by structure-based multisite micro-kinetic model. J Clean Prod 463. https://doi.org/10.1016/j.jclepro.2024.142480

Qadeer K, Hinai AA, Chuah LF, Sial NR, Muhtaseb AHA, Abri RA, Qyyum MA, Lee M (2023) Methanol production and purification via membrane-based technology: recent advancements, challenges and the way forward. Chemosphere 335. https://doi.org/10.1016/j.chemosphere.2023.139007

Rahmani F, Haghighi M, Estifaee P, Rahimpour MR (2012) A comparative study of two different membranes applied for auto-thermal methanol synthesis process. J Nat Gas Sci Eng 7. https://doi.org/10.1016/j.jngse.2012.04.001

Rajaee F, Guandalini G, Romano MC, Ritvanen J (2023) Techno-economic evaluation of biomass-to-methanol production via circulating fluidized bed gasifier and solid oxide electrolysis cells: a comparative study. Energy Convers Manag 301. https://doi.org/10.1016/j.enconman.2023.118009

Raso R, Tovar M, Lasobras J, Herguido J, Kumakiri I, Araki S, Menéndez M (2021) Zeolite membranes: comparison in the separation of $H_2O/H_2/CO_2$ mixtures and test of a reactor for CO_2 hydrogenation to methanol. Catal Today 364. https://doi.org/10.1016/j.cattod.2020.03.014

Reddy VJ, Hariram NP, Maity R, Ghazali MF, Kumarasamy S (2023) Sustainable e-fuels: green hydrogen, methanol and ammonia for carbon-neutral transportation. World Electr Veh J. https://doi.org/10.3390/wevj14120349

Rostovtseva V, Faykov I, Pulyalina A (2022) A review of recent developments of pervaporation membranes for ethylene glycol purification. Membranes 12. https://doi.org/10.3390/membranes12030312

Salkuyeh YK, Ashrafi O, Mostafavi E, Navarri P (2021) CO_2 utilization for methanol production; part I: process design and life cycle GHG assessment of different pathways. J CO_2 Util 50. https://doi.org/10.1016/j.jcou.2021.101608

Samimi F, Hamedi N, Rahimpour MR (2019) Green methanol production process from indirect CO_2 conversion: RWGS reactor versus RWGS membrane reactor. J Environ Chem Eng 7(1). https://doi.org/10.1016/j.jece.2018.102813

Sarp S, Hernandez SG, Chen C, Sheehan SW (2021) Alcohol production from carbon dioxide: methanol as a fuel and chemical feedstock. Joule 5(1). https://doi.org/10.1016/j.joule.2020.11.005

Schorn F, Breuer JL, Samsun RC, Schnorbus T, Heuser B, Peters R, Stolten D (2021) Methanol as a renewable energy carrier: an assessment of production and transportation costs for selected global locations. Adv Appl Energy 3. https://doi.org/10.1016/j.adapen.2021.100050

Schwarz JA, Contescu C, Contescu A (1995) Methods for preparation of catalytic materials. Chem Rev. https://doi.org/10.1021/cr00035a002

Scomazzon M, Barbera E, Bezzo F (2024) Alternative sustainable routes to methanol production: techno-economic and environmental assessment. J Environ Chem Eng 12(3). https://doi.org/10.1016/j.jece.2024.112674

Sen R, Goeppert A, Prakash GKS (2022) Homogeneous Hydrogenation of CO_2 and CO to methanol: the renaissance of low-temperature catalysis in the context of the methanol economy. Angew Chem Int Ed. https://doi.org/10.1002/anie.202207278

Seshimo M, Liu B, Lee HR, Yogo K, Yamaguchi Y, Shigaki N, Mogi Y, Kita H, Nakao S (2021) Membrane reactor for methanol synthesis using Si-rich LTA zeolite membrane. Membranes 11. https://doi.org/10.3390/membranes

Sheldon D (2017) Methanol production—a technical history. Johns Matthey Technol Rev 61(3). https://doi.org/10.1595/205651317X695622

Shi J, Zhu Y , Feng Y, Yang J, Xia C (2023) A prompt decarbonization pathway for shipping: green hydrogen, ammonia, and methanol production and utilization in marine engines. Atmosphere 14. https://doi.org/10.3390/atmos14030584

Shuit SH, Ong YT, Lee KT, Subhash B, Tan SH (2012) Membrane technology as a promising alternative in biodiesel production: a review. Biotechnol Adv 30. https://doi.org/10.1016/j.biotechadv.2012.02.009

Sollai S, Porcu A, Tola V, Ferrara F, Pettinau A (2022) Renewable methanol production from green hydrogen and captured CO_2: a techno-economic assessment. J CO_2 Util 68. https://doi.org/10.1016/j.jcou.2022.102345

Song G, Zhou W, Li C, Wang Z, Hu F, Wang T, Li Z, Tang A, Harold MP, Liu S, Kawi S (2023) Semi-hollow LTA zeolite membrane for water permeation in simulated CO_2 hydrogenation to methanol. J Memb Sci 678. https://doi.org/10.1016/j.memsci.2023.121666

Sonthalia A, Kumar N, Tomar M, Geo VE, Thiyagarajan S, Pugazhendhi A (2023) Moving ahead from hydrogen to methanol economy: scope and challenges. Clean Technol Environ 25(551–575). https://doi.org/10.1007/s10098-021-02193-x

Srivastava RK, Sarangi PK, Bhatia L, Singh AK, Shadangi KP (2022) Conversion of methane to methanol: technologies and future challenges. Biomass Convers Biorefin 12(1851–1875). https://doi.org/10.1007/s13399-021-01872-5

Šuhaj P, Husár J, Haydary J (2017) Modelling of syngas production from municipal solid waste (MSW) for methanol synthesis. Acta Chimica Slovaca 10(2). https://doi.org/10.1515/acs-2017-0019

Svitnič T, Beer K, Sundmacher K, Böcher M (2023) Optimal design of a sector-coupled renewable methanol production amid political goals and expected conflicts: costs vs. land use. Sustain Prod Consum 44. https://doi.org/10.1016/j.spc.2023.12.003

Szima S, Cormos CC (2018) Improving methanol synthesis from carbon-free H_2 and captured CO_2: a techno-economic and environmental evaluation. J CO_2 Util 24. https://doi.org/10.1016/j.jcou.2018.02.007

Tabibian SS, Sharifzadeh M (2023) Statistical and analytical investigation of methanol applications, production technologies, value-chain and economy with a special focus on renewable methanol. Renew Sustain Energy Rev 179. https://doi.org/10.1016/j.rser.2023.113281

Tanizume S, Maehara S, Ishii K, Onoki T, Okuno T, Tawarayama H, Ishikawa S, Nomura M (2021) Reaction of methanol to olefin using a membrane contactor on a silica substrate. Sep Purif Technol 254. https://doi.org/10.1016/j.seppur.2020.117647

Tian Z, Wang Y, Zhen X, Liu Z (2022a) The effect of methanol production and application in internal combustion engines on emissions in the context of carbon neutrality: a review. Fuel 320. https://doi.org/10.1016/j.fuel.2022.123902

Tian Z, Wang Y, Zhen X, Liu Z (2022b) The effect of methanol production and application in internal combustion engines on emissions in the context of carbon neutrality: a review. Fuel 320. https://doi.org/10.1016/j.fuel.2022.123902

Toyir J, Miloua R, Elkadri NE, Nawdali M, Toufik H, Miloua F, Saito M (2009) Sustainable process for the production of methanol from CO_2 and H_2 using Cu/ZnO-based multicomponent catalyst. Phys Proc 2. https://doi.org/10.1016/j.phpro.2009.11.065

Ullah A, Hashim NA, Rabuni MF, Mohd Junaidi MU (2023) A review on methanol as a clean energy carrier: roles of zeolite in improving production efficiency. Energies. https://doi.org/10.3390/en16031482

Varela H, Salazar EAP, Lima FHB, Eid K (2024) Renewable methanol and the energy challenge: the role of electrocatalysis. Curr Opin Electrochem https://doi.org/10.1016/j.coelec.2024.101539

Whittet DCB, Cook AM, Herbst E, Chiar JE, Shenoy SS (2011) Observational constraints on methanol production in interstellar and preplanetary ices. Astrophys J 742. https://doi.org/10.1088/0004-637X/742/1/28

Wojnarowski P, Janiga D (2024) Enhanced system for hydrogen storage and conversion into green methanol in a geothermal environment. Int J Hydrogen Energy 52. https://doi.org/10.1016/j.ijhydene.2023.10.133

Wolday AK, Ramteke M (2024) Surrogate model-based optimization of methanol synthesis process for multiple objectives: a pathway towards achieving sustainable development goals. Chem Eng Res Des 204(172–182). https://doi.org/10.1016/j.cherd.2024.02.021

Wu Z, Zhang Z, Xu G, Ge S, Xue X, Chen H (2024) Thermodynamic and economic analysis of a new methanol synthesis system coupled with a biomass integrated gasification combined cycle. Energy 300. https://doi.org/10.1016/j.energy.2024.131647

Yadav P, Athanassiadis D, Yacout DMM, Tysklind M, Upadhyayula VKK (2020) Environmental impact and environmental cost assessment of methanol production from wood biomass. Environ Pollut 265. https://doi.org/10.1016/j.envpol.2020.114990

Yang N, KangF, Zhang K, Zhou Y, Lin WF (2023) A strategy for CO_2 capture and utilization towards methanol production at industrial scale: an integrated highly efficient process based on multi-criteria assessment. Energy Convers Manag 293. https://doi.org/10.1016/j.enconman.2023.117516

Yue W, Li Y, Wei W, Jiang J, Caro J, Huang A (2021) Highly selective CO_2 conversion to methanol in a bifunctional zeolite catalytic membrane reactor. Angewandte Chemie Int Ed 60(33). https://doi.org/10.1002/anie.202106277

Zhang X, Zhang G, Song C, Guo X (2021) Catalytic conversion of carbon dioxide to methanol: current status and future perspective. Front Energy Res. https://doi.org/10.3389/fenrg.2020.621119

Zhang Z, Vo DN, Nguyen TBH, Sun J, Lee CH (2024) Advanced process integration and machine learning-based optimization to enhance techno-economic-environmental performance of CO_2 capture and conversion to methanol. Energy 293. https://doi.org/10.1016/j.energy.2024.130758

Chapter 4
Processes Used in the Production of Methanol

Ayat Gharehghani and Samareh Ahmadi

Abstract Methanol, as an alcohol, is a versatile chemical intermediate that is used in a wide range of industries. Since there are different methods for its production, according to the biomass production method, various processes are used in the production of methanol. Some processes may be common to different methods. So, its production may involve several processes that typically begins with the generation of syngas, a mixture of carbon monoxide and hydrogen and its preparation. This chapter delves into the various processes employed for methanol synthesis and production. The importance of methanolization process parameters including temperature, pressure, and gas composition as preparation processes of methanol synthesis reaction has been investigated. Furthermore, traditional methods such as steam reforming of natural gas, as well as emerging technologies and more recent methods like coal gasification, biomass gasification, biochemical and electrocatalytic methods, and so on are examined.

Keywords Methanol · Processes · Catalytic · Hydrogenation · Feedstock

4.1 Introduction

Reducing harmful emissions is crucial for addressing environmental concerns and air pollution. Sustainable fuel alternatives, such as natural gas, biodiesel, and various alcohols, offer promising solutions for mitigating emissions of carbon monoxide, carbon dioxide, hydrocarbons, and nitrogen oxides. Alcohols, including methanol, ethanol, butanol, bio-butanol, n-butanol, and iso-butanol, have gained significant attention as potential alternative fuels due to their convenient storage and easy handling (Srikrishnan et al. 2024).

A. Gharehghani (✉) · S. Ahmadi
School of Mechanical Engineering, Iran University of Science and Technology, Tehran, Iran
e-mail: ayat_gharehghani@iust.ac.ir

A. Bartl et al. (eds.), *Alcohol Production Processes and Their Utilisation in the Transport Sector*, Energy, Environment, and Sustainability,
https://doi.org/10.1007/978-981-96-7384-1_4

- Methanol (CH_3OH), as an alternative fuel, offers a cost-effective production process compared to other alcohols. However, it possesses lower energy density and exhibits higher toxicity. One significant advantage of methanol lies in its simple molecular structure, which enables easier steam reforming at relatively low temperatures (200–300 °C) with suitable catalysts (Sá et al. 2010).
- Ethanol (C_2H_5OH) is used as a solvent and a widely used biofuel, can be produced from various starchy or sugary feedstocks or fermentation or ethylene hydration. It offers several benefits, including non-toxicity, wide availability, easy transportation, and renewable energy source. Steam reforming of ethanol typically occurs at higher temperatures than as methanol (300–800 °C), but the process can be optimized with appropriate catalysts (Dybiński et al. 2023).
- Propanol (C_3H_7OH), often produced synthetically from propanal or propylene, is commonly used as a cost-effective substitute for ethanol or methanol. A common method for producing propanol involves the dehydration of propanal or propylene to yield propan-1-ol or propan-2-ol, respectively (Dybiński et al. 2023).
- Butanol (C_4H_9OH), used as a solvent and fuel, produced either from fossil fuels or through biomass fermentation or electrolysis of ethanol, exhibits lower water solubility compared to other alcohols. It can also be produced naturally through the ABE (acetone-butanol-ethanol) fermentation process (Lee et al. 2008). It exists in four isomeric forms with slightly varying properties such as solidification, boiling, and solubility parameter (Dybiński et al. 2023).

Methanol is a chemical substance with wide-ranging applications in energy, transportation, and industrial sectors. Its production needs various feedstocks and processes such as mechanical, chemical, biochemical, electrochemical, photochemical, etc., that include a series of catalytic reactions and transformations each offering distinct advantages and challenges (Snoeckx and Bogaerts 2017; Cifre and Badr 2007). Traditionally, methanol production has traditionally relied on fossil fuels, particularly utilizing natural gas through the process of steam methane reforming, followed by catalytic conversion to synthesize methanol. However, growing environmental concerns and the need for sustainable energy sources have spurred interest in alternative production methods using renewable feedstocks and innovative technologies (Deka et al. 2022; Marlin et al. 2018).

Methanol's physical characteristics, particularly its hydrogen bonding, enable it to stay in liquid form at room temperature, which simplifies storage and transportation compared to alternatives like methane or hydrogen, which require more specialized handling (Marlin et al. 2018). Methanol serves as an excellent alternative liquid fuel, capable of powering Otto engines and fuel cells with high thermodynamic efficiency while reducing environmental impact (Kotowski 2001). Therefore, the research about the processes and methods of its production is of special importance.

The processes used to produce methanol can be different based on the method and hydrocarbon source used. One such processes is gasification, in which biomass feedstocks are transformed into syngas—a mixture of carbon monoxide, carbon dioxide, and hydrogen (Gizaw et al. 2024). Biomass encompasses organic matter originating from plants or animals and can be utilized as an energy source. This includes materials like wood, crop remnants, agricultural and forestry byproducts, organic waste, and municipal waste (Vidhya et al. 2024). The resulting syngas is subsequently cleaned and purified to remove impurities before being catalytically converted into methanol. The gasification process can not only use abundant and renewable resources but also help in waste management and reduce greenhouse gas emissions (Bertau et al. 2014).

Another promising approach is the electrochemical reduction of carbon dioxide, which utilizes electrical energy, potentially sourced from renewables, to transform CO_2 into methanol. This method addresses the dual challenges of carbon capture and renewable energy utilization, presenting a sustainable pathway for methanol production (Deka et al. 2022; Osman et al. 2021a).

Advancements in catalytic technologies have significantly improved the efficiency and selectivity of methanol synthesis processes. Transition metal catalysts, such as copper-based systems, have been extensively studied and optimized to enhance methanol yield and reduce by-product formation (Bertau et al. 2014).

Photocatalytic and plasma-catalytic methods are also being explored for methanol production. These processes leverage the energy from light or plasma to drive the transformation of CO_2 and hydrogen into methanol, offering potential benefits in terms of lower operational temperatures and the utilization of renewable energy sources (Snoeckx and Bogaerts 2017; Bogaerts and Centi 2020). The integration of these innovative methods with traditional processes could pave the path for more impressive and eco-friendly methanol production systems.

In the following, the methods and processes used in the production of methanol will be discussed in detail.

4.2 Mechanical Processes

Methanol production consists of different processes but it is fundamentally a chemical process that involves a series of reactions and catalytic conversions. Some of these processes are necessary to establish the initial conditions for the reaction and to their execution and some others are applied to improve performance and achieving higher product and efficiency by spending less energy, cost and time.

4.2.1 Compression of Feed Gases

Compression is done to increasing the pressure of feed gases such as syngas (carbon monoxide, and hydrogen), natural gas, biogas, and carbon dioxide (Yang and Ge 2016; Van-Dal and Bouallou 2013) and is efficacious in the methanol synthesis, both in the preparation of feed gas and during the synthesis process within the methanol reactor. High-pressure conditions enhance the efficiency of catalytic reactions, leading to higher methanol yields (Sandvik et al. 2018). For this purpose, compressors are typically employed in this step (Bozzano and Manenti 2016; Lee et al. 2020).

In methanol production, feed gas typically comprises a mixture of carbon dioxide (CO_2), carbon monoxide (CO), and hydrogen (H_2), and sometimes methane (CH_4). These gases must be compressed to optimal pressures before entering the methanol synthesis reactor (Rasteiro et al. 2022). The primary purpose of feed gas compression is to ensure that the gas mixture achieves the required density and reactivity for efficient catalytic conversion. Compressing the feed gas to pressures between 50 and 100 bar significantly enhances the reaction rates and improves the overall efficiency of the methanol synthesis process (Sandvik et al. 2018; Dalena et al. 2018a; Yousuf et al. 2023). Higher pressures facilitate better contact between gas molecules and the catalyst, leading to higher conversion rates of CO_2 and CO into methanol (Deka et al. 2022).

Within the methanol synthesis reactor, maintaining high pressure is essential for maximizing methanol yield. The methanol synthesis reaction, typically conducted over copper-based catalysts, is exothermic and favored by high pressures (Deka et al. 2022).

Studies indicate that operating the methanol synthesis reactor at pressures ranging from 50 to 100 bar not only increases the reaction rate but also improves the selectivity toward methanol, reducing the formation of by-products such as dimethyl ether (DME) and higher alcohols (Bozzano and Manenti 2016; Lee et al. 2020; Mohanty et al. 2022; Riaz et al. 2013; Li and Tsotsis 2019). Furthermore, higher pressures contribute to better thermal management within the reactor, as the exothermic nature of the reaction generates significant heat (Yousuf et al. 2023; Riaz et al. 2013; Li and Tsotsis 2019).

The impact of pressure on the efficiency and yield of methanol production is well-documented. Increasing the pressure in the methanol synthesis reactor enhances the equilibrium conversion of CO_2 and CO to methanol due to Le Chatelier's principle. This principle states that higher pressures shift the equilibrium toward methanol formation, thus increasing yield (Yousuf et al. 2023). Moreover, higher pressures contribute to achieving higher space–time yields (STY), a measure of the reactor's productivity. Research shows that reactors operating at 100 bar pressure exhibit a 20–30% increase in STY compared to those operating at lower pressures. This increase in productivity is critical for industrial-scale methanol production, where maximizing output while minimizing operational costs is paramount (Leonzio 2018; Kotowicz et al. 2022).

4.2.2 Heating and Drying

Heating is an important process in methanol production, using to both the preparation of feed gases and the drying of raw materials such as biomass and enhancing reaction rates. In methanol production, feed gases are often heated to achieve the necessary reaction temperatures. This step is vital for processes like steam methane reforming

(SMR), where methane reacts with steam to produce syngas, a mixture of CO and H_2. Similarly, heating is required for the hydrogenation of CO_2 to facilitate the reaction kinetics. The methanol synthesis reaction is exothermic, meaning it releases heat. Maintaining an optimal temperature is essential to ensure the reaction proceeds at a desirable rate while avoiding the deactivation of catalysts. Typically, the reaction is conducted at temperatures between 200 and 300 °C. If the temperature is too low, the reaction rate decreases, leading to lower methanol yields. Conversely, excessively high temperatures can lead to the formation of unwanted by-products and catalyst degradation. Proper heat management, including the use of heat exchangers and temperature control systems, is crucial to maximize methanol production efficiency and maintain catalyst longevity (Oliveira et al. 2022a). Heating the feed gases ensures that the endothermic reactions proceed efficiently, maximizing the yield of syngas, which is subsequently converted to methanol (Yang and Ge 2016; Bozzano and Manenti 2016).

For biomass-based methanol production, drying the biomass feedstock is a critical step. Moisture in biomass can significantly reduce the efficiency of the gasification process. By heating the biomass to remove moisture, the gasification process can produce a higher quality syngas with better efficiency. The drying process typically involves heating the biomass to temperatures between 100 and 150 °C to evaporate the water content (Cifre and Badr 2007; Deka et al. 2022). Optimal heating conditions enhance reaction kinetics, improve catalyst performance, and increase the overall productivity of the methanol synthesis process (Hamelinck and Faaij 2002; Mignard and Pritchard 2008).

The result of a kinetic model was developed using Aspen Plus to simulate methanol production from syngas is presented in Fig. 4.1 (Adil and Rao 2022). The contour plots in Fig. 4.1 illustrate the influence of reaction temperature, operating pressure, and stoichiometric number on methanol production rate and yield. Altering the feed temperature within the 220–250 °C range has a minimal impact on methanol yield (Fig. 4.1a). While slightly lower feed temperatures exhibited a positive influence, so this effect was negligible, while the Fig. 4.1b indicates that both increasing pressure and stoichiometric number can enhance methanol yield (Adil and Rao 2022).

Where, stoichiometric number is defined as follows (Adil and Rao 2022):

$$\text{Stoichiometric number} = \frac{(H_2 - CO_2)}{(CO + CO_2)} \tag{4.1}$$

4.2.3 Gas Purification, Cleaning, and Separation

Gas purification is a critical step in the methanol production process, using in both the preparation of feed gases and the treatment of products post-synthesis. Purification of gases is critical to remove impurities that can poison catalysts or affect reaction

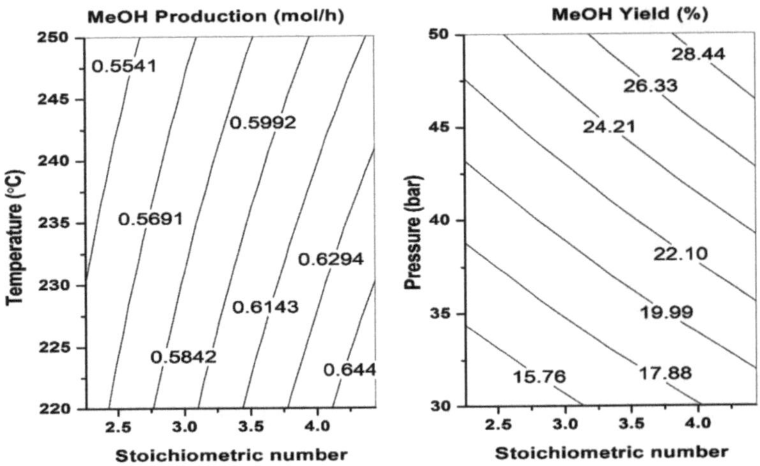

Fig. 4.1 Diagram of how methanol production rate (mol/h) and yield (%) based on the stoichiometric number and change with: **a** reaction temperature, **b** operating pressure (Adil and Rao 2022)

efficiency (Riaz et al. 2013). This process includes methods like adsorption, absorption, and membrane separation to eliminate contaminants such as sulfur compounds, CO_2, and other impurities from feed gases (Cifre and Badr 2007; Adil and Rao 2022; Qadeer et al. 2023). In the production of methanol, the gas purification process is used in two stages: Gas purification in feed gas preparation and gas purification post-synthesis.

Gas Purification and Cleaning in Feed Gas Preparation

Feed gases for methanol production typically include a mixture of carbon dioxide, carbon monoxide, and hydrogen. These gases often contain impurities such as sulfur compounds, nitrogen oxides, and other particulates that can poison catalysts and reduce the efficiency of the methanol synthesis process. Therefore, gas purification steps are employed to remove these impurities (Deka et al. 2022; Puig-Gamero et al. 2018; Shorokhov et al. 2023).

According to studies, purification steps can involve the use of materials like activated carbon and zeolites in adsorption units to capture impurities from the gas stream. The purified gas stream is essential for ensuring high catalytic activity and longevity of the catalysts used in methanol synthesis (Yousuf et al. 2023; Shorokhov et al. 2023).

Gas Purification of Post-synthesis and Separation

After methanol synthesis, the produced methanol often contains by-products such as water, higher alcohols, and unreacted gases. Purification and separation remove impurities, by-products, and unreacted feedstocks from the crude methanol mixture. Post-synthesis purification involves separation techniques such as distillation and condensation to isolate pure methanol from these by-products (Deka et al. 2022; Riaz et al. 2013). The primary purification method involves distillation, where the crude methanol is heated to separate it based on boiling points. Multi-stage distillation columns are typically employed to achieve high purity methanol. The purified methanol is then collected, and the by-products are either recycled or further processed (Bozzano and Manenti 2016; Riaz et al. 2013; Oliveira et al. 2022b). Effective gas purification ensures that the final methanol product meets purity standards required for its various applications. Among the purification methods, membrane-based separation technologies have been highlighted as efficient methods for methanol purification due to their ability to selectively permeate methanol over other components (Yousuf et al. 2023; Dalena et al. 2018b).

In addition to distillation, various methods are used in the purification process, some of them are (Deka et al. 2022; Marlin et al. 2018; Dalena et al. 2018a; Supp 1990; Olah et al. 2009):

- Membrane Separation: Utilizes selective membranes to separate gases based on their size and diffusivity. This method is effective in purifying CO_2 and H_2, enhancing the feed gas quality for methanol synthesis (Dalena et al. 2018a, b; Iulianelli et al. 2018).
- Pressure Swing Adsorption (PSA): PSA is used to separate specific gases from a mixture under pressure according to the molecular characteristics of the gases and their affinity for an adsorbent material. This method is often used to purify hydrogen from syngas mixtures before methanol synthesis (Supp 1990).
- Absorption: Involves the use of liquid solvents to selectively absorb specific gases from a mixture. In methanol production process, commonly used for removing CO_2 and sulfur compounds from natural gas or syngas (Cheng 1994).
- Catalytic Purification: In this method, catalysts are used to convert impurities into less harmful substances. Its application in methanol production is often to remove trace contaminants that can affect catalyst performance in the methanol synthesis reactor (Chinchen et al. 1987).

An example of the methanol synthesis loop and its purification is shown in Fig. 4.2 (Arnaiz del Pozo et al. 2022).

4.2.4 Cooling and Condensation

Cooling and condensation help manage the exothermic nature of methanol synthesis and separate methanol from by-products and unreacted gases. During methanol synthesis, the reaction between hydrogen and carbon oxides (CO and CO_2) is highly exothermic. Effective temperature management is crucial to prevent overheating, because it can lead to catalyst deactivation and reduced efficiency. Cooling systems, such as external heat exchangers and quenching, are used to remove excess heat from the reactor. For instance, traditional methanol reactors often utilize boiling

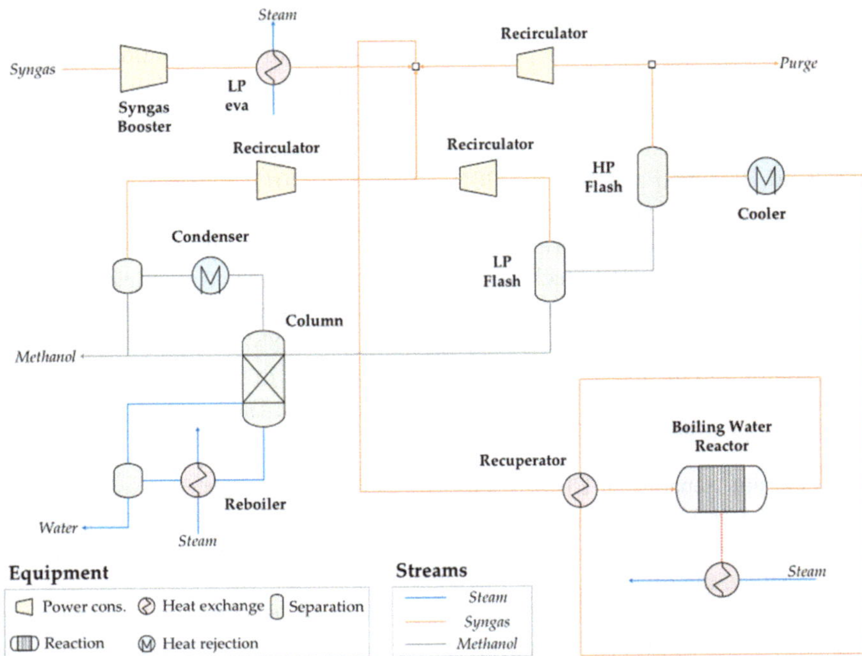

Fig. 4.2 Methanol synthesis and purification loop (Arnaiz del Pozo et al. 2022)

water reactors (BWRs), where water is boiled to absorb the heat generated during the reaction (Marlin et al. 2018; Oliveira et al. 2022b).

After the synthesis reaction, the product mixture typically contains methanol, water, unreacted gases, and by-products. Condensation is employed to separate methanol from this mixture. In intermediate condensation steps, methanol and water are condensed out of the gas phase, allowing for the removal of liquid methanol and water while recycling unreacted gases back into the reactor (Oliveira et al. 2022b; Palma et al. 2018; Keshavarz et al. 2020). This approach significantly increases the single-pass conversion efficiency and reduces the overall recycle flow, which can lower operating and investment costs (Deka et al. 2022; Sheldon 2017).

Proper heat management ensures that the reactor operates within the optimal temperature range, enhancing catalyst performance and prolonging catalyst life. Intermediate condensation steps, by efficiently removing methanol and water from the reaction mixture, help maintain higher conversion rates and reduce energy consumption (Deka et al. 2022; Marlin et al. 2018; Bozzano and Manenti 2016).

Table 4.1 summarizes the mechanical processes used in methanol production and the methanol production methods in which those processes are used (Deka et al. 2022; Marlin et al. 2018; Dalena et al. 2018a; Riaz et al. 2013; Olah et al. 2009).

Table 4.1 Mechanical processes used in methanol production methods

Mechanical process	Methanol production methods	Detailed explanation	Optimal conditions
Compression	Steam methane reforming (SMR), biomass gasification	Compression increases the pressure of feed gases to enhance reaction rates and efficiency	Pressure: 20–30 bar (SMR), 50–100 bar (biomass gasification)
Gas purification	Steam methane reforming (SMR), coal gasification, biomass gasification	Purification removes impurities from gases, typically using adsorption or absorption processes	Varies depending on impurities: 5–50 °C for adsorption
Heating and drying	All methods	Heating provides the necessary thermal energy to drive endothermic reactions. In addition, the gasification or production of syngas from biomass (drying and pyrolysis processes) are endothermic processes and heat must be given to raw materials	Temperature: 200–300 °C (SMR), 500–800 °C (biomass gasification)
Cooling and condensation	All methods	Cooling reduces the temperature of gases to condense out products or remove excess heat. Condensation involves converting vaporized products back into liquid form for collection	Temperature: 20–50 °C
Purification and separation	All methods	Purification further refines products to meet purity specifications and separation isolates desired products from by-products using techniques like distillation	Depends on method and product specifications Distillation (Methanol: 64.7 °C boiling point)

4.3 Chemical Processes

The production of methanol involves various processes especially chemical reactions and processes designed to convert raw materials like natural gas, biomass, or carbon dioxide into methanol. Central to this production are catalytic reactions, where catalysts like copper-based materials facilitate the conversion of synthesis gas (a mixture of CO, CO_2, and H_2) into methanol under specific temperature and pressure conditions. Methanol production is fundamentally a chemical process, and depending on the methods and raw materials determining the applicable processes (Deka et al. 2022; Bozzano and Manenti 2016; Chmielniak and Sciazko 2003). The various chemical processes involved in methanol production include hydrogenation, catalytic reactions, gasification, reforming, and oxidation, which are described below in the form of different methods of methanol production.

4.3.1 Direct Hydrogenation of CO_2

Heterogeneous catalytic hydrogenation of carbon dioxide involves the direct combination of CO_2 and H_2, which can lead to the production of methanol (Specht et al. 1999; Saito et al. 2000).

$$CO_2 + 3H_2 \rightarrow CH_3OH + H_2O \tag{4.2}$$

Producing methanol from carbon dioxide necessitates the addition of hydrogen. This method is illustrated in Fig. 4.3. Potential sources of CO_2 for this process encompass emissions from fossil fuel power plants, cement manufacturing facilities, fermentation operations, and water treatment plants. Additionally, carbon dioxide can be captured directly from the atmosphere (Cifre and Badr 2007).

CO_2 is a highly stable compound, and its C=O bond has a higher dissociation energy of ~ 750 kJ/mol compared to various other chemical bonds like C–H (~ 430 kJ/mol) and C–C (~ 336 kJ/mol), indicating that a substantial energy input is necessary for converting CO_2 (Xie et al. 2016; Kumar et al. 2012).

Typically, to convert CO_2 into methanol, a catalyst is necessary since CO_2 is a very stable molecule. Various catalysts can aid in the conversion of CO_2 to methanol, but the most commonly utilized ones are based on metals and/or metal oxides such as Cu, Zn, Cr, and Pd, which help facilitate the redox reaction (Azhari et al. 2022). Scientists are currently conducting research to discover methods for decreasing the release of CO_2 into the atmosphere. In essence, there are three potential approaches for lessening the accumulation of CO_2 in the atmosphere: decreasing the production of CO_2, utilizing CO_2, and storing CO_2 (Yang et al. 2008; Xiaoding and Moulijn 1996).

Methanol can be synthesized from CO_2 using two different approaches: a single-step process or a two-step process. The single-step method involves directly hydrogenating CO_2 into methanol. In the two-step method, CO_2 is initially converted into carbon monoxide through the Reverse Water Gas Shift (RWGS) reaction, and then

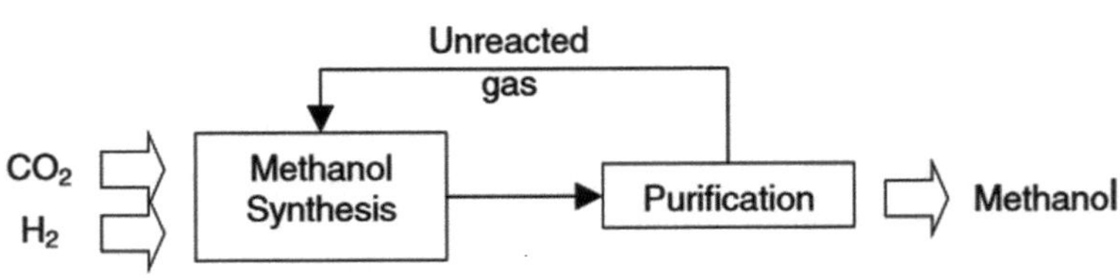

Fig. 4.3 Flowchart for methanol synthesis from CO_2 (Cifre and Badr 2007)

CO is hydrogenated to produce methanol. The CO_2 utilized in these processes can be sourced from the exhaust gases of thermal power plants (such as coal or natural gas plants) or from industries like steel manufacturing, cement production, and other significant CO_2 emitters (Van-Dal and Bouallou 2013).

The catalytic hydrogenation of CO_2 to produce methanol involves a principal reaction (Eq. 4.3) and two secondary reactions (Eqs. 4.4 and 4.5) as below. Equations 4.2 and 4.4 are exothermic reactions. The Reverse Water Gas Shift (RWGS) reaction, represented by Eq. (4.3), also takes place simultaneously (Rasteiro et al. 2022):

$$CO_2 + 3H_2 \rightarrow CH_3OH + H_2O \quad \Delta H_{298} = -49.7\,\text{kJ/mol} \tag{4.3}$$

$$CO_2 + H_2 \rightarrow CO + H_2O \quad \Delta H_{298} = 41.2\,\text{kJ/mol} \tag{4.4}$$

$$CO + 2H_2 \rightarrow CH_3OH \quad \Delta H_{298} = -90.5\,\text{kJ/mol} \tag{4.5}$$

The synthesis of methanol utilizing hydrogen and carbon dioxide offers significant environmental and economic benefits. This process becomes particularly advantageous when the consumption of carbon dioxide is higher compared to the energy and resources expended in generating the necessary hydrogen (Raudaskoski et al. 2009).

The $CuO/ZnO/Al_2O_3$ catalyst is widely utilized in industrial applications for facilitating the chemical reactions necessary for methanol production from CO_2 (Klier et al. 1982).

4.3.2 *Reforming of Methane and Biogas*

At present, over 75% of methanol is derived from natural gas (Roan et al. 2004). The production process involves three essential stages: reforming natural gas to generate synthesis gas (syngas), converting syngas into crude methanol, and distilling the crude methanol to reach the required purity levels (Roan et al. 2004; Rauch 2002; Hugill et al. 2001). Methane-derived syngas typically contains lower levels of impurities such as sulfur, halogenated compounds, and metals. This simplifies downstream purification and reduces the risk of catalyst poisoning in methanol synthesis reactors, which all utilize catalysts (Bozzano and Manenti 2016). The predominant method for syngas production is methane steam reforming, a highly endothermic process operating at 800–1000 °C and 20–30 atm (Ertl et al. 1997). This process generates CO and H_2, along with CO_2 from the water–gas shift reaction (Eq. 4.7) (Bozzano and Manenti 2016):

$$CH_4 + H_2O \rightarrow CO + 3H_2 \quad \Delta H_{298} = 49.1\,\text{kcal/mol} \tag{4.6}$$

$$CO + H_2O \rightarrow CO_2 + H_2 \quad \Delta H_{298} = -9.8\,\text{kcal/mol} \tag{4.7}$$

High temperatures and low pressures are conducive to syngas production. Elevating the process temperature diminishes the water–gas shift reaction's impact, thereby enhancing CO and H_2 generation. Heat is provided by combusting a portion of the natural gas fed to the reactor. However, coke and soot formation poses significant problems, as they can deactivate the catalyst and accumulate on the reformer's internal surfaces and downstream equipment.

According to researchers, three primary categories of catalysts are identified as the most efficient for converting methane to syngas (Iyer et al. 2004; Durante et al. 1990; Silva 2016; York et al. 2003):

- Nickel, cobalt, or iron catalysts supported on various substrates
- Supported noble metal catalysts
- Transition metal carbide catalysts.

Another established method for producing methanol involves the reforming of biogas, which is subsequently followed by the synthesis of methanol. Various catalysts have proven to be effective in this synthesis process, including $Cu/ZnO/Al_2O_3$, Pd/CeO_2, Cu/ZrO_2, $Mo(CO)_6$, $ZrO_2/CuZnO$, Cu/CeO_2, and Ni/Mod (Deka et al. 2022). Santos and colleagues conducted studies on methanol production from different biogas sources such as palm oil effluent, corncobs, landfill gases, and sorghum fermentation. They utilized mathematical modeling and simulation techniques to analyze the generation of methanol from these diverse sources (Santos et al. 2018).

4.3.3 Direct Methane Oxidation

Traditionally, a non-direct path is employed for converting natural gas into methanol. The process involves: (1) creating synthesis gas through steam reforming, and (2) using catalysis to convert synthesis gas into methanol. Syngas production is a highly energy-demanding procedure, achieving 65% thermodynamic efficiency at temperatures ranging from 800 to 1000 °C. Additionally, over 25% of the feed (natural gas) must be combusted to generate the necessary heat for the reaction. Therefore, for sustainable development reasons, converting methane directly into methanol is seen as a preferable alternative to current methods (Arutyunov 2014).

The partial oxidation of methane (POM) has emerged as the predominant method for producing methanol. This process, which was initially explored by Bone around 1935 (Bone 1935), has since gained widespread acceptance and popularity. POM involves the reaction of methane with a limited amount of oxygen to produce methanol, alongside other products (Bone 1935).

$$CH_4 + 0.5O_2 \rightarrow CH_3OH + H_2O \quad \Delta H_{298} = 128\,\text{kJ/mol} \tag{4.8}$$

Equation (4.8) illustrates the reaction for the partial oxidation of methane, conducted under high-pressure conditions ranging from 0.5 to 15 MPa, using a catalyst to achieve a high yield of target products (Gesser et al. 1985).

In addition, this method is also used to produce syngas. Thus, an alternative method for producing syngas involves the partial oxidation of methane (Bozzano and Manenti 2016):

$$CH_4 + 0.5O_2 \rightarrow CO + 2H_2 \quad \Delta H_{298} = -8.6\,\text{kcal/mol} \tag{4.9}$$

$$CO + 0.5O_2 \rightarrow CO_2 \quad \Delta H_{298} = -67.6\,\text{kcal/mol} \tag{4.10}$$

$$H_2 + 0.5O_2 \rightarrow H_2O \quad \Delta H_{298} = -67.6\,\text{kcal/mol} \tag{4.11}$$

All reactions in system are exothermic and operate at temperatures ranging from 1200 to 1500 °C, and typically produce syngas that is well-suited for methanol synthesis.

As shown in Fig. 4.4, a proposed mechanism for direct oxidation of methane to methanol over $CuMoO_4$ involves several steps. Initially, methane adsorbs onto the Cu sites of the catalyst, facilitating the breaking of a C–H bond. Subsequently, the resulting methyl group reacts with nearby oxygen species to form a methoxy group. Water plays a dual role, extracting the methoxy group to form methanol and methyl hydroperoxide (CH_3OOH), while also acting as a weak oxidizing agent to regenerate the catalyst's active sites. This cyclical process enables continuous methane oxidation and product formation (Wu et al. 2023).

4.3.4 *Gasification of Coal and Biomass*

Methanol production in fossil-based plants involves several key processes: (a)gasification, (b) gas cleaning, (c) hydrocarbon reforming, (d) water–gas shift reaction, (e) hydrogen addition and/or carbon dioxide removal, and (f) synthesis of methanol and filtration (Cifre and Badr 2007; Hamelinck and Faaij 2002).

The steps for producing methanol from coal are similar to those from natural gas, including syngas production via coal gasification, crude methanol synthesis, and purification (Williams et al. 1995; Assessment A 2003; Heydom et al. 1999). Although syngas from coal has lower hydrogen content, the Liquid Phase Methanol (LPMeOH) process is effective for this application (Heydorn et al. 2003) and can integrate with Integrated Gasification Combined Cycle (IGCC) power plants, utilizing excess syngas for methanol production (Heydorn et al. 2003). Methanol can be synthesized from almost any organic materials (Specht and Bandi 1999). One of the key sources is biomass, which is rich in carbon. This carbon-rich composition makes biomass particularly useful for creating carbon-based synthetic fuels (Rauch 2002). When biomass production is supplied sustainably, it creates a carbon–neutral system.

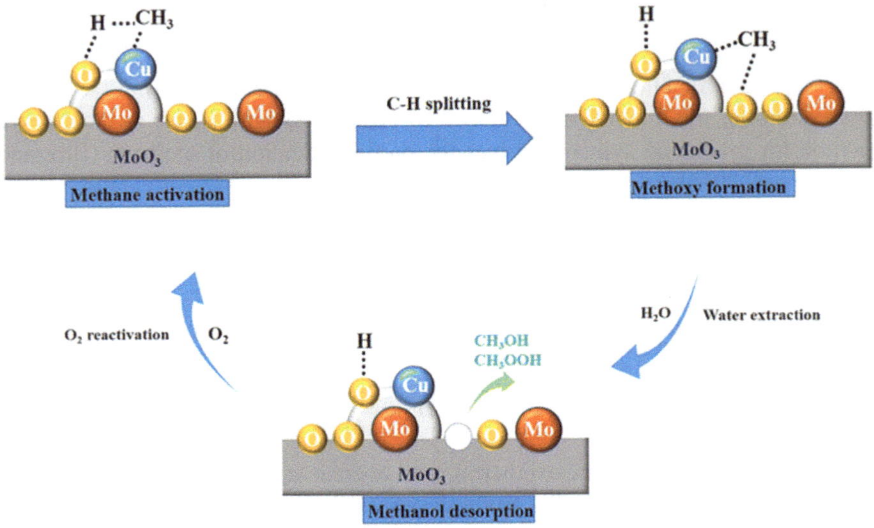

Fig. 4.4 Proposed mechanism for the oxidation of methane to methanol over a Cu/MoO_3 catalyst (Wu et al. 2023)

The process of converting biomass into methanol mirrors the steps used in natural gas and coal-based methods. Initially, synthesis gas is produced, and followed by the synthesis of methanol. The final stage involves purifying the methanol to achieve the desired quality (Cifre and Badr 2007; Specht et al. 1999; Rauch 2002; Specht and Bandi 1999).

For effective methanol synthesis, the ideal molar ratio of hydrogen to carbon dioxide in the syngas should exceed 2 (Specht and Bandi 1999). Traditional gasification methods used for biomass are not often successful to produce synthesis gas of the necessary quality for methanol production. Because, the gas produced from biomass gasification typically has a lower H_2 to CO_2 ratio than required. Unlike the syngas used for electricity generation, the syngas intended for methanol synthesis must have minimal inert gas components. Employing air as the gasification medium introduces a high nitrogen content into the syngas, which is undesirable for methanol synthesis (Cifre and Badr 2007). Therefore, ensuring the syngas composition is suitable for methanol production requires careful consideration of the gasification agents used, highlighting a significant difference between biomass gasification for fuel and power generation purposes. Specht et al. (1999) explored two main strategies to modify the hydrogen to carbon dioxide (H_2:CO_2) ratio in extracted syngas for methanol production:

- Carbon Dioxide Segregation: This method involves extracting excess carbon dioxide from the syngas. The removal is typically achieved through an acid gas separation process, which selectively eliminates CO_2, thereby increasing the relative concentration of hydrogen.

- Hydrogen Addition: In this approach, the syngas is enriched by adding the required amount of hydrogen. This additional hydrogen facilitates the hydrogenation of carbon dioxide, helping to achieve the desired H_2 ratio for efficient methanol synthesis.

By these methods, the syngas composition can be adjusted to meet the specific requirements for methanol production, ensuring a more efficient and effective process. A practical approach to utilizing biomass for energy is through the production and application of methanol as a motor fuel, derived from biomass gasification.

As studied by Zhang et al. (2010) and Asadieraghi and Daud (2015), gasification processes does not directly produced advanced liquid biofuels. In lieu of, these biofuels are created through an intermediate stage involving syngas, which is a mixture of hydrogen, carbon dioxide, carbon monoxide, and methane (Miao et al. 2014; Mahinpey and Gomez 2016; Maity 2015; McKendry 2002). Since 2001, there has been a significant rise in research and development in the field of biomass gasification across both industry and academia, driven by its potential advantages (Sikarwar et al. 2016). This process is particularly noted for being environmentally friendly, as it tends to result in lower emissions of greenhouse gases (Pauls et al. 2016).

Gasification processes produce syngas with high calorific value and a beneficial proportion of hydrogen (Higman et al. 2008). This syngas is a crucial feedstock for generating energy and manufacturing biochemical fuels (Parthasarathy and Narayanan 2014). It serves as a versatile precursor for the production of various valuable fuels, including methanol, hydrogen, and diesel (Carvalho et al. 2018; Kruse 2009; Tijmensen et al. 2002; Holmgren et al. 2012). Additionally, syngas can be directly combusted to produce electricity with using a Brayton–Rankine combined cycle (Ståhl and Neergaard 1998; Kirubakaran et al. 2009). The process involves three principal stages: first, pyrolysis, where biomass breaks down into tar, gas, and char; second, an oxidation phase that releases significant heat to facilitate the endothermic reactions; and third, gasification, which generates syngas (Moghadam et al. 2014). To improve the gasification reactions and minimize the formation of unwanted by-products, various catalysts are frequently employed. These include weak acid alkali metal salts such as potassium carbonate, sodium carbonate, potassium sulfide, and sodium sulfide, which help enhance the effectiveness of the process (Deka et al. 2022).

The syngas generated through biomass gasification primarily consists of hydrogen and carbon monoxide. However, it also contains other gases such as methane, carbon dioxide, water vapor, and nitrogen, along with various contaminants like ammonia, tars, and hydrogen sulfide (Abdoulmoumine et al. 2014). The overall quality of the produced syngas and the types of contaminants present are largely influenced by several factors. These factors include the properties of the biomass feedstock, the types of catalysts used, the gasifying agent employed, and the specific conditions within the reactor (Sikarwar et al. 2016; Farzad et al. 2016).

Under conditions of high temperature (250–300 °C) and pressure (5–15 MPa), syngas can be converted into methanol using an appropriate catalyst (Yang and Ge 2016). Various feedstocks, including biomass, biogas, and organic municipal waste, can be utilized to produce biomethanol, which is methanol derived from organic sources (Kasteren et al. 2016). The production process involves pretreating biomass (drying and chipping) before gasification to generate syngas. To enhance the production of hydrogen and carbon monoxide while reducing the generation of water and carbon dioxide, the oxygen supply is limited during the heating of feedstock at temperatures exceeding 700 °C (Deka et al. 2022). Subsequently, contaminants and impurities are removed from the product gas before it undergoes several conditioning steps to optimize its composition. The primary objective of syngas conditioning is to create a gas mixture with at least twice the amount of hydrogen compared to carbon monoxide. The optimal H_2/CO ratio depends on the initial composition of the syngas and the availability of hydrogen (Deka et al. 2022).

Figure 4.5 shows the stages of methanol production from biomass. Methanol can be produced through the gasification process, where a gasifier converts various biomass feedstocks, such as briquettes, forest residues, wood, crop residues, and waste biomass, into raw syngas. This raw syngas undergoes cleaning through cyclone, cooler, and scrubber units to generate clean syngas, which is subsequently fed into a methanol synthesis reactor. Following the condensation and separation of gases, methanol is synthesized from its constituent elements. The produced methanol has applications in transportation, energy production, industrial processes, and as a precursor for other fuels and chemicals (Deka et al. 2022).

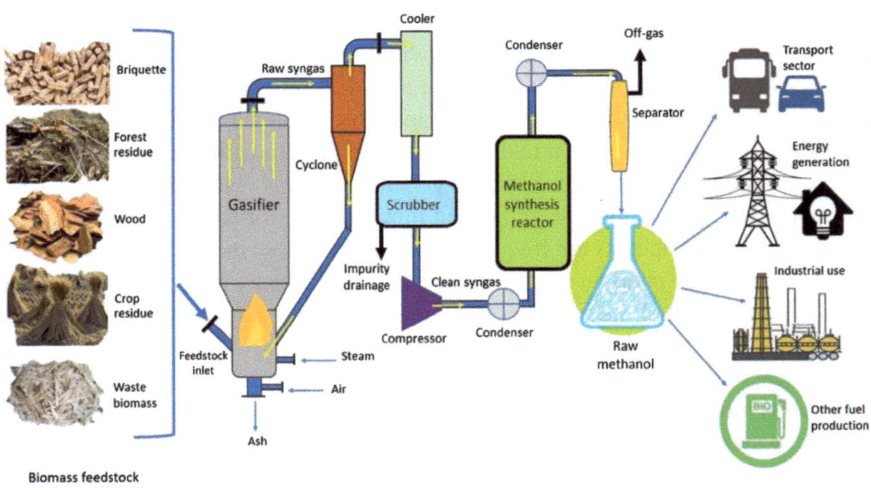

Fig. 4.5 Steps of methanol production via gasification process (Deka et al. 2022)

4.3.5 Synthesis Gas Conversion

Typically, methanol is manufactured on a large scale by utilizing synthesis gas, a mixture of hydrogen, carbon monoxide, and carbon dioxide, which is often generated through the gasification of coal or natural gas (Sheldon 2017; Ott et al. 2000). Currently, most industrial methanol is created by catalytically converting synthesis gas, a mix of CO, CO_2, and H_2. Though synthesis gas can be sourced from different carbon-rich materials, but natural gas is the primary source (Van-Dal and Bouallou 2013). The common catalyst used is $Cu/ZnO/Al_2O_3$, enabling methanol production under relatively "soft" conditions (210–270 °C and 50–100 bar) with a 99% selectivity toward Cox (Fiedler et al. 2000). The by-products include higher alcohols, methane, methyl formate, and DME (Van-Dal and Bouallou 2013).

4.4 Biochemical Methods

Methane can be transformed into methanol through indirect two-step methods. Initially, methane undergoes a reforming process to produce syngas, a mixture of carbon monoxide and hydrogen. This syngas is then catalytically converted into methanol using metal catalysts (Hwang et al. 2014). However, this indirect approach faces challenges such as low energy efficiency and substantial capital costs (Wood et al. 2012). Alternatively, methanol can be directly produced from methane through biological methods. The bioconversion process leverages the methane monooxygenase (MMO) enzyme or methanotrophic cells, operating under mild conditions. This method is advantageous due to its lower energy utilization, higher conversion rates, greater selectivity, and reduced capital costs compared to chemical methods (Conrado and Gonzalez 2014).

Enzymes can be anchored within membranes through multiple non-covalent interactions, including hydrogen bonding, entrapment, and hydrophobic or electrostatic adsorption (Luo et al. 2014).

Methane monooxygenase facilitates the activation of the resistant C–H bond in methane under ambient conditions. Methanotrophic organisms produce MMO to oxidize methane and used methane as their exclusive carbon and energy source under aerial situation (Hwang et al. 2014). A diagram outlining the production of methanol from waste biomass, purified biogas (using methane as a substrate for methanotrophic bacteria), is presented in Fig. 4.6.

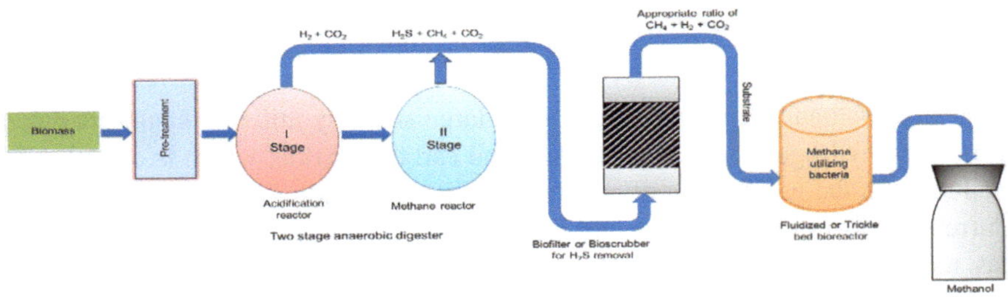

Fig. 4.6 Diagram illustrating the production of methanol by methanotrophic bacteria using waste biomass as a feedstock (Gautam et al. 2020)

4.4.1 Biological Conversion of Methane

Methane can be transformed into methanol through multiple techniques, and this methanol can then be used as a precursor for creating valuable chemicals via established chemical processes. The enzyme methane monooxygenase (MMO) is crucial in this conversion as it catalyzes the oxygenation of methane. Methanotrophic bacteria are capable of converting methane into methanol by inhibiting the action of methanol dehydrogenase (Hwang et al. 2014).

Transforming methane into methanol through biological processes presents an advantageous method for methanol production, offers a promising approach due to its high selectivity under mild conditions. However, challenges such as low conversion rates, yields, and enzyme stability hinder its industrial application. This method relies on two distinct groups of bacteria: ammonia-oxidizing bacteria and methanotrophic bacteria, which are effective in activating the carbon-hydrogen bonds in Deka et al. (2022). Ammonia-oxidizing bacteria utilize ammonia as an energy source, facilitating the partial oxidation of methane to produce methanol (Taher and Chandran 2013). Methanotrophic bacteria naturally produce methanol as an intermediate in their metabolic pathway, ultimately converting it into formaldehyde and other essential compounds. Methanotrophic bacteria are capable of converting methane into methanol by activating methane through the action of an enzyme called methane monooxygenase. This enzyme is essential for the bacteria to oxidize methane, which they use exclusively as their energy and carbon source (Han et al. 2013). This method not only leverages natural biological processes but also offers a potentially sustainable and efficient alternative for methanol production, minimizing the environmental impact associated with traditional chemical methods.

Methanotrophic bacteria have been investigated for their ability to biologically oxidize methane, particularly to address greenhouse gas emissions (Dunfield et al. 2007). Early studies, however, encountered limitations in terms of low conversion rates and bacteria intricate growth requirements (Hyman and Wood 1983).

Proposed strategy involved using these bacteria to oxidize methane in landfills using atmospheric oxygen (Scheutz et al. 2009). The natural metabolic process of methanotrophs involves the sequential, oxidation of methanol to formaldehyde

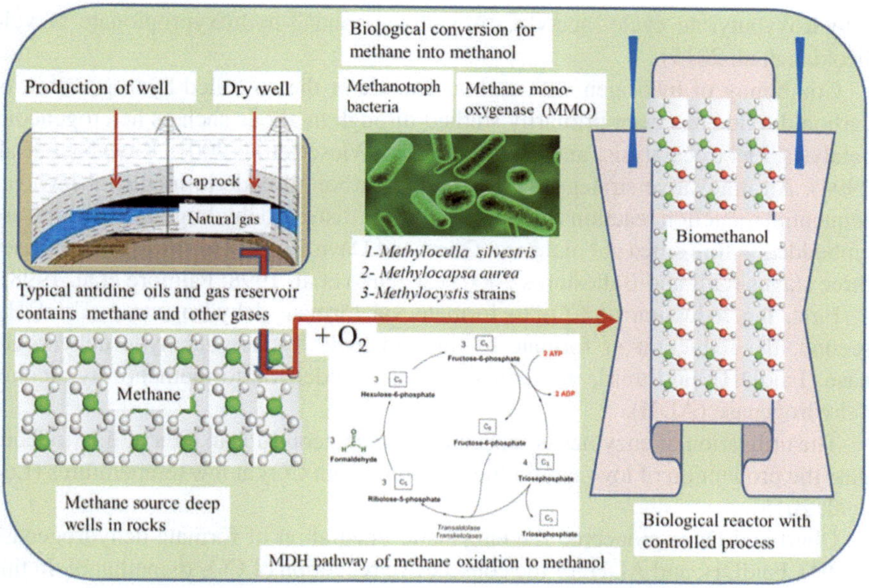

Fig. 4.7 Biological pathway of methanol production (Srivastava et al. 2022)

(CHOH) and the subsequent oxidation of formaldehyde to formate (CHOOH) and other metabolites, and ultimately, oxidation of methane to methanol that methanol is not the final product (Scheutz et al. 2009). It only a precursor of formaldehyde that is used to synthesize various metabolites that are essential to survival of bacteria (Park and Lee 2013). Biological pathway of methanol production from methane is shown in Fig. 4.7 (Srivastava et al. 2022).

Methanotrophic bacteria are known to oxidize methane to carbon dioxide through a series of enzyme-catalyzed reactions. These enzymes, including methane monooxygenase (MMO), methanol dehydrogenase (MDH), formaldehyde dehydrogenase (FaDH), and formate dehydrogenase (FDH), sequentially break down methane into CO_2 (Srivastava et al. 2022). Two primary types of methane monooxygenase exist: soluble (sMMO) and particulate (pMMO) (Park and Lee 2013).

4.4.2 Biocatalytic Conversion of CO_2

Bio-inspired materials offer numerous benefits, including biodegradability, low corrosiveness, easy disposal, and flexible design. Research indicates their potential as cost-effective catalysts for CO_2 reduction (Bhardwaj et al. 2021).

In general, there are six mechanisms for carbon dioxide reduction in nature, such as the reductive pentose phosphate cycle, citric acid cycle, dicarboxylate cycle,

4-hydroxybutyrate cycle, acetyl-CoA pathway, and 3-hydroxypropionate bicycle (Roldan et al. 2015).

Combining of hydrogen with carbon dioxide or the so-called hydrogenation of carbon dioxide has been primarily studied through methods such as heterogeneous catalysis, electrocatalysis, and photocatalysis (Wieser et al. 2001; Kuwabata et al. 1994). An innovative strategy for converting carbon dioxide to methanol involves sequential reduction reactions facilitated by three distinct dehydrogenases, which are embedded within silica gel matrices (Obert and Dave 1999). This process comprises three stages (Gill and Ballesteros 2000; Kutzenko et al. 1998; Palmore et al. 1998):

First, the reduction of CO_2 to formate via formate dehydrogenase (FateDH); second, the reduction of formate to formaldehyde by formaldehyde dehydrogenase (FaidDH); and third, the reduction of formaldehyde to methanol by alcohol dehydrogenase (ADH).

The utilization of enzymes is especially advantageous due to their ability to facilitate the production of low-carbon fuels straight from CO_2 at low temperatures (Luo et al. 2015).

Obert and Dave pioneered the enzymatic integration of formate dehydrogenase (FDH), FaidDH, and ADH for the consecutive reduction of CO_2 to methanol. In this process, reduced nicotinamide adenine dinucleotide (NADH) served as the ultimate electron donor in the enzymatic reaction (Obert and Dave 1999). This methodology involved the reversal of standard biological metabolic pathways (Luo et al. 2015).

Some biocatalysts that can reduce CO_2 to methanol include (Bhardwaj et al. 2021):

- Mineral greigite under ambient conditions.
- Graphitic carbon nitride with a honeycomb-like structure, prepared through thermal condensation.
- Multi-enzyme systems comprising formate dehydrogenase (FDH), alcohol dehydrogenase (ADH), and formaldehyde dehydrogenase (FaldDH).
- Iron (II) complexes for direct CO_2-to-methanol conversion.
- Hydrogen and iron (II) dichloro-scorpionate catalyst in solvent-free or amine-free.

4.5 Electrocatalytic Methods

At present, the electrocatalytic reduction of carbon dioxide (ERC) to selectively produce methanol remains in its infancy and is not yet ready for practical use. There is a scarcity of research on the process engineering design, economic viability, and environmental implications of ERC for methanol production (Wiranarongkorn et al. 2023). The initial study of ERC dates back over a century. In the first decade of the twentieth century, numerous investigations have utilizing metal electrodes, predominantly yielding formic acid. During the mid-1980s, researchers discovered the potential for methanol production from CO_2 using semiconductor materials like GaAs and InP, although the achieved current densities were inadequate (Canfield and Frese 1983; Frese and Leach 1985; Summers et al. 1986).

Hori et al. (1985) found that metallic copper could efficiently produce hydrocarbons, which was a significant breakthrough at the time and gave researchers hope for the development of more effective electrocatalysts for other products. Following that, the synthesis and performance of copper-based electrocatalysts, including copper nanostructures and copper alloys, as well as the application of advanced supports in conjunction with copper-based electrocatalysts, have been comprehensively reviewed (Wiranarongkorn et al. 2023).

Electrocatalytic reaction systems offer several advantages, including the ability to easily and precisely control the electrolytic conditions and the continuous in situ generation of reactive oxygen species through electrode reactions (Jiang et al. 2022).

4.5.1 Electrochemical Reduction of CO

Usually, the electrochemical reduction of carbon monoxide to produce methanol is conducted on copper-based electrodes (Bagger et al. 2019; Calle-Vallejo and Koper 2013; Montoya et al. 2015). The process involves the stabilization of key intermediates facilitated by the electrode surface and the electrolyte composition. Specific conditions, such as applied potential and the pH of the electrolyte, play critical roles in enhancing the efficiency and selectivity of methanol production from CO (Bagger et al. 2019; Hori et al. 1998).

Various electrodes are utilized for the electrochemical reduction of carbon monoxide to methanol such as: (Bagger et al. 2019; Hori et al. 1998; Li et al. 2014).

- Copper Electrodes (Cu): Copper electrodes are widely used due to their excellent electrocatalytic properties and their ability to stabilize key intermediates in the reduction of CO to methanol.
- Alloy Electrodes: Copper alloys with other elements such as zinc (Zn), silver (Ag), and gold (Au) are also employed to enhance the efficiency and selectivity of the CO reduction to methanol.
- Metal Oxide Electrodes: Certain metal oxides, such as copper oxide (Cu_2O) and zinc oxide (ZnO).

These electrodes are favored for their specific characteristics in stabilizing intermediates and improving the overall efficiency of the electrochemical reduction processes.

4.5.2 Direct Electrochemical Reduction of CO_2

Converting CO_2 into useful chemicals and fuels through electrochemical reduction is a promising way to reduce greenhouse gas emissions. But the commercial development of conventional catalytic hydrogenation of CO_2 for methanol production has been limited due to ample hydrogen consumption and is not widespread adoption used (Wiranarongkorn et al. 2023).

This process uses electricity to reduce CO_2 into valuable products like methanol (MeOH) through electrochemical catalysts (Malik et al. 2020). The process involves multiple steps, such as reducing CO_2 to CO, hydrogenating CO to formaldehyde (HCHO), and further hydrogenating CH_2O to MeOH. Additionally, electrochemical CO_2 reduction is capable of producing methanol along with other valuable chemicals including methane, methanol, formaldehyde, formic acid (HCOOH), syngas, and other hydrocarbons (Rumayor et al. 2019; Wyndorps et al. 2021). In comparison to catalytic hydrogenation, ERC operates under relatively mild conditions and offers favorable outcomes. It can utilize electricity prepared from any source, including renewable sources, to transform CO_2 into valuable products. As illustrated in Fig. 4.8, the anodic reaction involves water oxidation to produce O_2 and electrons, which travel through an external circuit to the cathode to produce H_2 or reduce CO_2. Simultaneously, protons migrate through an aqueous electrolyte to the cathode (Wiranarongkorn et al. 2023).

At the cathode, the interaction between electrons and protons results in the formation of H_2 or the reduction of CO_2 into various products. The allocation of electrons between hydrogen production and CO_2 reduction is influenced by the activation impediments of the respective reactions. The following reactions illustrate the potential reaction pathways along with their associated reduction potentials, relative to the standard hydrogen electrode (SHE) at 25 °C and pH 7 (Liang et al. 2020; Wu et al. 2017; Nguyen et al. 2020).

$$CO_2 + e^- \rightarrow CO + 2H_2 \quad E^0 = -1.90\,V \tag{4.12}$$

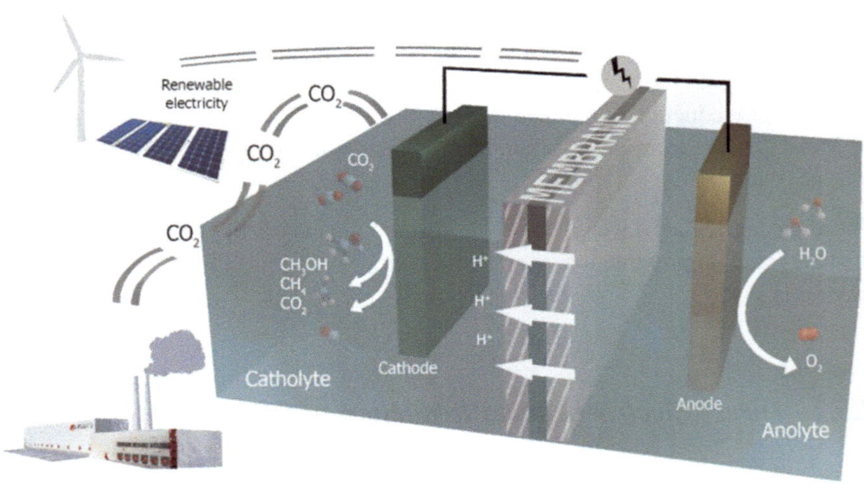

Fig. 4.8 Schematic of the standard arrangement for the electrochemical reduction of CO_2 (Wiranarongkorn et al. 2023)

$$CO_2 + 2H^+ + 2e^- \rightarrow HCOOH \quad E^0 = -0.61\,V \tag{4.13}$$

$$CO_2 + 2H^+ + 2e^- \rightarrow CO + H_2O \quad E^0 = -0.52\,V \tag{4.14}$$

$$CO_2 + 4H^+ + 4e^- \rightarrow CH_2O + H_2O \quad E^0 = -0.51\,V \tag{4.15}$$

$$CO_2 + 6H^+ + 6e^- \rightarrow CH_3OH + H_2O \quad E^0 = -0.38\,V \tag{4.16}$$

$$CO_2 + 8H^+ + 8e^- \rightarrow CH_4 + 2H_2O \quad E^0 = -0.24\,V \tag{4.17}$$

$$2H^+ + 2e^- \rightarrow H_2 \quad E^0 = -0.42\,V \tag{4.18}$$

At the cathode, the reduction of CO_2 (Eqs. 4.12–4.17) and the hydrogen evolution reaction (Eq. 4.18) take place, each comprising multiple elementary steps. Typically, the synthesis of more complex molecules like hydrocarbons or methanol requires more reaction steps compared to the production of simpler molecules such as CO and formic acid (Liang et al. 2020).

The ERC system consists of units dedicated to CO_2 capture, conversion of CO_2 to methanol, and subsequent methanol separation, depicted in Fig. 4.9 (Wiranarongkorn et al. 2023).

Key challenges in ERC include (Wiranarongkorn et al. 2023): (1) high overpotential due to CO_2 stability, (2) difficulty in finding efficient catalysts, (3) competing hydrogen evolution reactions, and (4) low CO_2 solubility in aqueous solutions, all affecting reaction rates and efficiency, and product yield.

Carbon dioxide can be electrochemically converted into carbon monoxide, formic acid, and various hydrocarbons using metal electrodes (Hori et al. 1985, 1986). Among them, Copper, in particular, is notable for its ability to efficiently reduce CO_2 to hydrocarbons and alcohols under standard temperature and pressure conditions (Hori et al. 1985, 1986, 1989). While numerous metals are capable of reducing CO_2 to CO, copper is distinguished by its ability to further reduce CO into methane

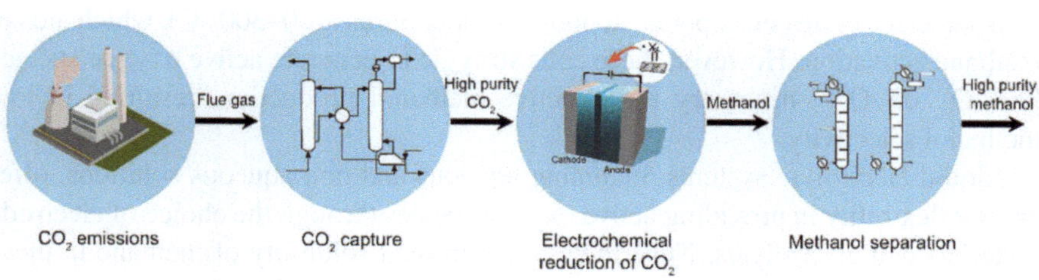

Fig. 4.9 Unit configuration for the electrochemical reduction system of CO_2 (Wiranarongkorn et al. 2023)

(CH_4), ethylene (C_2H_4), and alcohols at elevated current densities (Hori et al. 1989, 1987). Additionally, recent studies indicate that under specific conditions, silver (Ag) electrodes can also reduce CO_2 to CH_4, C_2H_4, and alcohols (Shiratsuchi and Nogami 1996).

Improved electrocatalysts are being developed to make the electrocatalytic reduction of carbon dioxide a practical method for producing methanol and other chemicals. Various materials based on their redox potentials, electron transfer rates, chemical reaction kinetics, and current efficiency were screened to find the best electrocatalyst. Many different metal-based electrocatalysts have been studied for ERC to produce methanol. These metal electrocatalysts can be categorized into four main groups: bulk monometallic, metal oxide, metal alloy, and nanostructured electrodes (Wiranarongkorn et al. 2023).

4.5.3 Direct Electrochemical Conversion of Methane to Methanol

The direct transformation of methane into methanol poses significant challenges due to the high stability of methane and the high reactivity of methanol. Transition metal oxides hold promise as potential catalysts for this electrochemical process (Shen et al. 2023). Methane features four identical C–H bonds arranged in a symmetrical tetrahedral structure, with a low electron affinity of 1.9 eV and polarizability of $2.8 \times 10^{-40}\ C^2\ m^2/J$, while exhibiting a high ionization energy of 12.6 eV. Breaking one of these C–H bonds require a substantial energy input of 439 kJ/mol (Jiang et al. 2022).

In stark contrast, under mild conditions methanol can be easily oxidized to formaldehyde or formic acid because of high reactivity. Consequently, when methanol is produced during the activation of methane, it tends to convert rapidly into other chemicals. This fundamental difference in reactivity between methane and methanol limits the efficiency and selectivity of converting methane to methanol in a single-step process.

Two primary electrochemical reaction systems have been utilized for the direct conversion of methane to methanol: solid electrolyte cells and liquid electrolyte cells.

Solid electrolyte cells operate at higher temperatures (100–300 °C), which aids in methane activation. However, these cells struggle to generate active oxygen species like $\cdot O^-$ or $\cdot O^{2-}$, necessary for selective methanol production, resulting in low methanol selectivity.

Liquid electrolyte systems, including aqueous and non-aqueous solutions, offer greater flexibility in providing active oxygen species through the choice of electrode materials and electrolytes. Nevertheless, the limited solubility of methane in these systems can restrict overall conversion efficiency (Jiang et al. 2022).

4.6 Photocatalytic Methods

Photocatalytic conversion involves the production of electrons through the action of light on a photocatalyst and the generation of hydroxyl radicals by the reaction between the produced electrons and water molecules. These hydroxyl radicals subsequently interact with methane molecules, breaking them down into methyl radicals. The methyl radicals then undergo further reactions with water molecules present in the system, leading to the formation of methanol and hydrogen gas. This innovative approach not only leverages renewable light energy but also integrates the use of water, making it a promising method for sustainable fuel production (Yang et al. 2014).

4.6.1 Photocatalytic Reforming of Methane

Photocatalysis is a gentle method for converting stable methane into high-value fuels like methanol, which has garnered significant interest (Zhou et al. 2024). Nonetheless, the initial C–H bond cleavage requires a substantial energy input of 439 kJ mol^{-1}, posing a significant energy barrier that limits the reaction's efficiency (Zhou et al. 2024; Wang et al. 2021; Huang et al. 2021). Additionally, these valuable fuels tend to be metastable (Zhou et al. 2024), and can easily be over-oxidized into less valuable CO or even CO_2 (Zhou et al. 2024; Shoji et al. 2020; Song et al. 2021).

Given these challenges, researchers have developed various photocatalysts and cocatalysts to enhance yield and selectivity by broadening the light-response spectrum, improving charge separation, or boosting catalytic C–H activation (Zhou et al. 2024). While semiconductor photocatalysts like TiO_2 have a strong oxidizing ability for methane activation, they often lead to over-oxidation, reducing selectivity for desired products (Feng et al. 2021). On the other hand, nitrides such as g-C_3N_4 display moderate oxidizing abilities but lack stability (Zhou et al. 2019). Recently, InGaN photocatalysts, with adjustable band edge potentials and surface band construction, have been synthesized by adjusting the indium content in the GaN segment (Zhou et al. 2019; Kibria et al. 2015). This advancement provides a suitable platform for controlling the reaction pathway in photocatalytic methane oxidation (Zhou et al. 2024).

4.6.2 Photocatalytic Reduction of CO_2

The process of CO_2 photoreduction includes complex multi-electron reactions that can result in a wide range of products, including gases like CO and CH_4, as well as higher hydrocarbons in the gas phase, and in the liquid phase, oxygen-containing compounds such as alcohols, aldehydes, and carboxylic acids (Hong et al. 2013).

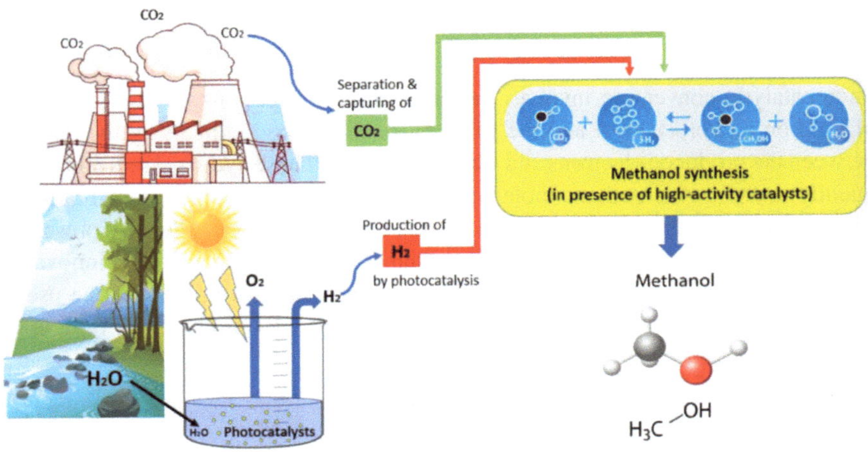

Fig. 4.10 Synthesis of methanol through photocatalysis using carbon dioxide captured from flue gas. In this process, water is split into oxygen and hydrogen via photocatalysis. The generated hydrogen then reacts with carbon dioxide, which has been captured from flue gas, to produce methanol (Deka et al. 2022)

Carbon is at its maximum oxidation state in CO_2, and lowering the oxidation state of CO_2 can result in a wide range of products with varying carbon oxidation levels, including gases like CO and CH_4, higher hydrocarbons in the gas phase, and liquid-phase oxygenates such as CH_3OH and HCOOH (Hong et al. 2013). In the process of photocatalytic conversion, hydroxyl radicals generated by light interaction produce methyl radicals from methane (Fig. 4.10). These methyl radicals then react with additional water molecules to form methanol and hydrogen (Yang et al. 2014).

Generally, there are three key essential phases in photocatalysis: capturing solar light, separating and transporting charges, and conducting surface reactions (Chang et al. 2016).

At first, (i), occurrence photons with energy equal to or exceeding the bandgap energy (E_g) of the photocatalyst are absorbed, leading to the creation of electron–hole ($e^- - h^+$) pairs. Next (ii), the photogenerated electrons and holes move independently toward the surface of the photocatalyst particles (in PC system) or the cathode and the anode (in PEC system) in the second step. In step three (iii), the electrons produced by light exposure convert CO_2 into fuels, while the holes facilitate the oxidation of H_2O into O_2. Throughout steps (ii) and (iii), a significant portion of the electrons and holes may be lost due to recombination (iv) (Chang et al. 2016).

As solar power is an essentially unlimited energy source, using it to convert CO_2 into chemicals and fuels like formic acid, methane, or methanol is viewed as a hopeful strategy for tackling the challenges of global warming and the energy storage (Lewis and Nocera 2006; Habisreutinger et al. 2013).

In general, there are two types of configurations of photoconversion systems for reducing CO_2. The initial setup of the photocatalytic system (PC) involves a suspension of photocatalyst particles in a solvent to reduce dissolved CO_2 (Fig. 4.11a).

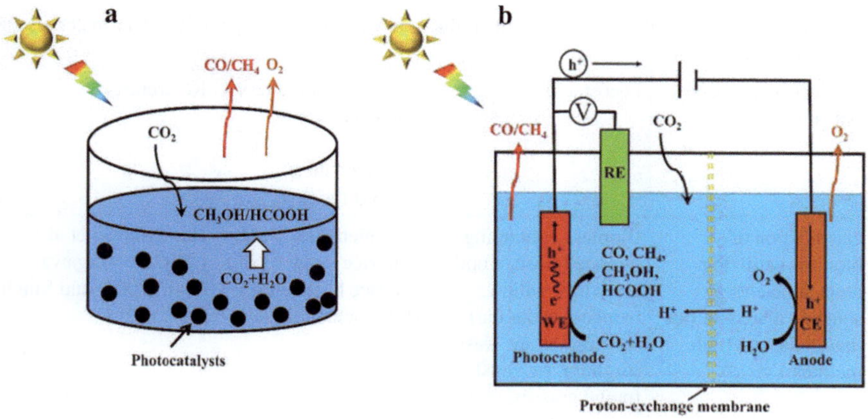

Fig. 4.11 **a** Photocatalyst particles dispersed in an electrolyte containing CO_2, facilitating both the reduction of CO_2 and the oxidation of water through photocatalytic reactions. **b** Photoelectrochemical cell with a photocathode serving as the working electrode (WE) for CO_2 reduction, a counter electrode (CE) for water oxidation, and a reference electrode (RE) immersed in a CO_2-containing electrolyte (Gao et al. 2020)

This type of system comprises simple equipment and widely available photocatalysts, with solar energy being the sole driving force for CO_2 reduction. Both the reduction and oxidation reactions take place on distinct surface sites of the identical photocatalyst particle, resulting in the admixture of all products (Chang et al. 2016; Ma et al. 2014; Gao et al. 2020).

The second configuration involves a photoelectrochemical (PEC) cell system, which comprises a semiconducting photoelectrode functioning as the working electrode, a counter electrode, and usually a reference electrode. The photoelectrode absorbs light to generate charge carriers and facilitates a half-cell reaction, typically occurring on the photocathode to reduce CO_2, while the counter electrode hosts the complementary half-cell reaction (Fig. 4.11b) (Chang et al. 2016; Gao et al. 2020).

Therefore, the products of reduction and oxidation can be physically detached in various half-cells using a proton exchange membrane (Nafion), which prevents the possibility of the re-oxidation of reduction products (Chang et al. 2016). Hence, conducting CO_2 photoreduction using a PEC system is advantageous, despite the complex preparation of photoelectrodes and the typically required external bias (White et al. 2015).

Table 4.2 shows some of the different processes used in the production of methanol with different raw materials and unique advantages, which were collected in the study of Deka et al. (2022).

Table 4.2 Different methods of methanol production with specified types of primary materials and present unique benefits

Production process (advantages)	Feedstocks	Practical parameters (methanol yield/ emission characteristics/ catalyst type)	References
Gasification of biomass (utilizing renewable energy sources, abundant raw materials, and high methanol yield)	Biomass containing lignocellulose, wood materials, willow, bamboo, cedar bark, sawdust, straw, waste, sorghum, rice husk, forage grasses, trees, Japanese larch, and agricultural residues	Low methanol yield: with rice straw (36%) and rice husk (39%) High methanol yield (55%)	Raudaskoski et al. (2009), Nakagawa et al. (2007) and Singh et al. (2014)
	Syngas	Integrated gasification combined cycle plant methanol production Reduction in sulfur oxides and nitrogen oxides	Osman et al. (2021b)
Indirect methods for conversion (high selectivity and productivity)	Biogas sourced from corncob fermentation, landfills, sorghum fermentation, and palm oil effluent	Most profitable source—palm oil Lowest producing source of methanol—landfill	Santos et al. (2018)
	Biogas derived from food, dairy, animal waste, and sludge of wastewater	Cu/Zn/Al/Zr catalyst used Biogas feed—1.8 × 10^6 Nm^3/year Maximum production of methanol = 2100 t/ day	Ghosh et al. (2019)
	Biogas	Enhancing methanol yield by regulation of hydrogen and supplementation gas composition	Patel et al. (2016)
	Hydrogen, syngas, and carbon dioxide captured from flue gas	Novel prepared $CuO/ZnO/ZrO_2$ catalyst used Economically and environmentally beneficial	Raudaskoski et al. (2009) and Jadhav et al. (2014)
	Carbon dioxide captured from industrial sources	Methanol costs 2.5 times high than usual	Atsonios et al. (2016)

(continued)

Table 4.2 (continued)

Production process (advantages)	Feedstocks	Practical parameters (methanol yield/ emission characteristics/ catalyst type)	References
	Carbon dioxide derived from fossil fuel combustion, and hydrogen resulting from renewable energy	–	Boretti (2013) and Anicic et al. (2014)
Partially oxidation of methane (high reaction rate and high conversion rate of methanol)	Methane	–	Gesser et al. (1985)
	Natural gas	–	Arutyunov et al. (1996)
	Methane	Selectivity = 40%, efficiency of conversion = 30%	Nozaki et al. (2011)
Photocatalytic conversion (high selectivity and low energy input)	Methane	Enhancing methanol yield by using silver-impregnated tungsten oxide Utilizing ultraviolet light	Hameed et al. (2014)
	Carbon dioxide sequestered from flue gas emissions	Producing methane with a lower environmental impact compared to methanol production	Trudewind et al. (2014)
	Methane, hydrogen peroxide	High methanol yield with the use of ultraviolet/hydrogen peroxide	López-Martín et al. 2017)
	Methane	Nickel oxide = 20% and Titanium dioxide = 21% Conversion of methanol selectivity with tungsten trioxide = 29%	Gondal et al. 2004)
	Methane, water	Maximum methanol yield = 43 g catalyst/ h, visible light using	Taylor and Noceti (2000)

(continued)

Table 4.2 (continued)

Production process (advantages)	Feedstocks	Practical parameters (methanol yield/ emission characteristics/ catalyst type)	References
Biological conversion processes (low energy requirement and high selectivity)	Methane	Utilizing ammonia-oxidizing bacteria, methanotrophs, and algae	Ge et al. (2014)
	Methane	Found advantageous in wastewater treatment plants Reduction of methanol cost	Taher and Chandran (2013)
	Methane	Utilizing nitrifying activated sludge Significant enhancements in the efficiency of wastewater treatment plants	Geng et al. 2014)

4.7 Hybrid Methods and Other Newer Approaches

4.7.1 Plasma-Catalytic Methanol Synthesis

Plasma-catalytic methanol synthesis offers a novel method for converting CO_2 and H_2 into methanol under milder conditions compared to traditional thermal techniques. This process leverages the distinctive properties of plasma to enhance catalytic reactions, presenting a sustainable option for methanol production (Cui et al. 2022).

Plasma technology is gaining significant attention in catalyst preparation due to its efficacy in oxidation, reduction, etching, doping, coating, and surface treatment (Zhang et al. 2021).

While low-temperature operations face dynamic limitations in CO_2 activation, high temperatures, despite facilitating CO_2 activation, primarily drive the reverse water–gas shift reaction (that leads to the production of carbon monoxide), which competes with methanol synthesis (Wang et al. 2018a).

Due to its capability to function at ambient temperature, non-thermal plasma (NTP) technology presents a viable alternative to thermal catalysis for overcoming the challenges associated with CO_2 activation (Wang et al. 2018a). Transition-metal catalysts produced using plasma-based technology exhibit increased activity at low temperatures (Yan et al. 2015; Wang et al. 2018b).

NTP excels in activating thermodynamically stable molecules like CO_2 at atmospheric pressure and low temperatures compared to the simple catalytic process (Snoeckx and Bogaerts 2017; Tu and Whitehead 2012, 2014; Mei et al. 2016; Yi et al. 2013). It generates numerous high-energy electrons with typical energies of 1 – 10 eV, has generated interest in employing these energetic electrons as alternative "catalysts" to activate reactants like CO_2 and H_2, transforming them into a range of reactive species such as radicals, excited atoms, ions, and molecules, which are crucial for initiating and propagating chemical reactions (Fridman 2008).

Plasma-catalytic methanol synthesis involves generating a plasma, which comprises species with high reactivity like ions, radicals, and excited molecules. These species interact with the feed gases (CO_2 and H_2), facilitating their conversion into methanol. The plasma generates excited species (e.g., CO_2, H_2) that interact with the catalyst to facilitate the formation of mediator material such as HCOO and CH_3O, eventually leading to methanol production (Zhang et al. 2023). Common catalysts include $Cu/ZnO/Al_2O_3$, MnO_x/ZrO_2, and CuO–MgO. These catalysts are selected for their ability to enhance the reaction rates and selectivity toward methanol (Snoeckx and Bogaerts 2017; Bogaerts and Centi 2020).

Non-thermal plasmas, including dielectric barrier discharge (DBD) (Appel et al. 2013) and radio-frequency (RF) plasmas, are frequently utilized because they can function effectively at near-ambient temperatures and pressures (Zhang et al. 2023). Figure 4.12 illustrates how non-thermal plasma catalysis operates to activate and transform CO_2 into various chemicals and fuels.

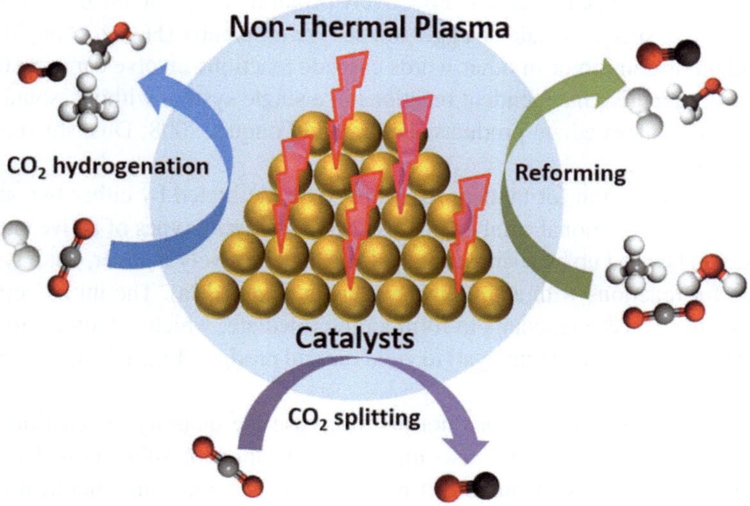

Fig. 4.12 Heterogeneous catalysis activated by NTP for CO_2 conversion (Appel et al. 2013)

4.7.2 *Bifunctional or Tandem Catalytic Methanol Synthesis*

In the context of advancing sustainable chemical science, there is a focus on novel technologies and streamlined procedures as effective tools for dealing with environmental concerns. Enhancing catalytic performance, minimizing waste, and employing eco-friendly substances are essential elements of eco-conscious chemistry (Felpin and Fouquet 2008). Within the framework of sustainable chemistry, cooperative catalysis and consecutive reactions represent resourceful and environmentally responsible approaches to chemical synthesis (Huang et al. 2017).

Synergistic catalysis, in particular, is a potent technique where multiple distinct active sites activate the reactants simultaneously, leading to a notable reduction in the reaction energy barrier (Felpin and Fouquet 2008). Therefore, this can lead to enhancements in both the efficiency and selectivity of current reactions, as well as the discovery of new chemical conversions (Huang et al. 2017).

Synergistic catalysts, which can be composed of two separate catalysts or one catalyst with multiple active sites, have found use in various fields such as organic catalysis (Allen and MacMillan 2012), photocatalysis (Martin et al. 2015), nano catalysis (You et al. 2013), electrocatalysis (Shao et al. 2016), and enzyme catalysis (Capon et al. 2005).

In the realm of synergistic catalysis, the cooperative interaction of the active sites elevates the energy level of the highest occupied molecular orbital (HOMO) of one reactant and decreases the energy level of the lowest unoccupied molecular orbital (LUMO) of the other reactant. By combining multifunctional active species thoughtfully, these resulting catalysts can effectively promote many chemical reactions that a single active species would struggle to catalyze efficiently (Huang et al. 2017).

Tandem or domino or in other words cascade reactions involve carrying out two or more consecutive independent reactions in a single system without isolating and purifying the intermediate products (Felpin and Fouquet 2008; Dhakshinamoorthy and Garcia 2014).

It is more common for tandem reactions to be facilitated by either two separate catalysts or a bifunctional catalyst possessing two or more types of active sites. The process can be sped up by a single catalyst in a consecutively manner, but it typically suitable for reactions with a similar mechanism (Fig. 4.13a). The initial active site (catalyst) triggers the reactants to form an intermediate, which is then catalyzed by second type of active site (catalyst) to yield the end product (Fig. 4.13b) (Huang et al. 2017).

As a result, tandem reactions not just decrease the quantity of reaction stages, energy usage, and waste, but also limit the utilization of solvents and reagents. Tandem reactions present numerous possibilities for enhancing chemical conversions, including Lew is acid–base catalysis, hydrogenation reactions, and alkene metathesis (Wasilke et al. 2005).

Metal–organic frameworks (MOFs) are complex, porous crystals made of metal ions and organic ligands, offering large surface area, tunable pores, and diverse structures, enabling them in gas separations, catalysis, sensing, and drug delivery. It

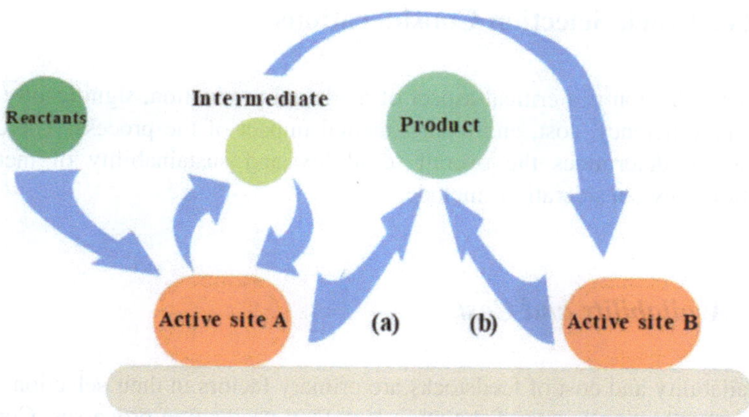

Fig. 4.13 Catalytic tandem reactions facilitated by either **a** a single active site or **b** multiple types of active sites

has been checked that the combination of different types of MOFs active sites, is a promising option as multifunctional materials for synergistic catalysis and tandem reactions (Huang et al. 2017).

4.7.3 Photoelectrochemical Catalytic (PEC) Methanol Synthesis

Photoelectrochemical systems combine photocatalysis and electrocatalysis to convert CO_2 into valuable products like methanol. However, finding a stable and efficient semiconductor material for photoelectrodes remains a significant challenge (Liang et al. 2017; Tursunov et al. 2017).

A key step in photoelectrochemical CO_2 reduction to methanol is the generation of electrons and protons through light absorption by a semiconductor photocathode. The absorbed light excites electrons from the valence band to the conduction band, which are then transferred to the cathode via an external circuit (Sarp et al. 2021). The semiconductor also separates electron–hole pairs near interface and sends into the electrolyte, driving the electrochemical conversion and oxidation processes.

A major limitation of PEC systems is the instability of n-type semiconductor materials, which are prone to surface oxidation. This issue hinders the long-term performance and efficiency of these systems (Hartadi et al. 2016).

4.8 Feedstock Selection Considerations

Feedstock selection is a critical aspect of methanol production, significantly influencing the efficiency, cost, and environmental impact of the process. The choice of feedstock determines the overall feasibility and sustainability of methanol production. Key considerations include.

4.8.1 Availability and Cost

The availability and cost of feedstocks are primary factors in their selection. Obviously, cost reduction is one of the main targets in production processes. Common feedstocks for methanol production include natural gas, coal, biomass, and carbon dioxide. Each feedstock has distinct supply chains and cost structures (Deka et al. 2022; Liu et al. 2018).

- Natural Gas: Widely available and relatively low-cost, natural gas is the predominant feedstock for methanol production. Its high hydrogen content makes it an efficient source for syngas production through steam methane reforming (Liu et al. 2018).
- Coal: Abundant and inexpensive, coal is another significant feedstock, particularly in regions with large coal reserves. However, coal gasification processes require extensive purification steps to remove impurities.
- Biomass: Renewable and sustainable, biomass feedstocks such as agricultural residues, wood, and municipal waste offer environmental benefits. However, their availability can be seasonal, and the cost of collection and transportation can be high.
- CO_2: Captured from industrial emissions or directly from the atmosphere, CO_2 is an emerging feedstock for methanol production. Utilizing CO_2 helps in carbon capture and storage (CCS) efforts, contributing to environmental sustainability (Liu et al. 2018).

Here are the current prices for some of the key feedstocks used in methanol production, based on data from November 2024: The price of natural gas, a primary feedstock for methanol production via steam methane reforming, varies by region (intratec 2024). As of recent data, the price in the United States is approximately $2.97 per million British thermal units (MMBtu). For coal gasification processes, the price of coal can also vary widely. In the United States, the price of thermal coal is around $60 per metric ton (Schorn et al. 2021). The cost of biomass feedstocks, such as wood chips or agricultural residues, can range from $30 to $70 per metric ton, depending on the type and source. The price of CO_2 for industrial use, such as in the direct hydrogenation of CO_2 to methanol, is approximately $30 to $50 per metric ton source. These prices are subject to fluctuations based on market conditions, regional availability, and other economic factors (intratec 2024; tradingeconomics 2024).

4.8.2 Environmental Impact

The environmental impact of feedstocks is a reasonable and attractive consideration, especially in the context of reducing greenhouse gas emissions and promoting sustainable practices: (Deka et al. 2022; Patil et al. 2024):

- Natural Gas: While efficient, natural gas is a fossil fuel, and its extraction and use contribute to CO_2 emissions. However, it produces fewer pollutants compared to coal.
- Coal: Coal has a high carbon footprint due to its significant CO_2 emissions during gasification. Advanced technologies and carbon capture and storage can mitigate some of these impacts.
- Biomass: Biomass is considered carbon–neutral, as the CO_2 released during its conversion is offset by the CO_2 absorbed during the growth of the biomass. It also helps in waste management by utilizing agricultural and municipal waste (Gautam et al. 2020).
- CO_2: Using CO_2 as a feedstock directly addresses carbon emissions by converting a greenhouse gas into a valuable product. This approach supports carbon capture and utilization (CCU) strategies (Meunier et al. 2020).

4.8.3 Process Compatibility

It is essential that the chosen production process for efficient methanol synthesis must be compatible with the feedstocks (Deka et al. 2022; Bertau et al. 2014; Sarvestani et al. 2024).

- Natural Gas: Ideal for steam methane reforming, natural gas produces a clean syngas with fewer impurities, reducing the need for extensive purification.
- Coal: Suitable for gasification processes, coal requires robust purification systems to remove sulfur and other contaminants before methanol synthesis.
- Biomass: Biomass gasification produces syngas, but the feedstock must be dried and pre-treated to ensure efficient conversion. The variability in biomass composition can affect process stability (Gautam et al. 2020).
- CO_2: Electrochemical and catalytic processes are well-suited for converting CO_2 into methanol. These methods often require renewable electricity, aligning with sustainable energy goals (Chiou et al. 2023).

4.8.4 Sustainability

The long-term sustainability of the feedstock is a key factor in ensuring continuous methanol production (Patil et al. 2024).

- Natural Gas and Coal: While abundant, these fossil fuels are finite resources. Their long-term use is unsustainable without significant advancements in CCS technologies.
- Biomass: Renewable and widely available, biomass offers a sustainable feedstock option. However, its use must be balanced with food production and land use considerations.
- CO_2: Utilizing CO_2 as a feedstock supports sustainability by recycling carbon emissions. This approach aligns with global efforts to reduce carbon footprints and combat climate change.

4.8.5 Technological Advancements

Advancements in technology can influence feedstock selection by improving the efficiency and feasibility of using various feedstocks (Patil et al. 2024; Sarvestani et al. 2024).

- Catalyst Development: Innovations in catalyst design can enhance the conversion efficiency of different feedstocks, making processes like CO_2 hydrogenation more viable.
- Process Optimization: Improved process designs and integration of renewable energy sources can reduce the environmental impact and operational costs of methanol production.
- Hybrid Systems: Combining traditional and emerging technologies can optimize feedstock utilization, such as integrating biomass gasification with CO_2 hydrogenation.

4.9 Future Prospects and Challenges

The future of methanol production is promising, driven by advancements in renewable feedstocks, catalyst innovations, and electrochemical methods. Utilizing renewable feedstocks such as biomass and captured CO_2 can significantly reduce the carbon footprint of methanol production, aligning with global sustainability goals. Innovations in catalyst materials, including transition metal carbides and bifunctional catalysts, are expected to enhance the efficiency and selectivity of methanol synthesis, reducing by-product formation and improving overall yields (Sahoo et al. 2023). Electrochemical methods, particularly those leveraging renewable electricity for CO_2 reduction, present a sustainable pathway that addresses both carbon capture and renewable energy utilization (Deka et al. 2022; Patil et al. 2024). The integration of traditional and emerging technologies, such as combining biomass gasification with CO_2 hydrogenation, offers potential for optimizing feedstock utilization and improving process efficiency. However, several challenges must be addressed to fully realize these prospects. The economic viability of renewable feedstocks and advanced catalysts remains a significant hurdle, as does the technological maturity of emerging methods like biochemical and photochemical processes. Ensuring the environmental sustainability of methanol production requires comprehensive life-cycle analyses to minimize the overall environmental footprint. Additionally, the transition to sustainable methanol production necessitates supportive infrastructure and policies, including investments in renewable energy, carbon capture and storage facilities, and regulatory frameworks (Liu et al. 2018; Patil et al. 2024; Sarvestani et al. 2024). Addressing these challenges is crucial to scaling methanol production sustainably and efficiently, reinforcing its role as a versatile and essential chemical and energy carrier in the future energy landscape.

4.10 Conclusion

In conclusion, this chapter has thoroughly examined the various methodologies and processes involved in methanol production, each contributing distinct advantages and challenges; the versatility of methanol as a critical chemical and energy carrier is evident through the integration of mechanical, chemical, electrochemical, biochemical, photochemical, and emerging technologies. Mechanical processes primarily focus on the preparation and purification of feedstocks, ensuring optimal conditions for subsequent reactions. Given that transformations in the feed materials occur via reactions, chemical processes are inherently involved in all methanol production methods. While some of these processes, like catalytic hydrogenation, generally exists in all methods, others are specific to the methanol production technique and may not be employed in other approaches. Electrochemical approaches, utilizing renewable electricity, present a sustainable alternative, particularly for CO_2 utilization. But it remains a nascent field with considerable challenges. Despite over a century of research, practical applications are still distant due to limitations such as low efficiency, competing side reactions, and the need for highly selective catalysts. However, recent advancements in catalyst design, reaction engineering, and understanding of reaction mechanisms offer hope for future breakthroughs. Biochemical and photochemical methods, although still in developmental stages, present promising pathways for methanol production with reduced environmental impact. Furthermore, emerging technologies continue to innovate by incorporating novel catalysts and reaction conditions aimed at improving methanol yield and purity. By integrating these diverse methodologies, methanol production can be optimized to meet both current and future energy demands, underscoring its potential as a versatile and sustainable energy carrier.

References

Abdoulmoumine N, Kulkarni A, Adhikari S (2014) Effects of temperature and equivalence ratio on pine syngas primary gases and contaminants in a bench-scale fluidized bed gasifier. Ind Eng Chem Res 53:5767–5777

Adil A, Rao L (2022) Methanol production from biomass: analysis and optimization. Mater Today Proc 57:1770–1775

Allen AE, MacMillan DWC (2012) Synergistic catalysis: a powerful synthetic strategy for new reaction development. Chem Sci 3:633–658

Anicic B, Trop P, Goricanec D (2014) Comparison between two methods of methanol production from carbon dioxide. Energy 77:279–289

Appel AM, Bercaw JE, Bocarsly AB, Dobbek H, DuBois DL, Dupuis M, Ferry JG, Fujita E, Hille R, Kenis PJA (2013) Frontiers, opportunities, and challenges in biochemical and chemical catalysis of CO_2 fixation. Chem Rev 113:6621–6658

Arnaiz del Pozo C, Cloete S, Jiménez Álvaro Á (2022) Techno-economic assessment of long-term methanol production from natural gas and renewables. Energy Convers Manag 266:115785. https://doi.org/10.1016/j.enconman.2022.115785

Arutyunov VS, Basevich VY, Vedeneev VI (1996) Direct high-pressure gas-phase oxidation of natural gas to methanol and other oxygenates. Russ Chem Rev 65:197

Arutyunov V (2014) Direct methane to methanol: foundations and prospects of the process. Elsevier

Asadieraghi M, Daud WMAW (2015) In-situ catalytic upgrading of biomass pyrolysis vapor: using a cascade system of various catalysts in a multi-zone fixed bed reactor. Energy Convers Manag 101:151–163

Assessment A (2003) Commercial-scale demonstration of the liquid phase methanol (LPMEOH^{TM}) process. DOE/NETL-2004/1199

Atsonios K, Panopoulos KD, Kakaras E (2016) Investigation of technical and economic aspects for methanol production through CO_2 hydrogenation. Int J Hydrogen Energy 41:2202–2214

Azhari NJ, Erika D, Mardiana S, Ilmi T, Gunawan ML, Makertihartha IGBN, Kadja GTM (2022) Methanol synthesis from CO_2: a mechanistic overview. Results Eng 16:100711. https://doi.org/10.1016/j.rineng.2022.100711

Bagger A, Arnarson L, Hansen MH, Spohr E, Rossmeisl J (2019) Electrochemical CO reduction: a property of the electrochemical interface. J Am Chem Soc 141:1506–1514. https://doi.org/10.1021/jacs.8b08839

Bertau M, Offermanns H, Plass L, Schmidt F, Wernicke H-J (2014) Methanol: the basic chemical and energy feedstock of the future. Springer

Bhardwaj R, Sharma T, Nguyen DD, Cheng CK, Lam SS, Xia C, Nadda AK (2021) Integrated catalytic insights into methanol production: Sustainable framework for CO_2 conversion. J Environ Manage 289:112468

Bogaerts A, Centi G (2020) Plasma technology for CO_2 conversion: a personal perspective on prospects and gaps. Front Energy Res 8:111

Bone WA (1935) Initial formation of methyl alcohol in the oxidation of methane. Nature 136:910

Boretti A (2013) Renewable hydrogen to recycle CO_2 to methanol. Int J Hydrogen Energy 38:1806–1812

Bozzano G, Manenti F (2016) Efficient methanol synthesis: perspectives, technologies and optimization strategies. Prog Energy Combust Sci 56:71–105

Calle-Vallejo F, Koper M (2013) Theoretical considerations on the electroreduction of CO to C_2 species on Cu (100) electrodes. Angew Chemie 125

Canfield D, Frese KW Jr (1983) Reduction of carbon dioxide to methanol on n-and p-GaAs and p-InP. Effect of crystal face, electrolyte and current density. J Electrochem Soc 130:1772–1773

Capon J-F, Gloaguen F, Schollhammer P, Talarmin J (2005) Catalysis of the electrochemical H_2 evolution by di-iron sub-site models. Coord Chem Rev 249:1664–1676

Carvalho L, Lundgren J, Wetterlund E, Wolf J, Furusjö E (2018) Methanol production via black liquor co-gasification with expanded raw material base—techno-economic assessment. Appl Energy 225:570–584

Chang X, Wang T, Gong J (2016) CO_2 photo-reduction: insights into CO_2 activation and reaction on surfaces of photocatalysts. Energy Environ Sci 9:2177–2196

Cheng W-H (1994) Methanol production and use. CRC Press

Chinchen GC, Denny PJ, Parker DG, Spencer MS, Whan DA (1987) Mechanism of methanol synthesis from CO_2/CO/H_2 mixtures over copper/zinc oxide/alumina catalysts: use of14C-labelled reactants. Appl Catal 30:333–338

Chiou H-H, Lee C-J, Wen B-S, Lin J-X, Chen C-L, Yu B-Y (2023) Evaluation of alternative processes of methanol production from CO_2: design, optimization, control, techno-economic, and environmental analysis. Fuel 343:127856. https://doi.org/10.1016/j.fuel.2023.127856

Chmielniak T, Sciazko M (2003) Co-gasification of biomass and coal for methanol synthesis. Appl Energy 74:393–403. https://doi.org/10.1016/S0306-2619(02)00184-8

Cifre PG, Badr O (2007) Renewable hydrogen utilisation for the production of methanol. Energy Convers Manag 48:519–527

Conrado RJ, Gonzalez R (2014) Envisioning the bioconversion of methane to liquid fuels. Science 343:621–623

Cui Z, Meng S, Yi Y, Jafarzadeh A, Li S, Neyts EC, Hao Y, Li L, Zhang X, Wang X (2022) Plasma-catalytic methanol synthesis from CO_2 hydrogenation over a supported Cu cluster catalyst: insights into the reaction mechanism. ACS Catal 12:1326–1337

Da Silva MJ (2016) Synthesis of methanol from methane: challenges and advances on the multi-step (syngas) and one-step routes (DMTM). Fuel Process Technol 145:42–61

Dalena F, Senatore A, Basile M, Knani S, Basile A, Iulianelli A (2018b) Advances in methanol production and utilization, with particular emphasis toward hydrogen generation via membrane reactor technology. Membranes 8:98

Dalena F, Senatore A, Marino A, Gordano A, Basile M, Basile A (2018) Methanol production and applications: an overview. Methanol 3–28

de Oliveira L, Campos B, John K, Beeskow P, Herrera Delgado K, Pitter S, Dahmen N, Sauer J (2022a) A detailed process and techno-economic analysis of methanol synthesis from H_2 and CO_2 with intermediate condensation steps. Processes 10:1535

Deka TJ, Osman AI, Baruah DC, Rooney DW (2022) Methanol fuel production, utilization, and techno-economy: a review. Environ Chem Lett 20:3525–3554. https://doi.org/10.1007/s10311-022-01485-y

Dhakshinamoorthy A, Garcia H (2014) Cascade reactions catalyzed by metal organic frameworks. Chemsuschem 7:2392–2410

dos Santos RO, de Sousa SL, Prata DM (2018) Simulation and optimization of a methanol synthesis process from different biogas sources. J Clean Prod 186:821–830

Dunfield PF, Yuryev A, Senin P, Smirnova AV, Stott MB, Hou S, Ly B, Saw JH, Zhou Z, Ren Y (2007) Methane oxidation by an extremely acidophilic bacterium of the phylum Verrucomicrobia. Nature 450:879–882

Durante VA, Walker DW, Gussow SM, Lyons JE (1990) Silicometallate molecular sieves and their use as catalysts in oxidation of alkanes. US Pat US4918249

Dybiński O, Milewski J, Szabłowski Ł, Szczęśniak A, Martinchyk A (2023) Methanol, ethanol, propanol, butanol and glycerol as hydrogen carriers for direct utilization in molten carbonate fuel cells. Int J Hydrogen Energy 48:37637–37653

Ertl G, Knözinger H, Weitkamp J (1997) Handbook of heterogeneous catalysis. VCH Weinheim

Farzad S, Mandegari MA, Görgens JF (2016) A critical review on biomass gasification, co-gasification, and their environmental assessments. Biofuel Res J 3:483–495

Felpin F, Fouquet E (2008) Heterogeneous multifunctional catalysts for tandem processes: an approach toward sustainability. Chem Sustain Energy Mater 1:718–724

Feng N, Lin H, Song H, Yang L, Tang D, Deng F, Ye J (2021) Efficient and selective photocatalytic CH_4 conversion to CH_3OH with O_2 by controlling overoxidation on TiO_2. Nat Commun 12:4652

Fiedler E, Grossmann G, Kersebohm DB, Weiss G, Witte C (2000) Methanol. In: Ullmann's encyclopedia of industrial chemistry. John Wiley & Sons, Ltd
Frese KW, Leach S (1985) Electrochemical reduction of carbon dioxide to methane, methanol, and CO on Ru electrodes. J Electrochem Soc 132:259
Fridman A (2008) Plasma chemistry. Cambridge University Press
Gao Y, Qian K, Xu B, Li Z, Zheng J, Zhao S, Ding F, Sun Y, Xu Z (2020) Recent advances in visible-light-driven conversion of CO_2 by photocatalysts into fuels or value-added chemicals. Carbon Resour Convers 3:46–59. https://doi.org/10.1016/j.crcon.2020.02.003
Gautam P, Neha, Upadhyay SN, Dubey SK (2020) Bio-methanol as a renewable fuel from waste biomass: current trends and future perspective. Fuel 273:117783. https://doi.org/10.1016/j.fuel.2020.117783
Ge X, Yang L, Sheets JP, Yu Z, Li Y (2014) Biological conversion of methane to liquid fuels: status and opportunities. Biotechnol Adv 32:1460–1475
Geng P, Yao C, Wei L, Liu J, Wang Q, Pan W, Wang J (2014) Reduction of PM emissions from a heavy-duty diesel engine with diesel/methanol dual fuel. Fuel 123:1–11
Gesser HD, Hunter NR, Prakash CB (1985) The direct conversion of methane to methanol by controlled oxidation. Chem Rev 85:235–244
Ghosh S, Uday V, Giri A, Srinivas S (2019) Biogas to methanol: a comparison of conversion processes involving direct carbon dioxide hydrogenation and via reverse water gas shift reaction. J Clean Prod 217:615–626
Gill I, Ballesteros A (2000) Bioencapsulation within synthetic polymers (part 1): sol–gel encapsulated biologicals. Trends Biotechnol 18:282–296
Gizaw DG, Periyasamy S, Redda ZT, Baskar G (2024) Industrial organic waste and byproducts as sustainable feedstock for bioenergy production. In: Circular bioeconomy perspectives in sustainable bioenergy production. Springer, pp 91–116
Gondal MA, Hameed A, Yamani ZH, Arfaj A (2004) Photocatalytic transformation of methane into methanol under UV laser irradiation over WO_3, TiO_2 and NiO catalysts. Chem Phys Lett 392:372–377. https://doi.org/10.1016/j.cplett.2004.05.092
Habisreutinger SN, Schmidt-Mende L, Stolarczyk JK (2013) Photocatalytic reduction of CO_2 on TiO_2 and other semiconductors. Angew Chemie Int Ed 52:7372–7408
Hameed A, Ismail IMI, Aslam M, Gondal MA (2014) Photocatalytic conversion of methane into methanol: performance of silver impregnated WO_3. Appl Catal A Gen 470:327–335. https://doi.org/10.1016/j.apcata.2013.10.045
Hamelinck CN, Faaij APC (2002) Future prospects for production of methanol and hydrogen from biomass. J Power Sour 111:1–22
Han J-S, Ahn C-M, Mahanty B, Kim C-G (2013) Partial oxidative conversion of methane to methanol through selective inhibition of methanol dehydrogenase in methanotrophic consortium from landfill cover soil. Appl Biochem Biotechnol 171:1487–1499
Hartadi Y, Widmann D, Behm RJ (2016) Methanol formation by CO_2 hydrogenation on Au/ZnO catalysts—effect of total pressure and influence of CO on the reaction characteristics. J Catal 333:238–250. https://doi.org/10.1016/j.jcat.2015.11.002
Heydom EC, Stein VE, Tijm PJA, Street BT, BoxSll PO, Komosky RM (1999) Commercial-scale demonstration of the liquid phase methanol (LPMEOH^{TM}) process: initial operating experience. In: Sixth clean coal technology conference. Citeseer, p 67
Heydorn EC, Diamond BW, Lilly RD (2003) Commercial-scale demonstration of the liquid phase methanol (LPMEOH) process. Air Products Liquid Phase Conversion Company, LP (US)
Higman C, van der Burgt M (2008) Chapter 5—gasification processes. In: Higman C, van der Burgt M (eds) Gasification, 2nd edn. Gulf Professional Publishing, Burlington, pp 91–191
Holmgren KM, Berntsson T, Andersson E, Rydberg T (2012) System aspects of biomass gasification with methanol synthesis—process concepts and energy analysis. Energy 45:817–828
Hong J, Zhang W, Ren J, Xu R (2013) Photocatalytic reduction of CO_2: a brief review on product analysis and systematic methods. Anal Methods 5:1086–1097

Hori Y, Kikuchi K, Suzuki S (1985) Production of CO and CH_4 in electrochemical reduction of CO_2 at metal electrodes in aqueous hydrogencarbonate solution. Chem Lett 14:1695–1698
Hori Y, Kikuchi K, Murata A, Suzuki S (1986) Production of methane and ethylene in electrochemical reduction of carbon dioxide at copper electrode in aqueous hydrogencarbonate solution. Chem Lett 15:897–898
Hori Y, Murata A, Takahashi R, Suzuki S (1987) Electroreduction of carbon monoxide to methane and ethylene at a copper electrode in aqueous solutions at ambient temperature and pressure. J Am Chem Soc 109:5022–5023
Hori Y, Koga O, Watanabe Y, Matsuo T (1998) FTIR measurements of charge displacement adsorption of CO on poly-and single crystal (100) of Cu electrodes. Electrochim Acta 44:1389–1395
Hori Y, Murata A, Takahashi R (1989) Formation of hydrocarbons in the electrochemical reduction of carbon dioxide at a copper electrode in aqueous solution. J Chem Soc Faraday Trans 1 Phys Chem Condens Phases 85:2309–2326
Huang Y-B, Liang J, Wang X-S, Cao R (2017) Multifunctional metal-organic framework catalysts: synergistic catalysis and tandem reactions. Chem Soc Rev 46:126–157
Huang W, Johnston-Peck AC, Wolter T, Yang W-CD, Xu L, Oh J, Reeves BA, Zhou C, Holtz ME, Herzing AA (2021) Steam-created grain boundaries for methane C–H activation in palladium catalysts. Science 373:1518–1523
Hugill JA, Overbeek JP, Spoelstra S (2001) A comparison of the eco-efficiency of the two-production routes for methanol
Hwang IY, Lee SH, Choi YS, Park SJ, Na JG, Chang IS, Kim C, Kim HC, Kim YH, Lee JW (2014) Biocatalytic conversion of methane to methanol as a key step for development of methane-based biorefineries. J Microbiol Biotechnol 24:1597–1605
Hyman MR, Wood PM (1983) Methane oxidation by *Nitrosomonas europaea*. Biochem J 212:31–37
intratec (2024) Methanol prices | current and forecast. https://www.intratec.us/chemical-markets/methanol-price
Iulianelli A, Ghasemzadeh K, Basile A (2018) Progress in methanol steam reforming modelling via membrane reactors technology. Membranes 8:65
Iyer MV, Norcio LP, Punnoose A, Kugler EL, Seehra MS, Dadyburjor DB (2004) Catalysis for synthesis gas formation from reforming of methane. Top Catal 29:197–200
Jadhav SG, Vaidya PD, Bhanage BM, Joshi JB (2014) Catalytic carbon dioxide hydrogenation to methanol: a review of recent studies. Chem Eng Res des 92:2557–2567
Jiang H, Zhang L, Han Z, Tang Y, Sun Y, Wan P, Chen Y, Argyle MD, Fan M (2022) Direct conversion of methane to methanol by electrochemical methods. Green Energy Environ 7:1132–1142
van Kasteren JMN (2016) 16—Production of bioalcohols via gasification. In: Luque R, Lin CSK, Wilson K, Clark J (eds) Handbook of biofuels production, 2nd edn. Woodhead Publishing, pp 495–507
Keshavarz A, Mirvakili A, Chahibakhsh S, Shariati A, Rahimpour MR (2020) Simultaneous methanol production and separation in the methanol synthesis reactor to increase methanol production. Chem Eng Process Intensif 158:108176
Kibria MG, Chowdhury FA, Zhao S, AlOtaibi B, Trudeau ML, Guo H, Mi Z (2015) Visible light-driven efficient overall water splitting using p-type metal-nitride nanowire arrays. Nat Commun 6:6797
Kirubakaran V, Sivaramakrishnan V, Nalini R, Sekar T, Premalatha M, Subramanian P (2009) A review on gasification of biomass. Renew Sustain Energy Rev 13:179–186
Klier K, Chatikavanij V, Herman RG, Simmons GW (1982) Catalytic synthesis of methanol from COH_2: IV. The effects of carbon dioxide. J Catal 74:343–360
Kotowicz J, Brzęczek M, Walewska A, Szykowska K (2022) Methanol production in the Brayton cycle. Energies 15:1480
Kotowski W (2001) Methanol and its derivatives as the strategic future engine fuel. Karbo 1:27–32
Kruse A (2009) Hydrothermal biomass gasification. J Supercrit Fluids 47:391–399

Kumar B, Llorente M, Froehlich J, Dang T, Sathrum A, Kubiak CP (2012) Photochemical and photoelectrochemical reduction of CO_2. Annu Rev Phys Chem 63:541–569

Kutzenko AS, Lamzin VS, Popov VO (1998) Conserved supersecondary structural motif in NAD-dependent dehydrogenases. FEBS Lett 423:105–109

Kuwabata S, Tsuda R, Yoneyama H (1994) Electrochemical conversion of carbon dioxide to methanol with the assistance of formate dehydrogenase and methanol dehydrogenase as biocatalysts. J Am Chem Soc 116:5437–5443

Lee SY, Park JH, Jang SH, Nielsen LK, Kim J, Jung KS (2008) Fermentative butanol production by Clostridia. Biotechnol Bioeng 101:209–228

Lee B, Lee H, Lim D, Brigljević B, Cho W, Cho H-S, Kim C-H, Lim H (2020) Renewable methanol synthesis from renewable H_2 and captured CO_2: How can power-to-liquid technology be economically feasible? Appl Energy 279:115827. https://doi.org/10.1016/j.apenergy.2020.115827

Leonzio G (2018) State of art and perspectives about the production of methanol, dimethyl ether and syngas by carbon dioxide hydrogenation. J CO_2 Util 27:326–354

Lewis NS, Nocera DG (2006) Powering the planet: chemical challenges in solar energy utilization. Proc Natl Acad Sci 103:15729–15735

Li CW, Ciston J, Kanan MW (2014) Electroreduction of carbon monoxide to liquid fuel on oxide-derived nanocrystalline copper. Nature 508:504–507

Li Z, Tsotsis TT (2019) Methanol synthesis in a high-pressure membrane reactor with liquid sweep. J Memb Sci 570–571:103–111. https://doi.org/10.1016/j.memsci.2018.09.071

Liang Z, Gao P, Tang Z, Lv M, Sun Y (2017) Three dimensional porous Cu-Zn/Al foam monolithic catalyst for CO_2 hydrogenation to methanol in microreactor. J CO_2 Util 21:191–199. https://doi.org/10.1016/j.jcou.2017.05.023

Liang S, Altaf N, Huang L, Gao Y, Wang Q (2020) Electrolytic cell design for electrochemical CO_2 reduction. J CO_2 Util 35:90–105

Liu W-C, Baek J, Somorjai GA (2018) The methanol economy: methane and carbon dioxide conversion. Top Catal 61:530–541

López-Martín Á, Caballero A, Colón G (2017) Photochemical methane partial oxidation to methanol assisted by H_2O_2. J Photochem Photobiol A Chem 349:216–223. https://doi.org/10.1016/j.jphotochem.2017.09.039

Luo J, Marpani F, Brites R, Frederiksen L, Meyer AS, Jonsson G, Pinelo M (2014) Directing filtration to optimize enzyme immobilization in reactive membranes. J Memb Sci 459:1–11

Luo J, Meyer AS, Mateiu RV, Pinelo M (2015) Cascade catalysis in membranes with enzyme immobilization for multi-enzymatic conversion of CO_2 to methanol. New Biotechnol 32:319–327

Ma Y, Wang X, Jia Y, Chen X, Han H, Li C (2014) Titanium dioxide-based nanomaterials for photocatalytic fuel generations. Chem Rev 114:9987–10043

Mahinpey N, Gomez A (2016) Review of gasification fundamentals and new findings: reactors, feedstock, and kinetic studies. Chem Eng Sci 148:14–31

Maity SK (2015) Opportunities, recent trends and challenges of integrated biorefinery: part I. Renew Sustain Energy Rev 43:1427–1445

Malik AS, Zaman SF, Al-Zahrani AA, Daous MA, Driss H, Petrov LA (2020) Selective hydrogenation of CO_2 to CH_3OH and in-depth DRIFT analysis for $PdZn/ZrO_2$ and $CaPdZn/ZrO_2$ catalysts. Catal Today 357:573–582. https://doi.org/10.1016/j.cattod.2019.05.040

Marlin DS, Sarron E, Sigurbjörnsson Ó (2018) Process advantages of direct CO_2 to methanol synthesis. Front Chem 6:446

Martin DJ, Liu G, Moniz SJA, Bi Y, Beale AM, Ye J, Tang J (2015) Efficient visible driven photocatalyst, silver phosphate: performance, understanding and perspective. Chem Soc Rev 44:7808–7828

McKendry P (2002) Energy production from biomass (part 2): conversion technologies. Bioresour Technol 83:47–54

Mei D, Zhu X, Wu C, Ashford B, Williams PT, Tu X (2016) Plasma-photocatalytic conversion of CO_2 at low temperatures: understanding the synergistic effect of plasma-catalysis. Appl Catal B Environ 182:525–532

Meunier N, Chauvy R, Mouhoubi S, Thomas D, De Weireld G (2020) Alternative production of methanol from industrial CO_2. Renew Energy 146:1192–1203. https://doi.org/10.1016/j.renene.2019.07.010

Miao Q, Zhu J, Barghi S, Wu C, Yin X, Zhou Z (2014) Model validation of a CFB biomass gasification model. Renew Energy 63:317–323

Mignard D, Pritchard C (2008) On the use of electrolytic hydrogen from variable renewable energies for the enhanced conversion of biomass to fuels. Chem Eng Res des 86:473–487

Moghadam RA, Yusup S, Azlina W, Nehzati S, Tavasoli A (2014) Investigation on syngas production via biomass conversion through the integration of pyrolysis and air-steam gasification processes. Energy Convers Manag 87:670–675

Mohanty P, Singh PK, Adhya TK, Pattnaik R, Mishra S (2022) A critical review on prospects and challenges in production of biomethanol from lignocellulose biomass. Biomass Convers Biorefinery 12:1835–1849. https://doi.org/10.1007/s13399-021-01815-0

Montoya JH, Shi C, Chan K, Nørskov JK (2015) Theoretical insights into a CO dimerization mechanism in CO_2 electroreduction. J Phys Chem Lett 6:2032–2037

Nakagawa H, Harada T, Ichinose T, Takeno K, Matsumoto S, Kobayashi M, Sakai M (2007) Biomethanol production and CO_2 emission reduction from forage grasses, trees, and crop residues. Japan Agric Res Q 41:173–180

Nguyen DLT, Kim Y, Hwang YJ, Won DH (2020) Progress in development of electrocatalyst for CO_2 conversion to selective CO production. Carbon Energy 2:72–98

Nozaki T, Ağıral A, Yuzawa S, Gardeniers JGEH, Okazaki K (2011) A single step methane conversion into synthetic fuels using microplasma reactor. Chem Eng J 166:288–293

Obert R, Dave B (1999) Enzymatic conversion of carbon dioxide to methanol: enhanced methanol production in silica sol−gel matrices. J Am Chem Soc 121:12192–12193

Olah GA, Goeppert A, Prakash GKS (2009) Chemical recycling of carbon dioxide to methanol and dimethyl ether: from greenhouse gas to renewable, environmentally carbon neutral fuels and synthetic hydrocarbons. J Org Chem 74:487–498

de Oliveira Campos B, John K, Beeskow P, Herrera Delgado K, Pitter S, Dahmen N, Sauer J (2022) A detailed process and techno-economic analysis of methanol synthesis from H_2 and CO_2 with intermediate condensation steps. Processes 10. https://doi.org/10.3390/pr10081535

Osman AI, Mehta N, Elgarahy AM, Al-Hinai A, Al-Muhtaseb AH, Rooney DW (2021a) Conversion of biomass to biofuels and life cycle assessment: a review. Environ Chem Lett 19:4075–4118

Osman AI, Hefny M, Abdel Maksoud MIA, Elgarahy AM, Rooney DW (2021b) Recent advances in carbon capture storage and utilisation technologies: a review. Environ Chem Lett 19:797–849

Ott J, Gronemann V, Pontzen F, Fiedler E, Grossmann G, Kersebohm DB, Weiss G, Witte C (2000) Methanol. Ullmann's Encycl Ind Chem

Palma V, Meloni E, Ruocco C, Martino M, Ricca A (2018) State of the art of conventional reactors for methanol production. Methanol 29–51

Palmore GTR, Bertschy H, Bergens SH, Whitesides GM (1998) A methanol/dioxygen biofuel cell that uses NAD+-dependent dehydrogenases as catalysts: application of an electro-enzymatic method to regenerate nicotinamide adenine dinucleotide at low overpotentials. J Electroanal Chem 443:155–161

Park D, Lee J (2013) Biological conversion of methane to methanol. Korean J Chem Eng 30:977–987

Parthasarathy P, Narayanan KS (2014) Hydrogen production from steam gasification of biomass: influence of process parameters on hydrogen yield—a review. Renew Energy 66:570–579

Patel SKS, Mardina P, Kim D, Kim S-Y, Kalia VC, Kim I-W, Lee J-K (2016) Improvement in methanol production by regulating the composition of synthetic gas mixture and raw biogas. Bioresour Technol 218:202–208

Patil T, Naji A, Mondal U, Pandey I, Unnarkat A, Dharaskar S (2024) Sustainable methanol production from carbon dioxide: advances, challenges, and future prospects. Environ Sci Pollut Res 31:44608–44648. https://doi.org/10.1007/s11356-024-34139-3

Pauls JH, Mahinpey N, Mostafavi E (2016) Simulation of air-steam gasification of woody biomass in a bubbling fluidized bed using Aspen Plus: A comprehensive model including pyrolysis, hydrodynamics and tar production. Biomass Bioenerg 95:157–166

Puig-Gamero M, Argudo-Santamaria J, Valverde JL, Sánchez P, Sanchez-Silva L (2018) Three integrated process simulation using aspen plus®: pine gasification, syngas cleaning and methanol synthesis. Energy Convers Manag 177:416–427. https://doi.org/10.1016/j.enconman.2018.09.088

Qadeer K, Al-Hinai A, Chuah LF, Sial NR, Ala'a H, Al-Abri R, Qyyum MA, Lee M (2023) Methanol production and purification via membrane-based technology: recent advancements, challenges and the way forward. Chemosphere 335:139007

Rasteiro LF, De Sousa RA, Vieira LH, Ocampo-Restrepo VK, Verga LG, Assaf JM, Da Silva JLF, Assaf EM (2022) Insights into the alloy-support synergistic effects for the CO_2 hydrogenation towards methanol on oxide-supported Ni_5Ga_3 catalysts: an experimental and DFT study. Appl Catal B Environ 302:120842

Rauch R (2002) Biomass gasification to produce synthesis gas for fuel cells, liquid fuels and chemicals. IEA Bioenergy Agreem

Raudaskoski R, Turpeinen E, Lenkkeri R, Pongrácz E, Keiski RL (2009) Catalytic activation of CO_2: use of secondary CO_2 for the production of synthesis gas and for methanol synthesis over copper-based zirconia-containing catalysts. Catal Today 144:318–323

Riaz A, Zahedi G, Klemeš JJ (2013) A review of cleaner production methods for the manufacture of methanol. J Clean Prod 57:19–37. https://doi.org/10.1016/j.jclepro.2013.06.017

Roan V, Betts D, Twining A, Dinh K, Wassink P, Simmons T (2004) An investigation of the feasibility of coal-based methanol for application in transportation fuel cell systems. Report USA Univ Florida

Roldan A, Hollingsworth N, Roffey A, Islam HU, Goodall JBM, Catlow CRA, de Leeuw NH (2015) King's research portal. Chem Commun 51:7501–7504

Rumayor M, Dominguez-Ramos A, Irabien A (2019) Innovative alternatives to methanol manufacture: carbon footprint assessment. J Clean Prod 225:426–434. https://doi.org/10.1016/j.jclepro.2019.03.015

Sá S, Silva H, Brandão L, Sousa JM, Mendes A (2010) Catalysts for methanol steam reforming—a review. Appl Catal B Environ 99:43–57

Sahoo KK, Katari JK, Das D (2023) Recent advances in methanol production from methanotrophs. World J Microbiol Biotechnol 39:360

Saito M, Takeuchi M, Fujitani T, Toyir J, Luo S, Wu J, Mabuse H, Ushikoshi K, Mori K, Watanabe T (2000) Advances in joint research between NIRE and RITE for developing a novel technology for methanol synthesis from CO_2 and H_2. Appl Organomet Chem 14:763–772

Sandvik P, Kathe M, Wang W, Kong F, Fan L-S (2018) High-pressure chemical looping reforming processes: system analysis for syngas generation from natural gas and reducing tail gases. Energy Fuels 32:10408–10420

Sarp S, Gonzalez Hernandez S, Chen C, Sheehan SW (2021) Alcohol production from carbon dioxide: methanol as a fuel and chemical feedstock. Joule 5:59–76. https://doi.org/10.1016/j.joule.2020.11.005

Sarvestani ME, Norouzi O, Di Maria F, Dutta A (2024) From catalyst development to reactor design: a comprehensive review of methanol synthesis techniques. Energy Convers Manag 302:118070

Scheutz C, Kjeldsen P, Bogner JE, De Visscher A, Gebert J, Hilger HA, Huber-Humer M, Spokas K (2009) Microbial methane oxidation processes and technologies for mitigation of landfill gas emissions. Waste Manag Res 27:409–455

Schorn F, Breuer JL, Samsun RC, Schnorbus T, Heuser B, Peters R, Stolten D (2021) Methanol as a renewable energy carrier: an assessment of production and transportation costs for selected global locations. Adv Appl Energy 3:100050

Shao M, Chang Q, Dodelet J-P, Chenitz R (2016) Recent advances in electrocatalysts for oxygen reduction reaction. Chem Rev 116:3594–3657

Sheldon D (2017) Methanol production-a technical history. Johnson Matthey Technol Rev 61:172–182

Shen K, Kumari S, Huang Y-C, Jang J, Sautet P, Morales-Guio CG (2023) Electrochemical oxidation of methane to methanol on electrodeposited transition metal oxides. J Am Chem Soc 145:6927–6943

Shiratsuchi R, Nogami G (1996) Pulsed electroreduction of CO_2 on silver electrodes. J Electrochem Soc 143:582

Shoji S, Peng X, Yamaguchi A, Watanabe R, Fukuhara C, Cho Y, Yamamoto T, Matsumura S, Yu M-W, Ishii S (2020) Photocatalytic uphill conversion of natural gas beyond the limitation of thermal reaction systems. Nat Catal 3:148–153

Shorokhov MM, Olabin VM, Klevs M, Zageris G, Geža V, Kharitonov VS (2023) Effect of CO_2 content on biogas-to-syngas conversion for methanol production. Theor Exp Chem 59:207–213

Sikarwar VS, Zhao M, Clough P, Yao J, Zhong X, Memon MZ, Shah N, Anthony EJ, Fennell PS (2016) An overview of advances in biomass gasification. Energy Environ Sci 9:2939–2977

Singh NB, Kumar A, Rai S (2014) Potential production of bioenergy from biomass in an Indian perspective. Renew Sustain Energy Rev 39:65–78

Snoeckx R, Bogaerts A (2017) Plasma technology—a novel solution for CO_2 conversion? Chem Soc Rev 46:5805–5863

Song S, Song H, Li L, Wang S, Chu W, Peng K, Meng X, Wang Q, Deng B, Liu Q (2021) A selective Au-ZnO/TiO_2 hybrid photocatalyst for oxidative coupling of methane to ethane with dioxygen. Nat Catal 4:1032–1042

Specht M, Bandi A (1999) The methanol-cycle–sustainable supply of liquid fuels. Ger Cent Sol Energy Hydrog Res, Stuttgart

Specht M, Bandi A, Baumgart F, Murray CN, Gretz J (1999) Synthesis of methanol from biomass/CO_2 resources. In: Greenhouse gas control technologies. Proceedings

Srikrishnan G, Shenbagamuthuraman V, Ağbulut Ü, Mishra I, Jain J, Balusamy S, Chinnadurai K, Chatterjee D, Shankar E, Shaik S (2024) Alcohol fuels in SI engines: a comprehensive state-of-the-art review on combustion, performance, and environmental impacts. J Therm Anal Calorim 1–63

Srivastava RK, Sarangi PK, Bhatia L, Singh AK, Shadangi KP (2022) Conversion of methane to methanol: technologies and future challenges. Biomass Convers Biorefin 12:1851–1875

Ståhl K, Neergaard M (1998) IGCC power plant for biomass utilisation, Värnamo, Sweden. Biomass Bioenerg 15:205–211

Summers DP, Leach S, Frese KW Jr (1986) The electrochemical reduction of aqueous carbon dioxide to methanol at molybdenum electrodes with low overpotentials. J Electroanal Chem Interfacial Electrochem 205:219–232

Supp E (1990) How to purify and condition methanol synthesis gas. In: How to produce methanol from coal. Springer, pp 44–101

Taher E, Chandran K (2013) High-rate, high-yield production of methanol by ammonia-oxidizing bacteria. Environ Sci Technol 47:3167–3173

Taylor CE, Noceti RP (2000) New developments in the photocatalytic conversion of methane to methanol. Catal Today 55:259–267. https://doi.org/10.1016/S0920-5861(99)00244-8

Tijmensen MJA, Faaij APC, Hamelinck CN, Van Hardeveld MRM (2002) Exploration of the possibilities for production of Fischer Tropsch liquids and power via biomass gasification. Biomass Bioenerg 23:129–152

tradingeconomics (2024) MethanolPrice-Chart-Historical Data-News. https://tradingeconomics.com/commodity/methanol

Trudewind CA, Schreiber A, Haumann D (2014) Photocatalytic methanol and methane production using captured CO_2 from coal-fired power plants. Part I—a life cycle assessment. J Clean Prod 70:27–37. https://doi.org/10.1016/j.jclepro.2014.02.014

Tu X, Whitehead JC (2012) Plasma-catalytic dry reforming of methane in an atmospheric dielectric barrier discharge: understanding the synergistic effect at low temperature. Appl Catal B Environ 125:439–448

Tu X, Whitehead JC (2014) Plasma dry reforming of methane in an atmospheric pressure AC gliding arc discharge: co-generation of syngas and carbon nanomaterials. Int J Hydrogen Energy 39:9658–9669

Tursunov O, Kustov L, Kustov A (2017) A brief review of carbon dioxide hydrogenation to methanol over copper and iron based catalysts. Oil Gas Sci Technol d'IFP Energies Nouv 72:30

Van-Dal ÉS, Bouallou C (2013) Design and simulation of a methanol production plant from CO_2 hydrogenation. J Clean Prod 57:38–45

Vidhya CV, Nandhini S, Sheela JM, Reenaa M (2024) Bioenergy-based sustainable bioeconomy—perspectives and challenges. In: Baskar G, Ashokkumar V, Rokhum SL, Moholkar VS (eds) Circular bioeconomy perspectives in sustainable bioenergy production. Springer Nature Singapore, Singapore, pp 5–36

Wang L, Yi Y, Guo H, Tu X (2018a) Atmospheric pressure and room temperature synthesis of methanol through plasma-catalytic hydrogenation of CO_2. Acs Catal 8:90–100

Wang Z, Zhang Y, Neyts EC, Cao X, Zhang X, Jang BW-L, Liu C (2018b) Catalyst preparation with plasmas: how does it work? ACS Catal 8:2093–2110

Wang Y, Hu P, Yang J, Zhu Y-A, Chen D (2021) C-H bond activation in light alkanes: a theoretical perspective. Chem Soc Rev 50:4299–4358

Wasilke J-C, Obrey SJ, Baker RT, Bazan GC (2005) Concurrent tandem catalysis. Chem Rev 105:1001–1020

White JL, Baruch MF, Pander JE III, Hu Y, Fortmeyer IC, Park JE, Zhang T, Liao K, Gu J, Yan Y (2015) Light-driven heterogeneous reduction of carbon dioxide: photocatalysts and photoelectrodes. Chem Rev 115:12888–12935

Wieser M, Yoshida T, Nagasawa T (2001) Carbon dioxide fixation by reversible pyrrole-2-carboxylate decarboxylase and its application. J Mol Catal B Enzym 11:179–184

Williams RH, Larson ED, Katofsky RE, Chen J (1995) Methanol and hydrogen from biomass for transportation. Energy Sustain Dev 1:18–34

Wiranarongkorn K, Eamsiri K, Chen Y-S, Arpornwichanop A (2023) A comprehensive review of electrochemical reduction of CO_2 to methanol: technical and design aspects. J CO_2 Util 71:102477

Wood DA, Nwaoha C, Towler BF (2012) Gas-to-liquids (GTL): a review of an industry offering several routes for monetizing natural gas. J Nat Gas Sci Eng 9:196–208

Wu J, Huang Y, Ye W, Li Y (2017) CO_2 reduction: from the electrochemical to photochemical approach. Adv Sci 4:1700194

Wu W, Li W, Wu M, Zhang H, Zhu C, Jiang Y (2023) Direct oxidation of methane to methanol using $CuMoO_4$. RSC Adv 13:5393–5404

Wyndorps J, Ostovari H, Von der Assen N (2021) Is electrochemical CO_2 reduction the future technology for power-to-chemicals? An environmental comparison with H_2-based pathways. Sustain Energy Fuels 5:5748–5761

Xiaoding X, Moulijn JA (1996) Mitigation of CO_2 by chemical conversion: plausible chemical reactions and promising products. Energy Fuels 10:305–325

Xie S, Zhang Q, Liu G, Wang Y (2016) Photocatalytic and photoelectrocatalytic reduction of CO_2 using heterogeneous catalysts with controlled nanostructures. Chem Commun 52:35–59

Yan X, Zhao B, Liu Y, Li Y (2015) Dielectric barrier discharge plasma for preparation of Ni-based catalysts with enhanced coke resistance: current status and perspective. Catal Today 256:29–40

Yang H, Xu Z, Fan M, Gupta R, Slimane RB, Bland AE, Wright I (2008) Progress in carbon dioxide separation and capture: a review. J Environ Sci 20:14–27

Yang L, Ge X (2016) Biogas and syngas upgrading. In: Advances in bioenergy. Elsevier, pp 125–188

Yang L, Ge X, Wan C, Yu F, Li Y (2014) Progress and perspectives in converting biogas to transportation fuels. Renew Sustain Energy Rev 40:1133–1152. https://doi.org/10.1016/j.rser.2014.08.008

Yi Y, Zhou J, Guo H, Zhao J, Su J, Wang L, Wang X, Gong W (2013) Safe direct synthesis of high purity H_2O_2 through a H_2/O_2 plasma reaction. Angew Chemie Int Ed 52:8446–8449

York APE, Xiao T, Green MLH (2003) Brief overview of the partial oxidation of methane to synthesis gas. Top Catal 22:345–358

You H, Yang S, Ding B, Yang H (2013) Synthesis of colloidal metal and metal alloy nanoparticles for electrochemical energy applications. Chem Soc Rev 42:2880–2904

Yousuf A, Hossain MS, Paul N, Shikder MW, Kumar D, Pirozzi D, Sakib AN, Kazempoor P (2023) Process integration approach to the methanol (MeOH) production variability from syngas and industrial waste gases. Energies 16:6557

Zhang L, Xu CC, Champagne P (2010) Overview of recent advances in thermo-chemical conversion of biomass. Energy Convers Manag 51:969–982

Zhang X, Zhang G, Song C, Guo X (2021) Catalytic conversion of carbon dioxide to methanol: Current status and future perspective. Front Energy Res 8:621119

Zhang X, Sun Z, Shan Y, Pan H, Jin Y, Zhu Z, Zhang L, Li K (2023) Boosting methanol production via plasma catalytic CO_2 hydrogenation over a MnO x/ZrO_2 catalyst. Catal Sci Technol 13:2529–2539

Zhou Y, Zhang L, Wang W (2019) Direct functionalization of methane into ethanol over copper modified polymeric carbon nitride via photocatalysis. Nat Commun 10:506

Zhou P, Tang S, Ye Z, Navid IA, Xiao Y, Sun K, Mi Z (2024) Water-promoted selective photocatalytic methane oxidation for methanol production. Chem Sci 15:1505–1510

Chapter 5
Methanol Production Processes in the Transport Sector

Nikolaos C. Kokkinos and Alexandros Psalidas

Abstract Methanol is one of the most consumed chemicals worldwide and its demand is constantly increasing due to the key role that possesses in the transport sector as an environmentally benign fuel. The processes for the production of methanol are divided into two categories: the chemical and biochemical processes. Currently, the chemical processes are those that are used at industrial scale, while biochemical processes are still at experimental stage. Specifically, the dominant chemical processes for methanol production are the natural gas to methanol and the coal to methanol processes. However, these processes are not sustainable, as they use fossil fuels as feedstock and have high levels of CO_2 emissions. As a result, more environmentally friendly processes have been researched. Such processes are the biomass to methanol and CO_2 to methanol, which are carbon neutral and have the potential to mitigate CO_2 emissions. Furthermore, the biochemical processes of methane to methanol and CO_2 to methanol are also remarkable, as they utilize microorganisms and enzymes as catalysts under mild conditions. In this book chapter, the aforementioned methanol production processes are thoroughly discussed.

Keywords Methanol · Methanol production processes · Industrial chemical processes · Biochemical processes · Natural gas to methanol · Coal gasification · Biomass gasification · CO_2 hydrogenation · Methanotrophs · CO_2 bioconversion

N. C. Kokkinos (✉) · A. Psalidas
Department of Chemistry, School of Sciences, Democritus University of Thrace, Kavala, Greece
e-mail: nikokkinos@mwpc.gr

N. C. Kokkinos
Petroleum Institute, Democritus University of Thrace, Kavala, Greece

Hephaestus Laboratory, School of Sciences, Democritus University of Thrace, Kavala, Greece

A. Bartl et al. (eds.), *Alcohol Production Processes and Their Utilisation in the Transport Sector*, Energy, Environment, and Sustainability,
https://doi.org/10.1007/978-981-96-7384-1_5

5.1 Introduction

Today, the established chemical processes for methanol production involve the conversion of fossil fuels to methanol. Specifically, at industrial scale methanol is produced from the conversion of natural gas and coal (Wagner et al. 2023; Ullah et al. 2023; Khan et al. 2023). Depending on the feedstock, there are different processes that are followed for the production of methanol. When natural gas is used as feedstock, the production of methanol is achieved through steam methane reforming (SMR), partial oxidation of methane (POX) or autothermal reforming (ATR) (Dalena et al. 2018; Seddon and Khan 2011; Kokkinos et al. 2024). On the other hand, methanol is produced from coal through the gasification process (Schubert 2020). Even though these processes are significantly different as far as it concerns the infrastructure needed, in general all of them carry out the methanol production with two main steps (Khan et al. 2023; Xiang et al. 2020; Park and Lee 2013; Collodi et al. 2017). In the first one, the feedstock, after the required pretreatment, is converted into syngas and in the second step, the syngas is converted into methanol through a group of reactions called methanol synthesis. The block diagrams of the natural gas to methanol process and coal to methanol process are presented at Figs. 5.1 and 5.2, respectively. Although these practices are the most dominant at industrial scale, their dependency on fossil fuels for feedstock and also their vast amount of carbon dioxide (CO_2) emissions, have driven the attention of the research to new sustainable pathways for methanol production (Park and Lee 2013; Khalafalla et al. 2020). Currently, research is focused on the biomass to methanol and the direct CO_2 hydrogenation processes, which have appeared to be the most potential for sustainable methanol production (Schubert 2020; Kokkinos et al. 2023). In the case of biomass to methanol production, biomass is converted to methanol through gasification similarly to coal to methanol process, but with many differences regarding the treatment needs of feedstock and syngas. On the other hand, direct hydrogenation of CO_2 is not connected to any of the previous technologies and does not involve the production of syngas. For CO_2 hydrogenation, hydrogen (H_2) and captured CO_2 are utilized for the production of methanol. Unlike the fossil fuel-based processes, these sustainable pathways are not industrially applied yet, but considerable progress has been achieved and some pilot plants have been established providing the first real data to evaluate these new processes.

In addition, there are biochemical processes for methanol production that leverage the metabolic pathways of microorganisms or the function of specific enzymes in order to achieve the conversion of feedstock to the desirable products. In contrast to chemical processes, biochemical processes take place under mild conditions (atmospheric pressure and temperatures around 30 °C), because at these conditions microorganisms and enzymes are functional. This fact gives a more environmentally friendly and sustainable character to them, as they are not energy intense processes (Park and Lee 2013; Abrar et al. 2023). Some biochemical processes for methanol production are the methanol production though methanotrophic bacteria and enzymatic CO_2 reduction. From the aforementioned pathways, the production of methanol

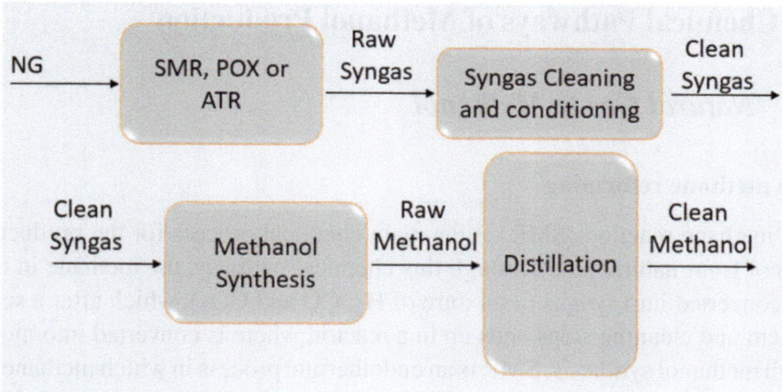

Fig. 5.1 Block diagram of natural gas to methanol process

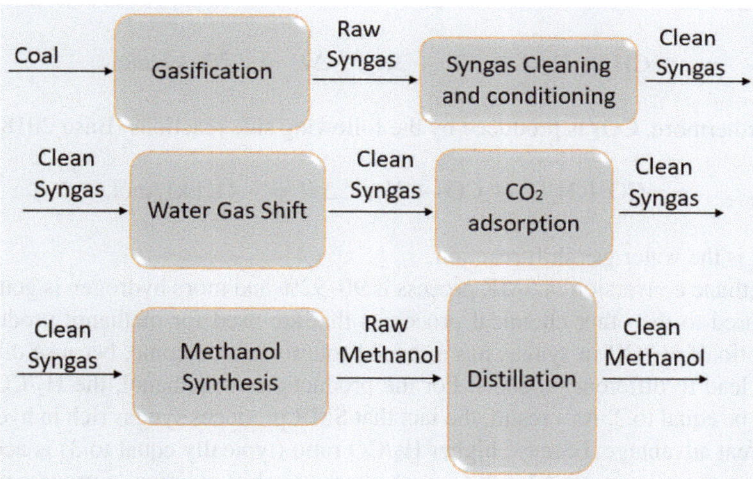

Fig. 5.2 Block diagram of coal to methanol process

with the use of methanotrophic bacteria is the one most researched and involves the conversion of methane to methanol, while all the other deal with the conversion of CO_2 to methanol. Even though they are very attractive as far as it concerns the sustainability aspect, they are not yet applicable to industrial scale.

5.2 Chemical Pathways of Methanol Production

5.2.1 *Natural Gas to Methanol*

Steam methane reforming

Steam methane reaction (SMR) is the main chemical process for the production of methanol from natural gas. Through this chemical pathway, the methane in natural gas is converted into syngas (a mixture of H_2, CO and CO_2), which after a series of treatment and cleaning steps ends up in a reactor, where is converted into methanol through methanol synthesis. SMR is an endothermic process in which methane reacts with steam in the presence of a metal-based catalyst (which is usually nickel-based) at high temperatures ranging from 800 to 1000 °C and pressures between 10 and 30 bar (Dalena et al. 2018; Park and Lee 2013), according to the following reaction (Basu 2018):

$$CH_4 + H_2O \rightarrow CO + 3H_2, \quad \Delta H = +206\,\mathrm{kJ/mol}, \tag{5.1}$$

Furthermore, CO_2 is produced by the following side reactions (Basu 2018):

$$CO + H_2O \rightarrow CO_2 + H_2, \quad \Delta H = -41.1\,\mathrm{kJ/mol}, \tag{5.2}$$

which is the water gas shift reaction.

Methane conversion in SMR process is 90–92% and more hydrogen is generated compared to the other chemical processes that are used for methanol production. The ratio of H_2/CO in syngas has a crucial role for the outcome, because different ratios lead to different products. For the production of methanol, the H_2/CO ratio has to be equal to 2. As a result, the fact that SMR produces syngas rich in hydrogen is a great advantage, because higher H_2/CO ratio (typically equal to 3) is achieved and thus there is no need for the use of water gas shift reaction, a process used to produce more hydrogen in order to increase the ratio (Xiang et al. 2020; Gupta et al. 2022; Tijm et al. 2001; Peña et al. 1996). In fact, this excess of H_2 has to be adjusted (Dalena et al. 2018). The main disadvantage of SMR is its strongly endothermic nature, which needs large amounts of energy to be consumed in order to fulfill its heating needs (Neumann and Veser 2005).

Partial oxidation of methane

Partial oxidation of methane (POM) is another route to convert methane to syngas and then to methanol, but unlike the SMR, methane in POM process reacts with limited supply of oxygen in order to achieve the partial oxidation. In contrast to SMR, POM is an exothermic reaction that takes place under temperatures ranging from 700 to 900 °C at atmospheric pressure and does not require the use of a catalyst (Muhammed et al. 2023; Ji and Wang 2021). Although, the reaction can also take place also at high temperatures between 1000 and 1500 °C (Ji and Wang 2021). POM reaction is as follows (Basu 2018):

$$CH_4 + 1/2O_2 \rightarrow CO + 2H_2, \quad \Delta H = -22.1\,\text{kJ/mol}, \tag{5.3}$$

POM produces less hydrogen than SMR and thus produces syngas with lower H_2/CO ratio (typically equal to 2) (Peña et al. 1996; York et al. 2003). Due to this fact, POM is also suitable for Fischer Tropsch synthesis and is less commonly used for methanol synthesis (Park and Lee 2013; Basu 2018; AlHumaidan et al. 2023). POM process being exothermic, is able to meet its heating requirements more economically compared to the strongly endothermic SMR process. In addition, POM reaction is faster and utilizes reactors which are smaller with higher productivity. The main drawback is the need of an oxygen separator unit, which increases the cost of POM plant (Peña et al. 1996; Ji and Wang 2021; York et al. 2003).

Autothermal reforming

Autothermal reforming (ATR) of methane is a process that combines both SMR and POM in a single reactor (Peña et al. 1996; Ji and Wang 2021; AlHumaidan et al. 2023). The reactor is separated in two zones, the combustion zone where methane and oxygen react and POM happens, and the reforming zone, where SMR reactions take place. In this way, the heat generated by the exothermic POM is utilized to fulfill the energy requirements of the endothermic SMR. ATR is taking place at temperature range of 900–1100 °C, pressure from 80 to 100 bar and in the presence of a catalyst. The process produces syngas with a molar ratio of H_2:CO equal to 2, but it can vary according to the needs, as ATR is a versatile process. The main advantages of this process are the energy efficiency, the compact reactor design and the versatility that it offers (Peña et al. 1996; Ji and Wang 2021; AlHumaidan et al. 2023). Although, the main disadvantages of the process are the relatively expensive processing control system in order to maintain stable operation and manage carbon and soot formation. A comparison of the three processes of natural gas to methanol is presented in Table 5.1.

5.2.2 *Coal to Methanol*

The production of methanol from coal is achieved through the gasification process. In this process, coal degrades into syngas that is further processed to produce methanol. Coal gasification takes place either at high or low temperatures in the presence of a gasifying agent. When low temperatures are used (lower than 1000 °C), a mixture of syngas along with some heavier hydrocarbons that were not converted are obtained. On the other hand, when high temperatures are employed (higher than 1200 °C), syngas of higher quality is produced, which has less hydrocarbons and more H_2 and CO content (Dalena et al. 2018; Basu 2018). Gasification process is conducted into four steps. The first stage is the drying stage in which the coal is heated to temperatures ranging from 100 to 150 °C, where the water is evaporated in order to

Table 5.1 Comparison of natural gas to methanol processes

Characteristics	SMR	POM	ATR
Operation conditions	Temperature: 800–1000 °C Pressure: 10–30 bar Catalyst: Nickel-based	Temperature: 700–900 °C Pressure: atmospheric pressure Catalyst: not required	Temperature: 900–1100 °C Pressure: 80–100 bar Catalyst: required
Reactants	Methane and steam	Methane and oxygen	Methane, oxygen and steam
Heat balance	Strongly endothermic process, requires external heat input	Mild exothermic process. It does not require external heat input	Balances endothermic and exothermic reactions
H_2/CO ratio of produced syngas	Typically equal to 3	Typically equal to 2 or lower	Typically equal to 2, but it can vary
Advantages	High hydrogen yield	Faster process, less energy intense process	Versatile process, energy efficient process
Disadvantages	Process with high energy consumption	Needs oxygen separator unit	Complex processing control system, needs oxygen separator unit

reduce the level of moisture content. In the second step, pyrolysis takes place. At this stage, the feedstock is heated at 400–600 °C in absence of oxygen in order to release its volatile compounds and result into a solid char rich in carbon content. The third stage is the partial combustion of the char and the volatile gases in the presence of controlled oxygen or air supply in order to raise the temperature sufficiently for the needs of the endothermic gasification reactions that take place in the next step. Furthermore, CO_2 and H_2O are produced from combustion, which are useful for the final step. In the fourth step, the gasification takes place, the temperature rises up to 800–1800 °C and the coal reacts with CO_2 and steam and syngas is produced (Shahabuddin and Alam 2022; Dai et al. 2023).

The result of the process relies on the gasifying agents and reactor type that are used and as a result it is important to address them. The gasifying agents that are used are air, O_2, steam and CO_2 (Gupta et al. 2022; Iglesias Gonzalez et al. 2011). When air is used, the costs for separating oxygen is saved, but the syngas produced has lower quality and heating value due to the presence of nitrogen. On the other hand, when O_2 is used the cost for its separation from air burden the production, but the syngas produced is more suitable for chemical synthesis. Steam increases the H_2 content of the syngas, as water gas shift reaction occurs at the fourth step of the gasification process, while CO_2 is useful for the adjustment of syngas's CO content (Gupta et al. 2022).

After syngas is produced it cannot be converted into methanol as easily as in natural gas to methanol process, because syngas produced from coal has more impurities and thus more cleaning and conditioning steps are required before it is used for methanol synthesis. This is because coal contains more impurities as feedstock and as a result the syngas produced is less clean and contains harmful compounds for the catalysts of methanol synthesis reactions. Specifically, raw syngas produced from coal gasification contains solid particulates (such as char and unconverted char), sulfur compounds (such as H_2S and COS), nitrogen compounds (such as NH_3, HCN) and trace metals (Schubert 2020; Gupta et al. 2022; Basu 2018; Iglesias Gonzalez et al. 2011). The removal of the solid particulates is achieved with the use of cyclones, filters or wet scrubbers. Desulfurization is contacted using zinc oxide beds or solvent absorption methods, such as MDEA, Selexol or Rectisol, while for nitrogen compounds and trace metal removal catalytic conversion or adsorption processes are used (Shahabuddin and Alam 2022; Dai et al. 2023; Mondal et al. 2011; Gupta et al. 2023).

In addition, after the treatment of syngas, another step is mandatory before the methanol synthesis. Water gas shift reaction is needed to adjust the H_2/CO ratio, because syngas from coal has lower H_2 content than is needed and H_2/CO ratio is commonly between 1 and 0.5 (Seddon and Khan 2011; Tijm et al. 2001). In water gas shift reaction CO reacts with steam according to Eq. (5.2).

The reaction can take place under different conditions depending on the catalyst that is used. Specifically, if copper-based catalysts are employed the reaction is carried out at lower temperatures ranging from 200 to 400 °C, while if iron-based catalysts are used the reaction takes place at higher temperatures between 400 and 500 °C (Basu 2018). As it can be seen at Eq. 5.2, the reaction is exothermic and as a result, its equilibrium is favored more at lower temperatures, producing syngas with significantly lower CO content, while in higher temperatures, the equilibrium is limited, but the reaction rate is increased. After the water gas shift reaction, the CO content is lower but the CO_2 content is significantly higher, which is a problem as CO_2 concentration affects the activity of the copper-based catalysts used for methanol production. When CO_2, is present in syngas at low concentrations it has positive impact on the catalysts because it inhibits the oxidation of Cu particles and the further catalyst deactivation. But when CO_2 is present in high concentrations it slows down the reaction rate (Sikarwar et al. 2017). Particularly, during coal to methanol process about 50% of coals carbon content is converted into CO_2 (Xiang et al. 2020). As a result, after the water gas shift reaction, an additional step of CO_2 adsorption and removal may be needed in order to obtain syngas with the desirable composition (Tijm et al. 2001; Arnaiz del Pozo et al. 2022).

After the necessary pretreatment, clean syngas is ready for the final step, its conversion to methanol. This is achieved through methanol synthesis which involves the following reactions (Xiang et al. 2020; Gupta et al. 2022):

$$\mathrm{CO} + 2\mathrm{H_2} \rightarrow \mathrm{CH_3OH}, \quad \Delta H = -90.8\,\mathrm{kJ/mol}, \tag{5.4}$$

$$\mathrm{CO_2} + 3\mathrm{H_2} \rightarrow \mathrm{CH_3OH} + \mathrm{H_2O}, \quad \Delta H = -49.2\,\mathrm{kJ/mol}, \tag{5.5}$$

$$\mathrm{CO_2} + \mathrm{H_2} \rightarrow \mathrm{CO} + \mathrm{H_2O}, \quad \Delta H = +41.6\,\mathrm{kJ/mol}, \tag{5.6}$$

which is the reverse water gas shift reaction

As it can be seen from the reactions above, methanol synthesis is a clearly exothermic process, as the main two reactions taking place are exothermic. For this reason, high temperatures cannot be used and as a result, the process temperature is ranging from 200 °C to 400 °C. In order to push the equilibrium toward the methanol, high pressures have to be employed (Schubert 2020; Abrar et al. 2023; Tijm et al. 2001; Sikarwar et al. 2017), for this reason pressure between 10–50 bar is commonly used; however, it is possible to reach 100 bar or even higher (Dalena et al. 2018; Gupta et al. 2022; Pandey et al. 2023). Nonetheless, the conversion of syngas is not high, so the use of a catalyst is mandatory. The most common catalyst used for the methanol synthesis is $Cu/ZnO/Al_2O_3$ (Schubert 2020), in which Cu is the active component for CO and CO_2 hydrogenation, while ZnO and Al_2O_3 are the catalyst support, ensuring the catalyst stability and functionality. These catalysts present significant selectivity, up to 99%, but the syngas conversion is about 50–80%. As a result, after the reaction, unconverted syngas is separated from the products and is recycled (Dalena et al. 2018; Schubert 2020; Sikarwar et al. 2017).

After the reaction, a mixture of methanol, water and other byproducts and impurities is obtained. Distillation is commonly used for separation and purification of methanol (Ullah et al. 2023).

5.2.3 Biomass Gasification

The production of methanol from biomass can be achieved through the gasification process in a similar way as coal gasification, but with some key differences in the individual steps of the process due to the different feedstock composition. Biomass has less carbon content than coal and higher moisture content and also faces the drawback that is less homogeneous (Gupta et al. 2022; Chmielniak and Sciazko 2003). Similar to coal gasification, the gasification of biomass can take place either at temperatures higher than 1200 °C (high temperatures) or at temperatures below 1000 °C (low temperatures) (Basu 2018).

The result of biomass gasification can be adjusted according to the desired outcome through appropriate selection of gasification agents and reactor type. Air, oxygen, carbon monoxide and steam are the main gasification agents used and affect the gasification's result as mentioned earlier in Sect. 5.2.2. Furthermore, the common reactors used for biomass gasification are the fixed bed, fluidized bed and entrained flow gasifiers. In the fixed bed gasifiers, the feedstock and the gasifying agents move either in co-current or in counter-current fashion. In the case of co-current, products contain less tar, but the reactor is more sensitive to moisture and ash content. On the other hand, fixed bed gasifiers with counter current are more preferable when there is feedstock with high moisture content, but syngas with high concentrations of tar is produced and as a result, syngas has to be cleaned more extensively before it is used. Fluidized bed gasifiers are suitable for small- and medium-scale applications and the efficiency of carbon conversion can reach up to 95%, but there is bed agglomeration and operational complexity. Finally, entrained flow gasifiers are able to operate at high temperatures up to 1500 °C, achieving higher carbon conversions rates, they are suitable for an economical large-scale application in combination with high-quality syngas production (Gupta et al. 2022).

Furthermore, syngas derived from biomass has higher content of impurities and harmful compounds for the equipment and the catalyst that have to be removed, as a result more intense cleaning and conditioning is needed. Common contaminants in the gasification products are sulfur-based compounds (COS, H_2S), nitrogen-based compounds (NH_3, HCN), hydrogen halides (HF, HCl), tars and trace metals (Na, K) (Abdoulmoumine et al. 2015). In addition, biomass derived syngas typically has a H_2/CO ratio lower than 1 and as a result the use of water gas shift reaction is mandatory (Gupta et al. 2022). When steam is used as a gasification agent, there is a significant increase in the H_2/CO ratio, which can be as high as 2.2 (Gao et al. 2023). This is because the use of steam prevents carbon deposition and at the same time supplies the system with a source of H_2 (Wang et al. 2017).

5.2.4 CO_2 Hydrogenation for Methanol Production

In recent years, hydrogenation of CO_2 has gained a lot of attention as a method to produce methanol with completely sustainable character. In contrast to previous chemical pathways, CO_2 hydrogenation does not involve syngas production from a fuel (either fossil or biofuels), nor multiple steps for syngas cleaning and conditioning. In CO_2 hydrogenation, captured CO_2 directly reacts with hydrogen gas at 200 °C and pressures ranging from 30 to 50 bar in the presence of copper-based catalysts. In particular, the most common catalyst used for CO_2 hydrogenation is $Cu/ZnO/Al_2O_3$ (Dalena et al. 2018; Schubert 2020). Furthermore, the main reaction is according to Eq. (5.5), while the CO hydrogenation and the reverse water gas shift reaction are also taking place as described at Eqs. (5.4) and (5.6) respectively. The reactions of Eqs. (5.4) and (5.5) may be the same as in methanol synthesis process, but the difference is that in CO_2 hydrogenation Eq. (5.5) is the main reaction and Eq. (5.4) a side reaction due to the side reaction of reversed water gas shift (Eq. 5.6). On the contrary, in methanol synthesis Eq. (5.4) is the primary reaction, while Eq. (5.5) is a secondary reaction which takes place between the excess of H_2 and CO_2, as CO_2 being more stable than CO, reacts more difficult (Basu 2018). In CO hydrogenation the activity of the catalysts is lower compared to that of syngas conversion to methanol and the conversion of CO_2 is about 35–45% (Schubert 2020). The unreacted gas is separated from the products and recycled. For the increase of process efficiency, the in situ removal of reaction products is suggested, as according to Le Chatelier's principle this would push the reaction equilibrium toward the products (Dalena et al. 2018). Specifically, the selective removal of water would be a useful application, because water causes thermodynamic limitations, hindering the efficient CO_2 conversion (Ullah et al. 2023). To address this problem zeolite-based membrane reactors are suggested to be used, as their hydrophilicity could facilitate the removal of water. Furthermore, the removal of water would not only positively affect the process efficiency, but would also lead to a cleaner product mixture and therefore less process intensive purification of methanol would be required (Ullah et al. 2023).

Furthermore, the needed CO_2 for the process can be obtained from many sources. Such sources are direct air capture (DAC) plants, carbon capture and storage (CCS) plants in industries with high levels of CO_2 emissions and also plants produce biogas, as a significant percentage of biogas CO_2 that needs to be removed. Moreover, H_2 for the process can be obtained from sustainable sources such as from water electrolysis using solar energy, wind energy and other forms of renewable energy (Schubert 2020). Thus, CO_2 hydrogenation process for methanol production has many advantages as it utilizes CO_2, and can be proved useful for the utilization of hydrogen, as methanol is considered a potential hydrogen carrier for its safe and convenient transportation and storage (Ullah et al. 2023; Dalena et al. 2018).

5.3 Biochemical Methods

5.3.1 Methanol Production Through Methanotrophic Bacteria

Methanotrophic bacteria are microorganisms that aerobically utilize methane as their source of carbon, oxidizing it into CO_2, under ambient conditions. Their metabolic pathway from methane to CO_2 is achieved within four steps as can be seen in Fig. 5.3.

In the first step, methane monooxygenase (MMO) oxidizes methane into methanol (Khan et al. 2023; Schubert 2020; Park and Lee 2013; Abrar et al. 2023; Taylor et al. 2018; Kulkarni et al. 2021). This ability has drawn the attention of researchers as it indicated that methanotrophs could potentially be used for methanol production through a biochemical process. In spite of the fact that methanol is an intermediate

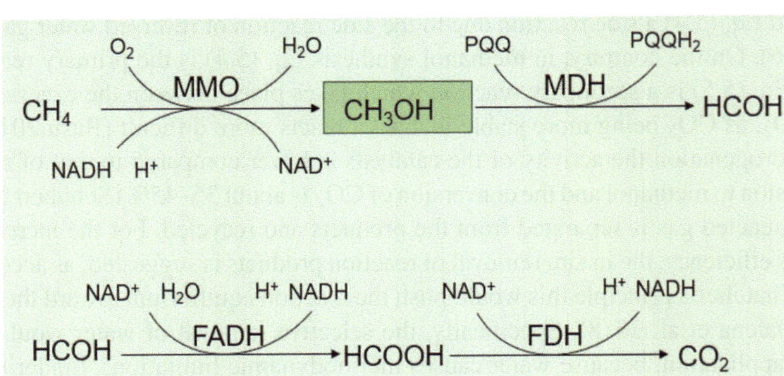

Fig. 5.3 Methanotroph's metabolic pathway of methane to CO_2

and not the final product of these bacteria's metabolic process, poses a problem as the produced methanol is oxidized into formaldehyde (Schubert 2020; Abrar et al. 2023; Taylor et al. 2018). To overcome this problem a possible solution is the use of compounds that inhibit the function of methanol dehydrogenase (MDH), which is responsible for methanol oxidation. Such compounds are cyclopropanol, EDTA, NaCl and phosphate (Taylor et al. 2018; Xin et al. 2004). However, the sole use of MDH inhibitors is not effective, because NADH molecules, which are necessary for the energy requirements of methane to methanol oxidation, are consumed and not regenerated as it would happen normally if methanol was let to be oxidized (Xin et al. 2004).For instance, Taylor et al. (2018) reported that by inhibiting the activity of MDH by 80%, the methanol production efficiency decreased from 82% on the first day of operation to 38% on the second day of operation. For this reason, some have suggested that the addition of formate is needed in order to ensure the regeneration of NADH. Notably, Sheets et al. (2016) found the addition of formate alone was able to achieve 25.5% conversion of methane to methanol, leading to 0.43 g/L methanol concentration, which was reported as a satisfying methanol production without the use of MDH inhibitors. On the other hand, Xin et al. (2004) found that CO_2 concentration affects the amount of methanol that is oxidized, as in high concentrations it can partially inhibit the MDH function. In this way, a part of the methanol produced is oxidized in order to ensure that the energy needs of methanotrophs are met and the other part of methanol is obtained as product.

The use of methanotrophs bacteria for the production of methanol is attractive because these bacteria operate with high selectivity under ambient conditions and thus it is not an energy intense process. Furthermore, methanotrophs can grow by feeding them biogas, which is a gaseous mixture derived from the anaerobic digestion of biomass and mainly consists of methane and CO_2 (the percentage of each can significantly vary) (Schubert 2020; Sheets et al. 2016; Patel et al. 2020). As a result, the production of methanol via methanotrophs is even more attractive, because it utilizes a sustainable feedstock (biogas), making the process completely environmentally friendly. The main challenges of this process are the poor mass transfer of methane to the bacteria, low yield and productivity of their cultivations which is an obstacle for the process's further development to industrial scale (Schubert 2020; Kulkarni et al. 2021; Patel et al. 2020). Patel et al. (2020) to address this problem immobilized methanotrophs on chitosan-APTES-GLA and observed great improvement of bacteria's productivity as it was increased from 3.79 to 4.90 mmol/L and noted that this approach has the potential to be up scaled for industrial application as the immobilized methanotrophs maintained their activity for 8 catalytic cycles. Furthermore, in order to achieve scalability, methanotrophs with specific characteristics have to be used. MMO enzymes in methanotrophs can be expressed in two forms, either soluble in the bacteria's cells (sMMO) or in particulate membrane bound form (pMMO) (Patel et al. 2020; Kim et al. 2019; Miyaji et al. 2020). Methanotrophs higher levels of pMMO expression, such as *Methylosinus trichosporium OB3b*, present higher productivity and therefore are more suitable for the development of this biochemical process (Miyaji et al. 2020).

5.3.2 *Bioconversion of CO_2 to Methanol*

In contrast to biochemical route for methane to methanol, the biochemical conversion of CO_2 to methanol is achieved by the use of enzymes and not microbial cultivations. The process involves a series of reduction reactions which are the reverse steps of methane to CO_2 metabolic pathway of methanotrophs presented earlier. Starting from CO_2, formic acid dehydrogenase (FDH) catalyzes the conversion of CO_2 into formic acid. Afterward, formaldehyde dehydrogenase (FALDH) converts the formic acid into formaldehyde. Finally, for the production of methanol, methanol dehydrogenase catalyzes the reduction of formaldehyde to methanol (Gomes et al. 2021; Shi et al. 2015; Zhang et al. 2021). As in the case of methanotrophs, these reactions are taking place under mild conditions between 25 and 40 °C, in neutral pH, making the process environmentally friendly. Although, challenges such as poor enzymes stability, sensitivity to the temperature and pH changes, the need for the regeneration of molecules such as NADH is the main obstacle for the scalability of this process. To address this problem, the immobilization of enzymes is suggested in order to improve the stability and reusability of the enzymes (Gomes et al. 2021; Shi et al. 2015; Zhang et al. 2021). Furthermore, for the scale up of the process, advanced bioreactor designs are being developed. Overall, enzymatic bioconversion of CO_2 to methanol is a novel sustainable process that has the potential to contribute to the mitigation of CO_2 emissions with further research.

5.4 Industrial Applications

Methanol is a valuable chemical with numerous uses both in industry, as platform chemical for the production of various petrochemical products, and in final products that people use in everyday life. For this reason, it is one of the most consumed primary petrochemicals. In fact, in 2022 the production of methanol reached 111 million tons and it accounted for the 16% of global consumption of bulk chemicals. In 2021, the global methanol market was worth over 37.4 billion USD and is expected to reach nearly at 61.7 billion USD by 2030. Saudi Arabia is the top methanol exporter, while China, United States, India and Netherlands are the major countries importing methanol, according to Statista.

From the aforementioned methanol production processes, only the chemical processes of natural gas to methanol and coal to methanol are currently applied at industrial scale (Wagner et al. 2023). Specifically, around 65% of methanol is produced from natural gas mostly with SMR, while the other 35% is produced through coal gasification process (Sollai et al. 2023). This difference is because generally the production of methanol from natural gas is more cost-effective. As mentioned before, methanol production via coal gasification requires more cleaning and conditioning steps of the intermediate syngas produced due to higher presence of impurities. Moreover, syngas produced from coal gasification has significant lower

H_2 content and as a result, there is a need for the use of water gas shift process in greater extent. All these in addition to the fact that natural gas is generally cheaper, make the production of methanol from natural gas more profitable, while coal gasification is preferable only at places where the coal is abundant such as China, where coal gasification is the dominant process of methanol production (Xiang et al. 2020; Li et al. 2023).

As far as it concerns biomass to methanol and CO_2 hydrogenation processes, they have not been applied to industrial scale yet. In the case of the biomass to methanol process, the technology required for its industrial-scale application exists, as the process has a technology readiness level (TRL) of 9, but the factor that hinders its scalability is that it is not yet a cost-effective way to produce methanol. On the other hand, the factor preventing the industrial application of methanol production by CO_2 hydrogenation is its technological immaturity, with a TRL of 2 (Harris et al. 2021). Nevertheless, there is significant effort toward their effective development and integration to methanol industry. Remarkably, according to Methanol Institute by July 2024 there are 170 renewable methanol projects worldwide, which are expected to achieve a total capacity 20.87 million tons by 2027. Successful example of green methanol pilot plant is the Leuna100 project which launched in 2023 in the Leuna Chemical Park with funding of 10.4 million Euros by the German Federal Ministry for Digital and transport. Leuna100 produces methanol from biomass and is the first cost-effective pilot plant in the world to do so. Furthermore, notable is the breakthrough that the Carbon Recycling International has made, converting in total 310,000 tons of CO_2 per year to 214,000 tons of methanol through the CO_2 hydrogenation process. An overall comparison of the chemical processes for methanol production is presented in Table 5.2.

5.5 Conclusions

Methanol is a valuable product that is expected to play an important role in the energy sector in the near future either for its potential to be used as drop in fuel or for platform chemical for the production of vast variety of sustainable fuel, such as renewable gasoline and other hydrocarbons. Moreover, methanol is considered as a suitable hydrogen carrier compound, which rises its future importance even more as it will contribute to the development of H_2-based applications. Considering all these and the fact that methanol's demand is already very high and it will keep rising, the study of its production methods is of great interest. Currently, the industrially established processes for methanol production are chemical processes that involve the conversion of natural gas and coal to methanol. Natural gas conversion to methanol is achieved through steam methane reforming, partial oxidation of methane and autothermal reforming, with the former being the most used among them for methanol production. On the other hand, coal to methanol pathway involves the gasification of coal. In all of these industrially established processes, for the production of methanol two main steps are needed: the conversion of feedstock to syngas, with the right

Table 5.2 Comparison of chemical production processes of methanol

Characteristics	Natural gas to methanol	Coal to methanol	Biomass to methanol	CO_2 hydrogenation to methanol
Feedstock	Methane from natural gas	Coal	Various types of biomasses (such as agricultural waste, forestry waste, domestic waste)	CO_2 and H_2
Global Share	~ 65%	~ 35%	< 1%	< 1%
Technological maturity	High maturity, well-established infrastructure	High maturity, well-established infrastructure	TRL:9	TRL:2
Environmental impact	High levels of CO_2 emissions, CH_4 leakage	High levels of CO_2 emissions	Carbon–neutral	Carbon neutral with potential for negative carbon emissions
Economic viability	High	High in regions rich in coal	Currently not viable due to high costs	Currently not viable due to high costs
Advantages	High process efficiency	Utilizes efficiently solid feedstocks	Renewable feedstock	Potential for mitigation of CO_2 emissions
Disadvantages	High energy consumption, high levels of CO_2 emissions	High energy consumption, high levels of CO_2 emissions	Feedstock variability	High costs due to H_2 requirements

H_2/CO ratio to be equal to 2, and the catalytic conversion of syngas into methanol through the methanol synthesis reaction. Typically, the use of natural gas as feedstock is preferable over coal, because coal gasification requires more steps for syngas cleaning and conditioning before the methanol synthesis. Although these processes are efficient for the conversion of their feedstock to the desired product, they both have high level of CO_2 emissions and they have a significant environmental impact. As a result, it is important for the growing methanol industry to move toward more sustainable production processes. For this reason, great efforts have been made for the development of biomass to methanol and CO_2 to methanol processes. Biomass to methanol is a carbon neutral process that converts various types of biomasses into methanol through gasification process in a similar fashion to coal to methanol pathway, but differences due to the different quality of feedstock and biomass derived syngas. CO_2 to methanol process involves the direct hydrogenation of CO_2 and would be even more preferable process, because it has the potential to be carbon negative. Both of these sustainable processes are currently being tested in pilot scale and significant progress has been made, but they have not introduced to industrial scale yet, because they are still not as cost-effective as they should be.

Furthermore, the biochemical processes of methanol production are of interest, too. There are two main biochemical processes, the conversion of methane into methanol with the use of methanotrophic bacteria and the conversion of CO_2 to methanol through a series of enzymatic reactions. These processes are attractive, because they do not require the use of high temperatures or pressures, as they take place under ambient conditions. Although, both face obstacles, such as low productivity and stability, which hinder their scaling up and therefore more research has to be made until they are suitable for industrial application.

References

Abdoulmoumine N, Adhikari S, Kulkarni A, Chattanathan S (2015) A review on biomass gasification syngas cleanup. Appl Energy 155:294–307. https://doi.org/10.1016/j.apenergy.2015.05.095

Abrar I, Arora T, Khandelwal R (2023) Bioalcohols as an alternative fuel for transportation: cradle to grave analysis. Fuel Process Technol 242. https://doi.org/10.1016/j.fuproc.2022.107646

AlHumaidan FS, Absi Halabi M, Rana MS, Vinoba M (2023) Blue hydrogen: current status and future technologies. Energy Convers Manage 283:116840. https://doi.org/10.1016/j.enconman.2023.116840

Arnaiz del Pozo C, Jiménez Álvaro Á, Cloete S (2022) Methanol from solid fuels: a cost-effective route to reduced emissions and enhanced energy security. Energy Convers Manage 270. https://doi.org/10.1016/j.enconman.2022.116272

Basu P (2018) Production of synthetic fuels and chemicals from biomass. In: Biomass gasification, pyrolysis and torrefaction: practical design and theory, pp 415–443

Chmielniak T, Sciazko M (2003) Co-gasification of biomass and coal for methanol synthesis. Appl Energy 74:393–403. https://doi.org/10.1016/S0306-2619(02)00184-8

Collodi G, Azzaro G, Ferrari N, Santos S (2017) Demonstrating large scale industrial CCS through CCU—a case study for methanol production, pp 122–138

Dai F, Zhang S, Luo Y, Wang K, Liu Y, Ji X (2023) Recent progress on hydrogen-rich syngas production from coal gasification. Processes 11:1765. https://doi.org/10.3390/pr11061765

Dalena F, Senatore A, Basile M, Knani S, Basile A, Iulianelli A (2018) Advances in methanol production and utilization, with particular emphasis toward hydrogen generation via membrane reactor technology. Membranes 8. https://doi.org/10.3390/membranes8040098

Gao Y, Wang M, Raheem A, Wang F, Wei J, Xu D, Song X, Bao W, Huang A, Zhang S, Zhang H (2023) Syngas production from biomass gasification: influences of feedstock properties, reactor type, and reaction parameters. ACS Omega 8:31620–31631. https://doi.org/10.1021/acsomega.3c03050

Gomes MZ V, Masdeu G, Eiring P, Kuhlemann A, Sauer M, Åkerman B, Palmqvist AEC (2021) Improved biocatalytic cascade conversion of CO_2 to methanol by enzymes Co-immobilized in tailored siliceous mesostructured cellular foams. Catal Sci Technol 11:6952–6959. https://doi.org/10.1039/D1CY01354H

Gupta PK, Kumar V, Maity S, Datta S, Kishore Gupta G (2022) A review on conversion of biomass to liquid fuels and methanol through indirect liquefaction route. Chem Select 7. https://doi.org/10.1002/slct.202203504

Gupta PK, Datta S, Saw SK, Kumari S, Sahu G, Singh RK, Chauhan V, Chavan PD (2023) Syngas cleaning technologies in gasification process for downstream applications. In: Sinha A, Sarkar BC, Mandal PK (eds) Proceedings of the 10th Asian mining congress 2023. Springer Nature Switzerland, Cham, pp 307–316

Harris K, Grim RG, Huang Z, Tao L (2021) A comparative techno-economic analysis of renewable methanol synthesis from biomass and CO_2: opportunities and barriers to commercialization. Appl Energy 303:117637. https://doi.org/10.1016/j.apenergy.2021.117637

Iglesias Gonzalez M, Kraushaar-Czarnetzki B, Schaub G (2011) Process comparison of biomass-to-liquid (BtL) routes Fischer-Tropsch synthesis and methanol to gasoline. Biomass Convers Biorefin 1:229–243. https://doi.org/10.1007/s13399-011-0022-2

Ji M, Wang J (2021) Review and comparison of various hydrogen production methods based on costs and life cycle impact assessment indicators. Int J Hydrogen Energy 46:38612–38635. https://doi.org/10.1016/j.ijhydene.2021.09.142

Khalafalla SS, Zahid U, Abdul Jameel AG, Ahmed U, Alenazey FS, Lee C-J (2020) Conceptual design development of coal-to-methanol process with carbon capture and utilization. Energies 13. https://doi.org/10.3390/en13236421

Khan S, Jain G, Srivastava A, Verma PC, Pande V, Dubey RS, Khan M, Haque S, Ahmad S (2023) Enzymatic biomethanol production: Future perspective. Sustain Mater Technol 38. https://doi.org/10.1016/j.susmat.2023.e00729

Kim HJ, Huh J, Kwon YW, Park D, Yu Y, Jang YE, Lee B-R, Jo E, Lee EJ, Heo Y, Lee W, Lee J (2019) Biological conversion of methane to methanol through genetic reassembly of native catalytic domains. Nat Catal 2:342–353. https://doi.org/10.1038/s41929-019-0255-1

Kokkinos NC, Emmanouilidou E, Zachopoulos F, Mittas N (2023) Application of artificial intelligence in modeling of methanol production from renewable sources. In: Reference module in chemistry, molecular sciences and chemical engineering. Elsevier

Kokkinos NC, Finos MA, Katsamperis I, Markou G, Tzevelekou P, Vasiakos A (2024) Process and reactor consideration for syngas production from natural gas bi-reforming. In: reference module in chemistry, molecular sciences and chemical engineering. Elsevier

Kulkarni PP, Khonde VK, Deshpande MS, Sabale TR, Kumbhar PS, Ghosalkar AR (2021) Selection of methanotrophic platform for methanol production using methane and biogas. J Biosci Bioeng 132:460–468. https://doi.org/10.1016/j.jbiosc.2021.07.007

Li S, Sun X, Liu L, Du J (2023) A full process optimization of methanol production integrated with co-generation based on the co-gasification of biomass and coal. Energy 267. https://doi.org/10.1016/j.energy.2022.126566

Miyaji A, Furuya D, Orita I, Baba T (2020) Growth-promoting effect of cyanocobalamin on *Methylosinus trichosporium* OB3b culture under high copper concentration for methanol synthesis from methane. Bioresour Technol Rep 11. https://doi.org/10.1016/j.biteb.2020.100473

Mondal P, Dang GS, Garg MO (2011) Syngas production through gasification and cleanup for downstream applications—recent developments. Fuel Process Technol 92:1395–1410. https://doi.org/10.1016/j.fuproc.2011.03.021

Muhammed NS, Gbadamosi AO, Epelle EI, Abdulrasheed AA, Haq B, Patil S, Al-Shehri D, Kamal MS (2023) Hydrogen production, transportation, utilization, and storage: recent advances towards sustainable energy. J Energy Storage 73:109207. https://doi.org/10.1016/j.est.2023.109207

Neumann D, Veser G (2005) Catalytic partial oxidation of methane in a high-temperature reverse-flow reactor. AIChE J 51:210–223. https://doi.org/10.1002/aic.10284

Pandey S, Srivastava VC, Kumar V (2023) Comparative assessment of different grades of coal for methanol production: simulation, optimization, environmental and economic analysis. Energy 284. https://doi.org/10.1016/j.energy.2023.129146

Park D, Lee J (2013) Biological conversion of methane to methanol. Korean J Chem Eng 30:977–987. https://doi.org/10.1007/s11814-013-0060-5

Patel SKS, Gupta RK, Kondaveeti S, Otari SV, Kumar A, Kalia VC, Lee J-K (2020) Conversion of biogas to methanol by methanotrophs immobilized on chemically modified chitosan. Bioresour Technol 315. https://doi.org/10.1016/j.biortech.2020.123791

Peña MA, Gómez JP, Fierro JLG (1996) New catalytic routes for syngas and hydrogen production. Appl Catal A 144:7–57. https://doi.org/10.1016/0926-860X(96)00108-1

Schubert T (2020) Production routes of advanced renewable C1 to C4 alcohols as biofuel components—a review. Biofuels Bioprod Biorefin 14:845–878. https://doi.org/10.1002/bbb.2109

Seddon D (2011) 12 - Methanol and dimethyl ether (DME) production from synthesis gas. In: Khan MR (ed) Advances in clean hydrocarbon fuel processing. Woodhead Publishing, pp 363–386

Shahabuddin M, Alam T (2022) Gasification of solid fuels (coal, biomass and MSW): overview, challenges and mitigation strategies. Energies 15:4444. https://doi.org/10.3390/en15124444

Sheets JP, Ge X, Li Y-F, Yu Z, Li Y (2016) Biological conversion of biogas to methanol using methanotrophs isolated from solid-state anaerobic digestate. Biores Technol 201:50–57. https://doi.org/10.1016/j.biortech.2015.11.035

Shi J, Jiang Y, Jiang Z, Wang X, Wang X, Zhang S, Han P, Yang C (2015) Enzymatic conversion of carbon dioxide. Chem Soc Rev 44:5981–6000. https://doi.org/10.1039/C5CS00182J

Sikarwar VS, Zhao M, Fennell PS, Shah N, Anthony EJ (2017) Progress in biofuel production from gasification. Prog Energy Combust Sci 61:189–248. https://doi.org/10.1016/j.pecs.2017.04.001

Sollai S, Porcu A, Tola V, Ferrara F, Pettinau A (2023) Renewable methanol production from green hydrogen and captured CO_2: a techno-economic assessment. J CO_2 Util 68:102345. https://doi.org/10.1016/j.jcou.2022.102345

Taylor A, Molzahn P, Bushnell T, Cheney C, LaJeunesse M, Azizian M, Semprini L (2018) Immobilization of *Methylosinus trichosporium* OB3b for methanol production. J Ind Microbiol Biotechnol 45:201–211. https://doi.org/10.1007/s10295-018-2010-z

Tijm PJA, Waller FJ, Brown DM (2001) Methanol technology developments for the new millennium. Appl Catal A 221:275–282. https://doi.org/10.1016/S0926-860X(01)00805-5

Ullah A, Hashim NA, Rabuni MF, Mohd Junaidi MU (2023) A review on methanol as a clean energy carrier: roles of zeolite in improving production efficiency. Energies 16. https://doi.org/10.3390/en16031482

Wagner N, Wen L, Frazão CJR, Walther T (2023) Next-generation feedstocks methanol and ethylene glycol and their potential in industrial biotechnology. Biotechnol Adv 69. https://doi.org/10.1016/j.biotechadv.2023.108276

Wang Z, Mai K, Kumar N, Elder T, Groom LH, Spivey JJ (2017) Effect of steam during Fischer-Tropsch synthesis using biomass-derived syngas. Catal Lett 147:62–70. https://doi.org/10.1007/s10562-016-1881-8

Xiang D, Li P, Yuan X, Cui P, Huang W (2020) Highly efficient carbon utilization of coal-to-methanol process integrated with chemical looping hydrogen and air separation technology: process modeling and parameter optimization. J Clean Prod 258. https://doi.org/10.1016/j.jclepro.2020.120910

Xin J-Y, Cui J-R, Niu J-Z, Hua S-F, Xia C-G, Li S-B, Zhu L-M (2004) Production of methanol from methane by methanotrophic bacteria. Biocatal Biotransform 22:225–229. https://doi.org/10.1080/10242420412331283305

York A, Xiao T, Green M (2003) Brief overview of the partial oxidation of methane to synthesis gas. Top Catal 22:345–358. https://doi.org/10.1023/A:1023552709642

Zhang Z, Li J, Ji M, Liu Y, Wang N, Zhang X, Zhang S, Ji X (2021) Encapsulation of multiple enzymes in a metal–organic framework with enhanced electro-enzymatic reduction of CO_2 to methanol. Green Chem 23:2362–2371. https://doi.org/10.1039/D1GC00241D

Chapter 6
Biomass Gasification and Design of Gasification Island for Methanol Synthesis Plant

Antti Pitkäoja

Abstract This chapter describes the production of syngas for methanol synthesis through biomass thermochemical conversion. The focus of the chapter is on the biomass gasification process and design of gasification island for methanol production. Synthesis of methanol requires syngas with a certain composition of CO, CO_2 and H_2 gases. The correct syngas feed composition for the synthesis reactor is essential for the process's high performance. Biomass gasification produces syngas enriched with CO and H_2 gases. The gas composition is close to the required composition for the methanol synthesis and can be modified to match the requirements. The overall process design of a methanol synthesis plant depends on the selected gasification technology. This chapter describes the biomass gasification process and reviews biomass gasification processes suitable for methanol synthesis. Reactor measurement data from the gasification processes are used to evaluate the process design and its alternatives for the gasification island of the biomass-to-methanol process, when steam-oxygen blown gasification and dual fluidised bed steam gasification are utilised.

Keywords Biomass-to-liquid · Gasification · Methanol · Biomass

A. Pitkäoja (✉)
Lappeenranta-Lahti University of Technology LUT, LUT School of Energy Systems, Lappeenranta, Finland
e-mail: antti.pitkaoja@lut.fi

A. Bartl et al. (eds.), *Alcohol Production Processes and Their Utilisation in the Transport Sector*, Energy, Environment, and Sustainability,
https://doi.org/10.1007/978-981-96-7384-1_6

6.1 Introduction

Biomass gasification is a thermochemical conversion process producing syngas suitable as feedstock for various chemical and fuel synthesis processes. Biomass-based products produced via biomass thermochemical conversion path can directly substitute chemicals and fuels refined from fossil-based raw materials. The role of biochemicals and biofuels is essential to lower fossil carbon emissions from the use of chemicals and fuels. Production of various chemicals and fuels is possible from biomass-based syngas produced with gasification. Some of the possible products are described in Fig. 6.1.

Methanol is a crucial commodity in the chemical sector today, predominantly used in producing other chemicals (e.g. formaldehyde, acetic acid, and various types of plastics) (IRENA and Methanol Institute 2021). Global methanol consumption reached 98.3 Mt in 2019 (IRENA and Methanol Institute 2021). The methanol use is expected to reach more than 120 Mt by 2025 (IRENA and Methanol Institute 2021). The methanol production in the present day is mainly based on fossil raw materials (natural gas and coal). Therefore, the development of alternative low-carbon emission methanol production technologies is essential to mitigate emissions from methanol production. Methanol can be produced using biomass-to-liquid (BtL) processes. In the process, biomass is first gasified, the syngas is purified and modified to be suitable for methanol synthesis and finally, methanol is synthesised from the syngas. Methanol is the largest chemical product manufactured by synthesis (Hannula and Kurkela 2013). Other production routes, e.g. biochemical conversion, are also possible (Speight 2008). Gasification is currently the main technology for syngas generation to produce second-generation biofuels, such as methanol, from non-edible biomass sources on a large scale. The methanol is synthesised from syngas containing CO, CO_2 and H_2 gases. Methanol synthesis requires a certain stoichiometric gas composition to operate the synthesis effectively. When the feed

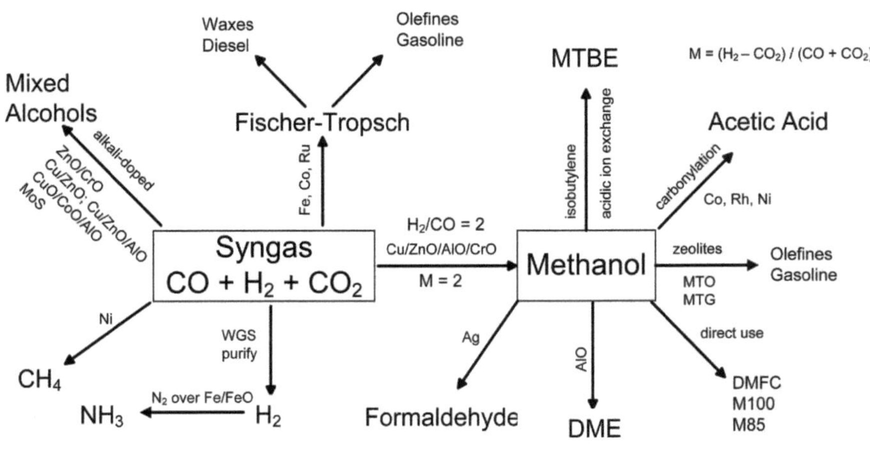

Fig. 6.1 Chemicals and fuels from syngas. Adapted from Spath and Dayton (2003)

is according to stoichiometry, the gasses are fully consumed without a slip of reactive gasses out from the reactor. The slip lowers the overall performance of the methanol synthesis reactor, and can lead to a larger reactor design than is necessary. For CO, CO_2 and H_2-based streams, stoichiometric compositions of $H_2/CO = 2$ or $M = \frac{H_2-CO_2}{CO+CO_2} = 2$ are preferred (Hannula and Kurkela 2013).

Syngas composition from a gasifier rarely satisfies the stoichiometric composition. Therefore, gas modification after the gasifier is required to obtain a suitable stoichiometric ratio for syngas. Depending on the gasifier type and overall process configuration H_2/CO or $M = \frac{H_2-CO_2}{CO+CO_2}$ is selected as the target composition. Typically, gas composition from biomass gasifiers is not very close to $M = 2$ ratio. Therefore, $H_2/CO = 2$ is a more common goal for gas modification after the gasifier. Catalytic gas conditioning processes and methanol synthesis require ultra-pure syngas. The impurities in syngas cause the deactivation of many catalysts. The impurities must be removed from the syngas before catalytic reactors. Gas purification steps must lower the concentration of impurities, such as sulphur compounds, to < 0.1 ppm level (Hannula and Kurkela 2013). Syngas purification is a multi-stage process where dedicated cleaning processes are used to remove impurities, such as the sulphur compounds.

This chapter aims to describe the biomass gasification-based biomass-to-methanol process with a focus on the gasification island of the synthesis plant. The gasification island includes process equipment for biomass gasification, biomass feedstock handling and syngas preparation for methanol synthesis. Gasification reactor measurement data from the literature are reviewed, and based on the data, a gasification island design is proposed for a methanol synthesis plant. Steam-oxygen blown gasification and dual fluidised bed steam gasification data are reviewed. Common process options are described to clean and modify syngas suitable for methanol synthesis with gasification technologies. Influence of the process options on the overall gasification island design is discussed.

6.2 Biomass Gasification Process

Biomass gasification converts carbonaceous material into gas species CO, CO_2, CH_4, H_2, various hydrocarbons and sulphur and nitrogen species. Gasification is a high-temperature process. The gasifier is typically operated in between 700 and 900 °C temperatures. The operation temperature varies by the gasifier type. Biomass gasification generally uses air, pure O_2, CO_2 and H_2O as gasification agents. Air is a typical gasification agent for power generation applications. Air is not used with the BtL processes suitable gasifiers since N_2 concentration in syngas must be minimised. The gases released from the biomass and formed during the gasification process react with a gasification agent, which results in high CO and H_2 concentrations for syngas. The gasification of biomass is a complex biomass conversion process. A simplified stepwise description of the process is shown in Fig. 6.2. When biomass is introduced in the gasifier, the biomass first dries. While heating up, pyrolysis products are

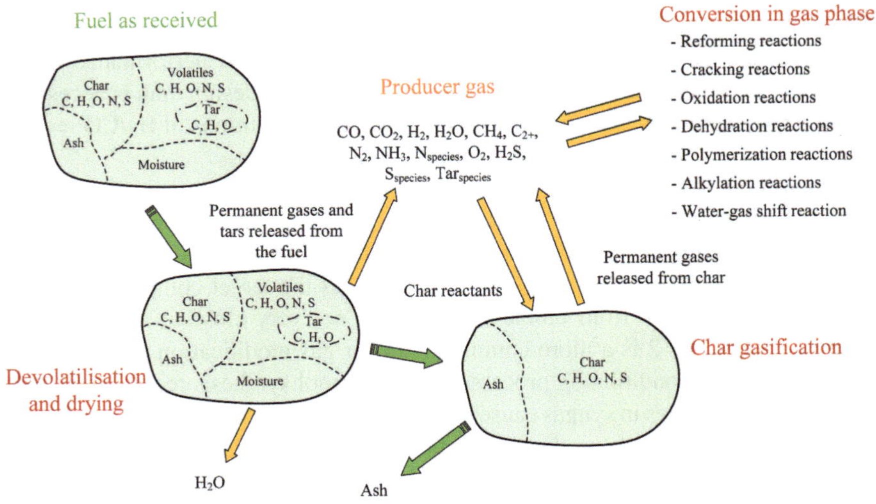

Fig. 6.2 Biomass conversion in gasifier

released from the dried biomass. The main pyrolysis products are permanent gases (CO, CO_2, H_2, CH_4 and C_2–C_4 hydrocarbons), char and tars (higher condensable hydrocarbons) (Dufour et al. 2009). The pyrolysis products are further converted by char gasification reactions, different reforming reactions and oxidation reactions. In reality, all the chemical reactions mentioned before are competitive and do not occur in steps as the description is given for the sake of simplicity. The main chemical reactions of biomass gasification are described in Table 6.1. Table 6.1 is a simplistic presentation of a complex chemical phenomenon.

Various types of biomass gasifiers have been proposed based on gas–solid contacting mode and gasifying agent (Basu 2010). Based on the gas–solid contacting mode, the gasifiers can be categorised broadly into three main categories:

- Fixed bed (updraft and downdraft)
- Entrained flow, and
- Fluidised bed.

The processes are shown in Fig. 6.3. The applicable thermal power range for each gasifier type varies. In general, the fixed bed gasifiers are mainly suitable for small-scale applications (< 10 MW), fluidised beds to intermediate scales (1–100 MW) and entrainment flow gasifiers for large capacities (> 50 MW) (Basu 2010). A more detailed description of gasification reactor types is available in De et al. (2018).

Fluidised bed gasifiers are often used for biomass gasification. The fluidised bed gasifier has a high tolerance for variation in fuel quality. The efficiency of the gasifier is higher than that of a fixed-bed gasifier (Basu 2010). The entrainment flow gasifier is not often considered for biomass gasification since the gasifier requires a small fuel particle size (Basu 2006). The entrainment gasifier is more typical for coal gasification, where the fuel is more homogeneous, and the coal can be more

Table 6.1 Chemical reactions in biomass gasification (Müller et al. 2024; Gómez-Barea and Leckner 2010)

Stoichiometry
Biomass pyrolysis
Biomass (CHONS) $\rightarrow$ char + tar + $CO + CO_2 + H_2 + CH_4 + C_{2+} + NH_3 + H_2S + \cdots$
Combustion reactions
$C + O_2 \rightarrow CO_2$
$C + 0.5O_2 \rightarrow CO$
$CO + 0.5O_2 \rightarrow CO_2$
$H_2 + 0.5O_2 \rightarrow H_2O$
$CH_4 + 2O_2 \rightarrow CO_2 + 2H_2O$
$C_xH_y + xO_2 \rightarrow xCO_2 + (y/2)H_2O$
Char gasification reactions
$C + H_2O \rightarrow CO + H_2$
$C + CO_2 \rightarrow 2CO$
$C + 2H_2 \rightarrow CH_4$
Reforming reactions
$H_2O + CO \leftrightarrow CO_2 + H_2$
$CH_4 + H_2O \leftrightarrow CO + 3H_2$
$C_xH_y + xCO_2 \rightarrow 2xCO + (y/2)H_2$
$C_xH_y + xH_2O \rightarrow xCO + (y/2 + x)H_2$
$C_xH_y + CO_2 \rightarrow C_{x-1}H_{-2} + 2CO + H_2$
$C_xH_y + H_2O \rightarrow C_{x-1}H_{-2} + CO + 2H_2$
$C_xH_y \rightarrow C + CH_4 + H_2$

easily milled into smaller particle sizes than biomass. Fluidised bed gasifiers can be bubbling fluidised bed (BFB) or circulating fluidised bed (CFB) hydrodynamics (see Fig. 6.4). The type of fluidisation depends on operating conditions, such as gas velocity within the gasifier (see more details from Kunii and Levenspiel (1991)). Biomass, by nature, is a commonly heterogeneous fuel, which makes the fluidised bed a good technological option. The fluidised bed allows for easy scalability of the gasification process to meet a specific output requirements.

Fluidised bed gasifiers can be designed as either autothermal or allothermal. The autothermal gasifier uses air or O_2 as a gasification agent to partially oxidase the fuel in the reactor, producing the heat the gasifier requires to operate. The autothermal gasifier can be either a BFB or a CFB gasifier. The allothermal gasifier produces heat required by the gasifier externally from the gasification reactor. Usually, interconnected fluidised beds are used where one reactor operates as a gasifier and the second as a combustor. The heat required by the gasifier is produced in the combustor and transported to the gasifier by bed material circulating between the reactors. The bed material circulation transports a char produced in the gasifier into the combustor, where it is combusted to produce heat. Typically, char alone is not enough to produce the heat required by the gasifier, and therefore, a support fuel is fed into the combustor. The support fuel can be fresh biomass or even off-gases from a refinery process. The

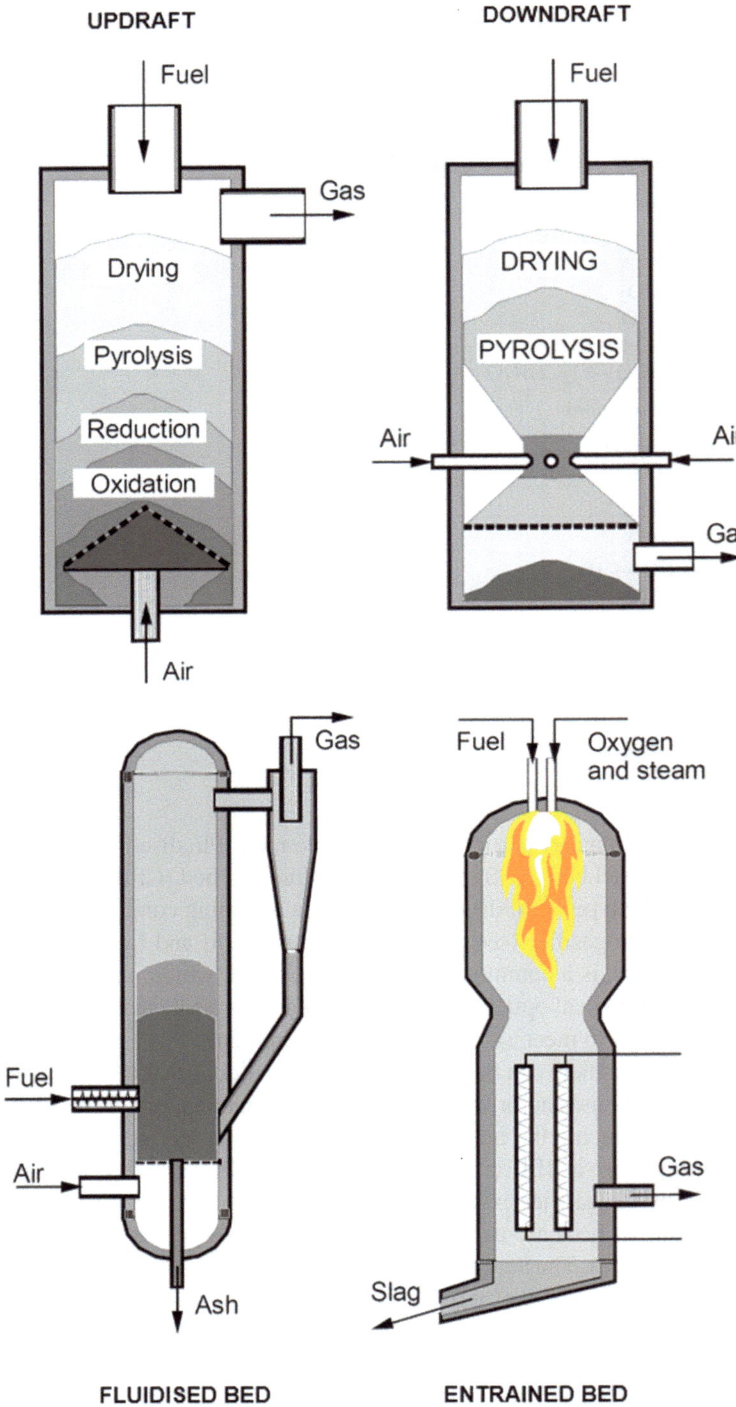

Fig. 6.3 Gasifier types. Adapted from Ohlström et al. (2001)

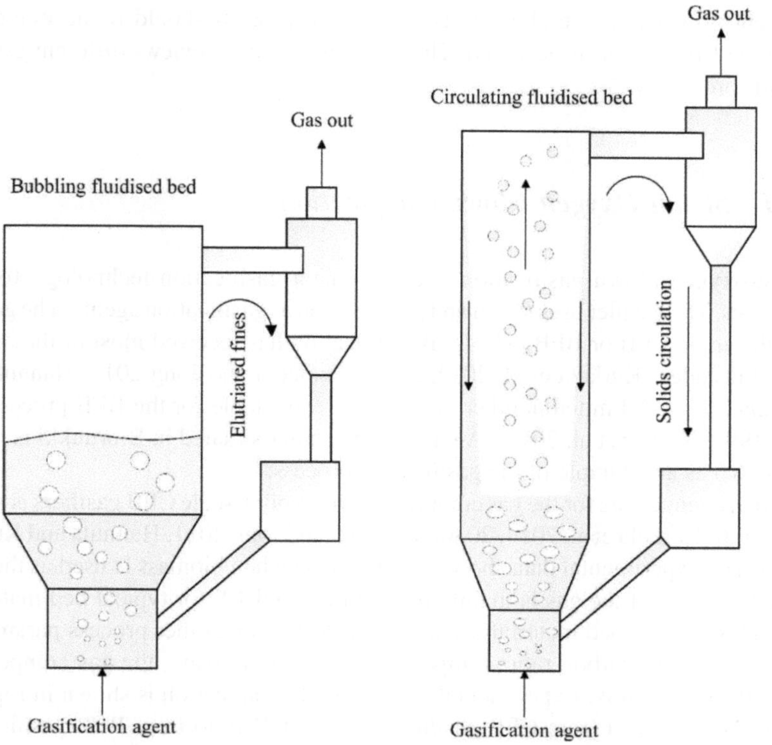

Fig. 6.4 Fluidised bed gasifiers

allothermal gasifier can be constructed by connecting BFB and CFB reactors together. The BFB is often operated as the gasifier (Fuchs et al. 2019). However, there is an example of applying a CFB reactor as a gasifier as well (Kurkela et al. 2019).

6.3 Fluidised Bed Gasifiers for BtL Processes

Fluidised bed gasifiers are currently employed with biomass gasification for power and heat generation (Patel et al. 2013; Tiilikka et al. 2014). The biofuel or chemical synthesis requires N_2-free syngas, for which various fluidised bed gasifier types are available. N_2 does not typically participate in chemical reactions to produce the synthesis product, and therefore, N_2 is an inert gas in most of the synthesis processes. Consequently, it only increases the size of the equipment and produces additional costs operationally and from an equipment design perspective. The common fluidised bed gasifiers proposed for the BtL process are steam-oxygen blown gasifiers and allothermal dual fluidised bed gasifiers (Kurkela et al. 2014; Schmid et al. 2021). Each gasifier type has its own characteristics, and the syngas produced by each have different properties. The characteristics of each process should be accounted for when the BtL process is designed. The following section reviews different gasifiers for BtL processes.

6.3.1 Steam-Oxygen Blown Gasification

Steam-oxygen blown gasification is a prominent gasification technology for BtL processes. The gasifier uses steam and pure oxygen as gasification agents. The gasifier design can be CFB or BFB. The CFB gasification has received most of the interest in past decades (Kurkela et al. 2014, 2016; Siedlecki and Jong 2011; Hannula and Kurkela 2012). A limited number of studies are available for the BFB process (Gil et al. 1997; Kitzler et al. 2011). A CFB pilot gasifier situated in Bioruukki is shown in Fig. 6.5 as an example of the gasification process.

Experimental data for the gas composition from pilot-scale CFB gasifiers is shown in Fig. 6.6 (Kurkela et al. 2014, 2016; Siedlecki and Jong 2011; Hannula and Kurkela 2012). The experimental data shows (see Fig. 6.7) when biomass is used as the fuel, an H_2/CO ratio of the gas is mainly between 0.6 and 1.8 for typical bed materials. Depending on the bed material used, reactor design, and other process parameters, such as steam-to-carbon ratio, temperature, and pressurising, the gas composition can vary significantly. Experimental data for BFB gasification is shown in Fig. 6.6, which includes data from CFB gasifiers. The H_2/CO ratio from BFB gasifier (Gil et al. 1997) varies between 0.3 and 0.8 in Fig. 6.7. The ratio is generally lower than in the CFB experiments. However, other experimental data show that the ratio can be

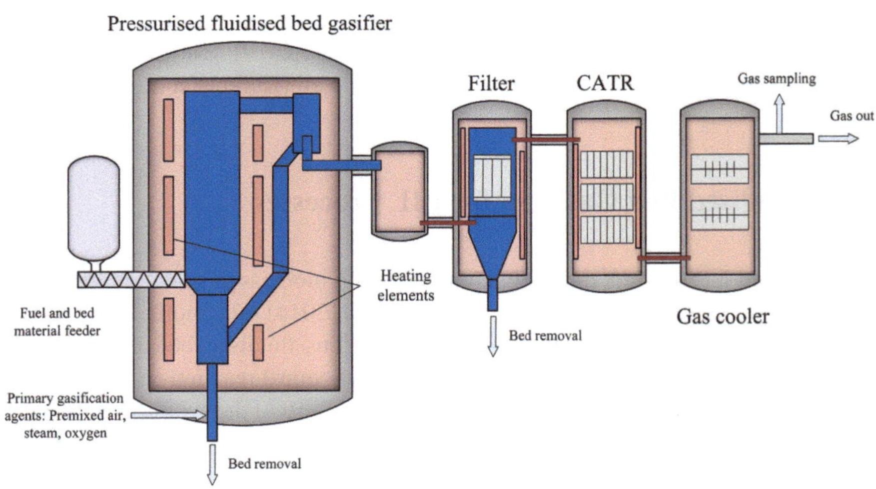

Fig. 6.5 Steam-oxygen blown gasifier situated in Bioruukki, VTT. Adapted from Hannula and Kurkela (2013)

even between 1.5 and 1.8 (Kitzler et al. 2011). Directly comparing all experiments is challenging since process conditions and reactor design vary significantly.

Besides the main gas species, various gas species considered impurities from the downstream processes' perspective are produced. The impurities are typically harmful to the downstream processes, and therefore, they must be removed from the gas. The impurities can cause catalyst deactivation in catalytic downstream processes.

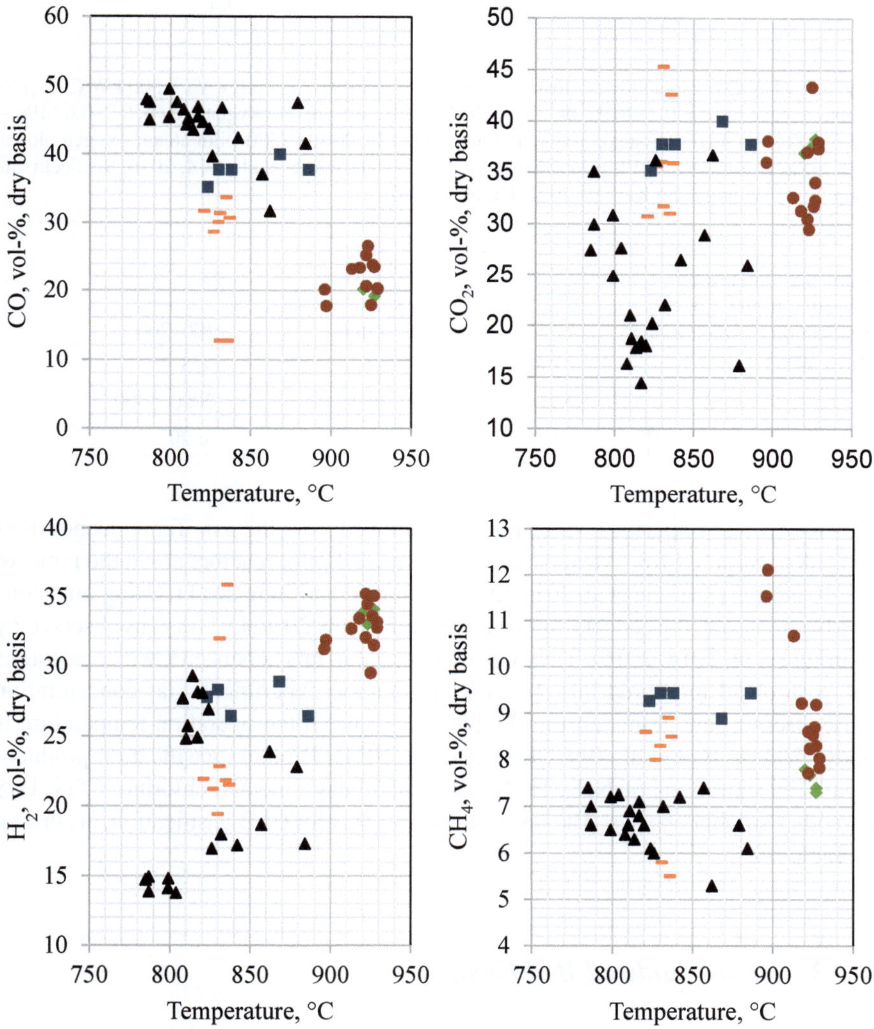

Fig. 6.6 Gas composition from steam-oxygen blown gasification. Legend: ◆ VTT 500 kW CFB pilot steam-oxygen blown, 0.25 MPa, ● VTT 500 kW CFB pilot steam-oxygen blown, 0.2–0.6 MPa, – Delft 100 kW CFB pilot steam-oxygen blown, ■ VTT 500 kW CFB pilot steam-oxygen blown, 0.25 MPa, ▲ Saragossa BFB pilot steam-oxygen blown (Kurkela et al. 2014, 2016; Siedlecki and Jong 2011; Hannula and Kurkela 2012; Gil et al. 1997)

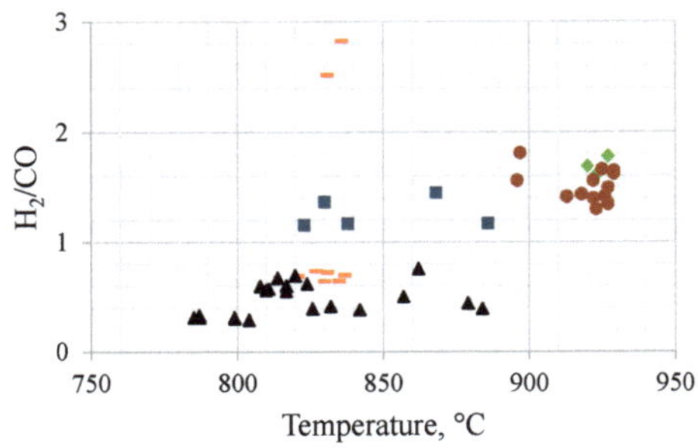

Fig. 6.7 H_2/CO composition of steam-oxygen blown gasifier. Legend: ◆ VTT 500 kW CFB pilot steam-oxygen blown, 0.25 MPa, ● VTT 500 kW CFB pilot steam-oxygen blown, 0.2–0.6 MPa, – Delft 100 kW CFB pilot steam-oxygen blown, ■ VTT 500 kW CFB pilot steam-oxygen blown, 0.25 MPa, ▲ Saragossa BFB pilot steam-oxygen blown (Kurkela et al. 2014, 2016; Siedlecki and Jong 2011; Hannula and Kurkela 2012; Gil et al. 1997)

Table 6.2 Impurities measured from CFB gasification (Kurkela et al. 2014, 2016)

Specie		
H_2S	ppm	100–200
COS	ppm	< 5
NH_3	ppm	2000–5000
HCN	ppm	< 20

Tars are typical impurities which must be removed from the gas. Tar concentrations typically measured from steam-oxygen blown CFB pilot reactors are in the range of 10–40 g/Nm^3 (Kurkela et al. 2014, 2016; Siedlecki and Jong 2011). Using limestone as a bed material, the concentration can be even below 20 g/Nm^3 (dry) as observed by (Kurkela et al. 2014, 2016). The main component in tars is benzene, which accounts for 70% of all measured tars. In BFB gasification experiments, tar concentration varies from a few grams up to 40 g/Nm^3 (dry) when the operation temperature is above 800 °C (Gil et al. 1997; Kitzler et al. 2011). Biomass gasification produces small concentrations of NH_3, HCN, alkali and heavy metals, chlorine, COS, H_2S and other sulphur species. The level of impurities is for CFB gasification, according to Table 6.2.

6.3.2 Dual Fluidised Bed Gasification

Dual fluidised bed (DFB) gasification is an oxygen-free alternative for steam-oxygen blown gasification for the BtL processes. The process consists of an interconnected gasifier and combustor. The gasifier is typically steam-blown, and the combustor is air-blown. Various concepts have been proposed for the system in past years (Corella

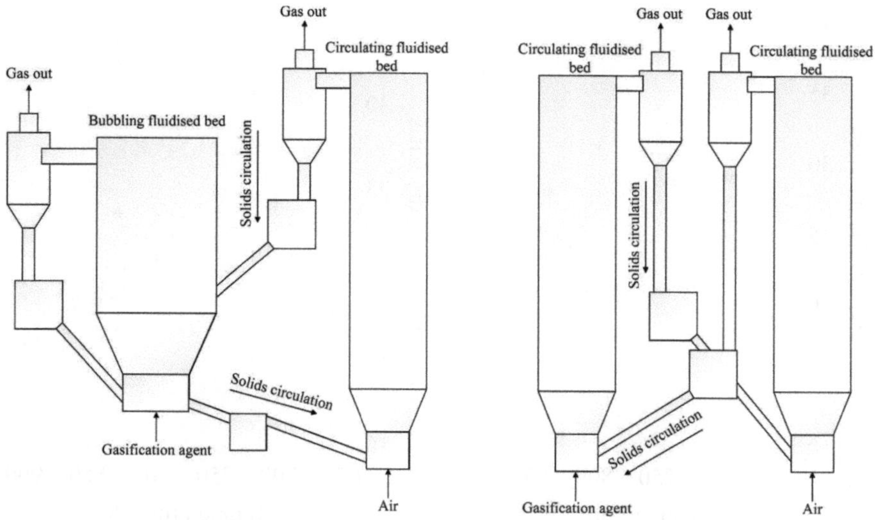

Fig. 6.8 Dual fluidised bed gasifiers

et al. 2007). The reactor configurations can be broadly divided into connected BFB-CFB and CFB-CFB reactors. The operation principles of both BFB-CFB and CFB-CFB reactors are, in principle, similar. The gasifier concepts are illustrated in Fig. 6.8. Experimental data for both gasifier types can be found in the literature (Kurkela et al. 2019; Schmid et al. 2021; Kirnbauer et al. 2013). A comparison of measurements from the pilot gasifiers is shown in Fig. 6.9. The data is for various biomasses, and it does not distinguish the influences of different operating parameters. The comparison shows that the CFB gasifier has smaller CO and higher CO_2 concentrations than the BFB gasifier. The H_2 concentrations are quite equal. When the H_2/CO ratio is plotted (Fig. 6.10), it can be observed that the CFB produces a higher H_2/CO ratio than BFB, which could cause a higher water–gas shift conversion in the gasifier. The difference in CO_2 concentrations also supports the observation.

Overall measured tar concentrations from the pilot gasifiers (Kurkela et al. 2019; Schmid et al. 2021; Kirnbauer et al. 2013) are shown in Fig. 6.11. VTT (Technical Research Centre of Finland) CFB gasifier and Technische Universität Wien (TUW) advanced gasifier produces quite similar total tar concentrations. Very low tar concentrations are achieved with the classic TUW gasifier. In each facility, the gasifier design is different. Also, the used bed material is different. A mixture of dolomite and sand was used in the VTT gasifier, while the TUW gasifier used olivine as the bed material. The dolomite catalytic activity for tar reforming is quite similar to olivine (Abu El-Rub et al. 2008). The experiments with Classic TUW gasifier were conducted with used olivine, which has enchanted catalytic properties compared to fresh olivine (Kirnbauer et al. 2013). This could be one possible explanation for the low tar concentrations in the TUW classic gasifier. Generally, the tar concentrations from the DFB gasifier are similar to steam-oxygen blown gasification.

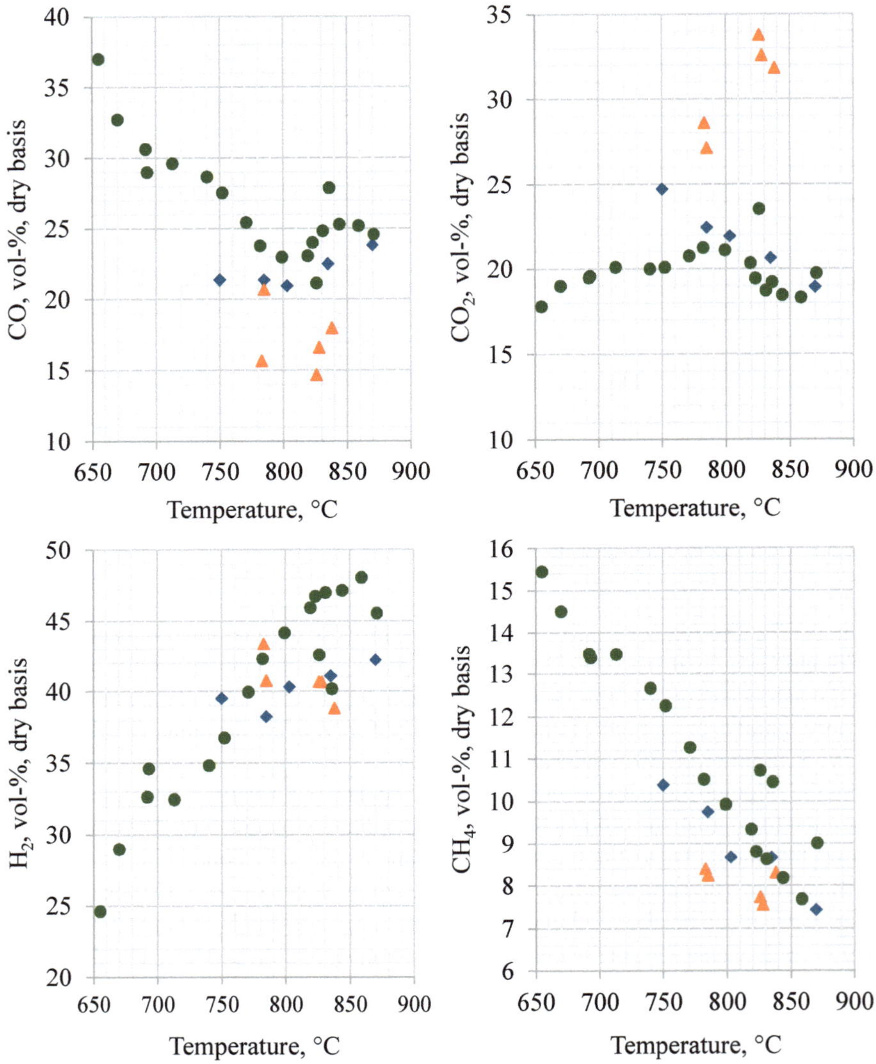

Fig. 6.9 Gas composition from DFB gasification experiments. Legend: ◆ TUW classic gasifier, ● TUW advanced gasifier, ▲ VTT CFB gasifier (Kurkela et al. 2019; Schmid et al. 2021; Kirnbauer et al. 2013)

DFB gasifiers show H_2S, NH_3 and COS forming during the gasification process (Kurkela et al. 2019; Pröll and Hofbauer 2008). The concentrations are similar to measurements from steam-oxygen blown gasification experiments shown earlier. The data shows concentrations to be dependent on the biomass used. NH_3 concentration for the CFB gasifier varied from 311 to 2242 mg/Nm^3 (dry). The use of bark as fuel produced higher NH_3 concentration than crushed wood pellets. H_2S and COS concentrations were 140–180 ppm and 3–5 ppm for bark, respectively. For crushed

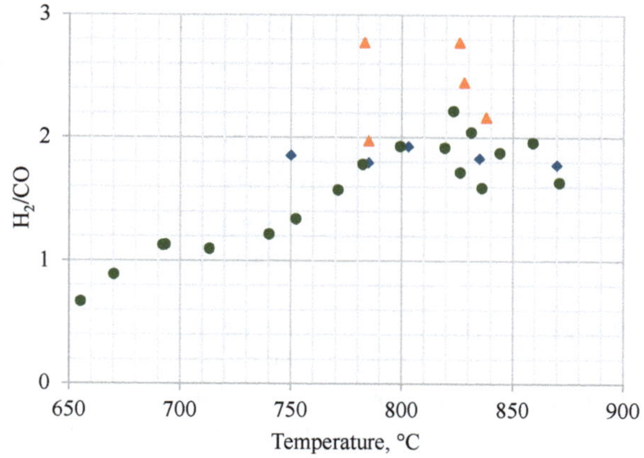

Fig. 6.10 H_2/CO ratio from DFB gasification experiments. Legend: ◆ TUW classic gasifier, ● TUW advanced gasifier, ▲ VTT CFB gasifier (Kurkela et al. 2019; Schmid et al. 2021; Kirnbauer et al. 2013)

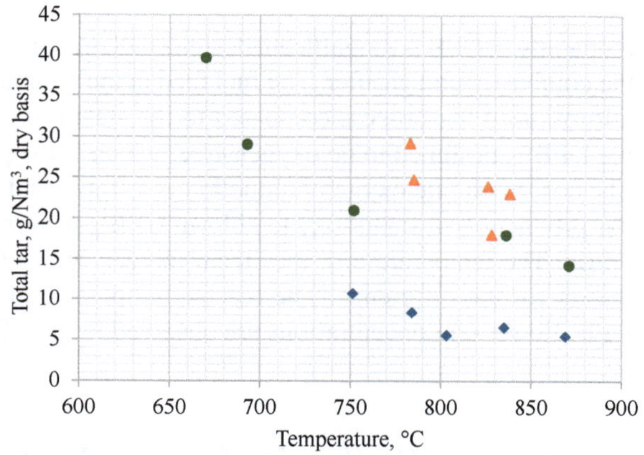

Fig. 6.11 Tar measurements from DFB gasifiers. Legend: ◆ TUW classic gasifier, ● TUW advanced gasifier, ▲ VTT CFB gasifier (Kurkela et al. 2019; Schmid et al. 2021; Kirnbauer et al. 2013)

wood pellets, H_2S concentration was 20 ppm and COS 0.5 ppm. For the Gussing 8 MW BFB gasifier (Kurkela et al. 2019) using wood chips as fuel, the concentrations of 130–170 ppm (dry) H_2S and 1100–1700 ppm (dry) NH_3 have been measured. The COS measurements are not available. The COS levels can still be expected to be similar to those measured in the other gasification processes with similar fuel.

6.3.3 Comparison of Gasification Technologies

Data presented in previous chapters is summarised in Table 6.3. The data shows DFB gasifiers to have an overall higher H_2/CO ratio for raw gas. The ratio can be around 2.0 or higher. A ratio of around 2.0 is beneficial as it reduces the need for syngas modification before many synthesis processes. The DFB gasifier, in general, is a more complex process than the steam-oxygen blown gasifier. Controlling solids flow

Table 6.3 Comparison of fluidised bed gasification reactor data

	Steam-oxygen blown gasification	Dual fluidised bed gasification
Temperature	800–900 °C	< 850 °C
Pressure	> 1 bar_a	1 bar_a
O_2 use	Used in gasifier	Can be used in combustor
Reactor	Gasification reactor	Gasification reactor + combustion reactor
Raw gas H_2/CO ratio	CFB: 0.6–1.8 BFB < 1.0	0.6–2.8
Raw gas tar concentration	CFB: 10–40 g/Nm^3 (dry) BFB: 5–40 g/Nm^3 (dry)	5–40 g/Nm^3 (dry) CFB: > 20 g/Nm^3 (dry)
Impurities	Similar	

efficiently in DFB is a significant design challenge. The DFB process requires fuel for the combustor and gasifier. The fuel feed in the combustor lowers the overall carbon utilisation of the process compared to the steam-oxygen blown gasifier. Typically, the DFB is designed to be oxygen-free, which is the greatest benefit of the process. The cost of on-site oxygen production is a significant additional operating expense. Tar production in steam-oxygen blown CFB gasifiers can be lower than that of the DFBs. The temperature of the steam-oxygen blown CFB gasifier can be the main reason for the difference. The reactor is operated at higher temperatures than the DFB. Lower tar concentrations are observed in the BFB gasifier than in the CFBs.

6.4 Biomass Gasification Island of Methanol Synthesis Plant

Production of synthetic fuels and chemicals requires a complex multi-stage process to produce the synthesis product. The gasification island is the main production line for preparing synthesis gas for the synthesis process. The gasification island's aim is to produce syngas with gas composition and quality criteria, such as concentration of impurities, are satisfied according to the requirements of the synthesis process. A general overview of the synthesis plant is shown in Fig. 6.12. The overall process consists of a gasification island, a power island, a synthesis island and auxiliary equipment. The gasification island includes a feedstock dryer, gasifier, tar removal section, filtration, gas cooler, gas modification (e.g. water–gas shift reactor, syngas hydrolysis reactors), acid-gas removal, syngas compression and various heat exchangers. The power island of the plant includes a steam cycle with a possible auxilliary boiler and steam turbine. The synthesis island includes a synthesis reactor and equipment related to finalising the synthesis product based on the quality criteria for the product. Auxiliary equipment required by the synthesis plant can be an air separation unit to

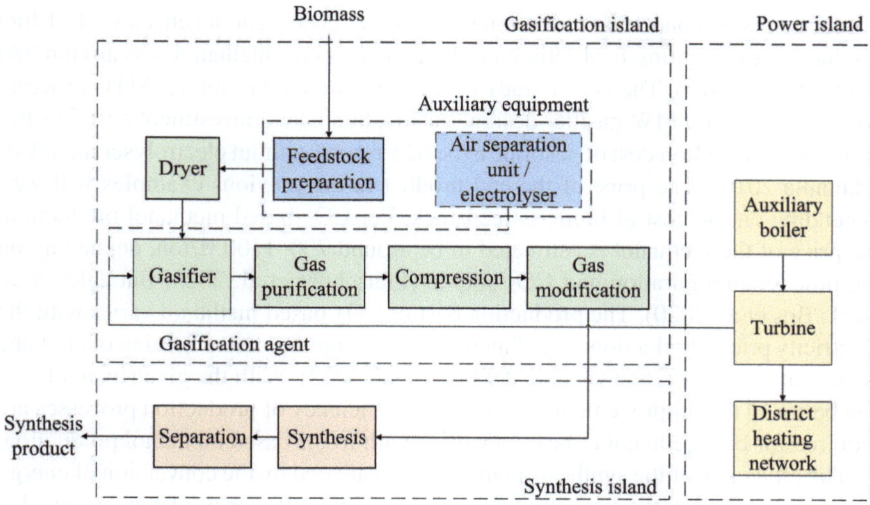

Fig. 6.12 BtL synthesis plant

produce oxygen, an electrolyser, feedstock handling equipment, etc., depending on the system's overall configuration. Gasification island design depends significantly on the selected gasification technology and the desired main equipment.

Synthesis of methanol occurs by hydrogenation of CO and CO_2:

$$CO + 2H_2 \leftrightarrow CH_3OH \tag{6.1}$$

$$CO_2 + 3H_2 \leftrightarrow CH_3OH + H_2O \tag{6.2}$$

The reactions take place in the presence of a copper oxide catalyst at 250–280 °C temperature and 60–80 bars (Hannula and Kurkela 2013). All commercially available modern catalyst systems are based on Cu–ZnO–Al_2O_3 or Cr_2O_3 with different additives and promoters. The modern catalyst allows over 99.9% selectivity for methanol. The process is the so-called low-pressure methanol synthesis, which is the main route for methanol synthesis. The syngas for the synthesis requires $M = 2$ or $H_2/CO = 2$ composition based on stoichiometry (Eqs. 6.1 and 6.2), and it must be free of sulphur ($H_2S < 0.1$ ppm) (Hannula and Kurkela 2013). Methane, argon and nitrogen are inert gases in the synthesis (Hannula and Kurkela 2013).

The economics of a large steam-oxygen gasification technology-based synthesis plant for methanol production was investigated previously for a plant with a 300 MW_{th} gasification unit (Hannula and Kurkela 2013). The investigation found that the total investment in the synthesis plant was 450 M€ (2013). The gasification island without oxygen production and biomass pretreatment equipment was 45% (150 M€) of the total capital investment costs of the synthesis plant. The share of synthesis island was 17%. The rest was from auxiliary equipment and power island. The oxygen production was alone 15% of the total investment costs. The levelised cost of fuel production (accounting for lifetime production costs) for methanol was around 380 €/ton (60 €/MWh). The newer study of methanol-derived gasoline (MTG process) with a smaller 100 MW gasifier design (gasification island investment cost 59 M€) shows the production cost of gasoline to be 623 €/ton (without electrolyser included) (Hannula 2016). The price of the end product in the previous examples will vary depending on the cost of biomass feedstock. For CO_2-based methanol production, the price of the methanol is estimated to be around 700–1100 €/ton, depneding on the process configuration and CO_2 source (Pratschner et al. 2023; Battaglia et al. 2021; Bos et al. 2020). The production cost of CO_2-based methanol varies with the electricity price. Production cost of methanol with autothermal reforming of methane is estimated as 269 €/ton (Arnaiz del Pozo et al. 2022). With the given examples, it can be stated that improvements in the overall efficiency of production processes are required for bio-methanol to be competitive with fossil-based methanol production.

The efficiency of the synthesis plant can be evaluated by the conversion of energy from feedstock into the final product and the carbon efficiency of the system (i.e. how much of the feedstock's carbon is converted into the final product). Typically, not all energy can be converted into the final product. A part of the energy is lost and can be used to produce utilities (heat and electricity) for the process itself or external consumers. Carbon conversion to the final product is an important parameter. Carbon efficiency often deteriorates if CO_2 produced by the gasifier is purged from the system. This is typically done when CO_2 cannot be utilised in the synthesis of the final product. Methanol synthesis can utilise CO_2 directly in the hydrogenation reaction (Eq. 6.2). It is therefore beneficial to set a goal to obtain an $M = 2$ ratio for the syngas where all carbon is used to produce the methanol. The following chapters will describe the typical gas modification options and the design of a gasification island based on gasifier's performance.

6.4.1 Process Design Alternatives for Gas Modification and Purification

The design of the downstream process for gas modification and purification depends on the gasification technology and the synthesis product. Generally, most of the synthesis processes have the same requirements for syngas purity. However, syngas modification is done based on the requirements of the synthesis. Heat integration should be considered when downstream processes are designed. The process equipment has different requirements for process temperature, which can allow heat extraction from the gasification island for the power island. After cleaning, the syngas should be ultra-pure. Where the concentrations of impurities are < 0.1 ppm and the syngas is particle-free.

The syngas is filtered after the gasifier to remove particles from it. High-temperature filtering also removes alkali and heavy metals, as well as, chlorides from the gas. The filtration temperature is typically 300–500 °C (Simell et al. 2014). The temperature is chosen based on the tar-loading of the gas. Low filtration temperatures can cause tar condensation, leading to a sticky dust cake that is difficult to clean.

The syngas tar concentration must be < 0.1 ppm. A catalytic reformer or oil washing could be employed to remove tars from syngas. The catalytic reformer allows tars to be removed at the ppm level in syngas. An absorbent reactor or equal is still needed to make the syngas virtually tar-free. The benefit of the catalytic reformer is that it could increase the H_2/CO ratio of syngas after the gasifier. Therefore, the catalytic reformer is a particularly suitable tar clean-up system for synthesis gas applications (Kurkela et al. 2019; Simell et al. 2014). The reformer can convert organic sulphur species to hydrogen sulphide (H_2S), which is a form of sulphur that is easy to remove downstream from syngas. Ammonia (NH_3) concentration of syngas is decreased by the reformer (Kurkela et al. 2019; Simell et al. 2014). The reformer also reforms 100% of C_{2+} hydrocarbons that are likely to cause coking of the synthesis catalysts later in the process if not removed (Kurkela et al. 2019; Simell et al. 2014). The reformer reforms a part of the methane in syngas, which is an inert gas in most of the synthesis process. Catalytic reformers allow for the reduction of syngas methane concentration even by 70%. High operating temperatures are required for the reformer. Therefore, it is often an autothermal reformer (ATR) using oxygen. The reformer using nickel catalyst requires operating temperatures > 900 °C to avoid catalyst deactivation (Simell et al. 2014). A catalytic reformer can be placed before the hot-gas filter, allowing the filter to operate at lower temperatures. The concept has been demonstrated in commercial operation in the Skive gasification plant (Patel et al. 2013). In this case, the catalysts must be monolithic type to avoid clogging by the fly ash and dust. Placing the reformer before the filter has a positive effect from a heat integration perspective. The design eliminates the temperature swing between the gasifier and the reformer caused by the syngas filtration. An option for catalytic reforming is oil washing of the syngas. Washing of syngas is conducted using oil, such as rapeseed methyl ester (RME) and equivalents. The washing has been demonstrated in the GoBiGas demonstration plant (Thunman et al. 2018). The use of washing requires processing steps for olefine removal downstream of washing, for example, by hydrogenation (Thunman et al. 2018).

In a conventional gasification island design, a water–gas shift reactor is used to adjust the H_2/CO ratio to 2.0. The water–gas shift reactor used can be a so-called sour shift reactor, where sulphur-tolerant catalysts are used, or a sweet shift reactor. The sour shift reactor can convert carbonyl sulphide (COS) and other organic sulphur compounds to hydrogen sulphide (H_2S) (Hannula and Kurkela 2013). Therefore, a separate COS hydrolysis reactor is not required before the sour shift reactor. The temperature of the sour shift reactor is 400 °C. When the sweet water–gas shift reactor is used, sulphur species removal from syngas is required before the reactor. The COS hydrolysis reactor and acid gas removal (AGR) unit are basically required before the sweet water–gas shift reactor. The AGR is used to remove H_2S and other sulphur species from syngas. Commercially, desulphurisation uses a regenerative absorber where solvents are used to separate H_2S from syngas. The commercially

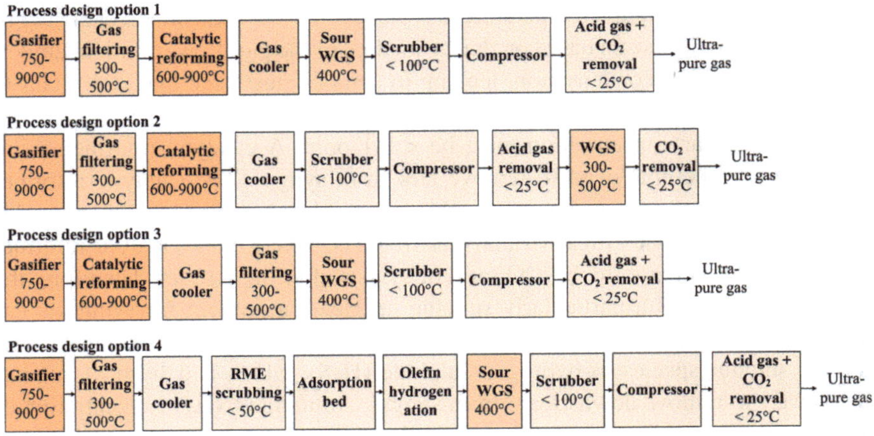

Fig. 6.13 Various process design options with the main unit processes

available AGR process, for example, is the Rectisol unit, which uses methanol as a solvent (Liquide 2017). Where H_2S, trace contaminants (tar, NH_3, and HCN), CO_2 and COS (trace concentrations) can be removed from syngas. Sulphur species can be removed to 0.1 ppm level and CO_2 to ppm level. The process design options for syngas cleaning are summarised in Fig. 6.13 with a several different process options.

The development of water electrolysis in recent years, along with increasing availability of renewable electricity, has made combining electrolysers with synthesis plants an interesting alternative. The H_2 produced by the electrolyser can be used to adjust the H_2/CO ratio downstream to the gasifier. The use of an electrolyser makes the water–gas shift reactor unnecessary. However, the electrolyser requires an external supply of electricity (higher than that of cryogenic separation of air (Hannula 2016)), which negatively affects the overall balance of the synthesis plant. Carbon utilisation from biomass to synthesis products can be enhanced (Hannula 2016). When designing the plant, the availability and prices of biomass and renewable electricity must be considered.

6.4.2 BtL Process with Steam-Oxygen Blown Gasification

The design of the BTL process for methanol production applying steam-oxygen blown gasification follows the principles shown in Fig. 6.12. The system includes a gasification island, a synthesis island and a power island. Oxygen on site can be produced via cryogenic air separation or electrolysis. Traditionally, a design with cryogenic separation has been the main choice. However, due to late developments in electrolysis, power-biomass-to-liquid (PBtL) has become an intriguing option.

A gasification island of the conventional BtL process is described in Fig. 6.14a. When a steam-oxygen blown gasifier is used, process's H_2/CO ratio and module (M) are below the requirements of the synthesis. The H_2/CO ratio is typically below 1.3, and M is very small. Therefore, syngas modification is required to adjust the composition to be suitable for the synthesis. The main unit processes of the BtL process with steam-oxygen blown gasifier are (1) Dryer, (2) Gasifier, (3) High-temperature filter, (4) Reformer, (5) Syngas cooler, (6) Sour shift, (7) Scrubber, (8) Compressor, (9) Acid gas removal unit. Impurities are cleaned with process steps 3, 4, 7 and 9.

In the PBtL concept (see Fig. 6.14b), the H_2/CO ratio is adjusted to a suitable level using the electrolyser. The electrolyser replaces the water–gas shift reactor and cryogenic air separation unit. The PBtL concept on the main parts is similar to the BtL process. In principle H_2/CO ratio can be adjusted to 2.0, or even significant quantities of H_2 could be produced to match the $M = 2$ ratio, which enhances carbon utilisation in methanol synthesis. However, the $M = 2$ ratio for syngas might require a significant-sized electrolyser.

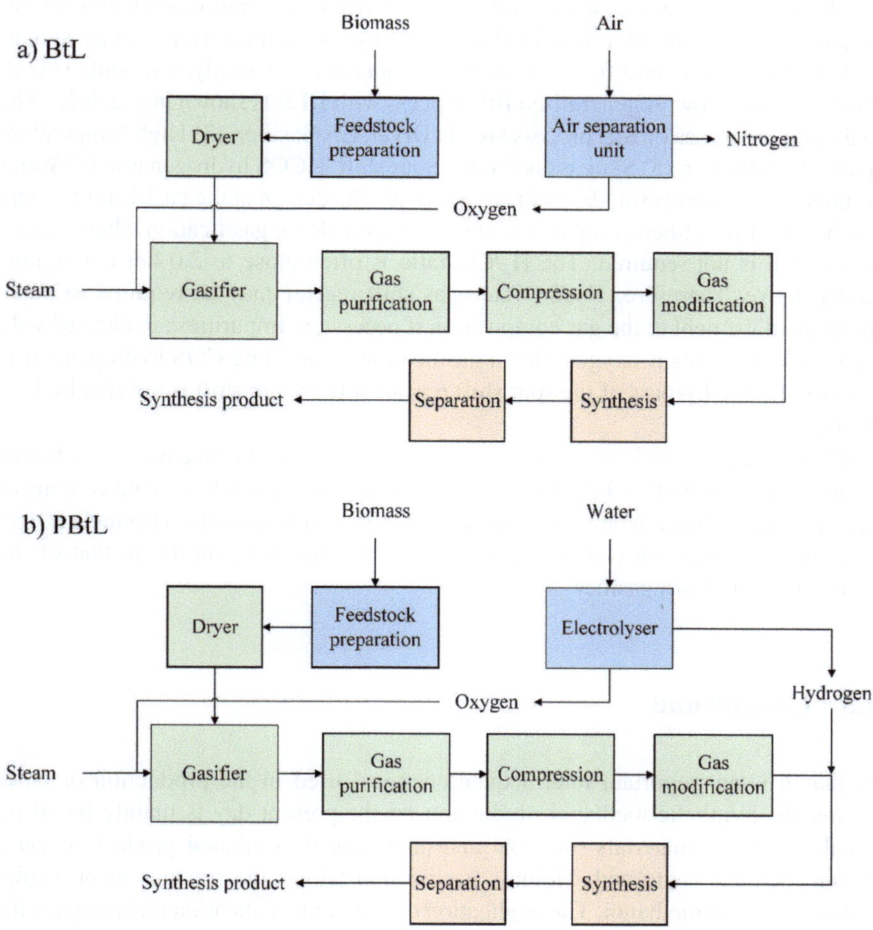

Fig. 6.14 Generalised **a**) BtL and **b**) PBtL processes with steam-oxygen blown gasifier

The conversion of energy to methanol for the conventional gasification island is around 60% of the feedstock LHV (lower heating value) (Hannula and Kurkela 2013). The efficiency stays similar if an electrolyser is included in the process (Hannula 2016). The CO-based methanol production can utilise 40% of the input carbon (Hannula 2016). However, a high carbon utilisation from feedstock to methanol is obtained with the electrolyser when $M = 2$ is the goal composition for syngas. The utilisation can be as high as 92% (Hannula 2016).

6.4.3 BtL Process with DFB Gasification

Methanol production employing a DFB process enables the simplification of the gasification island. The DFB does not require pure oxygen production for combustion. The DFB produces syngas with a higher H_2/CO ratio than steam-oxygen blown gasification. The gasifier with both CFB and BFB hydrodynamics can produce syngas with H_2/CO ratio over 2.0 at 800–900 °C temperatures. Usually, a module (M) of the gas remains low. A generalised BtL process with DFB is shown in Fig. 6.15. The main unit processes of BtL process are (1) Dryer, (2) Gasifier, (3) High-temperature filter, (4) Reformer, (5) Syngas cooler, (6) Sour shift + COS hydrogenator, (7) Water scrubber, (8) Compressor, (9) Acid gas removal. The design of the gasification island can be much simplified compared to steam-oxygen blown gasification when oxygen production is not required. The H_2/CO ratio is often close to 2.0 but not significantly above. Therefore, a sour water–gas shift reactor may be required to make the final adjustment to the gas composition if necessary. Impurities are cleaned with steps similar to steam-oxygen blown gasification. A separate COS hydrogenator is required for the by-pass of the sour shift reactor if the sourc shift is not required for the gas.

DFB gasifiers can have lower carbon efficiency than the steam-oxygen-blown gasifiers because part of the carbon supplied for the process is lost in the combustor flue gas. The carbon efficiency still around 30–40% can be expected (Hannula 2016). The energy conversion can be expected to be around 60%, similar to that of the steam-oxygen blown gasifier.

6.5 Conclusions

Methanol is an important intermediate chemical used in the production of other chemicals. Synthetic methanol production on the present day is mainly based on fossil-based raw materials (i.e. coal and natural gas). Methanol production via a thermochemical route with biomass gasification allows the production of a low-carbon emission methanol. The application of different gasification technologies for methanol production was reviewed in this chapter.

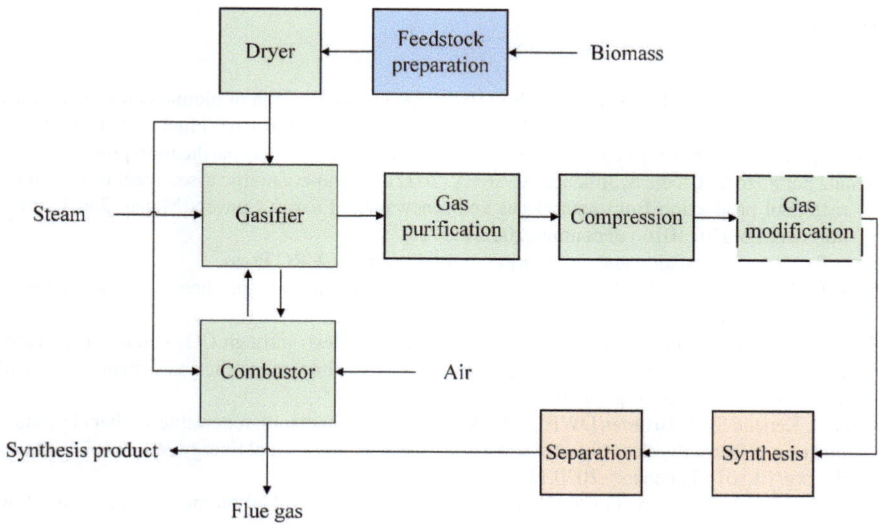

Fig. 6.15 Generalised BtL process with a DFB gasifier

The review concerned data from pilot-scale biomass fluidised bed gasification reactors. The fluidised bed is the most prominent gasifier type for biomass gasification in medium and large-scale applications. The review included steam-oxygen blown gasification and DFB gasifiers. The data from the gasifiers shows the H_2/CO ratio of DFB gasifiers to be higher than that of steam-oxygen blown gasifiers. Greater syngas modification is required for steam-oxygen blown gasifiers to make syngas suitable for methanol synthesis, compared to DFB gasifiers. Requirements for gas purification are similar in both gasifier types.

The design of the gasification island for the methanol synthesis plant was assessed based on available gasification technologies and performance data. The design of a gasification island with a DFB gasifier produces a simpler overall design of the system. However, the gasifier design itself is more complex than the steam-oxygen blown gasifier. Steam-oxygen blown gasification requires significant syngas modification and on-site oxygen production. Separate unit-processes are required for both, or alternatively, an electrolyser is required. A price range of methanol production with steam-oxygen blown gasification has been previously estimated for a 300 MW_{th} large-scale system without an electrolyser. The price of the methanol was around 400 €/ton. The production price of the methanol with the electrolyser on the site depends on the consumed electricity to produce the H_2 and the price and performance of the electrolysers in the future. Estimates for the methanol price with a DFB gasification are not available. However, the price can be estimated to be similar to steam-oxygen blown gasification. DFB gasification for methanol production can be recommended based on the system's overall complexity. The principles of design and operation of the DFB gasifier, as part of methanol synthesis, should be investigated to alleviate the uncertainty regarding the gasifier.

References

Abu El-Rub Z, Bramer EA, Brem G (2008) Experimental comparison of biomass chars with other catalysts for tar reduction. Fuel 87:2243–2252. https://doi.org/10.1016/j.fuel.2008.01.004

Air Liquide (2017) Lurgi Rectisol^{TM} the world's leading synthesis gas purification process

Arnaiz del Pozo C, Cloete S, Jiménez Álvaro Á (2022) Techno-economic assessment of long-term methanol production from natural gas and renewables. Energy Convers Manag 266:115785. https://doi.org/10.1016/j.enconman.2022.115785

Basu P (2006) Combustion and gasification in fluidized beds. CRC Press

Basu P (2010) Biomass gasification and pyrolysis: practical design and theory. Academic Press, Burlington, MA

Battaglia P, Buffo G, Ferrero D et al (2021) Methanol synthesis through CO_2 capture and hydrogenation: thermal integration, energy performance and techno-economic assessment. J CO_2 Util 44:101407. https://doi.org/10.1016/j.jcou.2020.101407

Bos MJ, Kersten SRA, Brilman DWF (2020) Wind power to methanol: renewable methanol production using electricity, electrolysis of water and CO_2 air capture. Appl Energy 264:114672. https://doi.org/10.1016/j.apenergy.2020.114672

Corella J, Toledo J, Molina G (2007) A review on dual fluidized-bed biomass gasifiers. Ind Eng Chem Res 6831–6839. https://doi.org/10.1021/ie0705507

De S, Agarwal AK, Moholkar VS, Thallada B (2018) Coal and biomass gasification. Springer Singapore, Singapore

Dufour A, Girods P, Masson E et al (2009) Synthesis gas production by biomass pyrolysis: effect of reactor temperature on product distribution. Int J Hydrogen Energy 34:1726–1734. https://doi.org/10.1016/j.ijhydene.2008.11.075

Fuchs J, Schmid JC, Müller S, Hofbauer H (2019) Dual fluidized bed gasification of biomass with selective carbon dioxide removal and limestone as bed material: a review. Renew Sustain Energy Rev

Gil J, Aznar MP, Caballero MA et al (1997) Biomass gasification in fluidized bed at pilot scale with steam−oxygen mixtures. Product distribution for very different operating conditions. Energy Fuels 11:1109–1118. https://doi.org/10.1021/ef9602335

Gómez-Barea A, Leckner B (2010) Modeling of biomass gasification in fluidized bed. Prog Energy Combust Sci 36:444–509. https://doi.org/10.1016/j.pecs.2009.12.002

Hannula I (2016) Hydrogen enhancement potential of synthetic biofuels manufacture in the European context: a techno-economic assessment. Energy 104:199–212. https://doi.org/10.1016/j.energy.2016.03.119

Hannula I, Kurkela E (2012) A parametric modelling study for pressurised steam/O_2-blown fluidised-bed gasification of wood with catalytic reforming. Biomass Bioenergy 38:58–67. https://doi.org/10.1016/j.biombioe.2011.02.045

Hannula I, Kurkela E (2013) Liquid transportation fuels via large-scale fluidised-bed gasification of lignocellulosic biomass. Espoo

IRENA and Methanol Institute (2021) Innovation outlook: renewable methanol. Abu Dhabi

Kirnbauer F, Wilk V, Hofbauer H (2013) Performance improvement of dual fluidized bed gasifiers by temperature reduction: the behavior of tar species in the product gas. Fuel 108:534–542. https://doi.org/10.1016/j.fuel.2012.11.065

Kitzler H, Pfeifer C, Hofbauer H (2011) Pressurized gasification of woody biomass—variation of parameter. Fuel Process Technol 92:908–914. https://doi.org/10.1016/j.fuproc.2010.12.009

Kunii D, Levenspiel O (1991) Fluidization engineering, 2nd edn. Butterworth-Heinemann

Kurkela E, Kurkela M, Hiltunen I (2014) The effects of wood particle size and different process variables on the performance of steam-oxygen blown circulating fluidized-bed gasifier. Environ Prog Sustain Energy 33:681–687. https://doi.org/10.1002/ep.12003

Kurkela E, Kurkela M, Hiltunen I (2016) Steam–oxygen gasification of forest residues and bark followed by hot gas filtration and catalytic reforming of tars: results of an extended time test. Fuel Process Technol 141:148–158. https://doi.org/10.1016/J.FUPROC.2015.06.005

Kurkela E, Kurkela M, Tuomi S et al (2019) Efficient use of biomass residues for combined production of transport fuels and heat. Espoo

Müller FJ, Fuchs J, Müller S, Winter F (2024) CO_2 conversion to CO by fluidized bed biomass gasification: Measuring CO_2 utilization via stable carbon isotope ratios. J CO_2 Util 83:102792. https://doi.org/10.1016/j.jcou.2024.102792

Ohlström M, Mäkinen T, Laurikko J, Pipatti R (2001) New concepts for biofuels in transport. In: Biomass-based methanol production and reduced emissions in advanced vehicles. VTT2074

Patel J, Salo K, Horvath A, Jokela P (2013) Andritz carbona biomass gasification process for power and biofuels. In: 21st European conference and exhibition. Copenhagen

Pratschner S, Radosits F, Ajanovic A, Winter F (2023) Techno-economic assessment of a power-to-green methanol plant. J CO_2 Util 75:102563. https://doi.org/10.1016/j.jcou.2023.102563

Pröll T, Hofbauer H (2008) H_2 rich syngas by selective CO_2 removal from biomass gasification in a dual fluidized bed system—process modelling approach. Fuel Process Technol 89:1207–1217. https://doi.org/10.1016/j.fuproc.2008.05.020

Schmid JC, Benedikt F, Fuchs J et al (2021) Syngas for biorefineries from thermochemical gasification of lignocellulosic fuels and residues—5 years' experience with an advanced dual fluidized bed gasifier design. Biomass Convers Biorefin 11:2405–2442. https://doi.org/10.1007/s13399-019-00486-2

Siedlecki M, de Jong W (2011) Biomass gasification as the first hot step in clean syngas production process—gas quality optimization and primary tar reduction measures in a 100 kW thermal input steam–oxygen blown CFB gasifier. Biomass Bioenergy 35:S40–S62. https://doi.org/10.1016/j.biombioe.2011.05.033

Simell P, Hannula I, Tuomi S et al (2014) Clean syngas from biomass—process development and concept assessment. Biomass Convers Biorefin 4:357–370. https://doi.org/10.1007/s13399-014-0121-y

Spath PL, Dayton DC (2003) NREL/TP-510-34929. Preliminary screening—technical and economic assessment of synthesis gas to fuels and chemicals with emphasis on the potential for biomass-derived syngas

Speight JG (2008) synthetic fuels handbook: properties, process, and performance. The McGraw-Hill Companies, Inc

Thunman H, Seemann M, Berdugo Vilches T et al (2018) Advanced biofuel production via gasification—lessons learned from 200 man-years of research activity with Chalmers' research gasifier and the GoBiGas demonstration plant. Energy Sci Eng 6:6–34. https://doi.org/10.1002/ese3.188

Tiilikka M, Bolhàr-Nordenkampf M, Honkola T (2014) Advanced gasification technologies for large scale energy production. In: Power-gen Europe. Cologne

Chapter 7
Production of Bioethanol from Second-Generation Feedstocks for the Transportation Sector

Shiladitya Ghosh, Rohit Das, and Tan Thakur

Abstract Petro-fuels-driven transportation sector (road, air, and water) is the largest emitter of greenhouse gases (GHGs), several other toxic gases and particulate matters causing rapid deterioration of the atmosphere. Displacing the major transportation fuels namely petrol/gasoline and diesel with suitable biofuels is viewed as the best alternative to bring down the emission/pollution index without imposing restrictions in the mobility of vehicles for communication, business and travel. The most promising alternative of liquid petro-fuels are alcohols, specifically ethanol, which has good fuel properties and compatibility as a blend-in liquid fuel for the existing designs of the internal combustion (IC) engines. While multiple routes of production of ethanol exist, the biochemical production processes are gaining increased popularity due to their potential waste-to-energy aspects and eco-friendly attributes. A few well-known microorganisms are dedicated as the industrial producers of ethanol which convert the fermentable carbohydrates to ethanol through specific solventogenic metabolism. Both edible carbohydrates and non-edible complex biomass can act as potential feedstocks for ethanol production, but the later ones are more favourable as it does not perturb food security. However, the margin of productivity and yield of the ethanol fermentation processes are limited by several factors and need to be elevated to secure its potential as the alternative fuel in the transportation sectors. This chapter presents the current state of the biochemical ethanol production processes and strategies for the improvement to facilitate its implementation as a dependable transportation fuel.

Keywords Ethanol · Lignocellulosic biomass · Second-generation platform · Blend-in fuel · Transportation sector

---

S. Ghosh (✉) · R. Das · T. Thakur
Amity Institute of Biotechnology, Amity University Kolkata, Kadampukur Village, Rajarhat, Newtown, Kolkata, West Bengal 700135, India
e-mail: sghosh10@kol.amity.edu

A. Bartl et al. (eds.), *Alcohol Production Processes and Their Utilisation in the Transport Sector*, Energy, Environment, and Sustainability,
https://doi.org/10.1007/978-981-96-7384-1_7

7.1 Introduction

Fossil fuels, such as coal, petroleum oils, and natural gas, have been used for primary sources of energy production in daily and industrial activities for a very long time now. The extreme reliance on these energy sources comes with numerous significant drawbacks, majorly concerning the environment. The combustion of these fossil fuels specifically coal and gasoline/diesel substantially contribute to greenhouse gas (GHG) emissions, with CO_2 emissions accounting for approximately 80% of the global total (Culaba et al. 2023). There is ample scientific evidence to support the theory that growing CO_2 concentrations in the atmosphere are mostly caused by non-renewable energy vectors, significantly contributing to global warming and climate anomalies of the present time. Before the industrial revolution, atmospheric CO_2 concentrations were stable at 280 parts per million (ppm), indicating 280 molecules per million molecules (US EPA, OA 2017). As of August, 2024, measurements at the Mauna Loa Observatory recorded a concentration of almost 419 ppm, marking over 50% increase attributed to the observed rise in global temperatures (US EPA, OA 2017; https://www.co2.earth/daily-co2). An estimated 7–9% of anthropogenic CO_2 emissions come from the iron and steel sector, one of the biggest industries that consumes fossil fuels and which will be increased by 25–30% by 2050 (Kim et al. 2022). During 2000–2023, global primary energy consumption was estimated to be around 620 exajoules (EJ), with fossil fuels comprising coal, natural gas and petroleum oils contributing 80–85% of this total (https://www.statista.com/statistics/265598/consumption-of-primary-energy-worldwide/). Fossil fuels are non-renewable resources and the finite reserves are projected to be severely depleted within the next 50 years. Use of alternatives such as biofuels and other renewable energy sources in place of fossil fuels is a promising way to remediate this concerning issue of future energy insecurity (Azarpour et al. 2022; Zohuri and McDaniel 2021).

Biofuels derived from biomass represent clean, environmentally sustainable, and efficient renewable energy sources (Mahapatra et al. 2021; Dahman et al. 2019; Jaiswal et al. 2022). They can be utilized for fuel and energy applications in solid (as native biomass), liquid (bioalcohols or biocrude/pyro-oil), or gaseous forms (hydrogen, methane, syngas) (Islas et al. 2019; Bajwa et al. 2018; Magda et al. 2021; Minteer 2011; Ghosh et al. 2019, 2017a; Gupta et al. 2024a; Abedinia et al. 2024; Fedyukhin et al. 2024; Roychowdhury and Ghosh 2024). Examples of liquid biofuels include biodiesel and bioalcohols which hold promise as alternatives to liquid fossil fuels (Magda et al. 2021; Minteer 2011; Thapa et al. 2018; Song et al. 2024). Substituting fossil fuels with biofuels could have substantial implications for the economy and the environment through potential reduction of greenhouse gas emissions which is the main driver of the energy transition (Oke et al. 2024; Rial 2024). As the transportation sector is the largest consumer of the liquid fossil fuels, the implementation of the replacement by convenient biofuels is urgently sought (Oke et al. 2024; Rial 2024; Nordin et al. 2024). The most promising alternative of liquid petro-fuels are alcohols, namely ethanol, butanol, and higher ones. Specifically, ethanol which shown good compatibility as a blend-in liquid fuel for gasoline/petrol has gained maximum popularity to be used in the existing designs of the internal combustion (IC) engines with efficiency (Abdellatief et al. 2024; Palani et al. 2024). Ethanol (C_2H_5OH) is a two-carbon organic compound which belongs to the alcohol group and has widespread use in alcoholic beverages and as a biofuel. Ethanol is used as a biofuel because it is renewable in nature and that it can be blended with gasoline in 5–85% range (Koyunoğlu 2024; Gupta et al. 2024b). The volumetric efficiency of gasoline is also enhanced on blending with ethanol due to its higher heat of vaporization, despite its energy content being lower compared to gasoline (66%–68% of pure gasoline). Due to this, ethanol has already been accepted and used as a blend-in additive for gasoline/petrol-based automobiles in many countries, including India (Koyunoğlu 2024; Azad et al. 2024).

Ethanol can be produced through chemical methods by using reactions like dimethyl ether with CO_2 and H_2 under specific reaction conditions and hydration of ethylene under activities of specific catalysts (Li et al. 2010; Feng et al. 2024). Biological fermentation process for ethanol produced by biochemical reactions, i.e., bioethanol, is the production route that is majorly used around the world due to its greener attributes compared to the chemical routes. Renewable materials which contain fermentable sugars and carbohydrates are the major sources of feedstocks for production of bioethanol. Based on the type of feedstock used, they can be majorly classified into three following types, Fig. 7.1. First-generation (1G) feedstocks consist of edible food crops like rice, corn, wheat, sugarcane, potato, cassava, etc. from where fermentative carbohydrates for ethanol are extracted (Vaskan et al. 2017; Rulli et al. 2016). Second-generation (2G) feedstocks consist of various lignocellulosic biomass (LCB) like rice straw, wheat straw, corn stover, sugarcane bagasse, sawdust, birchwood, pinewood switchgrass, etc. and many more (Gautam et al. 2024; Singh et al. 2023). Third generation (3G) feedstocks are mainly based on algal and aquatic weed biomass (Singh et al. 2023; Thakur and Singh 2024). A major drawback of 1G bioethanol is its competition with the food production which ultimately leads to the increase in prices of crops used in the production, also limiting their availability for food consumption (Vaskan et al. 2017). The major advantage of 2G bioethanol is that it addresses this drawback of 1G bioethanol. There is no lack of abundance of LCBs and no competition with food crops. Some 2G feedstocks, i.e., the energy crops, can also be grown with less water and on less fertile land than the agricultural crops of 1G platform (Vaskan et al. 2017; Halder et al. 2019). Main drawback of 2G bioethanol is the complexity and cost involved in the processing techniques which need to be done mandatorily (Gautam et al. 2024; Paliga et al. 2022; Ghosh et al. 2017b). This is not an issue with 1G bioethanol. 3G bioethanol is of interest to researchers due to its major advantage over 1G and 2G bioethanol mainly due to the cultivation cost of the source (algae/microalgae/aquatic weed) being low and conversion efficiency being high (Jambo et al. 2016). The major drawback of 3G bioethanol production is when produced on a large scale, billions of gallons of water might be necessary (Frankó et al. 2015). Moreover, there are stability issues unlike that of 1G and 2G because bioethanol sourced from algae is much less stable than the other two generations (Halder et al. 2019). Among the three generations, the 2G bioethanol is presently the major route of production of ethanol worldwide.

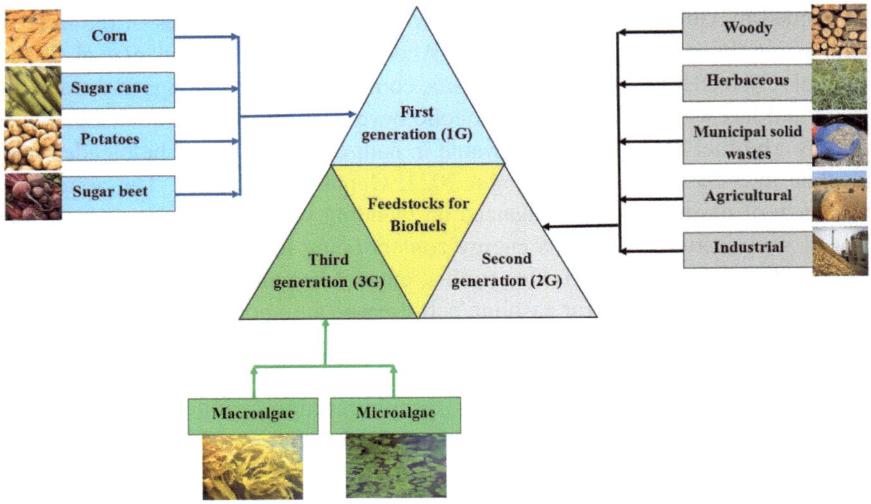

Fig. 7.1 Three generations of feedstocks for biofuel production (adapted from Singh et al. 2023)

The present chapter discusses bioethanol production technology particularly based on the prime feedstocks of the 2G platform and their processing for ethanol production. Various pretreatments used for different types of LCB originated from agro-processing or forestry have been mentioned to facilitate a broader understanding of the options. The hydrolysis of pretreated biomass and fermentative conversion of the hydrolysates are also discussed by focusing on the major biocatalysts involved in these, i.e., the hydrolytic enzymes and the ethanol producing microorganisms. Limitations associated with the production part have been highlighted along with promising solutions. Utilization of ethanol as the blend-in additive for gasoline/petrol as adopted by different countries and the factors influencing its progress are also discussed.

7.2 Feedstocks for Bioethanol Production: Types and Potential

Ethanol production goes back to a very long time in the history of mankind. In many of the ancient civilizations, ethanol was produced from commonly available biowastes and consumed as a beverage. Traditionally ethanol production is most commonly carried out by the brewer's yeast *Saccharomyces cerevisiae* which uses glucose as the main carbon source in its metabolic pathway for ethanol production (Silva Fernandes et al. 2022). The native strain of *S. cerevisiae* can only accept and use glucose (6-carbon sugar) as the carbon source and no 5-carbon sugar. Hence, the feedstock to be used for ethanol production should majorly contain glucose as the major carbon source. In this regard based on the source, the feedstocks for ethanol production can be classified into simple and complex ones.

7.2.1 First-Generation (1G) Feedstocks

The majority of the initial research on ethanol production used the native strain of *S. cerevisiae* and hence the monomeric sugars like glucose, fructose, and disaccharide sucrose (Atiyeh and Duvnjak 2001; Basso et al. 2011). These sugars are present in fruits and some vegetables and are considered simple feedstocks due to their structural simplicity. As the yeast can easily consume and directly assimilate these simple sugars, the rate and growth of yeast and fermentation were quite fast. However, all of these simple sugars are fit for human consumption and part of the edible food grade products which led to the food versus fuel conflict. Hence, the preference of the feedstocks shifted to the non-edible feedstocks in the later phase of research on ethanol production (Vaskan et al. 2017; Gautam et al. 2024; Singh et al. 2023). Those earlier studies significantly helped in establishing the technological foundation for standard ethanol fermentation processes and understanding the impact of different parameters like temperature, pH, dissolved oxygen, and soluble metabolite concentration on the yield and productivity of ethanol. Separation and recovery of ethanol from the fermentation effluent to produce low-water containing ethanol which is suitable for use as the blend-in fuel additive was also carried out in those studies.

Other than these, a significant number of studies have reported use of starchy substances and sugarcane juice for ethanol production (Li et al. 2022; Laluce et al. 2016). Sugarcane-based bioethanol is most popular and progressed in Brazil, which is the largest sugarcane producer followed by India. Sugarcane juice as well as the molasses is rich source of sucrose. Strains of *S. cerevisiae* possess invertase/sucrase due to which can take up sucrose as a precursor of glucose/fructose for ethanol production. Use of sugarcane juice directly in the ethanol fermentation might raise the margin of ethanol production to significantly high level but is unfavourable due the food versus fuel conflict. Molasses is the liquid byproduct of sugarcane refineries and is a good alternative source of sucrose for ethanol production (Bouallagui et al. 2013). Starch is a polysaccharide composed of glucose which can be hydrolysed to extract the glucose for fermentation. Starch is considered a complex polysaccharide but compared to the other plant polysaccharides like cellulose and hemicellulose the level of complexity is much less. Native strains of *S. cerevisiae* are not producers of amylase and hence cannot ferment starch directly without any additional enzymatic or chemical/thermal hydrolysis step. Starch-based feedstocks for bioethanol production are mainly sourced from corn, potato, barley, cassava, etc., majority of which are directly consumed for food production (Li et al. 2022; Ai et al. 2011). Hence, only organic wastes containing the starchy substances can be used as feedstocks for ethanol production. In general, all of the substances mentioned above are 1G type and hence unsustainable in the long run as feedstocks.

7.2.2 *Second-Generation (2G) Feedstocks*

The LCB comprises lignin and the polysaccharides cellulose and hemicellulose are considered as complex feedstocks due to their structural complexity. The entire 2G platform of biofuel production is dependent on the LCBs as the feedstocks (Gautam et al. 2024; Singh et al. 2023; Paliga et al. 2022; Ghosh et al. 2017b). Lignin removal is the first step of LCB conversion in the biochemical routes and it is conducted by various chemical, thermochemical, or biological methods (Ghosh et al. 2017b; Hassan et al. 2021). Lignin has no fermentable parts and must be removed from the LCB in order to retrieve the polysaccharides for extraction of fermentable sugars (Meshitsuka and Isogai 2017a). Cellulose is a homopolymer composed of glucan units connected throughout by β-1, 4-glycosidic bonds. It releases glucose upon enzymatic hydrolysis of the glycosidic bonds. Cellulose has a crystalline fraction and an amorphous fraction (Houfani et al. 2020). The crystalline fraction of the cellulose is more complex and resistant to chemical/biological hydrolysis compared to the amorphous part. The other polysaccharide hemicellulose is majorly composed of xylan and arabinan/mannan which release xylose and arabinose/mannose respectively after hydrolysis (Houfani et al. 2020). As mentioned in the previous section, the native strains of the yeast *S. cerevisiae* can only ferment glucose to ethanol but not the pentoses released from hemicellulose. This leads to loss of a fraction of carbon source present in the hydrolysate. To tackle this issue, strains of yeast other than *S. cerevisiae* and other ethanologenic microbes with pentose fermenting ability are being used in many studies. More recently, co-cultures of hexose and pentose fermenting microbes are used to ferment the mixed carbohydrates to ethanol (Chen 2011). The yield of ethanol varies based on the level of fermentable sugars available and the fermentative conversion efficiency of the microbes used for ethanol production.

For ethanol production in the 2G platform, the types of LCB majorly used are either various agro-wastes or forestry wastes. These wastes are generated directly or indirectly from the agricultural or forestry processing sectors. However, agro-wastes are more commonly available to many countries and popularly used as feedstocks for ethanol production compared to the forestry wastes. Comparatively shorter life span of the agricultural crops and regular cultivation for food production leads to continuous generation of agro-wastes. In terms of global abundance, the total amount of agro-wastes is much larger compared to forestry wastes. Lignin, cellulose, and hemicellulose contents greatly vary among different types of LCBs including the agro-wastes and forestry wastes, which determines the level of suitability of particular wastes as the feedstock for ethanol production. In the present chapter, the current state of the bioethanol production has been represented based on the latest reports available on the use of complex feedstocks.

7.3 Processes for Bioethanol Production: Current State and Limitations

In this section, studies reporting variety of agro-waste feedstocks for bioethanol production are considered which include rice straw or husk, wheat straw, banana peel, barley straw, corn stalk, corn stover, sugarcane bagasse, and sweet sorghum bagasse (Table 7.1). Some studies reporting forestry-based woody sources for ethanol production are also discussed. The lignocellulose composition, methods, and conditions of pretreatment and hydrolysis, details of the fermentation (microbe used, operating conditions and bioreactor), and performance indicators were tabulated. Table 7.1 summarizes the findings of the studies considered under the present chapter. Based on the information provided in Table 7.1, the current state of complex feedstock-based bioethanol production processes has been discussed and represented in the following section. The above mentioned feedstocks were locally sourced in different countries, including Spain, Korea, India, Bangladesh, China, Brazil, Turkey, Nigeria, Japan, Thailand, South Africa, USA, etc. for conducting bioethanol production. It indicates the effort exhibited by the global researchers towards ethanol production and also highlights the necessity of ethanol usage in the transportation sector (Balat and Balat 2009). The lignocellulose composition of the various feedstocks in their native state indicates the variability of the contents of lignin, cellulose, and hemicellulose among different types of agro-wastes or forestry wastes. Regardless of the type, it was clear to observe that all of the LCB feedstocks contained good amount of cellulose along with hemicellulose in their structure and indicated their potential to be used as the source of fermentable sugar extraction. Various pretreatment processes and hydrolysis techniques were employed to evaluate ethanol yield. Pretreatment or the delignification is carried out through various types of methods like chemical, physicochemical, thermal, and biochemical, whereas the hydrolysis is mainly conducted by the enzymatic route (Ghosh et al. 2017b).

Considering the technology of the second-generation platform, the processes are divided into three consecutive unit operations; 1. Pretreatment of the LCB, 2. Hydrolysis of the pretreated solid fraction of LCB, and 3. Fermentative conversion of hydrolysate to bioethanol, Fig. 7.2, (Paliga et al. 2022). After fermentation, the ethanol is recovered from the effluent by distillation or other methods. From the viewpoint of operation, the pretreatment step is most complex due to the variability of lignin content and composition in different LCBs. For efficient pretreatment various agents/catalysts as well as combinations of individual pretreatments are often used to enhance delignification of LCB. The second step is more or less conserved as a biological hydrolytic agent i.e., enzyme-driven process. Due to that, the operational variations are less compared to the pretreatment options. However, without proper conduction of pretreatment and hydrolysis, no LCB can be fermented to produce ethanol.

Table 7.1 Processing and conversion of various feedstocks for bioethanol production in 2G platform

Feedstock	Country of use	Raw LCB content	Pre-processing		Fermentation		Yield/ concentration
			Pretreatment/chemicals	Hydrolysis	Microorganisms	System	
Rice straw	Brazil (55)	35.3 ± 0.2% Cellulose, 23.8 ± 0.4% hemicellulose, 13.1 ± 0.7% acid-insoluble lignin, 4.4 ± 0.2% acid soluble lignin	NaOH, H_2SO_4	Dilute acid hydrolysis	*Kluyveromyces marxianus* NRRL Y-6860	SSF 125-mL Erlenmeyer flasks as reactors	20.4 g/L
	China (56)	Cellulose 38.82 ± 0.32%, hemicellulose 27.59 ± 0.81%, lignin 19.55 ± 0.37%	KOH, $CO(NH_2)_2$	Enzymatic hydrolyis	*Saccharomyces cerevisiae* Y2034	–	6.98 g/L
	China (57)	Cellulose 40.21%, hemicellulose 22.35%, lignin 17.15%	NaOH, H_2SO_4	Enzymatic hydrolysis	*Saccharomyces tanninophilus* CICC 1770	37 °C, 150 rpm	9.45 g/L
	India (58)	Cellulose 37 ± 0.4%, hemicellulose 24 ± 0.46%, lignin 28.8 ± 2.6%	NaOH, KOH, $NaClO_2$	Enzymatic hydrolysis	*Pichia stipitis* NCIM 3498, *Saccharomyces cerevisiae* OBC14	30 ± 0.5 °C, 150 rpm	26.12 ± 1.24 g/L
	India (59)	–	NaOH, H_2SO_4	Alkali/ enzymatic hydrolysis and acid/ enzyme hydrolysis	*Saccharomyces cerevisiae*	–	10 g/L
	India (60)	Cellulose 43%, hemicellulose 32% Lignin 25%	Na_2CO_3	Enzymatic hydrolysis	*Saccharomyces cerevisiae*	30 °C, pH 6.0, 150 rpm	18.07 g/L
	Brazil (61)	–	NaOH	Enzymatic hydrolysis	*Saccharomyces cerevisiae*	30 °C, 40 rpm	7.0 mg/mL
Banana peel	Mexico (62)	Cellulose 23.37%, hemicellulose 12.33%, lignin 12.33%	H_2SO_4	Enzymatic hydrolysis	*Kluyveromyces marxianus*	Rotatory shaker at 42 °C and 150 rpm	21 g/L
	India (63)	Cellulose 28.67%, hemicellulose 18.4%	–	Enzymatic hydrolysis	*Klebsiella* sp. SWET4	40 °C and pH 8	5.48 g/L

(continued)

Table 7.1 (continued)

Feedstock	Country of use	Raw LCB content	Pre-processing		Fermentation		Yield/ concentration
			Pretreatment/chemicals	Hydrolysis	Microorganisms	System	
	India (64)	Cellulose 15.14%, hemicellulose 11.32%, lignin 6.40%	H_2SO_4, ultrasonication	Enzymatic hydrolysis	*Saccharomyces cerevisiae*	37 °C with an agitation of 100 rpm	4.24 g/L
Wheat straw	Greece (65)	Cellulose 31.6%, xylan 24.4%, arabinan 3.7%, acid-insoluble lignin 19.1%, acid-soluble lignin 1%	H_2SO_4, ethanol, methanol, butanol, acetone, 2,2 oxydi(ethan-1-ol)	Enzymatic hydrolysis	*Saccharomyces cerevisiae*	100 mL Erlenmeyer flasks, 37 °C, pH 5.0	32.59 g/L
	India (66)	Cellulose 46.63 ± 4.16%, hemicellulose 27.47 ± 1.96% Lignin 15.20 ± 1.61%	H_2SO_4	Enzymatic hydrolysis	*Pichia stipitis* NCIM 3498	250 mL flasks, 30 °C	5.29% (v/v)
	China (67)	Cellulose 31.6%, hemicellulose 16.2%, lignin 21.2%	H_3PO_4, H_2O_2	Enzymatic hydrolysis	*Saccharomyces cerevisiae*	10 mL vials as reactor, 38 °C, 170 rpm	71.2 g/L
	Pakistan (68)	Cellulose 45%, hemicellulose 21.30%, lignin 19%	NaOH	Enzymatic hydrolysis	*Sacchromyces cervisiae*	30 °C, under anaerobic conditions	0.497 g/g
	China (69)	–	Subcritical water pretreatment	Enzymatic hydrolysis	*Saccharomyces cerevisiae* NX11424	30 °C, 150 rpm	37 g/L
	China (70)	Cellulose 42.8%, hemicellulose 25.0%, lignin 19.2%	NaOH, Na_2CO_3	Enzymatic hydrolysis	*Saccharomyces cerevisiae* SR8u	SSCF (Simultaneous saccharification and co-fermentation), 37 °C, pH 5, 200 rpm	31.1 g/L

(continued)

Table 7.1 (continued)

Feedstock	Country of use	Raw LCB content	Pre-processing		Fermentation		Yield/ concentration
			Pretreatment/chemicals	Hydrolysis	Microorganisms	System	
Forestry sources *Nothofagus pumilio* (40–60 years)	China (71)	45.0% Cellulose, 33.2% hemicellulose, 17.0% lignin	Ionic liquids: 1-ethyl-3-methylimidazolium-chloride, C2mimCl, 1-ethyl-3-methylimidazolium-acetate, C2mimOAc	Enzymatic hydrolysis	*Saccharomyces cerevisiae* Ethanol Red	SHF, SSF-reactor 250 mL Erlenmeyer flask, 40 °C, 250 rpm	0.173 g/g
Eucalyptus globulus Labill of 15 years	(71)	47.3% Cellulose, 39.3% hemicellulose, 17.3% lignin	Ionic liquids: 1-ethyl-3-methylimidazolium-chloride, C2mimCl, 1-ethyl-3-methylimidazolium-acetate, C2mimOAc	Enzymatic hydrolysis	*Saccharomyces cerevisiae* Ethanol Red	SHF, SSF—250 mL Erlenmeyer flask, 40 °C, 250 rpm	0.194 g/g
Betula pendula L.	Northern Sweden (72)	34.7% w/w Cellulose, 31.2% w/w Hemicellulose, 18.7% w/w Lignin	Ethanol, steam explosion	Enzymatic hydrolysis	*Saccharomyces cerevisiae* Ethanol Red	250 mL fasks, 35 °C, 180 rpm	80 g/L
Chinese fir sawdust	(73)	Cellulose 41.0%, hemicellulose 14.4%, lignin 36.5%	H_2SO_4	Enzymatic hydrolysis	*Saccharomyces cerevisiae*	100-mL glass flasks, 30 °C, 150 rpm	64.6 g/L
Barley straw	Spain (74)	Glucan 34.5 ± 0.04%; hemicelluloses 27.79 ± 0.39%; acid-insoluble lignin 17.22 ± 0.22%; acid-soluble lignin 2.45 ± 0.08%	Mechanical grinding chemical pretreatment-NaOH	Enzymatic hydrolysis	*Trametes villosa*	100 mL Erlenmeyer flasks	29.3%
	Spain (75)	32.9% Glucan, 26.1% hemicellulose, 18.8% total lignin, 10.4% extractives, 3.9% ash	NaOH	Enzymatic hydrolysis	*Saccharomyces cerevisiae* (co-fermenting yeast Cellu X™ 4	30 °C and 150 rpm	38 g/L

(continued)

Table 7.1 (continued)

Feedstock	Country of use	Raw LCB content	Pre-processing		Fermentation		Yield/ concentration
			Pretreatment/chemicals	Hydrolysis	Microorganisms	System	
	Korea (76)	–	NaClO	Enzymatic hydrolysis	*Saccharomyces cerevisiae* (KCCM 11304)	250 mL conical flask, 30oC, pH 7 ± 0.2, 120 rpm	370 mg/g
Cornstalk	Bangladesh (77)	Hemicellulose-33.55 ± 0.76, cellulose-38.70 ± 3.44, lignin-18.94 ± 2.98	NaOH	Enzymatic hydrolysis	*Saccharomyces cerevisiae*	250 mL flask, 30 °C, pH 5	43.8 g/L
	Bangladesh (78)	Hemicellulose-33.55 ± 0.76%, cellulose-38.70 ± 3.44%, lignin-18.94 ± 2.98%	CHW (compressed hot water)	Enzymatic hydrolysis	*Saccharomyces cerevisiae*	250 mL flask, 30 °C, pH 5	31.06 g/L
	China (79)	Cellulose-35.6 ± 0.3%, hemicellulose-28.3 ± 0.9%, lignin-19.5 ± 0.1%	NaOH	Enzymatic hydrolysis	*Saccharomyces cerevisiae*	30 °C, pH 5, 150 rpm	46.87 g/L
Corn stover	Nigeria (80)	Cellulose-72.87%, lignin-8.16%	NaOH	Enzymatic hydrolysis	*Saccharomyces cerevisiae*	SSF	0.045 L/kg
	China (81)	–	H_2SO_4	Enzymatic hydrolysis	*Saccharomyces cerevisiae* MT 3013	SSF, 37 °C, 180 rpm	3.53t/h
	China (82)	Glucan 59.6%, xylan 9.1%, arabinan 0.5%, lignin 26.2%, Ash 5.6%	SO_2	Enzymatic hydrolysis	*Saccharomyces cerevisiae* Ethanol Red	SSF, 118 mL glass bottles, 37 °C	76.8 g/L
	China (83)	–	Ethylenediamine	Enzymatic hydrolysis	*Saccharomyces cerevisiae*	SSF, Erlenmeyer flasks, 38–42 °C, 170 rpm	59.8 g/L

(continued)

Table 7.1 (continued)

Feedstock	Country of use	Raw LCB content	Pre-processing		Fermentation		Yield/ concentration
			Pretreatment/chemicals	Hydrolysis	Microorganisms	System	
	China (84)	Glucan 39.6% ± 0.7%, xylan 23.2% ± 0.3%, arabinan 3.5% ± 0.2%, acid-insoluble lignin 17.7% ± 0.4%, acid-soluble lignin 0.7% ± 0.1%	Ethylenediamine	Enzymatic hydrolysis	*Saccharomyces cerevisiae* SyBE00	250 mL Erlenmeyer flasks, 30 °C, 250 rpm	48.0 g/L
Rice husk	USA (85)	Cellulose 36.8%, Hemicellulose 19.7%, lignin 35.33%	H_2SO_4	Enzymatic hydrolysis	*Saccharomyces cerevisiae* NCYC 2826	SSF, 25 °C	7.55 ± 1.00%DW
	Nigeria (86)	–	$FeCl_3$	Enzymatic hydrolysis	*Saccharomyces cerevisiae*	SSF, 35 °C, pH 5, 50 rpm	3.802 ± 0.04%
	Japan (87)	Cellulose-29–36%, hemicellulose-12–29%, lignin-15–20%, ash-17–19%	NaOH	Enzymatic hydrolysis	*Escherichia coli* KO11, *Saccharomyces cerevisiae* NBRC 0224	Culture flask, 37 °C, 130 rpm	2.5 mg/mL
	Turkey (88)	Cellulose 35.86%, hemicellulose 18.2%, lignin 24.52%	H_2SO_4	Enzymatic hydrolysis	*Pichia stipitis* ATCC 58785	250 mL flasks, 37 °C, pH 5.6–6.2, 150 rpm	3.54 g/L
Sugarcane bagasse	Brazil (89)	Cellulose 38.72%, hemicellulose 23.01%, lignin 16.85%, extractives 5.27%, ash 3.72%	Imidazole solution	Enzymatic hydrolysis	*Saccharomyces cerevisiae* LPB 2705	125 mL Erlenmeyer flasks, 35 °C, pH 5.5, 150 rpm	10.0 ± 1.4 g/L

(continued)

Table 7.1 (continued)

Feedstock	Country of use	Raw LCB content	Pre-processing		Fermentation		Yield/ concentration
			Pretreatment/chemicals	Hydrolysis	Microorganisms	System	
	Brazil (90)	Cellulose 38.59 ± 3.45%, hemicellulose 38.59 ± 3.45%, lignin 17.79 ± 0.62	H_2O_2	Enzymatic hydrolysis	*Saccharomyces cerevisiae* PE-2, *Saccharomyces cerevisiae* CAT-1, *Kluyveromyces marxianus* CE025, *Kluyveromyces marxianus ATCC 36907*	125 mL Erlenmeyer flasks, 2L stirred tank reactor, 40 °C, pH 5.5, batch and fed-batch mode	29.45 g/L
	China (91)	Glucan-41.86 ± 0.38%, xylan-28.45 ± 0.49, lignin-20.07 ± 0.02%	NaOH	Enzymatic hydrolysis	*Saccharomyces cerevisiae* Y2034	30 °C, 150 rpm	22.37 g/L
	Brazil (92)	Cellulose-40.15%, hemicellulose-25.52%, lignin-26.17%, ash 8.16%	Hydrodynamic cavitation assisted NaOH	Enzymatic hydrolysis	*Scheffersomyces stipitis* NRRL-Y7124	SSF, Bubble column reactor, 30–50 °C, air flowrate at 0.26 vvm	17.26 g/L
Sweet sorghum bagasse	Thailand (93)	Cellulose 41.75 ± 0.25%, hemicellulose 25.40 ± 1.46%, lignin 10.88 ± 1.04%, ash 1.00 ± 0.05%	NaOH, H_2O_2	Enzymatic hydrolysis	*Saccharomyces cerevisiae*	SHF, 37 °C, 150 rpm	28.3 g/L
	China (94)	Cellulose 36.476 ± 0.089%, hemicellulose 21.927 ± 0.196%, acid-insoluble lignin 07.373 ± 0.050%, acid-soluble lignin 06.603 ± 0.135%, total lignin 13.98 ± 0.19%	NaOH, H_2O_2	Enzymatic hydrolysis	*Saccharomyces cerevisiae* CICC1308	36 ± 0.5 °C, pH 5, 100 rpm	19.332 ± 0.085 g/L

(continued)

Table 7.1 (continued)

Feedstock	Country of use	Raw LCB content	Pre-processing		Fermentation		Yield/concentration
			Pretreatment/chemicals	Hydrolysis	Microorganisms	System	
	Nigeria (95)	Starch 5.14a ± 0.54%, lignin 18.40a ± 0.3%	NaOH, H_2SO_4	Enzymatic hydrolysis	*Pichia tannophilus* NCYC614	SHF, 250 mL glass bottle, 32 °C, 130 rpm	23 g/L
	South Africa (96)	Cellulose-38 ± 5.09%, hemicellulose-19.6 ± 2.95%, Lignin-18 ± 0.03%, Ash-1.9 ± 0.05%	SO_2	Enzymatic hydrolysis	*Saccharomyces cerevisiae* MH1000	SSF, 5-L glass-jacketed bioreactor, 35 °C, pH 5.0	43.6 g/L

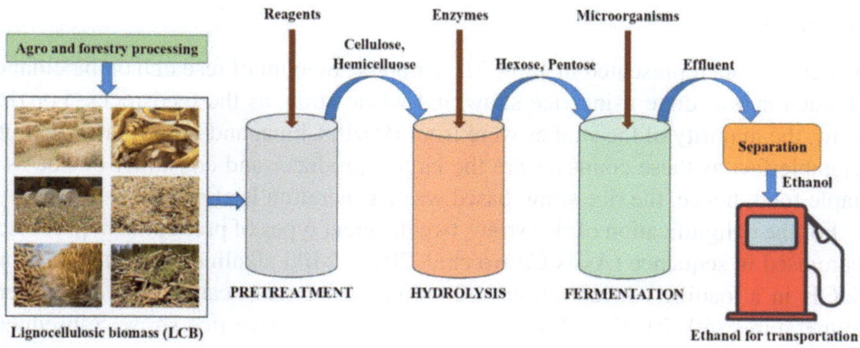

Fig. 7.2 Unit operations of the second-generation (2G) ethanol production technology

7.3.1 Pretreatment of LCB Feedstocks

The pretreatments of LCBs are commonly initiated with the reduction of the overall and particle size of the actually produced ones, i.e., straw, stalk, husk/hull, stem, etc. to smaller size like powdered form. For the size reduction, mechanical methods like grinding, milling, or extrusion are used to bring down the particle size in the range 0.1–4 mm. Afterwards, this powdered feedstock is mixed with the particular chemical agents and pretreatment is carried out under specific physical conditions to alter the LCB composition of the feedstock. The major aim of the pretreatment is to remove lignin and retain the polysaccharides in the remaining matrix (Gautam et al. 2024; Singh et al. 2023; Ghosh et al. 2017b). However, the selective removal of lignin also varies along with the method of pretreatment, and some amount of polysaccharide is degraded during pretreatment. Particularly, the hemicellulose and the amorphous fraction of the cellulose are affected during pretreatment. Removal of the mass fraction of the lignin and hemicellulose from the native LCB matrix creates pores and cracks over the surface of the feedstock and facilitates better binding of the hydrolytic enzymes in the subsequent step. In the following sections, the various pretreatment options are discussed for the different feedstocks considering common and unique types of pretreatments used, along with the details of the hydrolysis and fermentation systems Table 7.1 (Assis Castro et al. 2017; Wang et al. 2021, 2019; Jin et al. 2020; Banoth et al. 2017; Aziz et al. 2023; Singh and Kumar 2020; Lamb et al. 2018; Palacios et al. 2017; Sarkar et al. 2020; John et al. 2020; Salapa et al. 2017; Prasad et al. 2018; Qiu et al. 2017; Irfan et al. 2016; Chen et al. 2021; Yuan et al. 2018; Lienqueo et al. 2016; Matsakas et al. 2018; Ouyang et al. 2019; Coniglio et al. 2020; Duque et al. 2020; Sheikh et al. 2015; Rabeya et al. 2021; Adekunle et al. 2020; Zhao et al. 2019; Mensah et al. 2021; Zhang et al. 2020; Molaverdi et al. 2021; Zhu et al. 2020; Qin et al. 2018; Wu et al. 2018; Madu and Agboola 2018; Tabata et al. 2017; Germec et al. 2016; Valladares-Diestra et al. 2021; Araujo Guilherme et al. 2019; Hilares et al. 2017; Thanapimmetha et al. 2019; Cao et al. 2016; Nasidi et al. 2015; Bedzo et al. 2022).

Rice straw

As per the data represented in Table 7.1, a notable amount of research on bioethanol production was done using rice straw and wheat straw as the feedstocks. For rice straw, the majority of the studies were from Brazil, China, and India. This is a quite relatable fact as these countries are the largest producer and consumer of rice as a staple food; hence, the rice straw-based waste generation is also huge.

For the delignification of rice straw, two different types of pretreatment processes were used in sequence (Assis Castro et al. 2017). Mild alkaline pretreatment using NaOH in a loading concentration of 20–80 mg/g biomass carried out at different temperatures (50–70 °C) led to deacetylation of the native rice straw. Subsequent dilute acid pretreatment of the deacetylated rice straw with varying mild sulphuric acid concentration (0.5–1.5% w/v) for 30–90 min at 121 °C facilitates further lignin removal. The enzymatic hydrolysis of pretreated rice straw biomass under specific conditions of enzyme loading and cellulignin concentration, pH, temperature, and agitation using commercial enzyme cocktail (Cellic CTec2, Novozymes) leads to release of simple sugars (pentose and heoxe). Deacetylated biomass released higher amount of sugar after acid treatment compared to the only acid treated rice straw. Particularly, the deacetylation had a strong impact on the hydrolysis of hemicellulose and xylose enrichment of the hydrolysate (70 g/L) which led to maximum production of ethanol (20.4 g/L) during fermentation (Assis Castro et al. 2017).

Low temperature (< 100 °C) pretreatment of rice straw using mixture of KOH and urea also facilitates effective lignin removal and change in the crystallinity of cellulose and enhances release of sugar (Wang et al. 2021). KOH alone can also act as a delignification agent, but the effectiveness of combination of KOH and urea is more. In case of sole KOH pretreatment, a preprocessing of the straw by steam explosion at high temperature (170 °C) enhances the effectiveness of lignin removal and hydrolysis efficiency of the enzymes in the following step (Banoth et al. 2017). Oxidative method using sodium chlorite ($NaClO_2$) as the reagent is also used for rice straw pretreatment; however, the delignification is better for NaOH or KOH compared to $NaClO_2$ indicating the more effectiveness of the alkali pretreatment (Banoth et al. 2017). Commercial cellulase and xylanase mediated hydrolysis releases good amount of reducing sugar from the chemical pretreated rice straw for ethanol production.

Sodium carbonate (Na_2CO_3) which is also an alkaline reagent is used for pretreatment of rice straw as it also facilitates delignification. Due to the alkalinity, it has deacetylation property which cleaves glycosidic and ester bonds in the lignocellulose matrix and modifies the structure by solubilization of lignin and hemicellulose. In addition to lignin removal, Na_2CO_3 also facilitates silica removal from the polysaccharide part leading to better release of sugars from hydrolysis (Singh and Kumar 2020). Chemical pretreatments using acidic/alkaline reagents for solubilization of a fraction of lignin are most commonly used. Either the LCB is treated with mild acid or alkali in a single step or sometimes a combination of both in two steps is done. Generally, the acid mediated pretreatment is carried out under high value of temperature (> 100 °C). Sulphuric acid in dilute form is commonly used in acid pretreatment. Rice husk pretreatment with dilute sulphuric acid (0.5–2.5%) leads to better hydrolysis and ethanol production (Lamb et al. 2018).

Banana peel

Banana peel is considered as a suitable feedstock for ethanol production in Mexico and India (Palacios et al. 2017; Sarkar et al. 2020; John et al. 2020) due to its low lignin content and high polysaccharide content. Simple acid pretreatment of the mechanically ground banana peels with sulphuric acid at mild concentrations (0.5–1%) releases the majority of the hemicellulose fraction and a fraction of the lignin. Milling of the banana peel increases the reactive surface and facilitates more amount of hemicellulose fractionation. Mild acid pretreatment does not affect the lignin content in milled or pulverized banana peel (Palacios et al. 2017). Higher acid concentration, i.e., 2% combined with ultrasonication is more effective in delignification of banana peel. This combination selectively removes lignin and retains the polysaccharides in the biomass leading to better hydrolysis. Acid-assisted ultrasonication is more effective than sole acid pretreatment for banana peel due to the combined attack of the physical and chemical agents on the LCB. The sonication leads to formation of macro-radicals which facilitates bond breaking and enhances the acid mediated depolymerization of the LCB matrix.

Wheat straw

Like other feedstocks, alkali treatment with NaOH is also effective in delignification of wheat straw (68). Pretreatment with 2.5% NaOH under autoclaving condition increases the crystallinity index of wheat straw indicating removal of amorphous parts containing hemicellulose and lignin. Alkali is also used as reagents for pretreatment in two-step processes. NaOH and Na_2CO_3 mediated pre-extraction of wheat straw followed by addition of H_2O_2. The first step removes silica from the LCB matrix which enhances the efficiency of lignin and hemicellulose degradation. In this case, NaOH is a better chemical agent for delignification than Na_2CO_3 due to its higher alkaline potential (70).

Various solvents having different physicochemical properties could degrade the LCB matrix under specific process conditions and known as organosolv process. In the presence of a reaction catalyst (acid/alkali/salt), the solvents exhibit better solubilization of lignin than without catalysts. Wheat straw pretreatment using different solvents, i.e., ethanol, methanol, butanol, acetone, and diethylene glycol under high temperature (150–180 °C) shows effective delignification (65). Acetone and ethanol are better solvents for lignin removal compared to others. Methanol is not a good option of solvent to conduct organosolv pretreatment as it has low lignin solubilization potential. Partial hydrolysis of lignin monomer bonds and lignin-carbohydrate bonds is the major reaction responsible for selective solubilization of lignin (65).

Acid treatment combined with microwave heating is another effective pretreatment for wheat straw. In terms of sugar recovery, a lesser concentration of sulphuric acid, i.e., 2% is more effective when combined with microwave heating at 180 °C than 4% acid and 220 °C heating. At higher acid concentration and higher temperature, the sugars are converted to 5-hydroxymethylfurfural and furfural, which reduces the efficacy of fermentation (66). A different pretreatment process with mixture of phosphoric acid and hydrogen peroxide (PHP) also effectively removes lignin and hemicellulose from wheat straw. A highly concentrated solution (80% w/w) of phosphoric acid mixed with hydrogen peroxide (30% w/w) is used for pretreatment. Due to the self-exothermic nature, this pretreatment is operated at mild temperature (40.2 °C) unlike the high temperature pretreatment conducted with mild sulphuric acid. Lignin removal as high as 74.3% is obtained through this combined chemical pretreatment which enhances the polysaccharide hydrolysis. Phosphoric acid facilitates hemicellulose solubilization. Oxidative attack by peroxy-acids (e.g. peroxyacetic acid) and hydroxonium ions (HO^+) produced within the system during pretreatment are responsible for the delignification (67).

A chemical-free process, i.e., subcritical water pretreatment, is also effective in breaking the association of lignin and polysaccharides in the LCB matrix. Mainly, this pretreatment solubilizes the majority of hemicellulose fraction and partial fraction of lignin and preserves the cellulose (69). After the polysaccharides are removed by hydrolysis, the lignin can be recovered as a side product.

Forestry sources

Waste LCB produced during the processing of forestry and wood-based resources is also utilized as feedstocks for ethanol production in the countries where they are abundantly available. Forestry wastes from the mature plants of *Nothofagus pumilio* and *Eucalyptus globulus Labill* are considered as good feedstock sources. Ionic liquid pretreatment with 1-ethyl-3-methylimidazolium-chloride under high temperature of 150 °C enhanced the saccharification efficiency. The ionic liquid exhibited 'structure-disruptive' impact on the LCB matrix of the forestry wastes through breaking of inter- and intra-molecular hydrogen bonding leading to lignin removal (71). Bark free wood chips of silver birch pretreated through combined steam explosion and organosolv method. Under processing temperature of 200 °C in the steam explosion reactor, ethanol as the solvent and sulphuric acid as the catalyst show high delignification (77–86%) of the birch biomass. This is due to the cleavage of lignin-carbohydrate bonding and breaking of aryl-ether bonds of lignin by the combined action of ethanol and acid under high temperature (72). The concentration of the ethanol and acid, biomass to solvent ratio, and pretreatment time are important factors influencing the delignification efficiency.

Dilute sulphuric acid (0.2–06% w/w) treatment under 130–160 °C temperature followed by sodium chlorite (10% w/w) treatment at milder temperature (75 °C) of Chinese fir sawdust is also an effective LCB pretreatment method. The first stage of acidic pretreatment removes a major fraction of hemicellulose and a small amount of lignin. The second step shows effective removal of the remaining hemicellulose and lignin from the first step pretreated biomass (Ouyang et al. 2019). Delignification and hemicellulose removal reaches maximum values at 0.4% acid concentration in the first step. Higher concentration of acid, i.e., 0.6% facilitates formation of pseudo-lignin which causes restriction in the action of acidic chlorite in the second step and reduces the lignin and hemicellulose removal efficiency of the second step. Hence, using optimum process conditions for the first step is very important regarding the effectiveness of the subsequent step for such stepwise methods.

Barley straw

Extrusion is a method which is combined with chemical pretreatment to facilitate delignification of biomass. Barley straw is a potential LCB feedstock in Spain and Korea and used for bioethanol production. An alkali-extrusion pretreatment with 6% NaOH at mildly high temperature 70 °C facilitates delignification prior to enzymatic hydrolysis (Coniglio et al. 2020). The hydrolysis process conducted using Extruded barley straw with crude cell-free extract of *T. villosa* LBM 033's shows good saccharification yield from holocellulose (cellulose plus hemicellulose). A 29.3% release of total reducing sugar corresponding to 4.39 ± 0.05 mg/mL of sugars indicates the high saccharolytic efficiency of this microorganism which is beneficial for bioethanol production from barley straw (Coniglio et al. 2020). A higher alkali concentration (7.2% w/v) and maintenance of reaction temperature to 100 °C leads to more delignification and enhances the hydrolysis towards maximum. In the extrusion-based pretreatment, the mechanical action of the extruder blades exposes the interior parts of the compact LCB matrix to the chemical agent present in the system leading to more reaction. As a result, more amount of LCB biomass can be delignified by the chemical attack (Duque et al. 2020).

Combined chemical pretreatment with sodium hypochlorite (NaClO) and H_2O_2 of barley straw at 121 °C is also effective (Sheikh et al. 2015). Weight loss of barley straw biomass after pretreatment clearly indicates this. The weight loss of the biomass increases along with increase in concentration of NaClO and H_2O_2 mixture from 0.5 to 3% due to partial degradation of hemicellulose and lignification at higher concentrations. At high temperature, the reaction between NaClO and H_2O_2 forms several intermediates like singlet oxygen, superoxide, and hydroxyl radicals which have strong oxidative properties. Particularly, superoxide and hydroxyl radicals exhibit oxidative attack and causes degradation of lignin and polysaccharides by opening aromatic rings, cleaving aliphatic side chains, and ester bonds in the LCB matrix.

Cornstalk

Cornstalks from Bangladesh and China were pretreated using various chemical methods to enhance bioethanol production. Standard pretreatment of cornstalk with 0.5–2.5% NaOH at 70–100 °C shows delignification of the LCB matrix. However, 2% NaOH concentration is considered more suitable than 2.5%, as at this high concentration generation of lignin derived toxic chemicals are highest (Rabeya et al. 2021). Another physicochemical pretreatment used for cornstalk biomass is compressed hot water treatment using specially designed reactor. The pretreatment is carried out at 1 MPa pressure under nitrogen environment and varied temperature, i.e., 100, 120, 140, 160, 180, and 200 °C. Combined impact of temperature and pressure is responsible for breaking the structure of LCB. For cornstalk biomass, 180 °C is ideal as at higher temperature toxic compound generation becomes very prominent (Adekunle et al. 2020).

Comparison of multiple chemical pretreatments of cornstalk reveals important information reading suitability of a particular chemical as pretreatment agent. Individual chemical agents, i.e., acid (2% H_2SO_4, 2% HCI), alkali (2% NaOH, 2% KOH and 2% Na_2CO_3), oxidizing agent (2% H_2O_2), and combined chemical agents, i.e., acid combined with oxidizing agent (2% H_2SO_4 + 2% H_2O_2, 2% HCI + 2% H_2O_2) and alkali combined with oxidizing agent (2% NaOH + 2% H_2O_2, 2% KOH + 2% H_2O_2, 2% Na_2CO_3 + 2% H_2O_2) were used for the comparison of pretreatment efficiency. All the chemical pretreatment causes various amounts of lignin removal from the cornstalk biomass. Regarding delignification of cornstalk, the alkaline pretreatment with 2% Na_2CO_3 is least effective and combined pretreatment with alkali and oxidizing agent (2% NaOH + 2% H_2O_2) is most effective. This primarily indicates the specific interaction and impact a chemical agent might have on a particular biomass with specific LCB composition and variability of reactive action of different chemicals on the same biomass at same values and same reaction environment (Zhao et al. 2019).

Corn stover

Alkaline pretreatment of corn stover with 0.5–4% of NaOH at 121 °C shows a directly proportional relation between the extent of delignification and increasing load of concentration of NaOH. The rate of delignification follows first-order reaction kinetics at temperatures higher than 100 °C as the speed of the reaction depend mainly on the loading concentration of the alkali. The pretreatment increases the cellulose content in the pretreated biomass. Regarding the retention of the cellulose content, loading of alkali is a more influential parameter than the pretreatment time which has a minor effect. However, if the severity of pretreatment condition is increased, then chances of precipitation of the solubilized lignin over the surface of the pretreated biomass increase which might reduce the efficiency of the subsequent enzymatic hydrolysis (Mensah et al. 2021). The same pretreatment leads to higher lignin removal in cornstalk biomass than corn stover indicating the more complex structure of the latter.

Acid catalysed delignification is also attempted for pretreatment of corn stover with 2% (v/v) H_2SO_4 at 120 °C which leads to successful production of ethanol (Zhang et al. 2020). Corn stover pretreatment with sodium carbonate (Na_2CO_3) 100 °C also selectively removes lignin from the polysaccharides of the LCB matrix. A pre-step of SO_2 gas catalysed steam explosion treatment conducted at 190 °C further enhances the delignification by Na_2CO_3 and increases the glucan content. The combination of SO_2-catalysed steam and Na_2CO_3 removes the xylan-fraction of the hemicellulose and lignin from the corn stover biomass and increases the glucan fraction which leads to efficient release of glucose during enzymatic hydrolysis. The release of xylan and lignin from the LCB matrix in the combined pretreatment also increases the total crystallinity index which is beneficial for ethanol production (Molaverdi et al. 2021).

Ammonia and ethylenediamine are also used as chemical pretreatment agents. Stepwise aqueous ammonia and ethylenediamine mediated pretreatment of corn stover is very effective in lignin removal from corn stover. The xylan degradation is notably higher in case of the two-step pretreatment in comparison to the individual pretreatments with ammonia and ethylenediamine. Changes in functional group indicate alterations in the structures of lignin, cellulose, and hemicellulose in the LCB matrix (Zhu et al. 2020; Qin et al. 2018). These chemicals are considered advantageous over others like acid or alkali as the pretreatment conditions are less severe regarding the temperature or pressure and also the sugar loss due to conversion to toxic byproducts. Retention of majority of cellulose (particularly the crystalline fraction) enhances the effectiveness of the enzymatic hydrolysis and release of glucose.

Rice husk

Various methods are used for rice husk pretreatment. Hydrothermal pretreatment of rice husk is effective in modifying the lignocellulosic structure. The severity of the process regarding the temperature and time of exposure is an important factor for the delignification and formation of byproduct inhibitors. At higher exposure time and values of temperature, the toxic substances like hydroxy-methyl furfural (5-HMF), furfural, and acetic acid formation are higher (Wu et al. 2018). Hence, the values must be chosen in such a manner at which delignification is maximum and inhibitor formation is within a tolerable limit. Thermochemical pretreatment of rice husk carried out with 0.1 M solution of different catalysts like HCl, NaOH, and $FeCl_3$ and 121 °C leads to delignification. The comparison revels that $FeCl_3$ is an effective catalyst regarding the release of fermentable sugars and production of inhibitory byproducts during pretreatment. The inhibitor formation is least from $FeCl_3$ compared to HCl and NaOH which makes this more suitable as the pretreatment catalyst (Madu and Agboola 2018).

Steam explosion carried at 235–236 °C temperature and 3–4 MPa pressure is also an effective pretreatment of rice husk delignification. Saturated steam-driven explosion pretreatment facilitates subcritical oxidation of organic or oxidizable inorganic compounds present in the LCB matrix at elevated temperatures and pressure leading to mechanical breakdown. This makes the pretreated biomass more susceptible to enzymatic hydrolysis and release of fermentable sugars (Tabata et al. 2017). Dilute acid (1–5% H_2SO_4) mediated pretreatment of rice husk is an effective method for rice husk fractionation. The highest fermentable sugar release is facilitated by 3% acid concentration, at higher concentration of acid inhibitory compounds are formed at higher amount (Germec et al. 2016). Other than the acid concentration, temperature, biomass to liquid ratio, and hydrolysis time are also having influence on the release of sugars form the LCB matrix of rice husk.

Sugarcane bagasse

Green solvent mediated pretreatment is effectively used for delignification of sugarcane bagasse. Imidazole, as the green solvent, is used as the delignification agent in the 120–160 °C temperature which is indicated by the decrease of recovered solid fraction after treatment. In the pretreated solid biomass, increase of cellulose content indicates active lignin removal. Along with lignin, imidazole also affects the hemicellulose present in the LCB matrix and releases it as indicated by the removal of OH groups. Hemicellulose removal and delignification both are higher at more severe process temperature. Imidazole facilitates total solubilization of the lignin and prevents lignin regeneration or pseudo-lignin formation, making the lignin not recoverable after solubilization unlike acid or alkali pretreatment. Selectivity of the cellulose retention is due to the favourable degradation of LCB at higher temperatures through the breakage of the ester bonds binding lignin and hemicellulose and destabilization of the hydrogen bonds among lignin, cellulose and hemicellulose (Valladares-Diestra et al. 2021).

Pretreatment of sugarcane bagasse with multiple chemical and physicochemical methods indicates the suitability of a particular process. Mixed acid and alkali, alkali catalysed hydrothermal, sole alkali, and hydrogen peroxide pretreatment are used for sugarcane bagasse. Sole alkaline pretreatment shows higher lignin removal compared to the other pretreatments. Good hemicellulose reduction is obtained along with lignin removal through the synergistic effects of the combined pretreatments. The combined acid–alkali pretreatment maximizes the cellulose content in the post-treatment biomass (Araujo Guilherme et al. 2019). Generally, the acid or alkali mediated pretreatments of LCB are carried out at high temperatures, i.e., 100 °C of higher. Low temperature alkaline pretreatment of sugarcane bagasse with NaOH as the chemical agent conducted at 30–80 °C shows good effect on the deconstruction of the lignin and polysaccharide association leading to notable enzymatic hydrolysis (Wang et al. 2019). Different alkalis, i.e., NaOH, KOH, Na_2CO_3, and $Ca(OH)_2$, are used as pretreatment agents for sugarcane bagasse biomass. Use of these alkalis in a hydrodynamic cavitation system enhances their performance. Pretreatment in the hydrodynamic cavitation reactor is carried out at 60 °C. All of the alkali shows various levels of delignification and solid recovery after pretreatment. Among all alkalis used, NaOH performs best in the hydrodynamic cavitation pretreatment of sugarcane bagasse. The post-pretreatment black liquor of the NaOH hydrodynamic cavitation pretreatment has the potential to be reused in a sequential sugarcane bagasse pretreatment process, which reduces the chemical consumption and cost of pretreatment (Hilares et al. 2017).

Sweet sorghum bagasse

Sweet sorghum bagasse is used for bioethanol production in many countries like Thailand, China, Nigeria, and South Africa. NaOH (10% w/v), H_2O_2 (30%w/v), and sequential H_2O_2 and NaOH catalysed pretreatment show effective delignification of sorghum bagasse. NaOH pretreatment is carried out at higher temperature (121 °C) than the H_2O_2 pretreatment (25 °C). NaOH pretreatment substantially increases the cellulose content in the pretreated solid fraction of the biomass due to removal of lignin and hemicellulose from the LCB matrix of the sorghum bagasse. The sequential H_2O_2 and NaOH pretreatment further enhances the delignification and hemicellulose degradation (Thanapimmetha et al. 2019). Alkaline H_2O_2 pretreatment carried out at different temperature severity, i.e., 35 and 121 °C leads to different degree of lignin removal from the sorghum bagasse. At lower temperatures, the acid-soluble lignin fraction is mostly removed. At higher temperature, the combined agents can remove a larger amount of acid-insoluble lignin from the LCB matrix as well as hemicellulose. Cellulose enrichment of the biomass produces a higher amount of glucose from the enzymatic hydrolysis (Cao et al. 2016). Mild acid pretreatment catalysed by 2% v/v H_2SO_4 at 121 °C is effective in delignification of sweet sorghum bagasse and enhances the hydrolytic saccharification of the pretreated biomass (Nasidi et al. 2015).

Steam explosion pretreatment of sweet sorghum bagasse using water and SO_2 carried out at high temperature and pressure (7.4 MPa) is effective in degradation of the LCB matrix. Water catalysed steam explosion at 205 °C facilitates lesser sugar recovery from the sweet sorghum bagasse compared to the SO_2-catalysed steam explosion at 185 °C. Although higher sugar recovery is achievable at lesser severity, SO_2 as the catalyst generates more amount of inhibitors than water as catalyst (Bedzo et al. 2022).

7.3.2 Enzymatic Hydrolysis of Pretreated Feedstocks

The liquid slurry containing the solubilized lignin, hemicellulose, cellulose and other by products derived after pretreatment is separated from the undissolved solid fraction of the pretreated LCB. The solid fraction is washed to remove the traces of all substances produced during the pretreatment and dried before subjecting to enzymatic hydrolysis. As two different types of polysaccharides, namely hemicellulose and cellulose, are present which have different monomeric components and bonding patterns, specific enzymes are required to depolymerize the polysaccharides (Houfani et al. 2020; Meshitsuka and Isogai 2017b; Wyman et al. 2005). Hemicellulose is a heteropolymer composed of multiple pentose/hexose sugars including xylose, arabinose, mannose, glucose, galactose, 4-O-methyl glucuronic acid, and galacturonic acid residues. Hemicelluloses obtained from different plants vary greatly. The main sugar monomer of the backbone, i.e., xylans, mannans, and glucans, characterizes the hemicellulose of a specific plant species. Hemicellulose has lower molecular weight than cellulose and has a non-crystalline amorph structure which is more vulnerable to hydrolysis by physical agents like heat. Due to this, the high temperature pretreatment often degrades the hemicellulose along with lignin and reduces the selectivity of degradation. The other polysaccharide cellulose is a homopolymer composed of only glucose monomers bonded through β-1,4-Glycosidic bonds in unbranched chains. The majority of cellulose is present in a crystalline form with a low hydration value, which makes the cellulose resistant to degradation by physical agents like heat and mild chemical agents (Houfani et al. 2020; Meshitsuka and Isogai 2017b; Wyman et al. 2005).

Hydrolysis of the cellulose and hemicellulose rich pretreated biomass is carried out by the use of hydrolytic enzymes like cellulase and hemicellulose (Houfani et al. 2020). Due to the various types of monomeric units present in the polysaccharides of the pretreated LCBs combined through different chemical bonds, it cannot be hydrolysed by a single enzyme to release the monomers. A multienzyme mixture is generally used as an enzyme cocktail to hydrolyse polysaccharides. Particularly, cellulose offers more resistance towards hydrolysis due to its crystallinity and larger molecule size than hemicellulose. Hemicellulose is composed of xylan, arabinan, mannan, β-glucan, galactan, xyloglucan, arabinoxylan, galactomannan, galacturonan, etc. Hemicellulase comprises xylanases, β-mannanases, β-xylosidase, glucuronidases, acetyl xylan esterases, arabinofuranosidases, etc. as the major hydrolytic enzymes. Altogether these enzymes act upon a specific internal or external part of the hemicellulose molecules and break the glycosidic bonds to liberate the corresponding monomeric sugars (Houfani et al. 2020; Andlar et al. 2018).

The enzymes required for degradation of cellulose are:

- **Endoglucanase**: Endo-1,4-β-glucanases (carboxymethyl cellulase) randomly breaks the β-1,4-Glycosidic bonds on the internal sites of the cellulose and release smaller polysaccharide molecules like cellodextrin.
- **Exoglucanase**: Exo-1,4-β-D-glucanases (cellobiohydrolases) acts on the exoglucanases release cellobiose or glucose from the non-reducing end of the crystalline cellulose polymer.
- **B-glucosidase**: β-D-glucosidase (β-D-glucoside glucohydrolase) acts on the non-reducing ends of cellobiose or cellodextrin to release D-glucose. β-glucosidases complete hydrolysis by turning cellodextrin and cellobiose into glucose.

The hydrolytic enzyme cocktails used for saccharification are obtained from potential microorganisms with cellulolytic or hemicellulose degrading ability like *Trichoderma reesei* or *Aspergillus niger*. In the majority of the second-generation bioethanol production, the commercially available purified hydrolytic enzyme cocktails are used. Sometimes, enzyme cocktails are produced by culturing the hydrolytic microorganisms and the crude extract is used as produced (Ghosh et al. 2017b). Enzymatic hydrolysis is performed by mixing the pretreated LCB biomass as the substrate and enzyme cocktails in a proper loading value (generally 15–60 FPU/g biomass). Hydrolysis is favoured at a lower pH value like 5–5.5 and temperature within the range 50–55 °C. It takes around 24–48 h to complete the hydrolysis. The hydrolysate is collected and used in the fermentation step for production of bioethanol.

7.3.3 Fermentative Conversion of Hydrolysate

The hydrolysate obtained from the enzymatic hydrolysis of the pretreated LCB contains pentose sugars like xylose, arabinose released from the hemicellulose and glucose released from the cellulose. The hydrolysate is filtered to remove the particulate matter and insoluble parts of the biomass and neutralized to a pH value of 7. This hydrolysate is added to the specific fermentation media as the carbon source and inoculated with the fermentative ethanol producing microorganisms. Table 7.1 clearly shows that the yeast *Saccharomyces cerevisiae* is the most commonly used microorganism for ethanol production. High ethanol productivity and tolerance towards ethanol concentration are the favourable traits of this microorganism making it the most popular ethanol producer. Regarding the strategy of hydrolysis and fermentation, two types of bioprocesses are followed for ethanol production.

- **Separate hydrolysis and fermentation (SHF)**: SHF is a method in which the enzymatic hydrolysis of the pretreated solid feedstocks and ethanol fermentation are performed in a two-stage manner, Fig. 7.3 (Hage et al. 2023). In SHF, hydrolysis is carried out till completion and then ethanol producing microorganisms are used to ferment the sugars of the hydrolysate in separate reactors. Due to separate reactors used in SHF, hydrolysis and fermentation occur at their corresponding optimal values of operating parameters like temperature and pH. An inherent problem of SHF is that in the hydrolysis reactor, the increasing concentration of glucose or cellobiose often inhibits further hydrolysis action of the enzymes. Use of two separate reactors for enzymatic hydrolysis (stage 1) and fermentation (stage 2) increases the cost of operation. Disadvantages of SHF are chances of contamination during handling of hydrolysate before fermentation, formation of inhibitor byproducts, and long span of operation of separate reactor systems (Hage et al. 2023).
- **Simultaneous saccharification and fermentation (SSF)**: SSF is the alternative of SHF which consists of a single reactor in which both feedstock hydrolysis and fermentation occur simultaneously, Fig. 7.3 (Hage et al. 2023). The process starts with addition of the pretreated feedstock, hydrolytic enzymes, and ethanol producing microbes together in the reactor. In SSF, the glucose and cellobiose generated by hydrolysis are consumed by the microorganisms for ethanol production alongside their production. A major disadvantage of SSF is the fact that optimum temperature of hydrolytic enzyme activity (50–55 °C) differs from the optimal value of temperature (35–40 °C) supporting the growth of the ethanol producing microbe and ethanol production only one of which can be maintained in a single reactor. This reduces the efficiency of the other process which is operated under sub-optimal conditions. Use of a single reactor reduces the chances of hydrolysate contamination, time of operation, and overall cost of operation (Hage et al. 2023).

The hydrolysate must be free from any type of fermentation inhibitors like hydroxymethyl furfural (5-HMF), furfural, phenolic derivatives, acetic acid, etc. Besides these, the acidic or alkaline agents used in preprocessing must be neutralized or removed from the hydrolysate. To remove these substances, the hydrolysates are conditioned or detoxified by a variety of methods including membrane separation, adsorption, etc. and the pH is set at 7. The presence of inhibitors reduces the rate of growth and ethanol production by the ethanol producing microorganisms. Depending on the type of sugars and byproducts present in the hydrolysate after conditioning and based on the metabolic potential of the microorganism used, the ethanol productivity and yield vary in different systems, Table 7.1. Ethanol concentration as high as 80 g/L is produced by the specific yeast strains like *Saccharomyces cerevisiae* Ethanol Red (Matsakas et al. 2018).

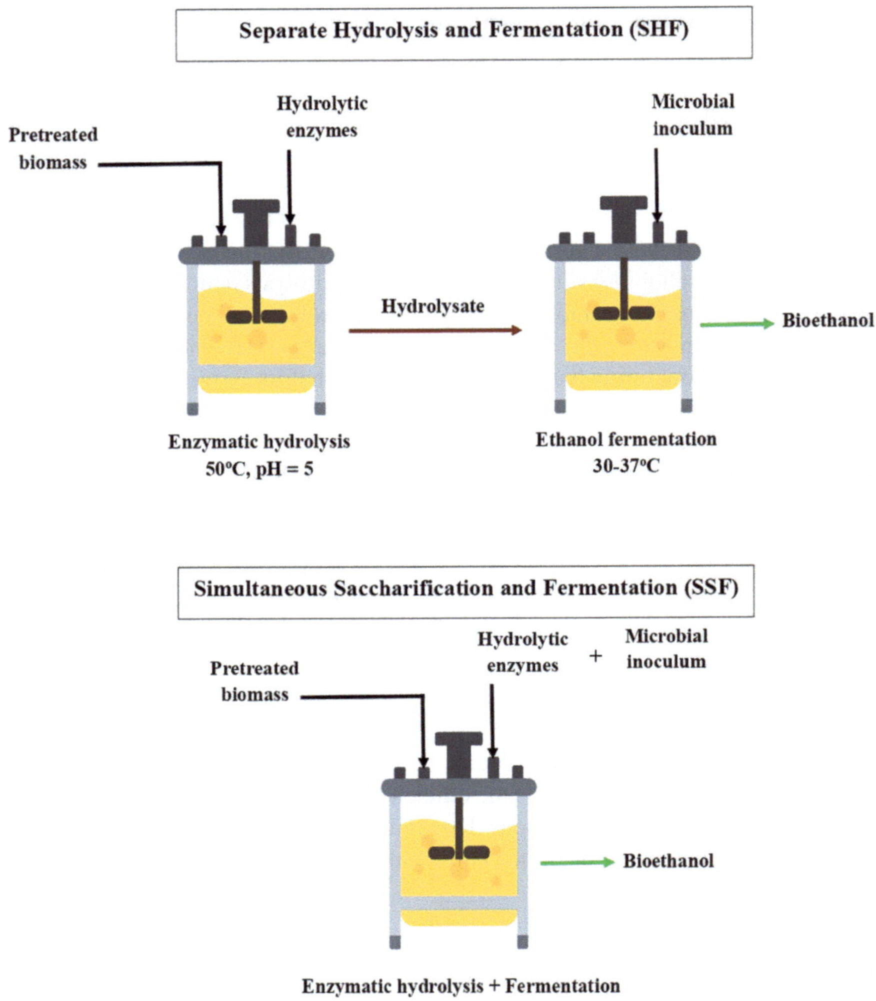

Fig. 7.3 Conversion of pretreated LCB to ethanol via SHF and SSF

7.3.4 Separation of Ethanol

After completion of the fermentation, the ethanol rich effluent is processed through specific separation method to extract ethanol from the highly diluted state (5–12 wt%) which cannot be used as a fuel. Distillation is one of the most commonly adopted separation processes for ethanol and also one of the most energy consuming separation process (Wang et al. 2024; Kiss 2014). For the separation of ethanol and water which acts as an azeotrope, high temperature processing is used in distillation. The requirement of high energy also increases the cost of separation and the produced ethanol to be used as fuel. Different types of alternative distillation methods like heat integrated distillation, membrane-based separation, pervaporation, etc. are designed to make the ethanol separation cost-effective and viable as a biofuel in the transportation sector (Wang et al. 2024; Kiss 2014; Khalid et al. 2019). The heat integrated distillation uses modified input of energy to the distillation system by recovering and using the exit heat from distillation columns to preheat the incoming feed. This technique is able to reduce the energy consumed for bioethanol production by $\approx$ 40%.

Membrane separation uses a mechanism based of higher partial pressure of ethanol in the ethanol–water mixture. Due to the difference, ethanol is transported to the membrane pores from the fermentation effluent. Pervaporation takes place under reduced pressure environment on the permeated side of the membrane, which facilitates enhanced transport of ethanol across the membrane due to the concentration gradient. Composition of the membrane (zeolites, polydimethylsiloxane, silicate, etc.) greatly influences the performance of this separation process. Combination of the membrane separation with vapour stripping is a better technique of ethanol separation from fermentation effluents. An ethanol content as high as 80 wt% can be achieved by this technique (Khalid et al. 2019). Combination of extractive and azeotropic distillation techniques is also effective in efficient recovery of ethanol from diluted effluents. Reduction in energy consumption compared to the traditional distillation is one advantage of the combined method. In the traditional fermentation systems, ethanol is present in a very diluted form in the fermentation broth. It only becomes suitable until it is converted to an anhydrous form through dehydration which is energy and cost intensive (Kumar et al. 2010; Vane 2008). As separation and purification is one of the major contributing steps to the overall cost of production, it must be minimized in order to make bioethanol a reliable biofuel for the transportation sector.

7.3.5 Existing Limitations of the Production Processes

Certain limitations can be outlined in the biochemical production of ethanol from lignocellulosic biomass feedstocks. In the preparative phase, the major limitation is caused by the presence of high amount of lignin in some feedstocks making their use difficult in the fermentation processes. Chemical pretreatments are material intensive (acid, alkali, other chemical agents) which increases the production cost. Also, the acidic or alkaline agents must be neutralized before the enzymatic hydrolysis or fermentation can take place, requiring more chemicals. On a large scale, the cost will become notably high. Biological/biochemical delignification is one cost-effective option, but the slow rate of lignin degradation limits its adoption for large-scale processes (Ghosh et al. 2017b). Conversion of fermentable sugars to toxic byproducts is another limiting factor associated with pretreatments of higher severity of temperature and heat. Detoxification of the pretreatment liquid adds additional cost input (Devi et al. 2021). For enzymatic hydrolysis, the cost of enzymes that are required for hydrolysis becomes much higher if they are purchased from the market in commercially available purified form. This cost can be reduced by the use of crude enzymes produced in house. However, it might not be feasible for every industrial plant for bioethanol production to produce large amounts of hydrolytic enzymes in house.

Saccharomyces cerevisiae, along with other standard industrial yeast strains, is the prime organism in biochemical ethanol production due to its high sugar tolerance ability, rapid growth, high fermentation rates, and substantial ethanol tolerance (Jetti et al. 2019; Dong et al. 2020). In both SHF and SSF, the monomeric sugar released from the pretreated LCB will be a mixture of pentoses and hexose which is a problem for *S. cerevisiae* due to its inability of pentose utilization. Xylose is the five-carbon sugar and besides glucose the second most abundant sugar in LCB derived hydrolysate. For this issue, ethanol producing strains belonging to other genera or microbial class is being used for ethanol fermentation. Table 7.1 shows that other than *S. cerevisiae*, ethanol producing yeasts namely *Kluyveromyces marxianus* and *Pichia stipitis* are used for ethanol fermentation from hydrolysate. These yeasts can use the pentoses like xylose, arabinose besides the hexoses and convert to ethanol. *Pichia stipitis*, also known as *Scheffersomyces stipitis*, can convert xylose to ethanol but suffers from low sugar and ethanol tolerance. Sometimes, a mixed culture of *S. cerevisiae* and one of these yeasts are used to maximize the conversion of all simple sugars present in the hydrolysate. Bacterial strains like *Klebsiella* sp. are also used in ethanol fermentation due to their higher ethanol tolerance and cellulolytic activity. The strains having cellulolytic and ethanol producing ability are potential microbes for the SSF type bioprocesses. Other than this, genetic modification of pentose non-fermenting strains with pentose metabolizing genes is effective in enabling pentose utilization. Genetic modification of *S. cerevisiae* with *Pichia stipitis* XYL1 xylose reductase and XYL2 xylitol dehydrogenase genes creates the recombinant strains of *S. cerevisiae*, YRH 396 and YRH 400 which has the xylose fermenting ability (Cortivo et al. 2018).

Due to its membrane solubilization property, ethanol is used as a common antimicrobial substance. This is an issue for its biochemical production as well. After its accumulation in the system beyond a certain value of concentration, it starts to inhibit its own production. Ethanol is produced as a soluble metabolite which acts as a stressor for its own producer. Ethanol at high concentration induces morphological changes, destabilization of cellular envelopes, structural protein damages, malfunction of organelles, and generation of reactive oxygen species which strongly inhibit the growth and metabolic activities of cells. Altogether these severely affect the physiological activity and fermentation performance and reduce the yield of ethanol. Although the ethanol producing microbes possesses, some degree of tolerance towards ethanol more tolerance would increase the margin of production. For that reason, many genetically modified strains are being developed with overexpression of the genes responsible for mediating the ethanol stress tolerance. These strains are capable of tolerating higher concentration of ethanol in the system and hence produce ethanol for an expanded period of time.

7.4 Present Use of Ethanol in Transportation Sector and Future Scope

Hydrous ethanol, characterized by a maximum oxygen concentration of 36.3%, presents a viable alternative to reduce engine knocking due to its higher octane number compared to gasoline. This property also enables a significant reduction in the emissions of pollutants such as particulate matter (PM), unburned hydrocarbons (HCs), and carbon monoxide (CO) (Sakthivel et al. 2018; Sydney et al. 2019; Paladugula et al. 2018). Internal combustion (IC) engine tested using predefined fuel blends, comprising gasoline and bioethanol in three different ratio (E20, E50, and E90) demonstrates that compared to gasoline, reduction as high as more than 50% can be achieved in the emission profiles of carbon monoxide and unburned hydrocarbons (HC) from the bioethanol-blended fuel mixtures. The nitrogen oxide content also reduces by over 30% when the E50 and E90 blends. Hence, the expected outcome of using ethanol in the transportation sector is quite achievable.

Among the various alcohols, ethanol is considered the best fuel for spark-ignition (SI) type engines. Ethanol has the highest octane number, flame speed, and improved thermal efficiency due to a higher compression ratio. It has higher heat of vaporization and broader flammability and shorter burning time compared to gasoline. These characteristics make it an excellent fuel for SI engines, either in pure form or blended with gasoline (Panoutsou et al. 2021). Three cylinders and four strokes SI engine operated with ethanol-blended gasoline with following ratio—E0 (100% gasoline), E5 (5% ethanol and 95% gasoline), E10 (10% ethanol and 90% gasoline), and E15 (15% ethanol and 85% gasoline) shows effectiveness of using ethanol. As ethanol-blended gasoline has a lower heating value per unit mass compared to neat gasoline, a greater amount of fuel is required to achieve the same energy output. Due to this, the total fuel consumption increases along with increase in the fraction ethanol in the blended fuel (E5 < E10 < E15). The emission of hydrocarbons and carbon monoxide in the exhaust gas decreases with an increase in ethanol fraction percentage in blended gasoline, while the emission of NO_X increased. Compared to unblended gasoline, use of ethanol gasoline blends i.e., E10 and E20 in SI engines with variable compression ratio of 4:1, 5.5:1, 7:1, 8.5:1 and 10:1 led to reduced emission of polluting gases. Ethanol-blended gasoline drastically reduces the hydrocarbon, CO, and NO emissions from engine run. At higher compression ratios carbon dioxide production increases. Brake specific fuel consumption in the E10 and E20 blends increased from 44 to 55% due to the less calorific value of ethanol. The requirement of a high amount of fuel is a potential drawback for blended gasoline if it is implemented in large scale, directly increasing costs and decreasing engine efficiency.

At present, many countries have mandated the blending of ethanol in the gasoline used in their transportation sector. Flex fuel engines have been developed to accommodate a higher fraction of ethanol in blended gasoline which is not suitable for the regular SI or IC engines. Brazil is the leading country in bioethanol production as well as implementation of it as the fuel for transportation sector. Using the flex-fuel type of vehicles, they have successfully adopted the use of 100% hydrous ethanol as fuel (Azhaganathan and Bragadeshwaran 2022). Next to Brazil, the USA is also using the flex-fuel vehicles run on E85 (85% ethanol:15% gasoline) blended fuel. Other countries like India, China, Europe, South Africa, etc. are using the E5–E25 blended fuel in their regular cars driven by conventional SI or IC engines. Ethanol blending in E5–E10 range does not require any modification of the design of the traditional engines. However, ethanol content in gasoline beyond 15–20 starts to reduce the efficiency of engines using blended fuel. Adoption of flex-fuel type engines is not very popular in a large part of world, unlike Brazil. Hence, biofuel policy and social acceptance plays vital role in the progress of implementation of the ethanol as a biofuel. Besides this, the high cost of production of the LCB-based bioethanol makes it less favourable as a competitive fuel to the fossil fuels. These factors affect the consistency of production and reliability of bioethanol in a negative manner. Due to the promising circular bioeconomy perspective of the LCB-to-bioethanol technology, it is expected that it will be implemented to the large-scale production systems in the near future (Baskar et al. 2024; Vidhya et al. 2024).

Overall ethanol has been well recognized as a potential clean fuel for combustion engines with low carbon emission characteristics which can notably contribute to renewable fuels-driven sustainable mobility (Agarwal and Valera 2022; Jena and Agarwal 2023; Prabakaran 2022; Sharma et al. 2023). Ethanol possesses the properties to become a dependable fuel for the sustainable transport system of the future (Sarıkoç 2022). Formation and implementation of new policies encouraging national bioethanol production efforts, developing successful prototypes to raise public awareness, and minimizing existing technical risks for adoption of ethanol as a reliable biofuel are required. Creation of a market incorporating ethanol as an integral part of primary transportation fuel mix is also essential. Globally, an effort has been initiated to increase the ethanol blending to at least 50% by the next decade in order to bring down the tailpipe emission notably. The ongoing progress in bioethanol production from small scale to larger scales is expected to support the implementation of ethanol as the major transportation fuel as per the projection.

7.5 Conclusions

This chapter introduces ethanol as a potential alternative fuel for the transportation sector and presents a detailed discussion on the current status of production through the biochemical processes. Various types of feedstocks for bioethanol production have been classified based on their native structure and the generation platform (1G or 2G) in which they are used. The present trend of usage clearly indicates that the feedstock preference for bioethanol production is majorly shifted from the edible 1G feedstocks to the non-edible LCB type 2G ones due to their promising sustainability aspects like abundance and renewability. The limitations existing in the 2G platform due to the inherent complexity of feedstock materials and multiple options of pretreatment methods used in small scales have been extensively discussed considering specific examples of major agro-wastes and forestry wastes. For large-scale production, only the pretreatment methods with lesser chemical, energy, and cost inputs will become viable options. Enzymatic hydrolysis of the pretreated LCB is also another decisive factor for large-scale 2G ethanol production due to the high cost of pure hydrolytic enzymes. From the economic viewpoint, the use of crude hydrolytic enzyme cocktails produced by specific microorganisms could be an alternative to reduce the cost of hydrolysis. In the fermentation step, the inability of the major ethanol producing microorganisms to consume and convert the C5 sugars is the biggest limitation. Microorganisms with the ability to ferment both C6 and C5 sugars are necessary to maximize the conversion of lignocellulose derived carbohydrates to ethanol. Genetic modification of the microorganisms or use of co-cultures are the strategies adopted to tackle this issue. The enzymatic hydrolysis and fermentation can be either performed in a sequential manner as SHF or in a combined manner as SSF. The SSF is economically better than SHF as it requires operation of a single reactor unit. However, separation of ethanol from fermentation broth remains a cost and energy intensive process if traditional distillation is adopted. Besides distillation, several other low-energy requiring processes like membrane separation are being attempted which might improve the ethanol separation in a cost-competitive manner and make the produced ethanol less costly. In the transportation sector, ethanol can be used in both IC and SI engines as blend-in fuel, but it suits more to the latter one as a fuel additive. However, the existing designs of the engines can only handle ethanol to a limited capacity or fraction which is not enough to bring down the emission significantly. Hence, engine modifications are also necessary to make the implementation of ethanol as the dependable blend-in fuel for the future transportation sector. Based on the overall progress of bioethanol production, it can be concluded that the technology will soon gain the readiness to be lifted towards commercial margins and make ethanol the reliable transportation fuel of tomorrow for majority of the countries.

Acknowledgements The corresponding author gratefully acknowledges the book editors for giving the opportunity to contribute this chapter to the book. The reviewers are gratefully acknowledged for providing their expert opinions and suggestions for the betterment of the chapter.

References

Abdellatief TM, Ershov MA, Kapustin VM, Chernysheva EA, Savelenko VD, Makhmudova AE et al (2024) A consensus novelty detection ensemble approach for formulating renewable gasoline-ethanol blends. Int J Hydrogen Energy 67:1308–1318

Abedinia O, Shakibi H, Shokri A, Sobhani B, Sobhani B, Yari M, Bagheri M (2024) Optimization of a syngas-fueled SOFC-based multigeneration system: enhanced performance with biomass and gasification agent selection. Renew Sustain Energy Rev 199:114460

Adekunle AE, Rabeya T, Jehadin F, Asad MA, Ayodele OO, Islam MS (2020) Compressed hot water pretreatment enhanced bioethanol production from corn stalk. Bioresour Technol Rep 12:100595

Agarwal AK, Valera H (2022) Introduction of potential and challenges of low carbon fuels for sustainable transport. In: Potential and challenges of low carbon fuels for sustainable transport. Springer Singapore, pp 3–6

Ai Y, Medic J, Jiang H, Wang D, Jane JL (2011) Starch characterization and ethanol production of sorghum. J Agric Food Chem 59(13):7385–7392

Andlar M, Rezić T, Marđetko N, Kracher D, Ludwig R, Šantek B (2018) Lignocellulose degradation: an overview of fungi and fungal enzymes involved in lignocellulose degradation. Eng Life Sci 18(11):768–778

Atiyeh H, Duvnjak Z (2001) Study of the production of fructose and ethanol from sucrose media by *Saccharomyces cerevisiae.* Appl Microbiol Biotechnol 57:407–411

Azad VK, De K, Majumder S (2024) Ethanol blending and its environmental impacts: a case study of India. Energy Sustain Dev 79:101385

Azarpour A, Mohammadzadeh O, Rezaei N, Zendehboudi S (2022) Current status and future prospects of renewable and sustainable energy in North America: progress and challenges. Energy Convers Manage 269:115945

Azhaganathan G, Bragadeshwaran A (2022) Critical review on recent progress of ethanol fuelled flex-fuel engine characteristics. Int J Energy Res 46(5):5646–5677

Aziz T, Shah Z, Sarwar A, Ullah N, Khan AA, Sameeh MY et al (2023) Production of bioethanol from pretreated rice straw, an integrated and mediated upstream fermentation process. Biomass Convers Biorefin 1–11

Bajwa DS, Peterson T, Sharma N, Shojaeiarani J, Bajwa SG (2018) A review of densified solid biomass for energy production. Renew Sustain Energy Rev 96:296–305

Balat M, Balat H (2009) Recent trends in global production and utilization of bio-ethanol fuel. Appl Energy 86(11):2273–2282

Banoth C, Sunkar B, Tondamanati PR, Bhukya B (2017) Improved physicochemical pretreatment and enzymatic hydrolysis of rice straw for bioethanol production by yeast fermentation. 3 Biotech 7:1–11

Baskar G, Ashokkumar V, Rokhum SL, Moholkar VS (2024) Introduction to circular bioeconomy perspectives in sustainable bioenergy production. In: Circular bioeconomy perspectives in sustainable bioenergy production. Springer Nature Singapore, Singapore, pp 1–4

Basso TO, de Kok S, Dario M, do Espirito-Santo JCA, Müller G, Schlölg PS et al (2011) Engineering topology and kinetics of sucrose metabolism in *Saccharomyces cerevisiae* for improved ethanol yield. Metab Eng 13(6):694–703

Bedzo OK, Dreyer CB, van Rensburg E, Görgens JF (2022) Optimisation of pretreatment catalyst, enzyme cocktail and solid loading for improved ethanol production from sweet sorghum bagasse. Bioenergy Res 15(2):1083–1095

Bouallagui H, Touhami Y, Hanafi N, Ghariani A, Hamdi M (2013) Performances comparison between three technologies for continuous ethanol production from molasses. Biomass Bioenerg 48:25–32

Cao W, Sun C, Qiu J, Li X, Liu R, Zhang L (2016) Pretreatment of sweet sorghum bagasse by alkaline hydrogen peroxide for enhancing ethanol production. Korean J Chem Eng 33:873–879

Chen Y (2011) Development and application of co-culture for ethanol production by co-fermentation of glucose and xylose: a systematic review. J Ind Microbiol Biotechnol 38(5):581–597

Chen J, Wang X, Zhang B, Yang Y, Song Y, Zhang F et al (2021) Integrating enzymatic hydrolysis into subcritical water pretreatment optimization for bioethanol production from wheat straw. Sci Total Environ 770:145321

Coniglio RO, Díaz GV, Fonseca MI, Castrillo ML, Piccinni FE, Villalba LL et al (2020) Enzymatic hydrolysis of barley straw for biofuel industry using a novel strain of *Trametes villosa* from Paranaense rainforest. Prep Biochem Biotechnol 50(8):753–762

Cortivo PRD, Hickert LR, Hector R, Ayub MAZ (2018) Fermentation of oat and soybean hull hydrolysates into ethanol and xylitol by recombinant industrial strains of *Saccharomyces cerevisiae* under diverse oxygen environments. Ind Crops Prod 113:10–18

Culaba AB, Mayol AP, San Juan JLG, Ubando AT, Bandala AA, Concepcion RS II et al (2023) Design of biorefineries towards carbon neutrality: a critical review. Biores Technol 369:128256
da Silva Fernandes F, de Souza ÉS, Carneiro LM, Alves Silva JP, de Souza JVB, da Silva Batista J (2022) Current ethanol production requirements for the yeast *Saccharomyces cerevisiae*. Int J Microbiol 2022(1):7878830
Dahman Y, Syed K, Begum S, Roy P, Mohtasebi B (2019) Biofuels: their characteristics and analysis. In: Biomass, biopolymer-based materials, and bioenergy. Woodhead Publishing, pp. 277–325
Daily CO_2. https://www.co2.earth/daily-co2
de Araujo Guilherme A, Dantas PVF, de Araújo Padilha CE, Dos Santos ES, de Macedo GR (2019) Ethanol production from sugarcane bagasse: use of different fermentation strategies to enhance an environmental-friendly process. J Environ Manage 234:44–51
de Assis Castro RC, Fonseca BG, dos Santos HTL, Ferreira IS, Mussatto SI, Roberto IC (2017) Alkaline deacetylation as a strategy to improve sugars recovery and ethanol production from rice straw hemicellulose and cellulose. Ind Crops Prod 106:65–73
Devi A, Singh A, Bajar S, Pant D, Din ZU (2021) Ethanol from lignocellulosic biomass: an in-depth analysis of pre-treatment methods, fermentation approaches and detoxification processes. J Environ Chem Eng 9(5):105798
Dong C, Qiao J, Wang X, Sun W, Chen L, Li S et al (2020) Engineering *Pichia pastoris* with surface-display minicellulosomes for carboxymethyl cellulose hydrolysis and ethanol production. Biotechnol Biofuels 13:1–9
Duque A, Doménech P, Álvarez C, Ballesteros M, Manzanares P (2020) Study of the bioprocess conditions to produce bioethanol from barley straw pretreated by combined soda and enzyme-catalyzed extrusion. Renew Energy 158:263–270
El Hage M, Louka N, Rezzoug SA, Maugard T, Sablé S, Koubaa M et al (2023) Bioethanol production from woody biomass: recent advances on the effect of pretreatments on the bioconversion process and energy yield aspects. Energies 16(13):5052
Fedyukhin AV, Semin DV, Dronov SA, Gusenko AG, Karasevich VA, Povernov MS (2024) Calculation of reconstruction volumes of hot-water boilers during the transition to a methane-hydrogen fuel. Int J Hydrogen Energy 49:681–689
Feng L, Guo J, Pang J, Yin M, Zhao Y, Wu P, Zheng M (2024) Nonenzymatic ethanol production in sustainable ways. Green Chem 26(15):8564–8582
Frankó B, Galbe M, Wallberg O (2015) Influence of bark on fuel ethanol production from steam-pretreated spruce. Biotechnol Biofuels 8:1–11
Gautam A, Pant M, Pant G, Kumar G (2024) Second-generation biofuels: concepts, applications, and challenges. In: Microbial applications for environmental sustainability, pp 277–304
Germec M, Kartal FK, Bilgic M, Ilgin M, Ilhan E, Güldali H et al (2016) Ethanol production from rice hull using *Pichia stipitis* and optimization of acid pretreatment and detoxification processes. Biotechnol Prog 32(4):872–882
Ghosh S, Dairkee UK, Chowdhury R, Bhattacharya P (2017a) Hydrogen from food processing wastes via photofermentation using Purple Non-sulfur Bacteria (PNSB)—a review. Energy Convers Manage 141:299–314
Ghosh S, Chowdhury R, Bhattacharya P (2017b) Sustainability of cereal straws for the fermentative production of second generation biofuels: a review of the efficiency and economics of biochemical pretreatment processes. Appl Energy 198:284–298
Ghosh S, Das S, Chowdhury R (2019) Effect of pre-pyrolysis biotreatment of banana pseudo-stem (BPS) using synergistic microbial consortium: role in deoxygenation and enhancement of yield of pyro-oil. Energy Convers Manage 195:114–124
Global primary energy consumption 2000–2023. https://www.statista.com/statistics/265598/consumption-of-primary-energy-worldwide/
Gupta P, Toksha B, Rahaman M (2024a) A Critical review on hydrogen based fuel cell technology and applications. Chem Rec 24(1):e202300295

Gupta S, Kanchan S, Kaur R, Sandhu SS (2024b) Investigation of performance and emission characteristics using ethanol-blended gasoline fuel as a flex-fuel in two-wheeler vehicle mounted on a chassis dynamometer. Clean Energy 8(3):174–193

Halder P, Azad K, Shah S, Sarker E (2019) Prospects and technological advancement of cellulosic bioethanol ecofuel production. In: Advances in eco-fuels for a sustainable environment. Woodhead Publishing, pp 211–236

Hassan MK, Chowdhury R, Ghosh S, Manna D, Pappinen A, Kuittinen S (2021) Energy and environmental impact assessment of Indian rice straw for the production of second-generation bioethanol. Sustain Energy Technol Assess 47:101546

Hilares RT, Ienny JV, Marcelino PF, Ahmed MA, Antunes FA, da Silva SS, Dos Santos JC (2017) Ethanol production in a simultaneous saccharification and fermentation process with interconnected reactors employing hydrodynamic cavitation-pretreated sugarcane bagasse as raw material. Biores Technol 243:652–659

Houfani AA, Anders N, Spiess AC, Baldrian P, Benallaoua S (2020) Insights from enzymatic degradation of cellulose and hemicellulose to fermentable sugars—a review. Biomass Bioenerg 134:105481

Irfan M, Asghar U, Nadeem M, Nelofer R, Syed Q, Shakir HA, Qazi JI (2016) Statistical optimization of saccharification of alkali pretreated wheat straw for bioethanol production. Waste Biomass Valoriz 7:1389–1396

Islas J, Manzini F, Masera O, Vargas V (2019) Solid biomass to heat and power. In: The role of bioenergy in the bioeconomy. Academic Press, pp 145–177

Jaiswal KK, Chowdhury CR, Yadav D, Verma R, Dutta S, Jaiswal KS, Karuppasamy KSK (2022) Renewable and sustainable clean energy development and impact on social, economic, and environmental health. Energy Nexus 7:100118

Jambo SA, Abdulla R, Azhar SHM, Marbawi H, Gansau JA, Ravindra P (2016) A review on third generation bioethanol feedstock. Renew Sustain Energy Rev 65:756–769

Jena A, Agarwal AK (2023) Understanding combustion in CI engines for adoption of renewable fuels. In: Renewable fuels for sustainable mobility. Springer Singapore, pp 317–347

Jetti KD, Gns RR, Garlapati D, Nammi SK (2019) Improved ethanol productivity and ethanol tolerance through genome shuffling of *Saccharomyces cerevisiae* and *Pichia stipitis*. Int Microbiol 22:247–254

Jin X, Song J, Liu GQ (2020) Bioethanol production from rice straw through an enzymatic route mediated by enzymes developed in-house from *Aspergillus fumigatus*. Energy 190:116395

John I, Yaragarla P, Appusamy A (2020) Production of bioethanol from banana peel using isolated cellulase from *Aspergillus niger*. In: Global challenges in energy and environment: select proceedings of ICEE 2018. Springer Singapore, pp 9–18

Khalid A, Aslam M, Qyyum MA, Faisal A, Khan AL, Ahmed F et al (2019) Membrane separation processes for dehydration of bioethanol from fermentation broths: recent developments, challenges, and prospects. Renew Sustain Energy Rev 105:427–443

Kim J, Sovacool BK, Bazilian M, Griffiths S, Lee J, Yang M, Lee J (2022) Decarbonizing the iron and steel industry: a systematic review of sociotechnical systems, technological innovations, and policy options. Energy Res Soc Sci 89:102565

Kiss AA (2014) Distillation technology–still young and full of breakthrough opportunities. J Chem Technol Biotechnol 89(4):479–498

Koyunoğlu C (2024) The economic case for blend fuels: a cost-benefit analysis in the European context. Sustain Technol Entrepren 3(2):100060

Kumar S, Singh N, Prasad R (2010) Anhydrous ethanol: a renewable source of energy. Renew Sustain Energy Rev 14(7):1830–1844

Laluce C, Leite GR, Zavitoski BZ, Zamai TT, Ventura R (2016) Fermentation of sugarcane juice and molasses for ethanol production In: Sugarcane-based biofuels and bioproducts, pp 53–86

Lamb CDC, Silva BMZD, de Souza D, Fornasier F, Riça LB, Schneider RDCDS (2018) Bioethanol production from rice hull and evaluation of the final solid residue. Chem Eng Commun 205(6):833–845

Li X, San X, Zhang Y, Ichii T, Meng M, Tan Y, Tsubaki N (2010) Direct synthesis of ethanol from dimethyl ether and syngas over combined H-mordenite and Cu/ZnO catalysts. Chemsuschem 3(10):1192–1199

Li J, Zhao R, Xu Y, Wu X, Bean SR, Wang D (2022) Fuel ethanol production from starchy grain and other crops: an overview on feedstocks, affecting factors, and technical advances. Renewable Energy 188:223–239

Lienqueo ME, Ravanal MC, Pezoa-Conte R, Cortínez V, Martínez L, Niklitschek T et al (2016) Second generation bioethanol from *Eucalyptus globulus* Labill and *Nothofagus pumilio*: ionic liquid pretreatment boosts the yields. Ind Crops Prod 80:148–155

Madu JO, Agboola BO (2018) Bioethanol production from rice husk using different pretreatments and fermentation conditions. 3 Biotech 8(1):15

Magda R, Szlovak S, Toth J (2021) The role of using bioalcohol fuels in sustainable development. In: Bio-economy and agri-production. Academic Press, pp 133–146

Mahapatra S, Kumar D, Singh B, Sachan PK (2021) Biofuels and their sources of production: a review on cleaner sustainable alternative against conventional fuel, in the framework of the food and energy nexus. Energy Nexus 4:100036

Matsakas L, Nitsos C, Raghavendran V, Yakimenko O, Persson G, Olsson E et al (2018) A novel hybrid organosolv: steam explosion method for the efficient fractionation and pretreatment of birch biomass. Biotechnol Biofuels 11:1–14

Mensah MB, Jumpah H, Boadi NO, Awudza JA (2021) Assessment of quantities and composition of corn stover in Ghana and their conversion into bioethanol. Sci Afr 12:e00731

Meshitsuka G, Isogai A (2017) Chemical structures of cellulose, hemicelluloses, and lignin. In: Chemical modification of lignocellulosic materials. Routledge, pp 11–33

Minteer SD (2011) Biochemical production of other bioalcohols: biomethanol, biopropanol, bioglycerol, and bioethylene glycol. In: Handbook of biofuels production. Woodhead Publishing, pp 258–265

Molaverdi M, Karimi K, Mirmohamadsadeghi S, Galbe M (2021) High efficient ethanol production from corn stover by modified mild alkaline pretreatment. Renew Energy 170:714–723

Nasidi M, Agu R, Deeni Y, Walker G (2015) Improved production of ethanol using bagasse from different sorghum cultivars. Biomass Bioenerg 72:288–299

Nordin I, Elofsson K, Jansson T (2024) Cost-effective reductions in greenhouse gas emissions: reducing fuel consumption or replacing fossil fuels with biofuels. Energy Policy 190:114138

Oke D, Dunn JB, Hawkins TR (2024) Reducing economy-wide greenhouse gas emissions with electrofuels and biofuels as the grid decarbonizes. Energy Fuels 38(7):6048–6061

Ouyang S, Qiao H, Xu Q, Zheng Z, Ouyang J (2019) Development of two-step pretreatment of Chinese fir sawdust using dilute sulfuric acid followed by sodium chlorite for bioethanol production. Cellulose 26:8513–8524

Palacios S, Ruiz HA, Ramos-Gonzalez R, Martínez J, Segura E, Aguilar M et al (2017) Comparison of physicochemical pretreatments of banana peels for bioethanol production. Food Sci Biotechnol 26:993–1001

Paladugula AL, Kholod N, Chaturvedi V, Ghosh PP, Pal S, Clarke L et al (2018) A multi-model assessment of energy and emissions for India's transportation sector through 2050. Energy Policy 116:10–18

Palani T, Esakkimuthu GS, Dhamodaran G, Seetharaman S (2024) Experimental study on dual oxygenates (ethanol, n-butanol) with gasoline on MPFI engine performance and emission characteristics. Int J Environ Sci Technol 21(1):245–254

Paliga LR, Warken AJ, Dalastra C, Rodrigues Soares ML, Kubeneck S, Correia Souza TR et al (2022) Feedstock for second-generation bioethanol production. Liquid biofuels: bioethanol. Springer International Publishing, Cham, pp 165–186

Panoutsou C, Germer S, Karka P, Papadokostantakis S, Kroyan Y, Wojcieszyk M et al (2021) Advanced biofuels to decarbonise European transport by 2030: markets, challenges, and policies that impact their successful market uptake. Energ Strat Rev 34:100633

Prabakaran B (2022) Engine performance and emissions of fuels from waste products for sustainable environment. In: Application of clean fuels in combustion engines. Springer Nature Singapore, pp 131–153

Prasad S, Malav MK, Kumar S, Singh A, Pant D, Radhakrishnan S (2018) Enhancement of bio-ethanol production potential of wheat straw by reducing furfural and 5-hydroxymethylfurfural (HMF). Bioresour Technol Rep 4:50–56

Qin L, Zhao X, Li WC, Zhu JQ, Liu L, Li BZ, Yuan YJ (2018) Process analysis and optimization of simultaneous saccharification and co-fermentation of ethylenediamine-pretreated corn stover for ethanol production. Biotechnol Biofuels 11:1–10

Qiu J, Ma L, Shen F, Yang G, Zhang Y, Deng S et al (2017) Pretreating wheat straw by phosphoric acid plus hydrogen peroxide for enzymatic saccharification and ethanol production at high solid loading. Biores Technol 238:174–181

Rabeya T, Jehadin F, Asad MA, Ayodele OO, Adekunle AE, Islam MS (2021) Alkali and intensified heat treatment of corn stalk for bioethanol production. Sugar Tech 23:643–650

Rial RC (2024) Biofuels versus climate change: exploring potentials and challenges in the energy transition. Renew Sustain Energy Rev 196:114369

Roychowdhury S, Ghosh S (2024) State-of-the-art of biochar amended dark fermentative hydrogen production: a sustainable coupling of decarbonization pathways towards low carbon future. J Clean Prod 141208

Rulli MC, Bellomi D, Cazzoli A, De Carolis G, D'Odorico P (2016) The water-land-food nexus of first-generation biofuels. Sci Rep 6(1):22521

Sakthivel P, Subramanian KA, Mathai R (2018) Indian scenario of ethanol fuel and its utilization in automotive transportation sector. Resour Conserv Recycl 132:102–120

Salapa I, Katsimpouras C, Topakas E, Sidiras D (2017) Organosolv pretreatment of wheat straw for efficient ethanol production using various solvents. Biomass Bioenerg 100:10–16

Sarıkoç S (2022) Some of the bio-fuels for internal combustion engines: alcohols and biodiesel. In: Potential and challenges of low carbon fuels for sustainable transport. Springer, Singapore, pp 9–31

Sarkar D, Prajapati S, Poddar K, Sarkar A (2020) Ethanol production by *Klebsiella* sp. SWET4 using banana peel as feasible substrate. Biomass Convers Biorefin 1–13

Sharma SK, Upadhyay RK, Kumar V, Valera H (2023) Introduction to sustainable transportation system. In: Transportation energy and dynamics. Springer Nature Singapore, Singapore, pp 3–6

Sheikh MMI, Kim CH, Park HH, Nam HG, Lee GS, Jo HS et al (2015) A synergistic effect of pretreatment on cell wall structural changes in barley straw (*Hordeum vulgare* L.) for efficient bioethanol production. J Sci Food Agric 95(4):843–850

Singh B, Kumar A (2020) Process development for sodium carbonate pretreatment and enzymatic saccharification of rice straw for bioethanol production. Biomass Bioenerg 138:105574

Singh A, Prajapati P, Vyas S, Gaur VK, Sindhu R, Binod P et al (2023) A comprehensive review of feedstocks as sustainable substrates for next-generation biofuels. Bioenergy Res 16(1):105–122

Song Y, Li R, Qian Y, Zhang L, Wang Z, Liu S et al (2024) Study on spray characteristics of biodiesel alternative fuels for in-cylinder environment of diesel engine. J Energy Inst 113:101507

Sydney EB, Letti LAJ, Karp SG, Sydney ACN, de Souza Vandenberghe LP, de Carvalho JC et al (2019) Current analysis and future perspective of reduction in worldwide greenhouse gases emissions by using first and second generation bioethanol in the transportation sector. Bioresour Technol Rep 7:100234

Tabata T, Yoshiba Y, Takashina T, Hieda K, Shimizu N (2017) Bioethanol production from steam-exploded rice husk by recombinant Escherichia coli KO11. World J Microbiol Biotechnol 33:1–7

Thakur P, Singh A (2024) Third-generation biofuels: a promising approach for environmental sustainability. In: Microbial applications for environmental sustainability. Springer Nature Singapore, Singapore, pp 305–327

Thanapimmetha A, Saisriyoot M, Khomlaem C, Chisti Y, Srinophakun P (2019) A comparison of methods of ethanol production from sweet sorghum bagasse. Biochem Eng J 151:107352

Thapa S, Indrawan N, Bhoi PR (2018) An overview on fuel properties and prospects of Jatropha biodiesel as fuel for engines. Environ Technol Innov 9:210–219

US EPA, OA (2017) Climate change indicators: atmospheric concentrations of greenhouse gasesl US EPA. https://www.epa.gov/climate-indicators/climate-change-indicators-atmospheric-concentrations-greenhouse-gases

Valladares-Diestra KK, de Souza Vandenberghe LP, Torres LAZ, Nishida VS, Zandoná Filho A, Woiciechowski AL, Soccol CR (2021) Imidazole green solvent pre-treatment as a strategy for second-generation bioethanol production from sugarcane bagasse. Chem Eng J 420:127708

Vane LM (2008) Separation technologies for the recovery and dehydration of alcohols from fermentation broths. Biofuels Bioprod Biorefin 2(6):553–588

Vaskan P, Ruiz Pachon EM, Gnansounou E (2017). Life cycle assessment of sugar crops and starch-based integrated biorefineries. In: Life-cycle assessment of biorefineries, pp 97–146

Vidhya CV, Nandhini S, Sheela JM, Reenaa M (2024) Bioenergy-based sustainable bioeconomy—perspectives and challenges. In: Circular bioeconomy perspectives in sustainable bioenergy production. Springer Nature Singapore, Singapore, pp 5–36

Wang Q, Wang W, Tan X, Chen X, Guo Y, Yu Q et al (2019) Low-temperature sodium hydroxide pretreatment for ethanol production from sugarcane bagasse without washing process. Biores Technol 291:121844

Wang W, Tan X, Imtiaz M, Wang Q, Miao C, Yuan Z, Zhuang X (2021) Rice straw pretreatment with KOH/urea for enhancing sugar yield and ethanol production at low temperature. Ind Crops Prod 170:113776

Wang X, Cui W, Guo W, Sun B, Huang M, Li J et al (2024) Separation techniques for manufacturing fruit spirits: from traditional distillation to advanced pervaporation process. Comprehen Rev Food Sci Food Safety 23(1):e13278

Wu J, Elliston A, Le Gall G, Colquhoun IJ, Collins SR, Wood IP et al (2018) Optimising conditions for bioethanol production from rice husk and rice straw: effects of pre-treatment on liquor composition and fermentation inhibitors. Biotechnol Biofuels 11:1–13

Wyman CE, Decker SR, Himmel ME, Brady JW, Skopec CE, Viikari L (2005) Hydrolysis of cellulose and hemicellulose. Polysacch Struct Div Funct Versatil 1:1023–1062

Yuan Z, Wen Y, Li G (2018) Production of bioethanol and value added compounds from wheat straw through combined alkaline/alkaline-peroxide pretreatment. Biores Technol 259:228–236

Zhang C, Chen H, Pang S, Su C, Lv M, An N et al (2020) Importance of redefinition of corn stover harvest time to enhancing non-food bio-ethanol production. Renew Energy 146:1444–1450

Zhao W, Zhao F, Zhang S, Gong Q, Chen G (2019) Ethanol production by simultaneous saccharification and cofermentation of pretreated corn stalk. J Basic Microbiol 59(7):744–753

Zhu JQ, Zong QJ, Li WC, Chai MZ, Xu T, Liu H et al (2020) Temperature profiled simultaneous saccharification and co-fermentation of corn stover increases ethanol production at high solid loading. Energy Convers Manage 205:112344

Zohuri B, McDaniel PJ (2021) Introduction to energy essentials: insight into nuclear, renewable, and non-renewable energies. Academic Press

Part III
Alcohol Fueled Engines

Chapter 8
Alcohol Fuels for Sustainable Energy Solutions

Ajeet Kumar Prajapati, Aditya Kashyap, and Rakesh Kumar

Abstract The mitigation of climate change requires a reduction in global CO_2 emissions, and this can only be achieved by moving away from fossil fuels. Because fossil fuels come from limited supplies, they are rapidly running out and seriously affecting ecosystems and the environment. This demonstrates how urgent it is to investigate alternative fuels, such as alcohol, biodiesel, vegetable oils, and gaseous fuels like hydrogen, natural gas, and biogas, that may be produced from locally accessible resources. Alcohol fuels stand out from the rest because they have performance, ignition, and combustion characteristics that are quite like conventional fuels derived from petroleum. Alcohols can be used directly in internal combustion (IC) engines as oxygenated fuels or combined with biofuels. Alcohol fuels are a good substitute for conventional fuels and help create a more sustainable energy future because of their intrinsic oxygen content, which also improves engine performance and lowers exhaust emissions.

Keywords Alcohol · Alternative fuels · Internal combustion engines · Oxygenated additives

Abbreviations

LPG	Liquefied petroleum gas
CDL	Coal-derived liquid fuels
B100	Biodiesel
FFVs	Flexible fuel vehicles
MTBE	Methyl tertiary butyl ether
ETBE	Ethyl tertiary butyl ether
LHOV	Latent heat of evaporation

A. K. Prajapati (✉) · A. Kashyap · R. Kumar
Department of Chemical Engineering and Biochemical Engineering, Rajiv Gandhi Institute of Petroleum Technology, Jais, Amethi, India
e-mail: 20ce0002@rgipt.ac.in

A. Bartl et al. (eds.), *Alcohol Production Processes and Their Utilisation in the Transport Sector*, Energy, Environment, and Sustainability,
https://doi.org/10.1007/978-981-96-7384-1_8

RVP	Reid vapour pressure
MON	Motor octane number
CO	Carbon monoxide
CO_2	Carbon dioxide
DI	Direct injection
RON	Research octane number
ECUs	Engine control units
BSFC	Brake-specific fuel consumption
FCVs	Fuel cell vehicles
H_2ICE	Hydrogen internal combustion engine
HC	Hydrocarbons
LHV	Lower heating value
HHV	Higher heating value
HEVs	Hybrid electric vehicles
MMO	Methane monooxygenase
ICEs	Internal combustion engines
ICEVs	Internal combustion engine vehicles
DME	Dimethyl ether
L-H_2ICE	Liquid hydrogen internal combustion engine
NO_x	Oxides of nitrogen
EBHVs	Ethanol and battery-powered hybrid vehicles
PPE	Personal protective equipment
CDC	Disease control and prevention
PM	Particulate matter
SI	Spark ignition
VOCs	Volatile organic compounds

8.1 Introduction

Energy plays a driving role in human development and improving living standards, essential for sustainable progress. Over recent decades, rapid global population growth has significantly heightened energy demands, particularly in the industrial and transportation sectors. This surge in demand has led to increased crude oil prices, closely linked to global economic activities (He et al. 2010). The increasing energy demands have prompted both the United States and China to position themselves as leading energy producers. In 2018, these two nations collectively accounted for 54% of the growth in energy production. Currently, fossil fuels dominate global primary energy consumption, constituting 82% of the total, with the transportation sector alone consuming 58% of these fuels (https://www.spglobal.com/commodityinsights/en/market-insights/latest-news/oil/062623-fossil-fuels-stubbornly-dominating-global-energy-despite-surge-in-renewables-energy-institute). Currently, more than 90% of energy consumed in the transportation sector is derived from oil products, reflecting only a 3.5% point reduction since the early 1970s. Consequently, the transportation sector is responsible for over one-third of global CO_2 emissions from end-use sectors (https://www.efginternational.com/insights/2023/the_essential_role_of_decarbonising_transportation_in_the_transition_to_a_regenerative_economy.html). Since 1973, worldwide primary energy demand has grown at an average annual pace of 2.0%. Despite this expansion, around one-third of the world's population continues to rely on non-commercial fuels. The widespread use of fossil fuels contributes to global warming, climate change, and various health issues. Major pollutants include CO_2, mostly produced by inefficient fuel burning, and NO_x, produced by both natural and manmade sources. Furthermore, SO_2 emissions from the combustion of fossil fuels like coal and petroleum cause environmental problems, such as acid rain (Ritchie et al. 2024). According to the National Energy Strategy in February 1992, the United States transportation sector alone utilized over 185 million buses, cars, and trucks, consuming 66.7% of the nation's oil. Similarly, in India, the proportion of oil consumption for transportation rose from 50% in 1991 to 61% by 2010, reflecting a rapid increase in oil usage and its environmental consequences. In June 1992, the United Nations Conference on Environment and Development (UNCED), commonly referred to as the Earth Summit, was held in Rio de Janeiro, Brazil. This significant gathering brought together global delegates to address the pressing issue of minimizing global warming and promoting sustainable development strategies (https://www.britannica.com/event/United-Nations-Conference-on-Environment-and-Development). Therefore, the search for viable alternatives to fossil fuels is increasingly crucial. Ideally, these alternatives should embody renewable, sustainable, efficient, and cost-effective energy sources with lower emissions. Significant global efforts are currently focused on developing and introducing gaseous fuels that can partially or fully substitute conventional fuels. Most of these gaseous fuels can be sourced from renewable resources and possess high self-ignition temperatures, making them well-suited for spark ignition engines. Key alternatives to internal combustion engines consist of natural gas, liquefied petroleum gas (LPG), hydrogen, acetylene, producer gas, alcohol, and vegetable oils. Among these, alcohol fuels are prominent contenders for alternative energy sources, suitable for use in internal combustion engines as well as synthetic fuels for transportation. Therefore, alcohol fuels are good alternatives to fossil fuels by enhancing energy security, reducing vehicle emissions, and ensuring economic stability for agricultural sectors through renewable sourcing and beneficial combustion properties. In conclusion, alcohol fuels offer significant potential benefits including increased energy supply security, decreased emissions from vehicles, and economic resilience for agricultural communities.

8.1.1 The Role of Alcohols in Alternative Fuel Solutions

This study defines alternative fuels to include alcohols such as ethanol (including blends with gasoline that exceed 85%), domestically produced natural gas and its liquefied forms, liquefied petroleum gas (LPG), coal-derived liquid fuels (CDL), hydrogen (H_2), biodiesel (B100), other bio-derived fuels, and non-petroleum fuels that provide notable energy security and environmental benefits.

Alcohol fuels are particularly favoured as alternative fuels due to several advantages over fossil fuels as follows:

i. Enhanced Power and Efficiency: When burned in internal combustion engines, alcohol fuels generate higher combustion pressures than gasoline, resulting in improved power output and thermal efficiency.
ii. Reduced Emissions: Alcohol fuels produce lower evaporative emissions and generate minimal ash during combustion due to their reduced carbon content, contributing to better fuel efficiency. Lower molecular weight alcohols can be produced from abundant and cost-effective indigenous energy sources such as biomass, coal, and natural gas.
iii. Sustainable Production: Lower molecular weight alcohols can be derived from plentiful and cost-effective domestic energy sources, including biomass, coal, and natural gas. They decrease the release of toxic gases into the environment and are miscible in water, facilitating easy cleanup in case of spills.
iv. Lower Greenhouse Gas Emissions: These fuels help decrease greenhouse gas emissions and typically have higher average octane ratings than gasoline, which boosts both power and fuel efficiency.
v. Environmental Safety: Alcohol fuels minimize the release of harmful gases into the atmosphere and are miscible in water, making cleanup easier in the event of spills.

In summary, alcohol fuels offer significant environmental and efficiency advantages compared to traditional fossil fuels, making them a promising option for reducing dependence on petroleum and mitigating environmental impacts.

8.1.2 Current Scenario Around the Globe

The origins of bioethanol can be traced back to 1899 when Ford Motor Company created an ethanol-powered vehicle, although this innovation was largely overlooked due to the low cost of gasoline at that time (https://lee-enterprises.com/a-short-history-of-biofuels/). The oil crisis in the 1970s revived interest in ethanol, leading to the introduction of gasohol, especially in Brazil. Currently, flexible fuel vehicles (FFVs) that can operate on both ethanol and gasoline are widely available. Additionally, methanol blending with gasoline, particularly the M15 blend (15% methanol and 85% gasoline), has been selectively implemented in China since the early 2000s. In the United States, bioethanol is mandated in transportation fuels, contributing to a significant reduction in crude oil dependency from 60 to 25%. There have been trials with near-neat fuels like M85 (85% methanol/15% gasoline). Methanol has a historical use in racing cars due to its cleaner combustion properties, preventing the emission of black smoke that can obstruct visibility.

Table 8.1 Concise overview of ethanol blending programmes across various countries

S/L No.	Countries adapting to blending programmes	Alcohol blending programme
1	Brazil	A required bioethanol blend of approximately 24% ± 2%
2	United States	A goal of 10% for the share of bioethanol in primary energy sources
3	European Union	The proportion of biofuel increased from 2% in 2005 to 5.75% in 2010, indicating a significant rise during this period
4	India	The existing requirement to blend 5% bioethanol will be raised to 20%
5	China	Compulsory integration at the regional government level
6	Colombia	Blending 10% bioethanol is required in urban regions
7	Argentina	Mandatory use of 5% bioethanol to be followed

Yun 2020; https://www.federalregister.gov/documents/2019/06/10/2019-11653/modifications-to-fuel-regulations-to-provide-flexibility-for-e15-modifications-to-rfs-rin-market

Brazil leads globally with most automobiles equipped with fuel-flex engine systems, enabling the vehicles to run on ethanol. Currently, the United States stands out as the world's largest producer and consumer of alcohol-based fuels (http://fastmarkets-prod-01.altis.cloud/insights/brazil-favours-ethanol-cars-flex-fuel-vehicles-overshadow-evs/). India also has committed to achieving a 20% ethanol blending in petrol (E20) by 2025 as part of its G20 pledges and national ethanol blending programme as shown in Table 8.1.

This chapter will provide a comprehensive and detailed discussion of the usage of alcohols (butanol, methanol, ethanol, and propanol) as fuels in spark ignition engines, which could potentially replace petroleum-based fuels.

8.2 Alcohol: Characteristics, Potential, and Obstacles

Alcohols are chemical compounds characterized by the presence of one or more hydroxyl (OH) groups attached to a carbon atom within an alkyl group. They are organic water derivatives that have an alkyl group in place of a hydrogen atom. Alcohols are common organic molecules with numerous applications, including sweets, perfumes, and chemical intermediates. The chemical industry creates a considerable number of alcohols. Ethanol and methanol are two major alcohols. Ethanol is utilized in personal care items, pharmaceuticals, fuel generation, and medical device sterilization. Methanol is used as a solvent, a precursor for formaldehyde and speciality resins, and in some fuels, antifreeze formulas, and metal cleaning procedures.

Despite the prevalent use of fossil fuels today, alcohol has historically served as a fuel source. One significant advantage of alcohol fuels is their high-octane rating, which improves fuel efficiency and compensates for their lower energy density compared to gasoline and diesel. This results in similar fuel economy measurements, such as kilometres per litre or miles per gallon. Among various alcohol fuels, biobutanol stands out for its energy density, which is comparable to gasoline, while its octane rating exceeds that of gasoline by more than 25% (Ayas 2018). However, biobutanol production is currently more complex than that of ethanol or methanol. Alcohol fuels have a common chemical formula of $C_nH_{2n+1}OH$. Most of the methanol is created from natural gas; however, comparable chemical techniques can be used to synthesize it from biomass. Ethanol, on the other hand, is primarily generated from biological materials through fermentation. The following sections will delve into the various types, characteristics, and aspects of alcohol fuels.

The following subsections will focus on the types, characteristics, and different aspects of alcohol fuels.

8.2.1 Types of Alcohols

Alcohol fuels comprise a diverse range of compounds, including methanol, ethanol, and various ethers such as MTBE and ETBE. They also include esters like biodiesel, which consists of methyl and ethyl esters derived from fatty acids found in vegetable oils and animal fats. Among these, methanol and ethanol are the most widely used. Their popularity stems from their capacity to be produced through both chemical and biological processes, along with their compatibility with current engine designs. Additionally, the focus is shifting towards the use of propanol and butanol, which are gaining attention for their unique properties and potential applications. Therefore, our focus will primarily be on these four alcohols: methanol, ethanol, propanol, and butanol.

Methanol

Methanol (CH_3OH), a fundamental alcohol with a single carbon atom per molecule, is a clear liquid widely utilized as a fuel. Its primary production from natural gas makes it accessible, but it can also be derived sustainably from sources like municipal solid waste and plant biomass, aligning with efforts to reduce carbon footprints. This versatility has positioned methanol as a viable option for enhancing fuel sustainability. In automotive applications, methanol has demonstrated significant advantages over gasoline. It has been chosen as the fuel for prestigious events such as the Indianapolis 500 for more than three decades, not only for its performance benefits but also for its enhanced safety profile (Ayas 2018). Methanol's lower flammability and reduced fire severity upon ignition contribute to its appeal as a safer alternative in motorsports and everyday use. Environmental considerations further bolster methanol's appeal. It biodegrades quickly when spilt and disperses rapidly in water, minimizing environmental impact compared to other fuels. Studies endorse ethanol as an alternative fuel due to its ability to lower urban air pollutant emissions, particularly compared to conventional gasoline. Methanol's typical usage involves blends like M85, consisting of 85% methanol and 15% gasoline, which balances performance and compatibility with existing fuel infrastructure. Beyond its role as a fuel, methanol holds promise as a hydrogen source. Establishing a methanol infrastructure is comparatively feasible to hydrogen, and methanol-powered fuel cell vehicles offer the potential for extremely low urban pollutant emissions (Thomas et al. 2000). Demonstrations by manufacturers like Daimler-Chrysler and Nissan showcase vehicles converting methanol directly into hydrogen for fuel cells, hinting at future applications in clean energy transportation (http://global.nissannews.com/en/releases/19990513-e_presskit). Commercially, methanol has been in use for over a century, finding applications in paints, plywood manufacturing, carpet production, and as an additive in gasoline and antifreeze (https://www.geeksforgeeks.org/uses-of-methanol/). Its versatility extends to serving as a precursor for essential chemicals such as formaldehyde and acetic acid, highlighting its role in various industrial processes.

Ethanol

Ethanol (C_2H_5OH) is a versatile alcohol derived from sources like corn, sugarcane, and grains, as well as from paper waste through fermentation and distillation processes (Hordeski 2020; Biomass for Renewable Energy 1999). Apart from its role in alcoholic beverages, ethanol serves as a significant fuel additive, commonly blended with gasoline to enhance fuel properties such as octane ratings and to reduce emissions. The use of ethanol as a fuel has gained traction globally due to its environmental advantages. When mixed with gasoline, it contributes to lower emissions, which is a key factor in reducing the overall carbon footprint of vehicles (Biomass for Renewable Energy 1999). Additionally, ethanol is known for its solubility in water, biodegradability, and ease of evaporation, making it less harmful in case of spills compared to petroleum products (Ayas 2018). This environmental compatibility has led to widespread acceptance by governments, automotive manufacturers, and consumers alike. Economically, ethanol production has historically supported agriculture by utilizing surplus grains and crops, thereby providing additional income to farmers and reducing federal farm programme costs. Ethanol production from surplus grain in the United States has significantly boosted farm incomes and reduced crude oil imports. In terms of production, ethanol is derived through fermentation processes, utilizing sugars from crops like corn in the United States and sugarcane in Brazil. This historical use of fermentation dates to ancient times and continues to be a cornerstone of ethanol production today. Ethanol's applications extend beyond fuel blending. It is used in various industrial processes, including the manufacture of personal care products, pharmaceuticals, and beverages (https://www.britannica.com/science/ethanol). Ethanol's role in personal care products includes its use as a solvent and preservative in cosmetics and skincare products. In pharmaceuticals, it acts as an antiseptic and solvent for various drugs. Moreover, ethanol is pivotal in the production of bioethanol, a renewable energy source that is increasingly integrated into transportation fuels. This integration not only helps in reducing greenhouse gas emissions but also supports the development of a sustainable energy economy. Ethanol's use in the production of biodiesel, known as ethyl ester, is another significant application, providing an alternative to conventional diesel fuels.

Propanol

Propanol (C_3H_7OH), specifically 1-propanol or *n*-propanol, is an alcohol distinguished by its three-carbon chain structure and hydroxyl (–OH) group. Its production typically involves the hydration of propylene from petroleum refining byproducts or fermentation processes utilizing specific bacteria strains (Walther and François 2016). Beyond its fundamental role as an industrial solvent, propanol is extensively utilized in pharmaceuticals, and cosmetics, and serves as a crucial chemical intermediate in various compound syntheses. Its efficacy as a solvent contributes significantly to applications in coatings, inks, and adhesives within industrial contexts. Propanol has recently gained attention as a possible biofuel due to its increased energy density and compatibility with current gasoline infrastructure, which may provide advantages over ethanol. Its application as a fuel additive or solo fuel shows promise in reducing greenhouse gas emissions and decreasing reliance on fossil fuels in the transportation sector (Jamrozik et al. 2022). The environmental appeal of propanol is further underscored by its biodegradability and relatively low toxicity, making it a preferred choice where environmental impact mitigation is a priority. Its versatile applications extend to disinfectants in healthcare settings and antiseptics in personal care products, highlighting its broad industrial applicability. In industrial and laboratory settings, isopropanol, a related alcohol, is often combined with acetone for its solvent properties. This blend is particularly effective in cleaning and degreasing applications, enhancing the dissolution of oils, resins, and stubborn contaminants on surfaces. This mixture finds extensive use in electronics cleaning and certain chemical formulations requiring potent solvent capabilities. Its integration into diverse industrial processes emphasizes its versatility and significance in advancing towards environmentally responsible solutions. In conclusion, propanol's diverse industrial applications, coupled with its potential as a renewable biofuel, highlight its growing importance in sustainable development initiatives.

Butanol

The two main types of butanol (C_4H_9OH), a four-carbon straight-chain alcohol, are isobutanol (*i*-butanol) and normal butanol (*n*-butanol). Butanol is colourless and combustible. These substances have good fuel qualities and can be used with ethers, ethanol, and other solvents to create a variety of combinations. In addition to being used as chemical intermediates and solvents in goods like pharmaceuticals, hydraulic fluids, and cosmetics, they are also essential in industrial processes like the synthesis of butyl and methacrylate (García et al. 2011). The structural variability of butanol allows for different isomers, including straight-chain and branched forms, each characterized by distinct properties such as octane number, boiling point, and viscosity. Despite these variations, all butanol isomers find common use as solvents, industrial cleaners, and potential additives in gasoline formulations (Liu et al. 2013). While most butanol is traditionally derived from fossil fuels using various methods, *n*-butanol can also be sourced from biomass sources, contributing to its potential as a renewable fuel option (Liu et al. 2013). In the context of fuel applications, butanol isomers are notable for being free of aromatics, olefins, and sulphur compounds, which can improve the overall composition of gasoline by diluting undesirable components. With an oxygen content of approximately 21.6 wt.%, butanol additives enhance combustion efficiency and contribute to reduced emissions during fuel combustion. In conclusion, butanol's versatility across various industrial applications, combined with its potential as a renewable fuel additive, underscores its pivotal role in advancing cleaner and more sustainable energy solutions. As research continues to explore and optimize its production and utilization, butanol remains a promising candidate for reducing environmental impact while meeting global energy demands.

Other Alcohols

Among other alcohols, fusel oil has recently garnered attention from researchers as a blending fuel in spark ignition (SI) engines. A blend of erratic, oily liquids created in tiny amounts when alcoholic beverages ferment is known as fusel oil. Fusel oil contains primarily *n*-propyl alcohol (15–20%), iso amyl alcohol (55–60%), isobutyl alcohol (6–8%), bits of *n*-butyl alcohol, and ethanol (Awad et al. 2017a). Before the industrial manufacture of synthetic amyl alcohols began, fusel oil was the principal commercial source of these chemicals, which are crucial components in lacquer solvents. The properties of fusel oil indicate that it has potential as an alternative fuel for spark ignition (SI) engines. When used as an additive in methanol-gasoline blends, fusel oil has been shown to enhance engine performance and emissions levels. Specifically, improvements in engine torque were observed across all loads and engine speeds compared to pure gasoline (Içingür and Calam 2012; Karaosmanoğlu et al. 1997; Calam et al. 2015a).

8.2.2 Properties of Alcohol Fuels

Chemical and physical properties are crucial in determining the quality of fuel used in engines, as they directly influence engine performance, combustion efficiency, and emission characteristics. The key properties of alcohol fuels are summarized in Table 8.2. A notable feature of alcohol fuels, especially methanol and ethanol, is their molecular structure, which includes oxygen atoms. When burning, this oxygen content significantly lowers the emissions of hydrocarbons (HC) and carbon monoxide (CO) (Saricks et al. 1999; Demirbaş 2005). Furthermore, the octane rating and other essential characteristics of alcohol fuels utilized in spark ignition engines are closely linked to their volatility. Alcohol fuels possess a latent heat of evaporation (LHOV) that is 3–5 times greater than that of gasoline, which significantly impacts the cold-start performance of SI engines. This higher LHOV contributes to improved volumetric efficiency and lower temperatures in the intake manifold. For example, methanol has an LHOV that is roughly 2.5 times that of gasoline, enhancing its performance in these applications (Canakci et al. 2013). The large molar expansion of alcohols especially methanol which produces additional pressure during chemical reactions without the need for additional heat is another important benefit of alcohols. However, compared to unleaded gasoline, alcohol fuels usually have a heating value of 20–40% lower, requiring 1.5–1.8 times more alcohol fuel to provide the same amount of energy. Alcohol fuels also help reduce engine emissions and exhaust temperatures while having a lower heating value to a large extent (Egolf and Jurs 1992). This also necessitates designing the engine fuelling system to include higher fuelling rates. However, alcohols offer the highest combustion efficiency, primarily due to their relative oxygen content, which enhances the completeness of combustion (Agarwal 2007; Nithyanandan et al. 2016). Vapour pressure is an important characteristic of engine fuels, as it affects the engine's ability to start in cold conditions and helps ensure compliance with evaporative emission standards. Alcohol fuels typically have a Reid vapour pressure (RVP) ranging from 17 to 55 kPa, which is lower than that of gasoline, which has an RVP of 53 to 60 kPa. This lower vapour pressure is considered advantageous for engine performance (Andersen et al. 2010).

Overall, alcohol fuels, particularly methanol and ethanol, aim to improve engine performance while reducing emissions in spark ignition engines. Their unique properties, including higher latent heat of vaporization, molar expansion, and oxygen content, contribute to these benefits, making them a viable alternative to traditional gasoline fuels. More details of alcohol fuels are described in the next sections.

Octane Number

When fuel is used in internal combustion engines that use the Otto cycle, such as gasoline engines or SI engines, the octane number indicates how resistant the fuel is to detonating during the combustion process. Fuel quality is evaluated by its octane number, which is derived from the combination of the motor octane number (MON) and the research octane number (RON). Under severe knocking conditions, engines can sustain damage in less than 50 h, and this damage builds up over time (Wang et al. 2017). Using high-octane fuels is essential to preventing engine knocking.

When alcohol fuels are mixed with gasoline, the octane number increases rapidly due to their high RON and MON properties. (Vallinayagam et al. 2017). These fuels have properties such as a high RON, MON, high oxygen content, and a single boiling point, making them viable alternatives for SI engines, as indicated in Table 8.2. Thus, fuels with higher octane ratings can serve as octane enhancers. Additionally, higher octane-rated fuels are desirable because they improve engine efficiency and reduce brake-specific fuel consumption (BSFC) by enabling higher compression ratios.

Calorific Value

The quantity of heat released per unit quantity during burning is indicated by a fuel's calorific or heating value. (Sheng and Azevedo 2005). This setting has a direct effect on the fuel consumption specific to brakes (BSFC) (Awad et al. 2017b) and influences engine power output and brake thermal efficiency, both of which depend on the fuel's energy content and the airflow rate into the engine (Ibrahim et al. 2017). Therefore, the calorific value (energy content) is pivotal in fuel selection. Alcohols often have a calorific value that is about three times lower than that of gasoline

Table 8.2 Properties of alcohols concerning gasoline (Awad et al. 2018)

	Gasoline	Butanol	Methanol	Ethanol	Fusel oil
Chemical formula	$C_{5-10}H_{12-22}$	$C_4H_{10}O$	CH_3OH	C_2H_5OH	$C_5H_{12}O$
CAS number	86290-81-5	35296-72-1	67-56-1	64-17-5	8013-75-0
Molecular Weight mass%	106.22	74.12	32.04	46.7	76.42
Carbon mass%	87.5	64.91	37.5	52.2	54.8
Hydrogen mass%	12.5	13.49	–	34.7	15
Oxygen mass%	0	21.6	49.93	34.7	30.32
Density (g/ml)	0.737	0.810	0.792	0.785	0.847
Boiling temperature (°C)	27–225	117.25	78	78.25	53.4–54.4
Reid vapour pressure (Kpa)	53–60	18.6	32.4	17	–
Research octane number	90–100	98	108.7	108.6–110	106.85
Motor octane number	82–92	78	86.6	92	103.72
LHOV (kJ/kg)	349	707.9	920	923	–
LHV (MJ/kg)	44.0	33.2	20.1	26.9	29.536
Freezing point (°C)	– 40	–	– 97.5	– 114	52
Viscosity (mm^2/s)	0.5–0.6	–	0.596	1.2–1.5	0.61
Flash point (°C)	– 45 to – 13	–	11	12–20	–
Auto-ignition temperature (°C)	257	385	423	425	41.6
Blend % of alcohol		35% (B35) 10–20% (*n*-B10, *n*-B20)	5–10% (M5, M10)	5–30% (E5, E10, E20, E30)	10–50% (F10, F20, F30, F50)
Effect on blended fuel RVP	–	– –	–	Increase by 10.33–10.99%	–
Effect on blended fuel octane number	–	– Dec. by 0.34–0.67%	–	Increase by 1.36–5. 56%	Increase by 1.4–15%

(continued)

Table 8.2 (continued)

	Gasoline	Butanol	Methanol	Ethanol	Fusel oil
Effect on blended fuel calorific value	–	Decrease by 3.54%	Decrease by 1.88–3.83%	Decrease by 4.76–14.7%	Decrease by 2–9%
		Decrease by 2.47–4.9%			

(see Table 8.2). A fuel's or biofuel's calorific value depends on several variables, including water content, fatty acids, carbon, hydrogen, and oxygen. Correlations between fuel characteristics such as density, viscosity, and calorific value have been noted by researchers. For biofuels, particularly alcohols, their lower calorific value stems from their reduced carbon content and increased oxygen content. Research, such as that by Demibras (Demirbas 2007), underscores oxygen's role in lowering a fuel's calorific value. Oxygen and water content tend to diminish the lower heating value (LHV) of fuel, whereas higher levels of hydrogen and sulphur can enhance it. Carbon content is notably significant, positively correlating with higher calorific values, which explains why fossil fuels generally have a higher heating value (HHV) compared to biofuels (Callejón-Ferre et al. 2011; Raveendran and Ganesh 1996). Moisture content adversely affects combustion efficiency, compromising engine performance (Calam et al. 2015b). To improve comprehension and predict calorific values, researchers have developed experimental correlations based on oxygen levels, carbon-hydrogen ratios, ash content, and water content data. These efforts aim to optimize fuel compositions and enhance their performance in various applications.

Auto-Ignition Temperature

Regular gasoline's auto-ignition temperature, also known as its fire point, can range from roughly 257 °C to as low as 221 °C (Yüksel and Yüksel 2004). Ethanol, in contrast, has an auto-ignition temperature of 425 °C, which is higher than that of gasoline. This characteristic is crucial for safety in transportation and storage, as it reduces the risk of spontaneous ignition in confined spaces due to vapour leaks. The auto-ignition temperature is defined as the temperature at which vapours ignite spontaneously in a confined environment. Therefore, the higher the auto-ignition temperature, much safer it is to transport. Factors such as turbulence can influence this temperature, with higher turbulence levels generally increasing the auto-ignition point (Egolf and Jurs 1992; Markides and Mastorakos 2011). Alcohol fuels, including methanol, exhibit combustion and ignition characteristics similar to traditional fuels.

Latent Heat of Vaporization

The latent heat of vaporization refers to the energy needed to transform a specific amount of liquid into vapour at a constant temperature and pressure, without altering the substance's temperature. The LHOV of ethanol (923 kJ/kg) is approximately 2.64 times the LHOV of gasoline (349 kJ/kg), thus indicating better heat-absorbing capacity. Fuels with high LHOV have also been found to result in reduced NO_X emissions (Karaosmanoğlu et al. 1997). Furthermore, it is thought that latent heat of vaporization (LHOV), which cools the air–fuel mixture as fuel vaporizes during the intake stroke in a direct-injection gasoline engine (DI SI), indirectly prevents engine knock. The familiar "cooling effect" occurs when ethanol in an internal combustion engine operates at a lower temperature, delaying the thermal ignition of the fuel in situations where knocking is likely to occur (Milpied et al. 2009).

Boiling Point and Volatility

Table 8.2 illustrates that alcohol typically has a lower boiling point than gasoline. One important factor that directly affects fuel volatility is the boiling point. Lower boiling point fuels evaporate more quickly and do not spill over time. Alcohols only have one boiling point, which has greatly aided in their increased release of energy (Milpied et al. 2009). Volatility refers to how readily alcohol fuels evaporate at lower temperatures. The carburetion or injection system cannot supply the proper air–fuel combination if vaporization does not take place within the engine's operating temperature range, which results in inefficient combustion. This considerably raises CO and HC emissions in addition to wasting gasoline. A fuel's flash point is also related to its volatility; highly evaporative compounds like gasoline pose hazards near sparks or open flames.

O_2 Content

Alcohols, distinguished by their high oxygen content, facilitate more complete and cleaner combustion, thereby reducing in-cylinder temperatures. This characteristic significantly enhances their combustion efficiency; however, the presence of oxygen limits fuel consumption efficiency by not contributing additional energy. The lower calorific value of alcohol and ether fuels compared to gasoline is due to their higher concentrations of oxygen and carbon (Demirbas 2004). Increased oxygen content in the fuel reduces heating value, while higher carbon content typically increases it (Awad et al. 2018). Understanding these relationships is crucial for optimizing fuel compositions and improving their combustion characteristics across diverse applications.

Phase Separation

Gasohol, a blend of alcohol, gasoline, and water, presents significant technical challenges, primarily stemming from the hygroscopic nature of alcohols (Osten and Sell 1983; French and Malone 2005). One of the most critical issues with using gasohol as a fuel is the phenomenon of phase separation, where the mixture divides into distinct liquid layers. This separation can be triggered by factors such as water content, temperature fluctuations, the specific type of alcohol used, and the composition of the gasoline. In ethanol-gasoline blends, water content emerges as the most significant technical hurdle. While gasoline itself has low water solubility and minor performance impacts, water in the fuel tank often accumulates at the bottom, leading to issues such as corrosion of engine components and clogging of fuel delivery systems. Tank bottom corrosion, attributed to the presence of water, is a notable concern in storage systems (Atadashi et al. 2012).

At low temperatures, water content poses additional risks, as it can freeze and obstruct fuel flow. Moreover, the presence of dissolved water reduces the calorific value of the fuel. If the water content is high enough, a separate water phase can form, displacing ethanol from the gasoline. This displacement disrupts the fuel's performance and efficiency, highlighting the necessity of maintaining dryness in alcohol-gasoline mixtures to avoid phase separation into water, alcohol, and hydrocarbon layers, especially under low-temperature conditions (Yanowitz et al. 2011).

To address these challenges, biodegradable surfactants such as Span 80 and Tween 80 can be incorporated into blends like hydrated bioethanol-gasoline and methanol-gasoline. These surfactants enhance the fuel's water tolerance, mitigate corrosive properties, and maintain fuel characteristics without significant alterations (Kassem et al. 2019). Additionally, the use of specific alcohols as co-solvents has been shown to improve blend stability by reducing the phase separation tendency (PST) of gasoline-alcohol mixtures. Such approaches represent a critical area of research and innovation aimed at overcoming the technical barriers associated with gasohol and enhancing its viability as a sustainable alternative fuel.

8.2.3 Technical Challenges: Solutions and Strategies

Challenges

The challenges of most used alcohol fuels are as follows:

i. **Toxicity and Corrosiveness**: Methanol is toxic and corrosive, requiring special fuel-handling systems (Deka et al. 2022).
ii. **Limited Greenhouse Gas Benefits**: Methanol derived from natural gas offers little to no net greenhouse gas benefits (Deka et al. 2022).
iii. **Engine Burns:** Ethanol can cause engine burns, necessitating ongoing research to improve its efficiency and reliability.
iv. **Large Land Requirement**: Producing ethanol on a large scale requires vast acres of land, leading to the potential destruction of natural habitats and additional costs for renting or leasing land (de Andrade Junior et al. 2019).
v. **Environmental Impact of Distillation**: The distillation process for ethanol production consumes significant amounts of heat, often generated from fossil fuels, resulting in high greenhouse gas emissions (Kiuchi et al. 2015).
vi. **Impact on Food Prices**: Increased demand for ethanol can drive up the price of corn, affecting the cost of ethanol production and other corn-dependent industries, potentially leading to higher food prices (https://www.epa.gov/environmental-economics/working-paper-impacts-ethanol-policy-corn-prices-review-and-meta-analysis).
vii. **Water Absorption**: Ethanol's hygroscopic nature leads to water contamination in fuel tanks which complicates its storage and use, as it can absorb moisture

from the environment, leading to engine problems, especially in marine applications (https://www.biobor.com/2020/05/06/the-challenges-associated-with-ethanol-blended-fuels/).

Solutions and Strategies

To address the challenges of alcohol fuels respectively, various research and technological advancements have proposed solutions:

i. Methanol's corrosive nature requires **specialized fuel-handling systems**. Using materials resistant to methanol corrosion, such as stainless steel and specific polymers, can mitigate this issue (Deka et al. 2022).
ii. While methanol from natural gas offers little net greenhouse gas benefits, **producing methanol from biomass or captured CO_2** significantly reduces its carbon footprint. Advanced methods include gasifying biomass and utilizing renewable electricity for methanol synthesis (Deka et al. 2022).
iii. Adding **corrosion inhibitors and modifying engine materials** can alleviate these issues. Research is also focused on developing ethanol-compatible engines and improving fuel formulations (Simon Araya et al. 2020).
iv. Large-scale ethanol production can increase land use and food prices. **Developing second-generation biofuels** from non-food biomass like agricultural residues and dedicated energy crops can reduce the impact on food supply and land use (Deka et al. 2022).
v. The distillation process for ethanol production is energy-intensive and typically relies on fossil fuels. Shifting to **renewable energy sources for distillation** and improving energy efficiency in the production process can mitigate its environmental impact (Simon Araya et al. 2020).
vi. Solutions to ethanol's hygroscopic nature include using additives to absorb water, improving storage systems, and refining ethanol to higher purity levels to minimize water content (Deka et al. 2022).
vii. The challenge of ethanol's difficulty in vaporizing in cold conditions can be effectively addressed through several methods, including increasing the quantity of injected fuel during startup and adjusting the Reid vapour pressure (RVP) of the blend. Research indicates that injecting about three times the usual amount of fuel or increasing the RVP to approximately 108 kPa can significantly enhance fuel vaporization and improve cold-start performance. Additionally, the use of fuel additives and blending ethanol with higher volatility fuels can further enhance its vaporization properties, ensuring reliable engine starts in cold environments (Aikawa et al. 2009).

These solutions are part of ongoing research and development aimed at making alcohol fuels more viable and sustainable as alternatives to traditional fossil fuels.

8.3 Sources and Production Methods

8.3.1 Sources of Alcohols

Many different types of organic materials, such as coal, biomass, natural gas, and organic wastes, can be converted into alcohol fuels. Historically, sugar cane and corn have been the main sources of alcohol fuels. The creation of alcohol fuels from non-food crops and agricultural waste, however, is now the main focus. For this reason, non-food lignocellulosic biomass which includes wastes, cellulosic residues, and energy crops has emerged as an essential source. A notable advancement has been the shift from first-generation, grain-based ethanol to second-generation, cellulosic ethanol, and more advanced cellulosic biofuels. After the 2000s, cellulosic ethanol became more well-known as a crucial metabolic pathway for turning biomass into fuels (Lynd et al. 2002). Third-generation feedstock derived from algae has also shown promise as a raw material for the alcohol fuel sector of the future. The main primary generations used to produce alcohol fuel are shown in Fig. 8.1.

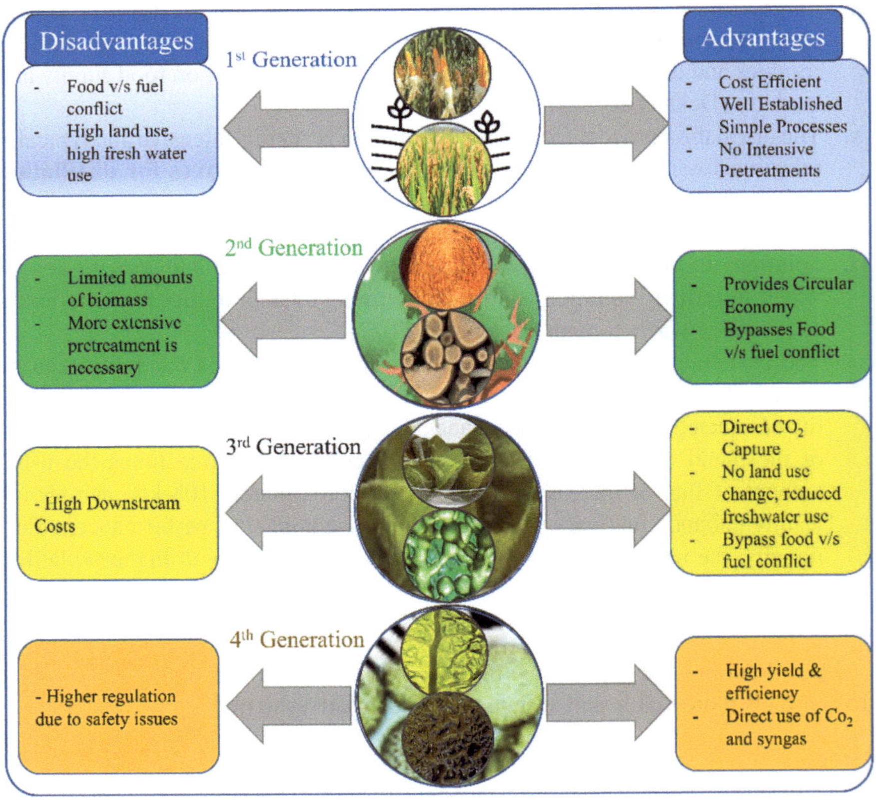

Fig. 8.1 Generation of sustainable renewable fuels from biomass

First Generation: Grain Feedstock

Traditionally, the primary raw materials used in the manufacturing of first-generation ethanol have been starch and carbohydrates (Zabed et al. 2017). In 2013, less than 10% of bioethanol originated from sources other than starch and carbohydrates, with corn, grain, and cassava being the primary crops utilized. This dependence on food crops for fuel raises significant issues, including environmental degradation linked to crop farming and ethanol production, as well as the ethical concerns surrounding the use of vital food resources for energy. So, in terms of producing biofuels, especially alcohol fuels, industrial powerhouses like the United States, Brazil, and China are at the forefront. Remarkably, corn starch accounts primary source of ethanol in the United States (Scully et al. 2021). In regions like Canada and the European Union, wheat and sugar beet also play crucial roles as first-generation feedstocks, capitalizing on the region's agricultural abundance despite facing similar sustainability and ethical challenges (Muir and Anderson 2022).

Second Generation: Lignocellulosic Biomass

In recent years, the production of bioethanol from grain-based sources has faced challenges, leading to a growing interest in second-generation bioethanol made from non-grain biomass. Traditionally, bioethanol has been produced from the starches of crops like corn and cassava, as well as from the sugars in sugar cane and sugar beets (Sanni et al. 2022). However, it can also be derived from lignocellulosic materials, such as agricultural waste (Mujtaba et al. 2023). While sugars and starches are easy to convert into bioethanol, they are often limited in supply and expensive. As a result, researchers are focusing on finding ways to produce bioethanol from more abundant and cost-effective lignocellulosic feedstocks. One major advantage of lignocellulosic crops over traditional sources is their high growth productivity, which is crucial for making bio-alcohol production economically viable (Silva et al. 2024). However, a key challenge is extracting valuable components from the complex structures of lignin, hemicellulose, and cellulose. This extraction process requires advanced technologies and can be costly. Additionally, despite their productivity advantages, lignocellulosic crops may need substantial land and water resources, raising potential environmental concerns (Haberzettl et al. 2021).

Third Generation: Algae Species

Third-generation biofuels primarily use algae as their raw material, which grow much faster than most land-based plants, making them a promising source of alcohol fuels. Unlike terrestrial crops, algae do not require extensive farmland, and non-edible species are particularly well-suited for bioethanol production. Efficient algae cultivation primarily depends on a direct supply of CO_2, which can be sourced from industrial emissions or captured directly from the atmosphere. In traditional cultivation systems, approximately 70% of the CO_2 provided is utilized for photosynthesis, leading to biomass production (Doucha and Lívanský 2006). Figure 8.2 shows the basic processes involved in biofuel production from third-generation feedstocks.

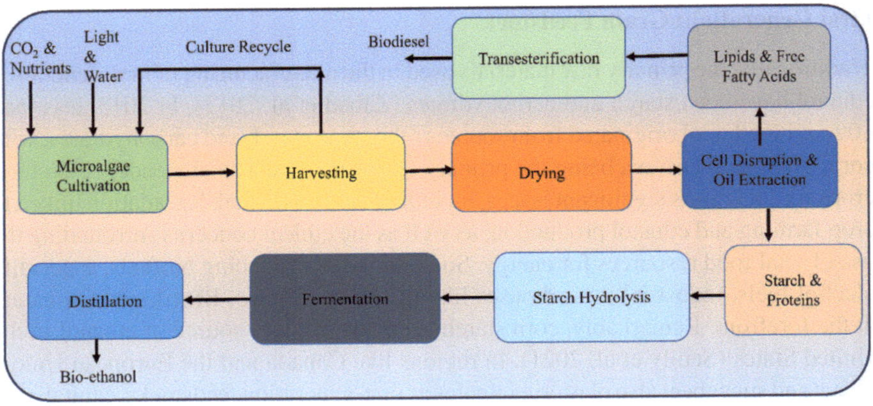

Fig. 8.2 Ethanol production from algae. Reproduced from the study by Scott-Kerr et al. (2010)

Third-generation biofuel production fully dissociates itself from the need for agricultural land, in contrast to second-generation biofuels. Furthermore, oil produced from algae may be produced more frequently than oil from higher plants. Unlike higher plants, which usually accumulate lipids in certain areas, like rape seeds, algae may produce large amounts of lipids in each cell, making their mass-production process more effective (Cavelius et al. 2023). Despite their potential, algae-based biofuels encounter considerable obstacles to commercial viability. The high costs of large-scale cultivation and harvesting pose significant economic challenges, necessitating technological advancements for efficient algae cultivation and processing. Consequently, further technological improvements in scaling up and reducing production costs are crucial for commercialization.

Fourth Generation: Genetic Engineering

The latest generation of biofuel production utilizes genetically engineered microalgae as the feedstock, aiming to overcome the limitations of earlier generations by using non-food biomass sources and improving yield and efficiency (Cavelius et al. 2023; Shokravi et al. 2022). This method is relevant to a range of characteristics, such as the use of various sugars (like pentoses and hexoses), increased lipid production, and enhanced photosynthesis and carbon fixation. For model organisms such as *Escherichia coli* and *Saccharomyces cerevisiae*, there are many tools available for genetically modifying the regulation of existing pathways or for introducing new pathways (Cavelius et al. 2023). One key example is the production of biobutanol from genetically modified algae. Butanol has several advantages over traditional biofuels like bioethanol, including higher energy density, lower corrosiveness, and the ability to be blended with gasoline at higher concentrations (Shokravi et al. 2022). Researchers are working to further optimize the production of biobutanol and other alcohols from engineered algal strains. However, the use of genetically modified organisms in biofuel production also comes with risks that need to be carefully managed, such as potential environmental impacts and socioeconomic effects. Ongoing research aims to develop robust mitigation strategies to enable the safe and sustainable scale-up of fourth-generation biofuels derived from genetically engineered algae.

8.3.2 *Production Methods of Alcohols*

Methanol Production Routes

Conventional Methanol Production

The traditional process for methanol production generally consists of four key steps: purification of the feed gas, steam reforming, methanol synthesis, and distillation. Figure 8.3 shows the process flow diagram of conventional methanol production.

The process descriptions are as follows:

i. **Feed Gas Purification**

The initial step in methanol production is the purification of the natural gas feedstock. This process involves the removal of impurities, particularly sulphur compounds, which can contaminate and deactivate the catalysts used in subsequent steps. Effective purification is essential to maintain catalyst activity and overall process efficiency (Ayas 2018).

ii. **Steam Reforming**

Following purification, the feed gas undergoes steam reforming. In this procedure, hydrocarbons are subjected to temperatures generally between 250 °C and 300 °C with steam, resulting in the formation of synthesis gas (syngas), a blend of carbon monoxide (CO) and hydrogen (H_2) (Ayas 2018). The general reaction

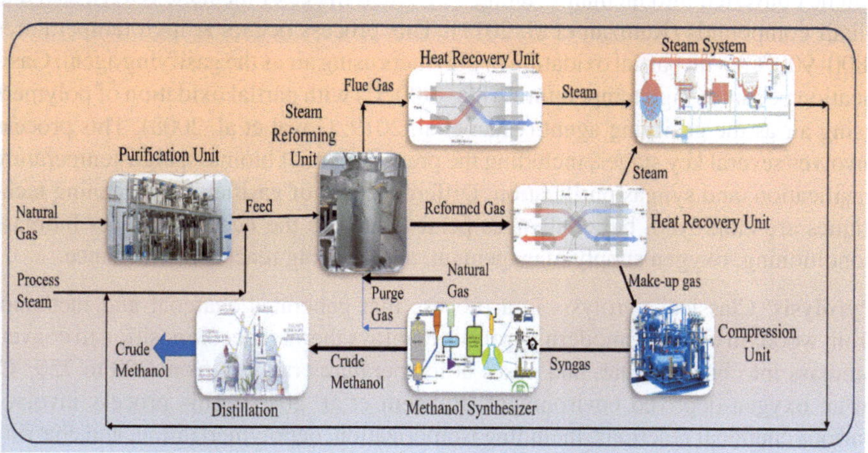

Fig. 8.3 Conventional production of methanol. Reproduced from the study by Ayas et al. (2018)

for steam reforming can be represented as follows:

$$C_nH_m + nH_2O \rightarrow nCO + (n + 2m)H_2.$$

iii. **Methanol Synthesis**

Methanol is then produced by passing the syngas over a catalyst bed at high pressures and temperatures. The key reactions involved are:

$$CO + 2H_2 \rightarrow CH_3OH, \quad \Delta H_{298K} = -21.7 \text{ kcal/mol}$$

$$CO_2 + 3H_2 \rightarrow CH_3OH + H_2O, \quad \Delta H_{298K} = -11.9 \text{ kcal/mol.}$$

iv. **Distillation**

The final step in the methanol production process is distillation. The crude methanol undergoes a two-step purification process to remove water and residual ethanol, ensuring that the final product meets stringent purity standards. The desired purity level is typically 99.85%, following ASTM D-1152/97 specifications. This high level of purity is essential for methanol's applications in various industries, including as a fuel and a chemical feedstock (Ayas 2018).

Advance methanol production route

i. **Thermochemical Process**

This route of methanol production involves four key steps: gasification of biomass, pyrolysis, liquefaction, and methanol synthesis from Syngas.

Gasification of Biomass: The gasification process transforms biomass waste into syngas, primarily composed of hydrogen (H_2) and carbon monoxide (CO), which can be converted into methanol; along with small traces of methane (CH_4), ash, and alkali compounds (Ramzan et al. 2011). This process occurs at high temperatures (800–900 °C) with partial oxidation of polymers using air as the gasifying agent. Gasification occurs at high temperatures (800–900 °C) with partial oxidation of polymers using air as the gasifying agent (Chan et al. 2019; Goyal et al. 2008). This process involves several key stages, including the pretreatment of biomass, high-temperature gasification, and syngas purification. Different types of gasifiers and cleaning techniques are employed, but challenges persist, such as the need for proper biomass conditioning, oxygen supply management, and ongoing reactor maintenance.

Pyrolysis: Classical pyrolysis is a process that generates charcoal and methanol from wood. In contrast, modern pyrolysis employs thermal decomposition to convert biomass into bio-oil, char, and gases at temperatures ranging from 200 to 750 °C, in an oxygen-deprived environment (Raheem et al. 2015). This process involves various chemical reactions, including isomerization, depolymerization, and charring (Collard and Blin 2014; Lange 2007). However, this bio-oil requires further catalytic upgrading before it can be converted into methanol.

Liquefaction: Liquefaction is a process that transforms wet biomass into liquid fuels by subjecting it to low temperature (200–500 °C) and high pressure (5–20 bar) conditions (Goyal et al. 2008). It involves reactions like dehydrogenation and decarboxylation, and it produces bio-oil, liquid tar, and condensable organic vapours (Raheem et al. 2015; Verma et al. 2012).

Synthesis from Syngas: Methanol production from syngas, primarily composed of H_2 and CO, is enhanced through catalytic conversion, yielding hydrocarbons and alcohols depending on catalytic conditions (Göransson et al. 2011). Catalysts like ZnO/Cr_2O_3 and $Cu/ZnO/Al_2O_3$ are used, with alkali doping producing higher alcohols (Yang and Ge 2016). Optimal methanol yield requires high pressure (300 bar) and temperature (200–400 °C). Adjusting H_2/CO ratios and removing CO_2 improves yield (Gerber et al. 2007).

ii. **Biochemical Process**

Biofuels can be produced efficiently from both fresh and waste biomass through biochemical processes such as anaerobic digestion, fermentation, and photo-biological conversion. While methane (60–70%), carbon dioxide, ammonia, hydrogen sulphide, and other trace gases are the main components of biogas produced by anaerobic digestion, ethanol is produced from fermentable biomass through alcoholic fermentation. This biogas is frequently used to generate energy or heat. However, biochemical techniques can be used to convert biogas into higher-value fuels like methanol with the right processing (Pittman et al. 2011; Voloshin et al. 2016; Lee et al. 2019).

Through a sequence of metabolic events, methane is transformed into carbon dioxide by methanotrophs, bacteria that use methane as their only carbon source. Methane monooxygenase (MMO) is an enzyme that is essential to this process and helps convert methane to methanol (Qiu et al. 2023; Guimarães et al. 2023; Gebreeyessus et al. 2019). This method presents a promising alternative for producing methanol from waste biomass (Vohra et al. 2014). Figure 8.4 gives an overview of the methanol production routes.

Ethanol Production Routes

Cellulosic Ethanol Production

The production of cellulosic ethanol involves several key steps that transform ligno-cellulosic biomass into ethanol. Figure 8.5 depicts the process flow diagram of cellulosic ethanol production. Here is a detailed overview of the production methods:

Feedstock Preparation: The process begins with the selection and collection of ligno-cellulosic biomasses, which includes agricultural residues such as corn stover, wheat straw, and rice straw (Chundawat et al. 2011). This biomass is characterized by low mass and energy densities, making it more suitable for distributed production facilities located near the feedstock supply to minimize logistics costs.

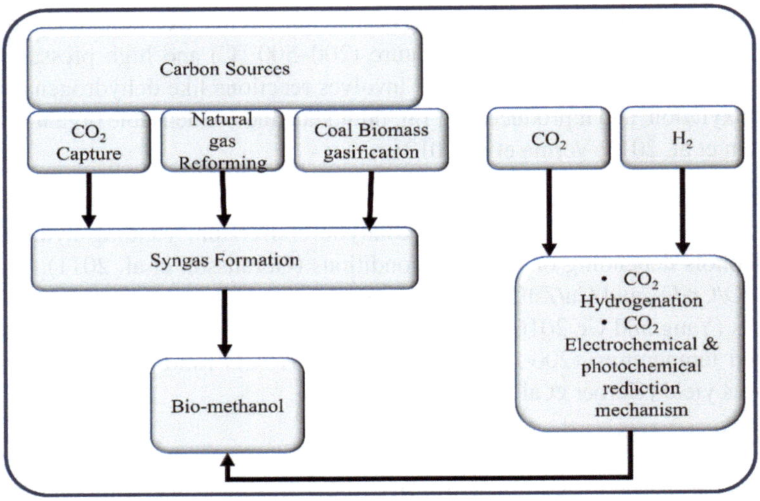

Fig. 8.4 Routes of methanol production

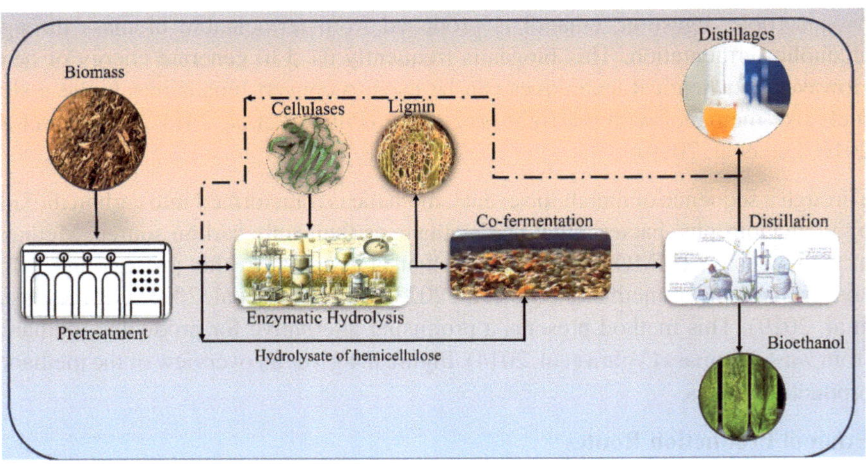

Fig. 8.5 Cellulosic bioethanol production. Reproduced from the study of Scott-Kerr et al. (2010)

Pretreatment: The complicated structures of cellulose, hemicellulose, and lignin make lignocellulosic biomass intrinsically resistant. It takes pretreatment to break down this structure and make cellulose more accessible for enzymatic hydrolysis. Lignin-carbohydrate complexes (LCCs) are broken down and hemicellulose and cellulose are released using a variety of pretreatment techniques, including physical, chemical, and biological treatments. The goal is to produce less inhibitory byproducts and improve sugar yields from subsequent enzymatic hydrolysis (Liu et al. 2019).

Enzymatic Hydrolysis: After pretreatment, the cellulose component undergoes enzymatic hydrolysis, where cellulase enzymes convert cellulose into glucose. This step is critical, as the efficiency of hydrolysis directly affects the overall yield of ethanol (Hahn-Hägerdal et al. 2006). The presence of lignin can hinder this process by occupying reactor space and adsorbing enzymes, leading to lower hydrolysis rates (Yuan et al. 2021).

Fermentation: The glucose obtained from hydrolysis, along with pentose sugars released from hemicellulose, must be fermented to produce ethanol. However, traditional yeast strains like *S. cerevisiae* can only ferment hexose sugars (like glucose) and not pentose sugars (like xylose) (Qiu et al. 2023). Therefore, engineered microorganisms or co-fermentation strategies are needed to efficiently convert both hexose and pentose sugars into ethanol. This may involve using recombinant strains that can metabolize pentose sugars or employing a two-step fermentation process.

Distillation: Once fermentation is complete, the resulting broth contains ethanol, water, and residual biomass. Distillation is employed to separate and purify ethanol from this mixture. However, the presence of inhibitors and the complexity of the feedstock can increase energy consumption during this stage, making it less efficient compared to first-generation ethanol production from sugar or starch (Guimarães et al. 2023).

Stillage Treatment: The leftover material after distillation, known as stillage, must be treated properly to minimize environmental impact. This can involve anaerobic digestion or other methods to recover energy or nutrients from the waste (Gebreeyessus et al. 2019).

In summary, the production of cellulosic ethanol involves the preparation of lignocellulosic biomass, pretreatment to enhance sugar availability, enzymatic hydrolysis to release glucose, fermentation to produce ethanol and subsequent distillation and treatment of byproducts. Ongoing research focuses on overcoming challenges related to efficiency, cost, and environmental impact to make cellulosic ethanol a viable alternative to conventional ethanol production.

Synthetic Ethanol Production

The main process used to create chemically manufactured ethanol is the hydration of ethylene, a byproduct of oil refinement. This process is efficient and widely used in industrial applications. Below is a detailed overview of the production method:

Ethylene Production: Ethylene (C_2H_4) is generated from the cracking of hydrocarbons, typically derived from petroleum or natural gas. This involves heating crude oil or natural gas to high temperatures and breaking down larger hydrocarbon molecules into smaller ones, including ethylene (Küüt et al. 2019).

Hydration of Ethylene: The core of the synthesis process is the hydration of ethylene to produce ethanol (C_2H_5OH). There are two primary methods for this reaction (Küüt et al. 2019):

Direct Catalytic Hydration: Ethylene is reacted with water vapours in the presence of a catalyst, usually phosphoric acid. The reaction can be represented as:

$$C_2H_4(g) + H_2O(g) \leftrightarrow C_2H_5OH(g) \quad \Delta H = -43.4 \text{ kJ/mol}.$$

This reaction occurs at temperatures between 250 and 300 °C and pressures of 5–8 MPa. The molar ratio of ethylene to water is typically maintained at 1:1.

Indirect Hydration (Sulphuric Acid Process): In this method, ethylene is absorbed in concentrated sulphuric acid to form ethyl hydrogen sulphate, which is then hydrolysed to produce ethanol. The steps are as follows:

a. Ethylene reacts with sulphuric acid:

$$C_2H_4 + H_2SO_4 \rightarrow C_2H_5OSO_3H.$$

b. Hydrolysis of ethyl hydrogen sulphate produces ethanol:

$$C_2H_5OSO_3H + H_2O \rightarrow C_2H_5OH + H_2SO_4.$$

Purification: After hydration, the resulting mixture contains ethanol, water, and unreacted ethylene. Distillation is used to separate and purify ethanol. By volume, the azeotropic combination is roughly 95.57% ethanol and 4.43% water (Küüt et al. 2019). Further dehydration is required to produce anhydrous ethanol; this can be done with techniques like further distillation or molecular sieves.

The production of chemically synthesized ethanol primarily involves the hydration of ethylene, either directly or through an indirect method using sulphuric acid. This process is efficient and allows for large-scale production of ethanol, making it a vital component in various industrial applications (Küüt et al. 2019).

Propanol Production Routes

Microbial Production of Propanol

The microbial production of propanol is gaining traction as a sustainable method to generate this valuable chemical from renewable resources. This process involves several key steps, which are detailed below.

Selection of Microbial Strains: Effective propanol production relies on the choice of microbial strains. Commonly used organisms include *E. coli*, various *Clostridium* species, and *Thermobifida fusca* (Walther and François 2016). These microorganisms can be genetically modified to enhance their propanol production capabilities, making them suitable for industrial applications.

Metabolic Pathways for Propanol Production: Several synthetic pathways facilitate propanol production. The acetone-dependent pathway uses acetone as an intermediate to generate isopropanol. The succinate-dependent pathway converts succinate into propanol through a series of enzymatic reactions (Walther and François 2016). The threonine pathway employs threonine as a precursor for propanol synthesis.

Genetic Engineering: Genetic engineering is vital to this process. It involves introducing and overexpressing specific genes that encode enzymes necessary for converting substrates into propanol. For example, in *E. coli*, overexpressing genes such as *methylmalonyl-mutase, arginine kinase, and methylmalonyl-CoA decarboxylase* can significantly increase propanol yields (Walther and François 2016). Incorporating the bifunctional butyraldehyde/alcohol dehydrogenase (adhE) gene from *Clostridium acetobutylicum* also enhances production.

Selection of Substrates: The selection of appropriate substrates is crucial. Renewable feedstocks like glucose, glycerol, and lignocellulosic biomass are commonly used. Glycerol, a byproduct of biodiesel production, is advantageous due to its additional NADH aiding in reduction processes (Walther and François 2016). Using enzymatic hydrolysis, lignocellulosic biomass can be transformed into glucose.

Optimization of Fermentation Conditions: Optimizing fermentation conditions is essential for maximizing yield. Key factors include temperature (higher temperatures enhance microbial tolerance and productivity), pH (maintained at optimal levels for microbial activity), aeration (depending on aerobic or anaerobic requirements), and continuous product removal to prevent toxicity and inhibit microbial growth (Walther and François 2016).

Downstream Processing: After fermentation, downstream processing separates and purifies propanol. Distillation is commonly used to separate propanol from the fermentation broth, while gas stripping can recover volatile products like butanol, applicable to propanol.

By focusing on these areas, microbial production of propanol offers a viable alternative to traditional chemical synthesis methods, contributing to a more sustainable chemical industry.

Butanol Production Routes

Biological Production Process

Biobutanol can be produced from lignocellulosic biomass (as shown in Fig. 8.6) through a series of steps involving pretreatment, hydrolysis, and fermentation.

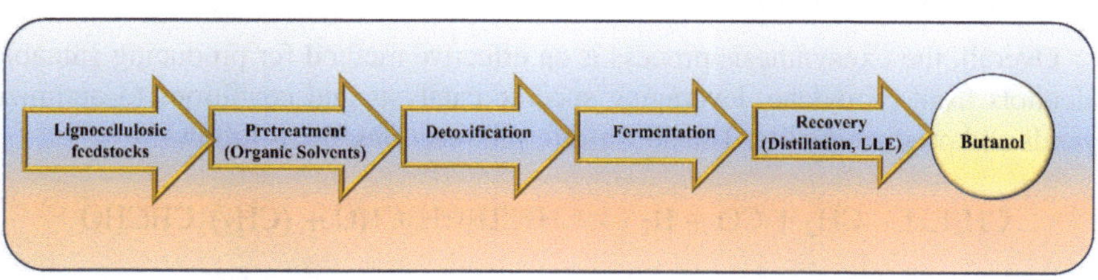

Fig. 8.6 Biological route of butanol production. Reproduced from the study of Liu et al. (2013)

The biomass is first pretreated to increase its accessibility for subsequent processing. A variety of pretreatment techniques, including steam explosion, hydrothermal, alkaline, and diluted acid pretreatment, can be used. These techniques aid in the disintegration of lignocellulose's intricate structure, increasing the hydrolysis susceptibility of cellulose and hemicellulose. After pretreatment, the biomass undergoes hydrolysis, which can be carried out using acids or enzymes. The fermentation step involves the conversion of the sugars into biobutanol using microorganisms, particularly from the Clostridium genus. These anaerobic bacteria can ferment a wide range of sugars to produce a mixture of butanol, acetone, and ethanol (Liu et al. 2013). After fermentation, the desired product (butanol) is separated and purified from the fermentation broth.

Chemical Production Processes

Butanol can be produced through various chemical technologies, including oxosynthesis and aldol condensation, in addition to the biological fermentation processes.

1. Oxosynthesis Process

   The oxosynthesis process entails the reaction of propylene with carbon monoxide and hydrogen, facilitated by a catalyst, usually cobalt or rhodium. This reaction produces a mixture of *n*-butyraldehyde and isobutyraldehyde, which can then be hydrogenated to generate *n*-butanol and isobutyl alcohol (Ayas 2018).

   Key Steps of the Oxosynthesis Process.

   i. Reaction of propylene, carbon monoxide, and hydrogen:

      At pressures between 10 and 20 MPa and temperatures between 130 and 160 °C, the reaction takes place when cobalt is employed as the catalyst. Consequently, there is an about 3:1 ratio of *n*-butyraldehyde to isobutyraldehyde (Ayas 2018). In contrast, using rhodium as the catalyst allows the reaction to take place at lower pressures (0.7–3 MPa) and temperatures (80–120 °C), significantly improving the *n*-butyraldehyde to isobutyraldehyde ratio to between 8 and 16:1 (Ayas 2018).

   ii. Hydrogenation of the Aldehyde Mixture:

      Nickel or copper catalysts in the gas phase or nickel catalysts in the liquid phase can be used to carry out the hydrogenation step. High pressure and temperature applied during this phase improve the final butanol product's purity (Ayas 2018).

Overall, the oxosynthesis process is an effective method for producing valuable alcohols from propylene, leveraging specific catalysts and conditions to optimize yields and product quality. The above-referred reactions are as follows:

$$CH_3CH - CH_2 + CO + H_2 \rightarrow CH_3CH_2CH_2CHO + (CH_3)_2CHCHO$$

$$CH_3CH_2CH_2CHO + H_2 \rightarrow CH_3CH_2CH_2CH_2OH$$

$$(CH_3)_2CHCHO + H_2 \rightarrow (CH_3)_2CHCH_2OH.$$

2. Aldol Condensation Process

   Another chemical route for butanol production is the aldol condensation process, which involves the following steps (Ayas 2018):

   i. Condensation and dehydration of two molecules of acetaldehyde to form crotonaldehyde.
   ii. Hydrogenation of crotonaldehyde to *n*-butanol at 180 °C and 0.2 MPa.

   The reaction is as follows:

$$CH_3CH = CHCHO + 2H_2 \rightarrow CH_3CH_2CH_2CH_2OH.$$

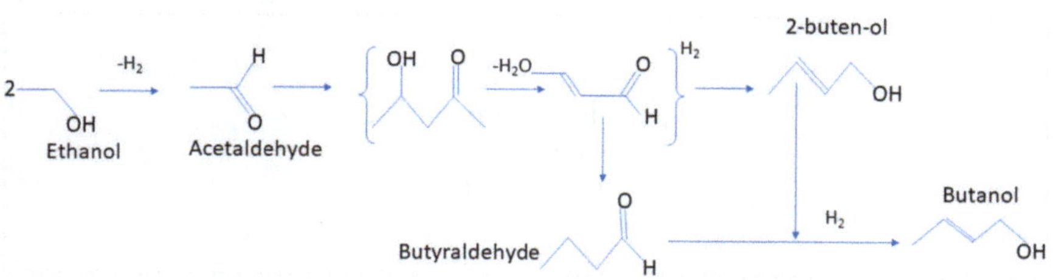

The reaction mechanism of the aldol condensation reaction to produce *n*-butanol from ethanol in a liquid phase is as follows (Mahmudul et al. 2016):

Compared to the oxo synthesis process, the aldol condensation route has the advantage of using more readily available raw materials (acetaldehyde). However, the oxosynthesis process is generally preferred for industrial-scale *n*-butanol production due to its overall advantages (Ayas 2018).

The choice between chemical and biological (fermentation) routes for butanol production also depends on factors such as raw material availability, process efficiency, and environmental considerations.

8.4 Alcohol Compatibility in IC Engines

Currently, internal combustion engines are extensively utilized in several sectors including industrial, agriculture, transportation, and marine applications. The compatibility of internal combustion engines is enhanced when they can operate on numerous fuels without requiring significant adjustments, while also generating electricity with low environmental consequences (Yue and Liu 2023). As depicted in the diagram (Fig. 8.7), any alternative fuels are deemed appropriate provided they demonstrate sustainability across various crucial factors, including

Fig. 8.7 Combability of alcohol on internal combustion engine

fuel compatibility, materials, emission control, mechanical aspects, and technological compatibility (Muthuraman et al. 2022).

8.4.1 Alcohol Fuel for CI Engine

Diesel engines are widely utilized in industries, agriculture, marine applications, and transportation due to their advantageous characteristics, including higher brake thermal efficiency, brake power, adaptability, reliability, and cost-effectiveness. Meanwhile, diesel fuels are automobiles believed to be significant contributors to environmental pollution and are considered a major factor in the development of energy shortages. To mitigate the energy crisis and comply with the new emission laws, alternative fuels are considered one of the most efficient alternatives (Wei et al. 2015; Aydin and İlkılıç 2010). Alcohols, which are alternative fuels, have been extensively utilized in CI engines as additives or directly to the engine. While alcohols offer economic advantages in CI engines compared to traditional diesel and can be used as a single fuel or additive without any engine modifications (Çelebi and Aydın 2018), some challenges need to be addressed in their utilization. These challenges include

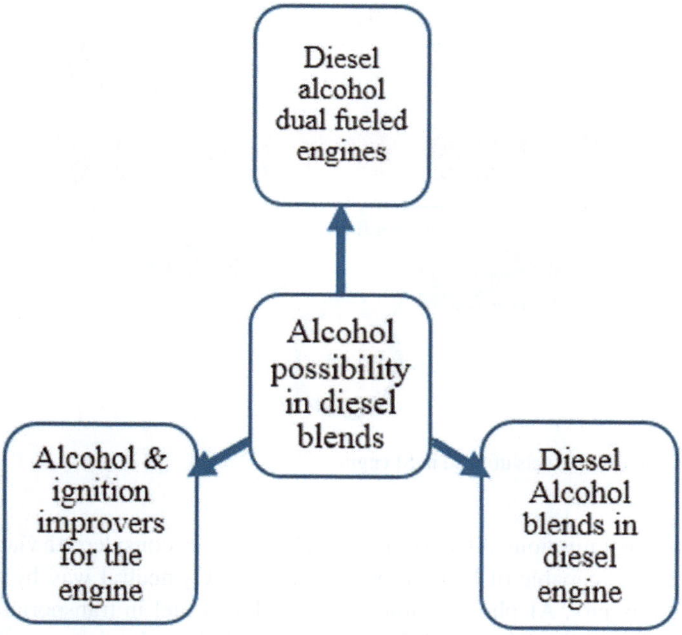

Fig. 8.8 Steps for alcohol possibility in diesel blends

blending with diesel fuel, low lubricity, difficulty in vaporization, and high auto-ignition temperature. These challenges can be resolved by several approaches, such as including chemicals or elevating the intake air temperature (Yilmaz et al. 2005; Yilmaz and Donaldson 2007). Combining alcohol with diesel fuels is an alternative method for addressing issues related to lubrication and vaporization. However, alcohols and diesel fuel do not mix and need the assistance of co-solvents (1-Dodecanal, Heptanol, octanol etc.) or emulsifiers to combine these liquids as depicted in Fig. 8.8. Nevertheless, these techniques incur high costs and the procedure of combining the ingredients may provide challenges such as blending, separation, splashing, heating, or other procedural processes (Theinnoi et al. 2021). While alcohols do not mix with diesel fuel, they do mix with different biodiesel. Under specific constraints, alcohols, diesel, and biodiesel can be used in a mixture without encountering any issues of miscibility (Mahmudul et al. 2016).

8.4.2 *Alcohol Fuel for SI Engine*

Alcohols are appealing alternative fuels because of their ability to be derived from both natural and synthetic sources. To tackle rising pollution levels and achieve sustainability goals, it is necessary to enhance the utilization of alternative fuels,

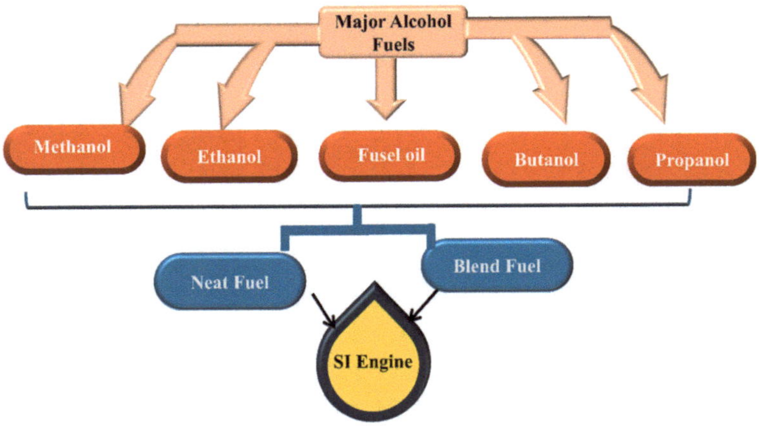

Fig. 8.9 Major alcohol fuels utilized in SI engine

which have zero emissions. Alcohol-based fuels have been considered a viable renewable alternative, capable of being used in a nearly CO_2-neutral way by efficiently converting biomass. Alcohol has long been used as a fuel in transportation, especially in small vehicles (Yusri et al. 2017; White 1907), but it has just started to attract major global attention. Apart from conventional fuels like diesel and gasoline, alcohols including ethanol, butanol, methanol, and fusel oil (as shown in Fig. 8.9), as well as ethers such as methyl tertiary butyl ether (MTBE) and dimethyl ether (DME), are utilized as fuels in internal combustion engines (Theinnoi et al. 2021; Mahmudul et al. 2016; Pourkhesalian et al. 2010; Semelsberger et al. 2006). Historically, the initial types of alcohol used in Spark Ignition (SI) engines have primarily consisted of mixtures of gasoline and ethanol. Current fuel quality regulations generally permit ethanol to be added to gasoline at levels ranging from 5 to 20%.

8.4.3 Alcohol Fuel for Hybrid Flex Fuel Engine

Alcohol-powered hybrid vehicles as shown in Fig. 8.10 offer a promising blend of renewable energy and advanced automotive technology (Zacharias et al. 2023). These vehicles combine the high efficiency of hybrid powertrains with the environmental benefits of alcohol fuels like ethanol and methanol. Ethanol and battery-powered hybrid vehicles (EBHVs) are potential alternatives to conventional fossil fuel-based vehicles. These vehicles combine an ethanol engine with an electric motor and a battery, allowing them to switch between ethanol and electricity as needed. They could offer a range of benefits for the sugar industry, the environment, and the transportation sector. But they also face some challenges and limitations (Patil 2024). These cars can profit from the first-ever plug-in electric hybrid flex fuel

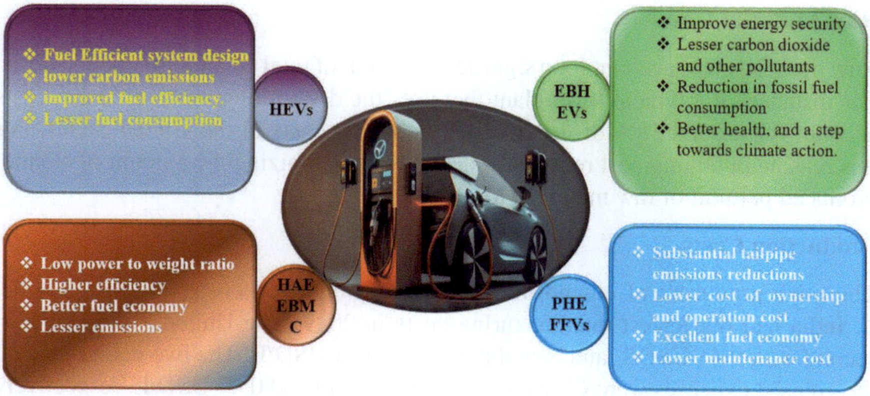

Fig. 8.10 Hybrid flex fuel engine

vehicle in the world's economic and environmental domains (PHEFFV) (https://ethanolrfa.org/media-and-news/category/news-releases/article/2024/06/new-study-shows-plug-in-hybrid-flex-fuel-vehicle-has-major-economic-climate-advantages-over-battery-evs).

A redesigned Miller-cycle engine with regenerative air preheating and auto-ignition similar to a diesel engine would be the internal combustion subsystem of a proposed hybrid (internal combustion/electric) vehicular engine system (Jahanbakhshi et al. 2021). To achieve full combustion, the fuel ethanol would be burned lean. Even though the suggested engine would have a lower power-to-weight ratio than the majority of current engines, this would not be an issue if the engine were utilized in a non-hybrid system because hybrid systems require significantly less power, allowing for smaller engines than those that are only powered by internal combustion engines (https://www.techbriefs.com/component/content/article/1559-lar16365).

8.5 Economic and Environmental Sustainability

8.5.1 Economic Aspects of Alcohol-Based Fuels

Ethanol production has significant economic implications, both at the individual producer level and the broader market scale. The cost of producing ethanol varies across countries due to differences in raw material costs, production methods, and economies of scale.

Raw Material Costs

Raw material costs account for a significant portion of total ethanol production costs. For instance, at a typical Iowa plant that uses the dry-milling method, maize costs account for around 58% of the overall production costs in the United States. For this reason, obtaining cost reductions depends on optimizing the amount of ethanol produced per unit of raw material (Mizik 2021).

Production Costs

The main producers of ethanol have different production costs. With 0.37 USD/L, India had the lowest manufacturing costs in 2019. The United States came in second with 0.41 USD/L and Brazil third with 0.44 USD/L. The two countries with the greatest expenses were China and the EU, at 0.49 and 0.47 USD/L, respectively (Mizik 2021). However, when adjusted for ethanol's lower energy content than gasoline, India's production cost was still 1.82% higher than the gasoline price (Mizik 2021).

Competitiveness with Gasoline

At current oil prices, ethanol production is generally not competitive with gasoline on a purely cost basis. In 2019, the West Texas Intermediate crude oil price averaged 57 USD/barrel, equivalent to 0.36 USD/L. Even the most efficient producer, India, had a production cost higher than the gasoline price (Mizik 2021). This suggests that without subsidies, tax credits, or mandatory blending rates, ethanol production would be unprofitable worldwide (Mizik 2021).

Market Concentration

The top two producers, the United States and Brazil accounted for 74% of the world's ethanol output in 2019. This indicates that the ethanol industry is extremely consolidated. Brazil produced 36,238 million litres (28% share), compared to the US 59,809 million litres (46% of total production). The other major producers were China (8%), the EU (5%), and India (3%) (Mizik 2021).

Coproducts and Biorefineries

The use of coproducts and the development of integrated biorefineries can improve the economic viability of ethanol production. Coproducts such as corn germ oil, fibre, and animal feed from wet-milling processes can provide additional revenue streams (Mizik 2021). Integrated biorefineries that combine first-generation (1G) and advanced technologies, such as lignocellulosic or algae-based production, can also enhance economic performance (Mizik 2021).

In conclusion, while ethanol production has significant economic implications, it currently faces challenges in terms of competitiveness with gasoline. Reducing raw material costs, improving yields, and developing integrated biorefineries that utilize coproducts can help improve the economic viability of ethanol production. However, government support through subsidies, tax credits, and mandatory blending rates may still be necessary to ensure the continued growth of the ethanol industry.

8.5.2 Environmental Aspects of Alcohol-Based Fuels

Compared to conventional fossil fuels, alcohol-based fuels like methanol, butanol, and ethanol have major environmental advantages. Since the transportation industry contributes significantly to air pollution and greenhouse gas (GHG) emissions, adopting alternative fuels is essential to reducing climate change and enhancing air quality.

Reduced Greenhouse Gas Emissions

The process of producing and burning ethanol leads to a more balanced carbon cycle. The amount of CO_2 taken by plants during their growth is approximately equal to the amount emitted after the combustion of ethanol. Studies show that compared to gasoline, ethanol made from grains can reduce greenhouse gas emissions by 44–52%. It is projected that new technology will raise this decrease to roughly 70% in the upcoming years (Dees et al. 2023).

Butanol, another promising alcohol-based fuel, can also significantly lower tailpipe emissions and environmental impacts in the fuel use stage compared to gasoline. A life cycle assessment study found that using a farmed wood butanol-gasoline blend as an alternative transport fuel for passenger cars has the potential to reduce greenhouse gas emissions (Feinauer et al. 2021).

Lower Tailpipe Emissions

Fuels combined with alcohol have been shown to dramatically lower tailpipe emissions of air toxics such as benzene, exhaust hydrocarbons, carbon monoxide, and fine particulate matter from vehicles. According to studies, using ethanol-blended fuels reduces the amount of harmful emissions that cars emit into the air and poses a lesser health risk to people than using conventional gasoline (Mueller et al. 2021; Suarez-Bertoa et al. 2015).

Butanol and isobutanol have also been shown to lower particulate matter emissions from direct-injection spark ignition engines during laboratory and on-road tests compared to gasoline (Vojtisek-Lom et al. 2015).

Methanol blends can reduce emissions of carbon monoxide and hydrocarbons significantly. However, methanol has a lower energy density than gasoline, which can lead to slightly higher fuel consumption and emissions of CO_2 per mile driven. Despite this, the overall reduction in pollutants like carbon monoxide and particulates can still provide substantial environmental benefits.

Biodegradability and Reduced Spill Impacts

Alcohols like ethanol, butanol, and methanol are biodegradable and break down into harmless substances when spilt, making them less harmful to the environment than fossil fuels. However, they are flammable and require careful handling. Fuel-grade ethanol often contains denaturants to prevent consumption, and methanol is toxic, posing serious health risks if ingested. Despite their benefits, these alcohols still present safety concerns that must be addressed.

Potential Drawbacks and Considerations

While alcohol-based fuel production offers environmental benefits, it is not without its challenges. The conversion of natural land into cropland to support increased production can lead to a temporary increase in GHG emissions due to the release of carbon stored in soil and vegetation. This concept is known as carbon debt. Over time, however, the use of these fuels can still lead to lower overall emissions compared to continued reliance on fossil fuels. Additionally, the production of alcohol-based fuels requires significant amounts of water and energy, which can have indirect environmental impacts. The choice of feedstock and production methods can also influence the overall environmental impact. Specifically, methanol production from natural gas can result in higher GHG emissions if the natural gas is not sourced sustainably.

8.5.3 Sustainability

The sustainability of using alcohol-based fuels, such as methanol, ethanol, and butanol, encompasses a range of environmental, economic, and social factors. These alcohols are considered renewable alternatives to fossil fuels, with varying degrees of sustainability depending on their sources and production methods. Methanol, produced from natural gas or biomass, offers a renewable fuel option with significant greenhouse gas reduction potential, especially when using CO_2 capture. Butanol, with higher energy density than ethanol, can be produced from lignocellulosic biomass or algae, avoiding food competition and enhancing sustainability. Despite higher costs due to complex processes, technological advancements are reducing production costs. In conclusion, while first-generation alcohol-based fuels provide a viable short-term solution for reducing greenhouse gas emissions and dependence on fossil fuels, they pose economic and social sustainability challenges. Advanced alcohol technologies, although currently more expensive, offer a more sustainable long-term solution by utilizing non-food biomass and innovative production methods. A comprehensive sustainability analysis, considering environmental, economic, and social impacts, is essential for the future development and adoption of alcohol-based fuels like methanol, ethanol, and butanol.

8.6 Vulnerability of Alcohol-Based Fuels

Alcohol-based fuels, predominantly ethanol and methanol, are gaining traction as alternatives to traditional fossil fuels. While they offer benefits such as reduced greenhouse gas emissions and potential for renewable production, they also present significant safety and health risks. Understanding these vulnerabilities is crucial for safe handling and effective use as shown in Fig. 8.11.

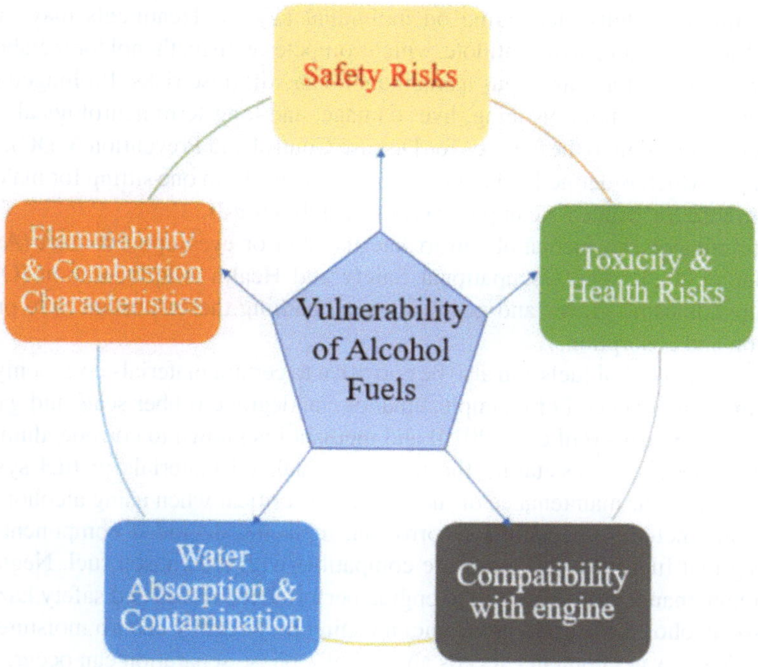

Fig. 8.11 Risks of alcohol fuel usage

Alcohol-based fuels, predominantly ethanol and methanol, are gaining traction as alternatives to traditional fossil fuels. While they offer benefits such as reduced greenhouse gas emissions and the potential for renewable production, they also present significant safety and health risks. Understanding these vulnerabilities is crucial for safe handling and effective use.

Alcohol-based fuels are highly flammable, with ethanol having a flash point of approximately 13 °C (55 °F) and methanol at about 11 °C (52 °F). This places them in the Class IB flammable liquid category, indicating a high risk of ignition. According to the National Fire Protection Association (NFPA), alcohol fuels can ignite easily from sparks, static electricity, or open flames, necessitating stringent safety measures during storage and handling.

The vapours of alcohol-based fuels can pose significant health risks. Methanol vapours are particularly toxic, with exposure leading to symptoms such as dizziness, headache, and nausea. A study found that inhalation of methanol vapours at concentrations as low as 200 ppm can cause acute health effects (Ernstgård et al. 2005). Prolonged exposure may result in severe respiratory distress and neurological damage, underscoring the need for adequate ventilation and personal protective equipment (PPE).

Methanol ingestion can lead to metabolic acidosis, central nervous system depression, and visual disturbances, potentially resulting in permanent blindness or death (Ashurst and Nappe 2024). The lethal dose of methanol is estimated to be 30 to 240 mL for adults, depending on individual factors. Treatments may include ethanol administration as an antidote, which competes with methanol for metabolism. Ethanol, while not as hazardous as methanol, can still pose risks. Prolonged exposure can cause alcohol poisoning, liver damage, and long-term neurological consequences. According to the Centres for Disease Control and Prevention (CDC), binge drinking—which is defined as having five or more drinks in one sitting for males and four or more for women—can pose serious health hazards.

Both ethanol and methanol can irritate the skin or eyes upon contact (Ashurst and Nappe 2024). The Occupational Safety and Health Administration (OSHA) recommends using gloves and goggles when handling these substances to prevent irritation and absorption.

Moreover, alcohol fuels can also be corrosive to certain materials commonly used in engine components. For example, ethanol can degrade rubber seals and gaskets, leading to fuel leaks (Gil et al. 2010) and methanol is known to corrode aluminium and other metals, necessitating the use of specialized materials in fuel systems. Therefore, regular maintenance of fuel systems is critical when using alcohol-based fuels. This includes checking for corrosion, replacing degraded components, and ensuring that fuel lines and tanks are compatible with the chosen fuel. Neglecting these maintenance tasks can lead to engine performance issues and safety hazards.

Also, alcohol fuels are hygroscopic, meaning they readily absorb moisture from the air. When water content exceeds about 0.5%, phase separation can occur, where water and fuel separate into distinct layers. This can lead to engine misfires and reduced efficiency. Phase separation in ethanol-blended fuels can lead to a decrease in energy content and negatively impact vehicle performance, as the water that separates from the fuel can dilute the fuel mixture. Additionally, water contamination can promote microbial growth, leading to the formation of biofilms and sludge in fuel tanks. This can clog filters and injectors, causing performance issues and increasing maintenance costs. The National Renewable Energy Laboratory (NREL) estimates that microbial contamination can lead to fuel-related failures in diesel engines using biodiesel blends.

Alcohol-based fuels have different combustion characteristics compared to gasoline. Ethanol has a higher octane rating (around 100 compared to gasoline's average of 87–93), which can improve engine performance and reduce knocking. However, its higher heat of vaporization can lead to cold-start difficulties, particularly in colder climates. While alcohol fuels can produce fewer greenhouse gas emissions, their combustion can generate harmful byproducts. Incomplete combustion may lead to higher emissions of formaldehyde and other volatile organic compounds (VOCs). According to the Environmental Protection Agency (EPA), ethanol can produce more acetaldehyde than gasoline, contributing to air quality issues.

Hence, it can be concluded that alcohol-based fuels present both opportunities and challenges in the transition to sustainable energy sources. While they can reduce reliance on fossil fuels and lower carbon emissions, their safety, toxicity, and health risks must be managed effectively. By implementing proper handling procedures, ensuring material compatibility, and maintaining fuel systems, users can mitigate the risks associated with alcohol-based fuels.

8.7 Safety Considerations for Alcohol-Fuelled IC Engine

Alcohol-powered internal combustion (IC) engines offer an alternative to conventional fossil fuels, offering potential advantages in terms of reduced emissions and the utilization of renewable resources (Kasibhatta 2019). Nevertheless, there are safety factors that need to be considered to guarantee secure functioning. These issues encompass a range of factors, including the proper handling and storage of alcohol fuels, as well as the construction and maintenance of the engine.

8.7.1 Fuel Handling and Storage

Proper fuel handling and storage of alcohol, such as ethanol, is essential for safety and efficiency. One should store alcohol in well-ventilated places, away from heat sources and direct sunlight, in authorized containers that are clearly labelled as flammable (https://www.gexcon.com/blog/ethanol-associated-hazards-and-safety-measures/). Using non-sparking tools and properly grounded equipment is essential to prevent static discharge during handling. It is important to routinely inspect storage areas for any leaks or damage, and to have spill containment measures in place along with readily accessible fire extinguishers. Hence, the employees should be trained on safe handling practices and emergency protocols. To mitigate the risks associated with the flammability of alcohol fuels, compliance with local regulations and standards must be ensured.

Alcohol fuels possess a significantly higher level of flammability compared to diesel, necessitating stringent safety measures during storage and handling (Yusri et al. 2017). To mitigate the risk of unintentional ignition, storage facilities should incorporate flame arresters and pressure relief valves on fuel tanks, reducing the likelihood of explosions (Vaivads et al. 1994). Designated handling areas must be marked as non-smoking zones and equipped with appropriate fire suppression systems, such as foam or dry chemical extinguishers, to address potential fire hazards effectively.

Additionally, alcohol fuels exhibit corrosive effects on certain metals and rubber components commonly used in fuel systems (Kumar and Ashok 2023). To ensure durability and safety, materials in contact with alcohol fuels should be selected carefully, favouring stainless steel or specially treated metals for mechanical components and resistant rubber or plastics for seals and hoses. Hence, regular inspections and maintenance of fuel lines, tanks, and seals are essential to detect and address potential leaks or material degradation.

8.7.2 Engine Design and Modifications

Alcohol-based fuels exhibit distinct combustion properties in contrast to gasoline or diesel. Engines must undergo tuning and calibration to maintain optimal combustion and minimize knocking, which has the potential to harm engine parts (Jacob and Ashok 2020). Modifications to fuel injection systems may be necessary to accommodate the varying viscosity and energy content of alcohol fuels. The operating heat level of the engine may be affected by the fact that alcohol-based fuels normally burn at a lower temperature than gasoline. Make sure that the cooling system has enough capacity to keep the engine running at its optimal temperature and avoid overcooling or overheating. Alcohol can eliminate lubricants, increasing friction and damage to engine components (Costa et al. 2023). Consequently, employing lubricants specifically designed for alcohol-powered engines ensures sufficient lubrication and reduces friction.

8.8 Future Prospect and Research Directions

The transition towards sustainable energy sources has prompted significant interest in alcohol fuels as viable alternatives to traditional fossil fuels. Advances in alcohol fuel technology, combined with the potential for renewable alcohol production, along with the policy and incentive frameworks present promising pathways for achieving energy sustainability.

Recent advancements in alcohol fuel technology have highlighted the potential of various alcohols, including methanol, ethanol, propanol, and butanol, as effective diesel substitutes. Research indicates that blending these alcohols with diesel can enhance engine performance while reducing emissions. For instance, studies have shown that ethanol–diesel blends can increase combustion efficiency and reduce soot emissions, although they may also lead to higher carbon monoxide (CO) emissions at certain blend ratios (Kim et al. 2020). Furthermore, the development of catalysts and additives to improve the combustion properties of these blends is an active area of research, aiming to optimize performance and minimize environmental impact (Çelebi and Aydın 2019; Kumar et al. 2013).

The economic viability of bioethanol production hinges on the development of cost-effective technologies that can efficiently convert low-energy density feedstocks into alcohol fuels. Countries in Africa and South America, with their favourable climates and abundant agricultural resources, are well-positioned to increase bioenergy production, including alcohol fuels. The potential for large-scale production of bioethanol and other alcohols from locally available feedstocks could significantly contribute to energy security and economic development in these regions (Kumar et al. 2013).

Apart from the above parameters, for alcohol fuels to realize their full potential, supportive policy and incentive frameworks are essential. Governments can play a pivotal role by implementing regulations that promote the use of renewable fuels, such as tax incentives for biofuel production and mandates for blending alcohol fuels with conventional diesel. Therefore, investment in research and development is crucial to overcoming the technological barriers associated with alcohol fuel production and utilization. Policies that encourage public–private partnerships and funding for innovative technologies can accelerate progress in this field, making alcohol fuels more competitive with traditional fossil fuels (Kumar et al. 2013).

8.9 Conclusion

This chapter discusses the potential of alcohol fuels as sustainable alternatives to traditional fossil fuels, emphasizing their role in enhancing energy security, reducing emissions, and driving economic resilience, particularly in agricultural regions. Alcohol fuels, including methanol, ethanol, propanol, and butanol, are distinguished by their higher octane ratings, renewable availability, and lower environmental impact. They can be produced from a variety of feedstocks such as lignocellulosic biomass and waste materials, reducing competition with food crops and contributing to waste management.

Regions with abundant agricultural resources are uniquely positioned to capitalize on the large-scale production of alcohol fuels like bioethanol, offering opportunities for job creation and energy independence. By utilizing local resources, communities can reduce their reliance on volatile fossil fuel markets and foster economic growth, as demonstrated by the successful implementation of alcohol fuel programmes in countries such as the United States, Brazil, and China. These case studies illustrate the practicality of flexible fuel vehicle (FFV) adoption and blending initiatives, which have driven widespread acceptance of alcohol fuels in transportation systems.

From a technological perspective, alcohol fuels have inherent advantages due to their high oxygen content, which improves combustion efficiency and lowers emissions. Studies have shown that alcohol fuels produce significantly lower levels of pollutants, including carbon monoxide (CO), hydrocarbons (HC), nitrogen oxides (NO_x), and particulates. Moreover, advancements in internal combustion engine design and innovative fuel management systems have enabled modern engines to operate effectively on alcohol blends with minimal modifications.

However, challenges such as phase separation in alcohol-fossil fuel blends, handling safety concerns, and the need for engine recalibration remain. Addressing these issues requires sustained research and development, supported by effective policy frameworks and incentives. Regulatory measures that promote the sustainable production and adoption of alcohol fuels, coupled with public–private research collaborations, are vital to overcoming these barriers.

Looking to the future, continued innovation in alcohol fuel technologies and supportive policies will likely accelerate their integration into existing energy systems. Their ability to offer cleaner-burning alternatives, reduce reliance on finite fossil fuel resources, and stimulate rural economic development makes them a key component of a sustainable energy strategy. By embracing these renewable energy solutions, the global community can take significant steps towards mitigating climate change and building a resilient energy future.

In conclusion, alcohol fuels exemplify the intersection of environmental sustainability, technological innovation, and economic growth, underscoring their importance in shaping a balanced and sustainable energy policy.

Acknowledgements The project is supported by the Rajiv Gandhi Institute of Petroleum Technology (RGIPT), Jais, which is gratefully acknowledged.

CRediT Authorship Contribution Statement Ajeet Kumar Prajapati: Conceptualization, Writing—original draft, Methodology, Formal analysis, Data curation, Conceptualization.

Orcid Id: 0000-0003-1717-0762

Aditya Kashyap: Writing—original draft, Methodology, Formal analysis; Rakesh Kumar: Writing—review and editing, Supervision, Resources

Data Availability Data will be made available on request.

Declaration of Competing Interest The authors declare that they have no known competing financial interests or personal relationships that could have appeared to influence the work reported in this chapter.

References

Agarwal AK (2007) Biofuels (alcohols and biodiesel) applications as fuels for internal combustion engines. Prog Energy Combust Sci 33:233–271. https://doi.org/10.1016/j.pecs.2006.08.003

Aikawa K, Sakurai T, Hayashi A, Aikawa K, Sakurai T, Hayashi A (2009) Study of ethanol-blended fuel (E85) effects under cold-start conditions. https://doi.org/10.4271/2009-01-0620

Andersen VF, Anderson JE, Wallington TJ, Mueller SA, Nielsen OJ (2010) Vapor pressures of Alcohol−Gasoline blends. Energy Fuels 24:3647–3654. https://doi.org/10.1021/ef100254w

Ashurst JV, Nappe TM (2024) Methanol toxicity. In: StatPearls. StatPearls Publishing, Treasure Island (FL)

Association RF. New study shows plug-in hybrid flex fuel vehicle has major economic, climate advantages over battery EVs. https://ethanolrfa.org/media-and-news/category/news-releases/article/2024/06/new-study-shows-plug-in-hybrid-flex-fuel-vehicle-has-major-economic-climate-advantages-over-battery-evs

Atadashi IM, Aroua MK, Abdul Aziz AR, Sulaiman N.M.N.: The effects of water on biodiesel production and refining technologies: a review. Renew Sustain Energy Rev 16:3456–3470. https://doi.org/10.1016/j.rser.2012.03.004

Awad OI, Ali OM, Mamat R, Abdullah AA, Najafi G, Kamarulzaman MK, Yusri IM, Noor MM (2017a) Using fusel oil as a blend in gasoline to improve SI engine efficiencies: a comprehensive review. Renew Sustain Energy Rev 69:1232–1242. https://doi.org/10.1016/j.rser.2016.11.244

Awad OI, Mamat R, Ibrahim TK, Hagos FY, Noor MM, Yusri IM, Leman AM (2017b) Calorific value enhancement of fusel oil by moisture removal and its effect on the performance and combustion of a spark ignition engine. Energy Convers Manag 137:86–96. https://doi.org/10.1016/j.enconman.2017.01.009

Awad OI, Mamat R, Ali OM, Sidik NAC, Yusaf T, Kadirgama K, Kettner M (2018) Alcohol and ether as alternative fuels in spark ignition engine: a review. Renew Sustain Energy Rev 82:2586–2605. https://doi.org/10.1016/j.rser.2017.09.074

Ayas N (2018) 2.13 Solvent materials. In: Dincer I (ed) Comprehensive energy systems. Elsevier, Oxford, pp 368–395

Aydin H, İlkılıç C (2010) Effect of ethanol blending with biodiesel on engine performance and exhaust emissions in a CI engine. Appl Therm Eng 30:1199–1204. https://doi.org/10.1016/j.applthermaleng.2010.01.037

Biomass for renewable energy, fuels, and chemicals. Choice Reviews Online, 36-4508-36-4508 (1999)

Brazil favours ethanol cars as 2022 flex-fuel vehicles continue to overshadow EVs (2023). http://fastmarkets-prod-01.altis.cloud/insights/brazil-favours-ethanol-cars-flex-fuel-vehicles-overshadow-evs/

Calam A, Solmaz H, Uyumaz A, Polat S, Yilmaz E, İçingür Y (2015a) Investigation of usability of the fusel oil in a single cylinder spark ignition engine. J Energy Inst 88:258–265. https://doi.org/10.1016/j.joei.2014.09.005

Calam A, İçingür Y, Solmaz H, Yamık H (2015b) A comparison of engine performance and the emission of fusel oil and gasoline mixtures at different ignition timings. Int J Green Energy 12:767–772. https://doi.org/10.1080/15435075.2013.849256

Callejón-Ferre AJ, Velázquez-Martí B, López-Martínez JA, Manzano-Agugliaro F (2011) Greenhouse crop residues: energy potential and models for the prediction of their higher heating value. Renew Sustain Energy Rev 15:948–955. https://doi.org/10.1016/j.rser.2010.11.012

Canakci M, Ozsezen AN, Alptekin E, Eyidogan M (2013) Impact of alcohol–gasoline fuel blends on the exhaust emission of an SI engine. Renew Energy 52:111–117. https://doi.org/10.1016/j.renene.2012.09.062

Cavelius P, Engelhart-Straub S, Mehlmer N, Lercher J, Awad D, Brück T (2023) The potential of biofuels from first to fourth generation. PLOS Biol 21:e3002063. https://doi.org/10.1371/journal.pbio.3002063

Çelebi Y, Aydın H (2018) Investigation of the effects of butanol addition on safflower biodiesel usage as fuel in a generator diesel engine. Fuel 222:385–393. https://doi.org/10.1016/j.fuel.2018.02.174

Çelebi Y, Aydın H (2019) An overview on the light alcohol fuels in diesel engines. Fuel 236:890–911. https://doi.org/10.1016/j.fuel.2018.08.138

Chan YH, Cheah KW, How BS, Loy ACM, Shahbaz M, Singh HKG, Yusuf NR, Shuhaili AFA, Yusup S, Ghani WAWAK, Rambli J, Kansha Y, Lam HL, Hong BH, Ngan SL (2019) An overview of biomass thermochemical conversion technologies in Malaysia. Sci Total Environ 680:105–123. https://doi.org/10.1016/j.scitotenv.2019.04.211

Charting the course: the essential role of decarbonising transportation in the transition to a regenerative economy—EFG International. https://www.efginternational.com/insights/2023/the_essential_role_of_decarbonising_transportation_in_the_transition_to_a_regenerative_economy.html

Chundawat SPS, Beckham GT, Himmel ME, Dale BE (2011) Deconstruction of lignocellulosic biomass to fuels and chemicals. Annu Rev Chem Biomol Eng 2:121–145. https://doi.org/10.1146/annurev-chembioeng-061010-114205

Collard F-X, Blin J (2014) A review on pyrolysis of biomass constituents: mechanisms and composition of the products obtained from the conversion of cellulose, hemicelluloses and lignin. Renew Sustain Energy Rev 38:594–608. https://doi.org/10.1016/j.rser.2014.06.013

Costa HL, Cousseau T, Souza RM (2023) Current knowledge on friction, lubrication, and wear of ethanol-fuelled engines—a review. Lubricants 11:292. https://doi.org/10.3390/lubricants11070292

de Andrade MAU, Valin H, Soterroni AC, Ramos FM, Halog A (2019) Exploring future scenarios of ethanol demand in Brazil and their land-use implications. Energy Policy 134:110958. https://doi.org/10.1016/j.enpol.2019.110958

Dees J, Oke K, Goldstein H, McCoy ST, Sanchez DL, Simon AJ, Li W (2023) Cost and life cycle emissions of ethanol produced with an oxyfuel boiler and carbon capture and storage. Environ Sci Technol 57:5391–5403. https://doi.org/10.1021/acs.est.2c04784

Deka TJ, Osman AI, Baruah DC, Rooney DW (2022) Methanol fuel production, utilization, and techno-economy: a review. Environ Chem Lett 20:3525–3554. https://doi.org/10.1007/s10311-022-01485-y

Demirbas A (2004) Combustion characteristics of different biomass fuels. Prog Energy Combust Sci 30:219–230. https://doi.org/10.1016/j.pecs.2003.10.004

Demirbas A (2007) Effects of moisture and hydrogen content on the heating value of fuels. Energy Sources Part Recovery Util Environ Eff 29:649–655. https://doi.org/10.1080/009083190957801

Demirbaş A (2005) Bioethanol from cellulosic materials: a renewable motor fuel from biomass. Energy Sources 27:327–337. https://doi.org/10.1080/00908310390266643

Doucha J, Lívanský K (2006) Productivity, CO_2/O_2 exchange and hydraulics in outdoor open high density microalgal (Chlorella sp.) photobioreactors operated in a Middle and Southern European climate. J Appl Phycol 18:811–826. https://doi.org/10.1007/s10811-006-9100-4

Egolf LM, Jurs PC (1992) Estimation of autoignition temperatures of hydrocarbons, alcohols, and esters from molecular structure. Ind Eng Chem Res 31:1798–1807. https://doi.org/10.1021/ie00007a027

Ernstgård L, Shibata E, Johanson G (2005) Uptake and disposition of inhaled methanol vapor in humans. Toxicol Sci 88:30–38. https://doi.org/10.1093/toxsci/kfi281

Ethanol|Definition, formula, uses, & facts|Britannica. https://www.britannica.com/science/ethanol

Ethanol: associated hazards and safety measures. https://www.gexcon.com/blog/ethanol-associated-hazards-and-safety-measures/

Feinauer M, Ehrenberger S, Buchgeister J (2021) Life cycle assessment of a farmed wood butanol-gasoline blend as an alternative transport fuel for passenger cars. Fuel 306:121651. https://doi.org/10.1016/j.fuel.2021.121651

French R, Malone P (2005) Phase equilibria of ethanol fuel blends. Fluid Phase Equilib 228–229:27–40. https://doi.org/10.1016/j.fluid.2004.09.012

García V, Päkkilä J, Ojamo H, Muurinen E, Keiski RL (2011) Challenges in biobutanol production: how to improve the efficiency? Renew Sustain Energy Rev 15:964–980. https://doi.org/10.1016/j.rser.2010.11.008

Gebreeyessus GD, Mekonnen A, Alemayehu E (2019) A review on progresses and performances in distillery stillage management. J Clean Prod 232:295–307. https://doi.org/10.1016/j.jclepro.2019.05.383

Gerber MA, White JF, Stevens DJ (2007) Mixed alcohol synthesis catalyst screening. Pacific Northwest National Lab (PNNL), Richland, WA (United States)

Gil P, Fernandes J, Neves MM, Reis LG (2010) Suitability of corkrubber gaskets in gasoline-ethanol blends. Mater Sci Forum 636–637:266–272. https://doi.org/10.4028/www.scientific.net/MSF.636-637.266

Göransson K, Söderlind U, He J, Zhang W (2011) Review of syngas production via biomass DFBGs. Renew Sustain Energy Rev 15:482–492. https://doi.org/10.1016/j.rser.2010.09.032

Goyal HB, Seal D, Saxena RC (2008) Bio-fuels from thermochemical conversion of renewable resources: a review. Renew Sustain Energy Rev 12:504–517. https://doi.org/10.1016/j.rser.2006.07.014

Group SM. Hybrid automotive engine using ethanol-burning miller cycle. https://www.techbriefs.com/component/content/article/1559-lar16365

Guimarães CEC, Neto FS, de Castro Bizerra V, do Nascimento JGA, Valério RBR, de Sousa PG, de Sousa Braz AK, Melo RLF, de França Serpa J, de Lima RKC, Guimarães AP, de Souza MCM, Lopes AAS, de Sousa Rios MA, Desai AS, Bilal M, Smułek W, Jesionowski T, dos Santos JCS (2023) Sustainable bioethanol production from first- and second-generation sugar-based feedstocks: advanced bibliometric analysis. Bioresour Technol Rep 23:101543. https://doi.org/10.1016/j.biteb.2023.101543

Haberzettl J, Hilgert P, von Cossel M (2021) A critical review on lignocellulosic biomass yield modeling and the bioenergy potential from marginal land. Agronomy 11:2397. https://doi.org/10.3390/agronomy11122397

Hahn-Hägerdal B, Galbe M, Gorwa-Grauslund MF, Lidén G, Zacchi G (2006) Bio-ethanol—the fuel of tomorrow from the residues of today. Trends Biotechnol 24:549–556. https://doi.org/10.1016/j.tibtech.2006.10.004

He Y, Wang S, Lai KK (2010) Global economic activity and crude oil prices: a cointegration analysis. Energy Econ 32:868–876. https://doi.org/10.1016/j.eneco.2009.12.005

Hordeski MF (2020) Megatrends for energy efficiency and renewable energy. River Publishers, New York

Ibrahim Tk, Mohammed MK, Awad OI, Rahman MM, Najafi G, Basrawi F, Abd Alla AN, Mamat R (2017) The optimum performance of the combined cycle power plant: a comprehensive review. Renew Sustain Energy Rev 79:459–474. https://doi.org/10.1016/j.rser.2017.05.060

Içingür Y, Calam A (2012) The effects of the blends of fusel oil and gasoline on performance and emissions in a spark ignition engine. J Fac Eng Archit Gazi Univ 27:143–149

Jacob A, Ashok B (2020) An interdisciplinary review on calibration strategies of engine management system for diverse alternative fuels in IC engine applications. Fuel 278:118236. https://doi.org/10.1016/j.fuel.2020.118236

Jahanbakhshi A, Karami-Boozhani S, Yousefi M, Ooi JB (2021) Performance of bioethanol and diesel fuel by thermodynamic simulation of the miller cycle in the diesel engine. Results Eng. 12:100279. https://doi.org/10.1016/j.rineng.2021.100279

Jamrozik A, Tutak W, Grab-Rogaliński K (2022) Effects of propanol on the performance and emissions of a dual-fuel industrial diesel engine. Appl Sci 12:5674. https://doi.org/10.3390/app12115674

Karaosmanoğlu F, Işiğigür A, Aksoy HA (1997) Methanol-unleaded gasoline blends containing fusel oil fraction as spark ignition engine fuel. Energy Sources 19:567–577. https://doi.org/10.1080/00908319708908873

Kasibhatta S (2019) Alcohol fuels as an alternative fuels—bringing new heights in sustainability. In: alcohol fuels—current technologies and future prospect. IntechOpen

Kassem MGA, Ahmed A-MM, Abdel-Rahman HH, Moustafa AHE (2019) Use of Span 80 and Tween 80 for blending gasoline and alcohol in spark ignition engines. Energy Rep 5:221–230. https://doi.org/10.1016/j.egyr.2019.01.009

Kim HY, Ge JC, Choi NJ (2020) Effects of ethanol-diesel on the combustion and emissions from a diesel engine at a low idle speed. Appl Sci 10:4153. https://doi.org/10.3390/app10124153

Kiuchi T, Yoshida M, Kato Y (2015) Energy saving bioethanol distillation process with self-heat recuperation technology. J Jpn Pet Inst 58:135–140. https://doi.org/10.1627/jpi.58.135

Kumar TS, Ashok B (2023) Corrosion behaviour analysis of SI engine components for ethanol-gasoline blends in flex fuel vehicular application. Fuel Process Technol 240:107574. https://doi.org/10.1016/j.fuproc.2022.107574

Kumar S, Cho JH, Park J, Moon I (2013) Advances in diesel–alcohol blends and their effects on the performance and emissions of diesel engines. Renew Sustain Energy Rev 22:46–72. https://doi.org/10.1016/j.rser.2013.01.017

Küüt A, Ritslaid K, Küüt K, Ilves R, Olt J (2019) Chapter 3—State of the art on the conventional processes for ethanol production. In: Basile A, Iulianelli A, Dalena F, Veziroğlu TN (eds) Ethanol. Elsevier, Amsterdam, pp 61–101

Lange J-P (2007) Lignocellulose conversion: an introduction to chemistry, process and economics. Biofuels Bioprod Biorefining 1:39–48. https://doi.org/10.1002/bbb.7

Lee SY, Sankaran R, Chew KW, Tan CH, Krishnamoorthy R, Chu D-T, Show P-L (2019) Waste to bioenergy: a review on the recent conversion technologies. BMC Energy. 1:4. https://doi.org/10.1186/s42500-019-0004-7

Liu C-G, Xiao Y, Xia X-X, Zhao X-Q, Peng L, Srinophakun P, Bai F-W (2019) Cellulosic ethanol production: progress, challenges and strategies for solutions. Biotechnol Adv 37:491–504. https://doi.org/10.1016/j.biotechadv.2019.03.002

Liu H, Wang G, Zhang J (2013) The promising fuel-biobutanol. In: Fang Z (ed) Liquid, gaseous and solid biofuels—conversion techniques. InTech

Lynd LR, Weimer PJ, Van Zyl WH, Pretorius IS (2002) Microbial cellulose utilization: fundamentals and biotechnology. Microbiol Mol Biol Rev 66:506–577. https://doi.org/10.1128/MMBR.66.3.506-577.2002

Mahmudul HM, Hagos FY, Mamat R, Abdullah AA, Awad OI (2016) Experimental investigation of the impact of using alcohol- biodiesel-diesel blending fuel on combustion of single cylinder CI engine. IOP Conf Ser Mater Sci Eng 160:012038. https://doi.org/10.1088/1757-899X/160/1/012038

Markides CN, Mastorakos E (2011) Experimental investigation of the effects of turbulence and mixing on autoignition chemistry. Flow Turbul Combust 86:585–608. https://doi.org/10.1007/s10494-010-9268-1

Milpied J, Jeuland N, Plassat G, Guichaous S, Dioc N, Marchal A, Schmelzle P (2009) Impact of fuel properties on the performances and knock behaviour of a downsized turbocharged DI SI engine—focus on octane numbers and latent heat of vaporization. https://doi.org/10.4271/2009-01-0324

Mizik T (2021) Economic aspects and sustainability of ethanol production—a systematic literature review. Energies 14:6137. https://doi.org/10.3390/en14196137

Modifications to fuel regulations to provide flexibility for E15; modifications to RFS RIN market regulations. https://www.federalregister.gov/documents/2019/06/10/2019-11653/modifications-to-fuel-regulations-to-provide-flexibility-for-e15-modifications-to-rfs-rin-market

Mueller S, Dennison G, Liu S (2021) An assessment on ethanol-blended gasoline/diesel fuels on cancer risk and mortality. Int J Environ Res Public Health 18:6930. https://doi.org/10.3390/ijerph18136930

Muir BM, Anderson AR (2022) Development and diversification of sugar beet in Europe. Sugar Tech 24:992–1009. https://doi.org/10.1007/s12355-021-01036-9

Mujtaba M, Fernandes Fraceto L, Fazeli M, Mukherjee S, Savassa SM, Araujo de Medeiros G, do Espírito Santo Pereira A, Mancini SD, Lipponen J, Vilaplana F (2023) Lignocellulosic biomass from agricultural waste to the circular economy: a review with focus on biofuels, biocomposites and bioplastics. J Clean Prod 402:136815. https://doi.org/10.1016/j.jclepro.2023.136815

Muthuraman VS, Patel A, Shreya V, Vaidyanathan A, Reshwanth KNGL, Karthick C, Jan Gęca M, Ashok B, Sivagami K, Nanthagopal K (2022) Progress on compatibility issues of alcohols on automotive materials: kinetics, challenges and future prospects—a comprehensive review. Process Saf Environ Prot 162:463–493. https://doi.org/10.1016/j.psep.2022.04.022

Nissan begins driving tests of a methanol fuel cell vehicle. http://global.nissannews.com/en/releases/19990513-e_presskit

Nithyanandan K, Zhang J, Li Y, Wu H, Lee TH, Lin Y, Lee CF (2016) Improved SI engine efficiency using Acetone–Butanol–Ethanol (ABE). Fuel 174:333–343. https://doi.org/10.1016/j.fuel.2016.01.001

Osten DW, Sell NJ (1983) Methanol-gasoline blends: blending agents to prevent phase separation. Fuel 62:268–270. https://doi.org/10.1016/0016-2361(83)90079-0

Patil D (2024) How ethanol and battery powered hybrid vehicles can fuel a greener future. https://www.chinimandi.com/how-ethanol-and-battery-powered-hybrid-vehicles-can-fuel-a-greener-future/

Perkins R, Edwardes-Evans H. Fossil fuels "stubbornly" dominating global energy despite surge in renewables: energy Institute. https://www.spglobal.com/commodityinsights/en/market-insights/latest-news/oil/062623-fossil-fuels-stubbornly-dominating-global-energy-despite-surge-in-renewables-energy-institute

Pittman JK, Dean AP, Osundeko O (2011) The potential of sustainable algal biofuel production using wastewater resources. Bioresour Technol 102:17–25. https://doi.org/10.1016/j.biortech.2010.06.035

Pourkhesalian AM, Shamekhi AH, Salimi F (2010) Alternative fuel and gasoline in an SI engine: a comparative study of performance and emissions characteristics. Fuel 89:1056–1063. https://doi.org/10.1016/j.fuel.2009.11.025

Qiu Y, Wu M, Bao H, Liu W, Shen Y (2023) Engineering of *Saccharomyces cerevisiae* for co-fermentation of glucose and xylose: current state and perspectives. Eng. Microbiol. 3:100084. https://doi.org/10.1016/j.engmic.2023.100084

Raheem A, Wan Azlina WAKG, Taufiq Yap YH, Danquah MK, Harun R (2015) Thermochemical conversion of microalgal biomass for biofuel production. Renew Sustain Energy Rev 49:990–999. https://doi.org/10.1016/j.rser.2015.04.186

Ramzan N, Ashraf A, Naveed S, Malik A (2011) Simulation of hybrid biomass gasification using Aspen plus: A comparative performance analysis for food, municipal solid and poultry waste. Biomass Bioenergy 35:3962–3969. https://doi.org/10.1016/j.biombioe.2011.06.005

Raveendran K, Ganesh A (1996) Heating value of biomass and biomass pyrolysis products. Fuel 75:1715–1720. https://doi.org/10.1016/S0016-2361(96)00158-5

Ritchie H, Rosado P, Roser M (2024) Energy mix. Our World Data

Sanni A, Olawale AS, Sani YM, Kheawhom S (2022) Sustainability analysis of bioethanol production from grain and tuber starchy feedstocks. Sci Rep 12:20971. https://doi.org/10.1038/s41598-022-24854-7

Saricks C, Santini D, Wang M (1999) Effects of fuel ethanol use on fuel-cycle energy and greenhouse gas emissions. Argonne National Laboratory, Argonne, IL (United States)

Scott-Kerr C, Johnson T, Johnson B, Kiviaho J (2010) Bioethanol—status report on bioethanol production from wood and other lignocellulosic feedstocks

Scully MJ, Norris GA, Falconi TMA, MacIntosh DL (2021) Carbon intensity of corn ethanol in the United States: state of the science. Environ Res Lett 16:043001. https://doi.org/10.1088/1748-9326/abde08

Semelsberger TA, Borup RL, Greene HL (2006) Dimethyl ether (DME) as an alternative fuel. J Power Sources 156:497–511. https://doi.org/10.1016/j.jpowsour.2005.05.082

Sheng C, Azevedo JLT (2005) Estimating the higher heating value of biomass fuels from basic analysis data. Biomass Bioenergy 28:499–507. https://doi.org/10.1016/j.biombioe.2004.11.008

Shokravi H, Heidarrezaei M, Shokravi Z, Ong HC, Lau WJ, Din MFM, Ismail AF (2022) Fourth generation biofuel from genetically modified algal biomass for bioeconomic development. J Biotechnol 360:23–36. https://doi.org/10.1016/j.jbiotec.2022.10.010

Silva TAL, Varão LHR, Pasquini D (2024) Lignocellulosic biomass. In: Handbook of biomass. Springer, Singapore, pp 1–39

Simon Araya S, Liso V, Cui X, Li N, Zhu J, Sahlin SL, Jensen SH, Nielsen MP, Kær SK (2020) A review of the methanol economy: the fuel cell route. Energies 13:596. https://doi.org/10.3390/en13030596

Suarez-Bertoa R, Zardini AA, Keuken H, Astorga C (2015) Impact of ethanol containing gasoline blends on emissions from a flex-fuel vehicle tested over the worldwide harmonized light duty test cycle (WLTC). Fuel 143:173–182. https://doi.org/10.1016/j.fuel.2014.10.076

Support T. A short history of biofuels. https://lee-enterprises.com/a-short-history-of-biofuels/

The challenges associated with ethanol blended fuels—biobor fuel additives (2022). https://www.biobor.com/2020/05/06/the-challenges-associated-with-ethanol-blended-fuels/

Theinnoi K, Sawatmongkhon B, Wongchang T, Haoharn C, Wongkhorsub C, Sukjit E (2021) Effects of diesel–biodiesel–ethanol fuel blend on a passive mode of selective catalytic reduction to reduce NO_x emission from real diesel engine exhaust gas. ACS Omega 6:27443–27453. https://doi.org/10.1021/acsomega.1c04365

Thomas CE, James BD, Lomax FD, Kuhn IF (2000) Fuel options for the fuel cell vehicle: hydrogen, methanol or gasoline? Int J Hydrog Energy 25:551–567. https://doi.org/10.1016/S0360-3199(99)00064-6

United Nations Conference on Environment and Development (UNCED)|History & Facts|Britannica. https://www.britannica.com/event/United-Nations-Conference-on-Environment-and-Development

US EPA O. Working paper: impacts of ethanol policy on corn prices: a review and meta-analysis of recent evidence. https://www.epa.gov/environmental-economics/working-paper-impacts-ethanol-policy-corn-prices-review-and-meta-analysis

Uses of methanol. https://www.geeksforgeeks.org/uses-of-methanol/

Vaivads R, Bardon M, Rao V, Battista V (1994) Flammability of alcohol-gasoline blends in fuel tanks. Fire Saf Sci 4:575–586. https://doi.org/10.3801/IAFSS.FSS.4-575

Vallinayagam R, Vedharaj S, Roberts WL, Dibble RW, Sarathy SM (2017) Performance and emissions of gasoline blended with terpineol as an octane booster. Renew Energy 101:1087–1093. https://doi.org/10.1016/j.renene.2016.09.055

Verma M, Godbout S, Brar SK, Solomatnikova O, Lemay SP, Larouche JP (2012) Biofuels production from biomass by thermochemical conversion technologies. Int J Chem Eng 2012:542426. https://doi.org/10.1155/2012/542426

Vohra M, Manwar J, Manmode R, Padgilwar S, Patil S (2014) Bioethanol production: feedstock and current technologies. J Environ Chem Eng 2:573–584. https://doi.org/10.1016/j.jece.2013.10.013

Vojtisek-Lom M, Beranek V, Stolcpartova J, Pechout M, Klir V (2015) Effects of n-butanol and isobutanol on particulate matter emissions from a Euro 6 direct-injection spark ignition engine during laboratory and on-road tests. SAE Int J Engines 8:2338–2350. https://doi.org/10.4271/2015-24-2513

Voloshin RA, Rodionova MV, Zharmukhamedov SK, Nejat Veziroglu T, Allakhverdiev SI (2016) Review: biofuel production from plant and algal biomass. Int J Hydrog Energy 41:17257–17273. https://doi.org/10.1016/j.ijhydene.2016.07.084

Walther T, François JM (2016) Microbial production of propanol. Biotechnol Adv 34:984–996. https://doi.org/10.1016/j.biotechadv.2016.05.011

Wang Z, Liu H, Reitz RD (2017) Knocking combustion in spark-ignition engines. Prog Energy Combust Sci 61:78–112. https://doi.org/10.1016/j.pecs.2017.03.004

Wei L, Yao C, Wang Q, Pan W, Han G (2015) Combustion and emission characteristics of a turbocharged diesel engine using high premixed ratio of methanol and diesel fuel. Fuel 140:156–163. https://doi.org/10.1016/j.fuel.2014.09.070

White TL (1907) Alcohol as a fuel for the automobile motor. Presented at the Pre-1964 SAE Technical Papers Jan 1

Yang L, Ge X (2016) Chapter three—biogas and syngas upgrading. In: Li Y, Ge X (eds) Advances in bioenergy. Elsevier, Amsterdam, pp 125–188

Yanowitz J, Christensen E, McCormick RL (2011) Utilization of renewable oxygenates as gasoline blending components. National Renewable Energy Lab. (NREL), Golden, CO (United States)

Yilmaz N, Donaldson AB (2007) Modeling of chemical processes in a diesel engine with alcohol fuels. J Energy Resour Technol 129:355–359. https://doi.org/10.1115/1.2794771

Yilmaz N, Donaldson AB, Johns A (2005) Some perspectives on alcohol utilization in a compression ignition engine. SAE International, Warrendale, PA

Yuan Y, Jiang B, Chen H, Wu W, Wu S, Jin Y, Xiao H (2021) Recent advances in understanding the effects of lignin structural characteristics on enzymatic hydrolysis. Biotechnol Biofuels 14:205. https://doi.org/10.1186/s13068-021-02054-1

Yue Z, Liu H (2023) Advanced research on internal combustion engines and engine fuels. Energies 16:5940. https://doi.org/10.3390/en16165940

Yüksel F, Yüksel B (2004) The use of ethanol–gasoline blend as a fuel in an SI engine. Renew Energy 29:1181–1191. https://doi.org/10.1016/j.renene.2003.11.012

Yun Y (2020) Alcohol fuels: current status and future direction. In: Alcohol fuels—current technologies and future prospect. IntechOpen

Yusri IM, Mamat R, Najafi G, Razman A, Awad OI, Azmi WH, Ishak WFW, Shaiful AIM (2017) Alcohol based automotive fuels from first four alcohol family in compression and spark ignition engine: a review on engine performance and exhaust emissions. Renew Sustain Energy Rev 77:169–181. https://doi.org/10.1016/j.rser.2017.03.080

Zabed H, Sahu JN, Suely A, Boyce AN, Faruq G (2017) Bioethanol production from renewable sources: Current perspectives and technological progress. Renew Sustain Energy Rev 71:475–501. https://doi.org/10.1016/j.rser.2016.12.076

Zacharias LGL, Oliveira LDB, Alves VH, Guichet X, Peyerl D (2023) The future of diesel: paths and new alternatives to energy security and sustainability. In: Peyerl D, Relva S, Da Silva V (eds) Energy transition in Brazil. Springer Nature Switzerland, Cham, pp 173–192

Chapter 9
Alcohol for Fueling IC Engines

Sudhir Kumar Pathak, Pravin Omprakash Sharma, and Har Mohan Singh

Abstract Alcohol fuels have been increasingly considered for internal combustion (IC) engines due to their potential to address rising energy costs and reduce greenhouse gas emissions. Ethanol and methanol are the primary alcohol fuels studied for this purpose. Ethanol is valued for its higher-octane number, which allows for increased compression ratios and improved thermal efficiency in IC engines. Its oxygen content promotes more complete combustion, thereby reducing emissions of carbon monoxide and unburned hydrocarbons. On the other hand, methanol is noted for its high-octane rating and potential for use in various engine types. Both ethanol and methanol are produced from renewable sources, supporting their role as sustainable alternatives to traditional fossil fuels. The combustion properties of alcohol fuels, such as their lower energy density and higher viscosity compared to gasoline and diesel, are also examined. Challenges related to the use of alcohol fuels include their impact on engine performance and fuel system materials. Ongoing research aims to address these challenges and enhance the viability of alcohol fuels in IC engines, contributing to a cleaner and more sustainable energy future.

Keywords IC engines · Alcohols · Alternative fuels · Sustainability · Transportation

S. K. Pathak (✉)
Research and Development Division, PLUSS Advanced Technologies Limited, Bawal, Rewari, Haryana, India
e-mail: sudhirpathak95@gmail.com

P. O. Sharma (✉)
Mechanical Engineering Department, Dr. D. Y. Patil Institute of Technology, Pimpri, Pune, Maharashtra, India
e-mail: pravin.sharma@dypvp.edu.in; pravinosharma@gmail.com

H. M. Singh
Department of Physics and Environmental Sciences, Sharda School of Engineering & Science, Sharda University, Uttar Pradesh, Greater Noida, India
e-mail: har.singh@sharda.ac.in

A. Bartl et al. (eds.), *Alcohol Production Processes and Their Utilisation in the Transport Sector*, Energy, Environment, and Sustainability,
https://doi.org/10.1007/978-981-96-7384-1_9

Nomenclature

CI	Compression ignition
CO_2	Carbon dioxide
CO	Carbon monoxide
CN	Cetane number
BS	Bharat stage
EGR	Exhaust gas recirculation
HC	Hydrocarbons
HVO	Hydrotreated vegetable oil
IC	Internal combustion
ON	Octane number
NO_x	Nitrogen oxides
RPM	Revolutions per minute
SI	Spark ignition
THC	Thermal hydrocarbons

9.1 Introduction of Alcohol Fuel for IC Engine

Various fossil fuels, i.e., oil, coal, and natural gas sum up to almost 90% of carbon emissions globally. Despite the adversities and depletion of fossil fuels, heavily industrialized nations have struggled to curb their dependency on such fuels (Das et al. 2022). Meanwhile, it was in the year 2022 that a notable rise was witnessed in renewable energy investments. Yet, the emissions caused by the continued combustion of fossil fuel kept increasing at an alarming rate, and the subsidies for fossil fuel consumption reached a whopping $1 trillion. The need for discovering alternative fuels for internal combustion (IC) engines has arisen due to increasing energy demands throughout the globe and for reducing greenhouse gas emissions. Alcohol, as such, has emerged as a potential candidate owing to its ability to be produced from natural resources and significantly favorable combustion characteristics. This section aims to introduce and provide a comprehensive overview of alcohol as a substitute fuel in IC engines. Various attributes, such as the benefits, challenges and current state-of-the-art research and development in alternative fuels with a special focus on alcohol, are discussed here (Percy and Edwin 2024).

Figure 9.1 shows global energy consumption based on fuel type. It is evident from Fig. 9.1 that the world heavily relied upon oil as a prime energy source. Using alcohol as a fuel is not a novel concept at all. Still, due to the cheap and abundant availability of petroleum-based fuels in the mid-twentieth century, alcohol unknowingly got rejected by the research community. The role of alcohol hence existed only lasted as a secondary fuel. Interestingly, the world had to wait for an oil crisis and other environmental issues in the 1970s to attract researchers' interest in alcohol fuels (Percy and Edwin 2023).

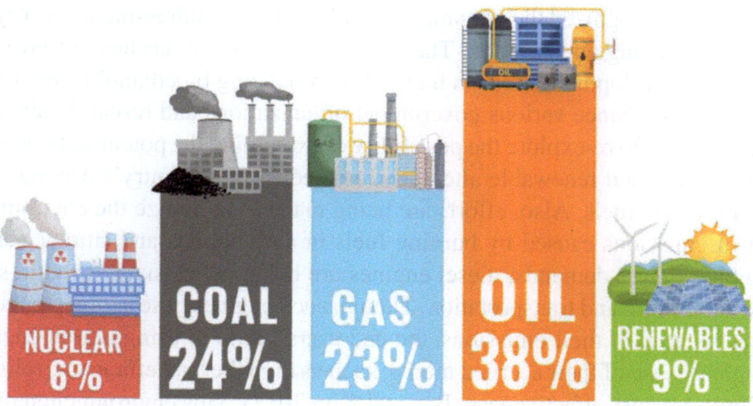

Fig. 9.1 Global energy consumption as per the fuel type (https://fpif.org/how-to-rapidly-reduce-fossil-fuel-use/)

The global transition to alcohol fuels, particularly ethanol and methanol, is driven by the urgent need for sustainable energy solutions and reduced emissions. Various nations have articulated clear visions and strategies to incorporate alcohol fuels into their transportation sectors, reflecting a commitment to environmental sustainability, energy security, and economic development. Brazil has been a leader in the production and utilization of bioethanol derived from sugarcane. The country's extensive flex-fuel vehicle market allows cars to operate on high-ethanol blends, significantly reducing CO_2 emissions. Brazil's National Alcohol Program, which was initiated in the 1970s, has established a robust infrastructure for biofuel production and distribution, making it a model for other nations. The government continues to support the expansion of ethanol use through tax incentives and blending mandates, aiming for a sustainable energy future.

The Renewable Fuel Standard (RFS) program of the United States plays a crucial role in promoting ethanol production from corn. The RFS mandates increasing volumes of renewable fuel to be blended into gasoline, which has bolstered domestic agriculture and reduced reliance on imported oil. Additionally, various states have implemented their own policies encouraging the use of ethanol in transportation. Other agencies like the U. S. Department of Agriculture (USDA) supports research and development initiatives aimed at improving biofuel production technologies and expanding feedstock sources. The European Union has set targets for renewable energy consumption in transport as part of its broader climate action strategy. The EU aims to reduce emissions of all kinds from transport by at least 14% by promoting sustainable biofuels and ensuring that they meet sustainability criteria. India is actively pursuing an ethanol blending program, aiming for 20% ethanol blending in petrol by 2025. This initiative is part of India's broader strategy to enhance energy security, reduce air pollution, and support farmers through increased demand for sugarcane. The Indian government has introduced policies to facilitate the production of ethanol from various feedstocks, including agricultural residues and molasses.

China, too has recognized the potential of alcohol fuels in addressing its energy challenges and reducing air pollution. The Chinese government has heavily invested in research and development projects focused on producing bioethanol from non-food biomass sources. Since various government organizations and research labs worldwide have begun to re-explore the possibility of examining the potentiality of alcohol as a sustainable and renewable alternative for reducing a country's dependence on petroleum-based fuels. Also, efforts are being made to minimize the environmental impact of emissions caused by burning fuels in automobiles and similar vehicles. Due to high power demands, diesel engines are being aggressively used in sectors such as agriculture and transportation. Though such engines pose a threat to the environment because of their emissions, the end users' trend in using search engines is very high because of their accrued merits, such as high thermal efficiency, reliability, versatility, and cost-effectiveness (Bae and Kim 2017). This environmental damage can be minimized by exploring research avenues contributing to the use of substitute fuels like as alcohols. Alcohol use can mitigate unwanted environmental pollution effectively through stringent compliance with revised emission regulations. Alcohols have been extensively used as additives in diesel engines due to their economic benefits compared to pure diesel. They can be used directly as a single fuel or an additive in diesel engines without major engine modifications (Verger et al. 2022).

Figure 9.2 shows the energy used by transportation modes and fuel types worldwide. So, to be less dependent on oil, prevent any undue oil crisis event, and, most importantly, avoid any further harm to the environment, exploring and globally promoting alternative fuels in IC engines has become vital. This alarming scenario is now leading to promotion of research that aims to achieve the desired power output in the existing engines without having to modify them structurally and be able to operate in various conditions while being able to minimize the cost and the environmental impact. The research community must take care of parameters such as cost-effectiveness, domestic sourcing and compatibility with existing fuel systems with fewer or no modifications for broader adoption of alternative fuels (Göktaş et al. 2021).

Various challenges still exist in using alcohol as alternative fuel. These include issues with blending with diesel fuel, low lubricity, difficulty in vaporization, and high auto-ignition temperatures. All such challenges can be mitigated by using additives or increasing the intake air temperature. The alcohols are not directly miscible with diesel fuel; thus, co-solvents or emulsifiers are needed. However, blending alcohols with diesel fuel can assist resolve lubricity and vaporization difficulties. Various procedure, which includes mixing, separating, splashing, and heating, can be expensive and complex. Within a specific limit, alcohols can be used in diesel/biodiesel/alcohol combinations without experiencing miscibility issues since they are compatible with biodiesel despite their immiscibility with diesel (Verhelst et al. 2019). Methanol, propanol, and butanol are commonly used alcohols as alternative fuels in CI engines. The adoption of alcohol fuels offers numerous advantages that align with the visions of these countries:

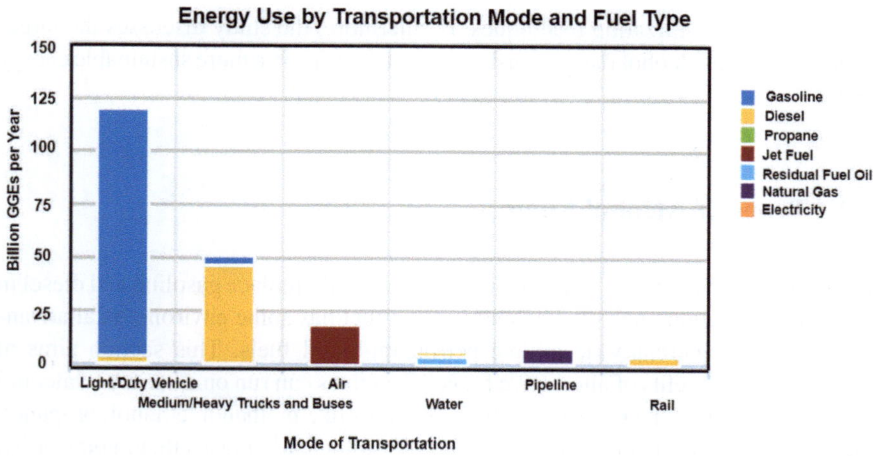

Fig. 9.2 Energy usage based on the transportation mode and fuel type (https://afdc.energy.gov/data)

- Alcohol fuels are derived from biomass or waste materials, making them renewable sources of energy that can be replenished through sustainable agricultural practices.
- Alcohol fuels contribute to lower carbon emissions compared to conventional fossil fuels by recycling carbon dioxide through plant growth.
- They generally have a higher-octane rating than gasoline, leading to improved engine performance and efficiency.
- Alcohol fuels produce fewer harmful emissions such as particulate matter and sulfur oxides, contributing to better air quality and public health outcomes.
- The production and use of alcohol fuels stimulate rural economies by creating jobs in agriculture and biofuel production sectors.
- Alcohol fuels can be blended with gasoline or used in dedicated vehicles without requiring significant modifications to existing infrastructure.

Technological advancements in the IC engine, such as fuel formation and renewable fuel production, have helped trigger research and interest in using alcohol as an alternative fuel. To name a few, researchers are focused on designing and optimizing internal combustion engines to reap the benefits of alcohol fuels, developing cost-efficient and effective methods for producing alcohol, and addressing any challenges associated with their use. In addition, using alcohol as fuel with the option of hybridization and flex-fuel vehicles will reduce the environmental impact cost for the transportation sector alone.

The potential of alcohol fuels for IC engines is thoroughly examined in this research paper. For this various chemical and physical properties, combustion characteristics, and environmental effects of each alcohol are considered. It also examines the most recent developments in engine modifications, pollution control techniques, and alcohol fuel utilization techniques. Furthermore, the study discusses the large-scale adoption of alcohol fuels, focusing on transitioning to a more sustainable energy future.

9.2 Types of Alcohol Fuels

The term "alternative fuel" refers to any fuel that could replace gasoline and diesel in internal combustion engines. Each alternative fuel has some environmental advantages over commercially accessible petroleum-based fuels. This section aims at discussing the benefits of alternative fuels. IC engines can run on a variety of alcohol kinds. The fact that the primary aliphatic alcohols like methanol, ethanol, propanol, and butanol can be created chemically and physiologically makes them noteworthy. These alcohols are helpful for internal combustion engines because of specific characteristics; the basic formula for alcohol fuels is $C_nH_{2n+1}OH$. The most popular and potential fuels are discussed in the subsequent paragraphs.

9.2.1 *Methanol (CH_3OH)*

Methanol is primarily obtained from natural gas, yet biomass waste can also be used to produce methanol through chemical methods. On the other hand, ethanol can be produced through the fermentation of biological materials. Similarly, biobutanol offers various advantages for internal combustion engines as it has energy density values that are very close to gasoline and offers 25% higher octane ratings than simple alcohols. Despite the advantages offered, biobutanol has various challenges during its production compared to ethanol or methanol. It should be noted that there is no clear distinction between the alcohols produced biologically and those produced using chemicals. Methanol is even seen as an alternative to hydrogen fuel. Historical evidence of methanol being used in racing vehicles in northern America has been reported in the literature. The ease of methanol being produced from syngas derived from biomass will make it a potent biofuel option in future (Fig. 9.3).

9.2.2 *Ethanol (C_2H_5OH)*

Ethanol, the second member of the alcohol family, has two carbon hydroxyl groups in its chemical structure. It needs to be stored, handled, and transported with care due to its lower flashpoint compared to fossil fuels. With a high-octane number, ethanol can be used in spark-ignition (SI) engines either partially or entirely in place of gasoline. The biological production process of ethanol leads to low toxicity. It contains about 34% oxygen by weight, has lower density and viscosity, and demonstrates better

Fig. 9.3 Chemical structure of methanol fuel

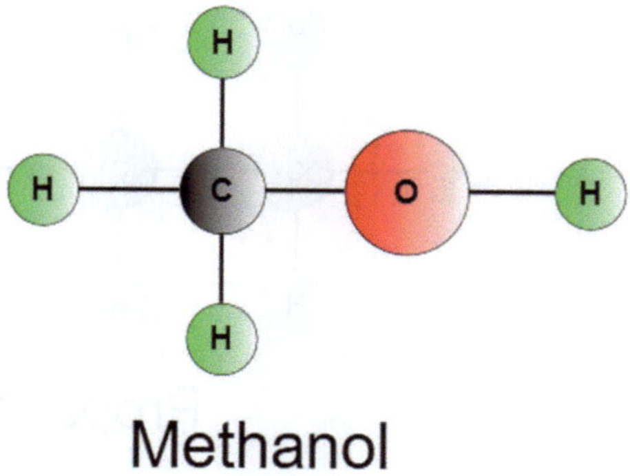

Fig. 9.4 Chemical structure of ethanol fuel

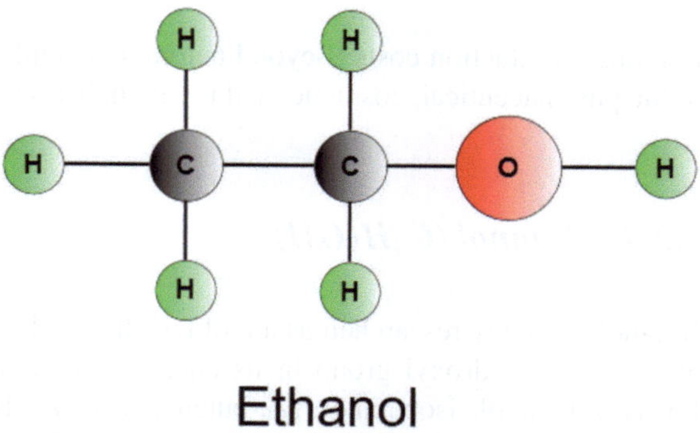

cold flow properties than gasoline. Ethanol is already widely used as a fuel additive and is increasingly used as a primary fuel or in blends with gasoline. Ethanol is less corrosive and non-toxic than methanol, although it produces some toxic emissions (Fig. 9.4).

9.2.3 *Propanol (C_3H_7OH)*

Propanol, a member of the alcohol family, has three carbon hydroxyl groups in its chemical structure and exists in two isomeric forms: *n*-propanol and isopropanol. It is considered a more desirable alternative fuel than methanol and ethanol due to its higher cetane number, lower corrosiveness, and better calorific value because of its higher carbon content. Propanol is also considered safer because it has a higher energy density and flashpoint compared to methanol and ethanol. It has a lower autoignition temperature than other alcohols, which reduces the ignition delay time, even though it requires more heat to vaporize than diesel. However, the exorbitant expense of producing propanol prevents it from being widely used as an automotive fuel. While limited research exists on propanol's use in engines, ongoing efforts aim

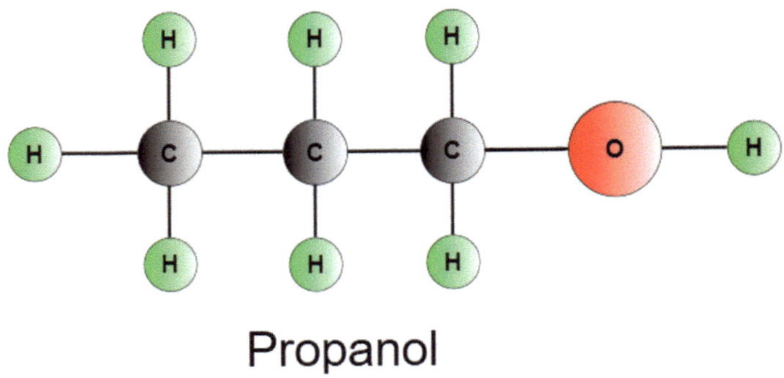

Fig. 9.5 Chemical structure of propanol fuel

to reduce production costs. Beyond automotive applications, propanol is also used in the pharmaceutical, cosmetic, and textile industries (Fig. 9.5).

9.2.4 *Butanol (C_4H_9OH)*

Butanol is widely researched as a biologically produced alcohol. It has four carbon atoms and a hydroxyl group in its chemical structure. It exists in four isomeric forms: *n*-butanol, isobutanol, *sec*-butanol, and *tert*-butanol, each with the chemical formula C_4H_9OH. Although these isomers have similar heat capacities, their combustion properties vary due to their unique molecular structures. For instance, *n*-butanol, isobutanol, and *sec*-butanol have similar burning rates but produce flames with different structures. Butanol is considered compatible with current fuel systems due to its physicochemical characteristics, which are more similar to gasoline and diesel than lower alcohols. These characteristics include a higher flashpoint, lower toxicity, and reduced corrosiveness. Additionally, butanol's viscosity is similar to that of diesel, which enhances lubrication. When combined with diesel, butanol offers various benefits, such as a higher cetane number, higher heating value, reduced latent heat of evaporation, lower viscosity, and improved stability. Due to its lower heat of evaporation, butanol requires less heat and air temperature at the intake compared to lower alcohol fuels. The requirement of butanol in an internal combustion engine is lower than that of methanol and ethanol because of its higher heating value. Therefore, butanol is considered a prominent contender for future biofuel use (Fig. 9.6).

9.2.5 *Merits and Demerits of Use of Alcohol Fuels*

The use of internal combustion (IC) engines with alcohol as fuel offers several merits and demerits, which are listed below:

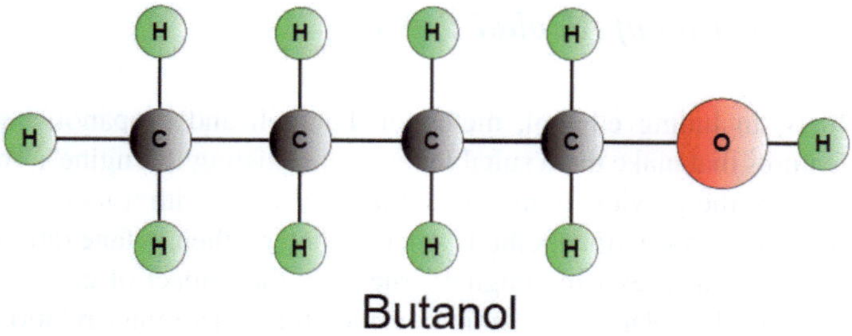

Fig. 9.6 Chemical structure of butanol fuel

9.2.5.1 Merits

- Alcohol fuels can lead to lower emissions of certain pollutants. Ethanol and methanol burn cleaner than gasoline, reducing levels of hydrocarbons (HC), carbon monoxide (CO), and particulate matter (soot).
- Ethanol is primarily produced from biomass, making it a renewable fuel source. Such use helps reduce fossil fuel dependence and supports agricultural industries.
- Alcohols have a higher specific heat capacity and can absorb more heat, which may enhance engine efficiency under certain operating conditions.
- Alcohol fuels can provide better mixing and more uniform combustion, leading to smoother engine operation.

9.2.5.2 Demerits

- The energy content available per liter of alcohol fuels is less compared with gasoline or diesel. For the same energy output, more alcohol fuel is required, which can lead to increased fuel consumption.
- Alcohols can be corrosive to certain metals and materials used in fuel systems, such as aluminum and rubber components. This necessitates modifications to engine components and fuel system parts.
- Alcohol fuels are more hygroscopic, meaning they absorb water from the environment. This can lead to phase separation and microbial contamination, affecting fuel quality and performance.
- The use of alcohol fuels can lead to higher levels of unwanted oxides such as NO*x*, which pollute and lead to smog development and adversely affect the environment and health.
- To accommodate higher concentrations of alcohol fuels, engines must be adjusted to match the fuel injection timing, altering the compression ratios, and other parameters, which can increase maintenance and operational costs.

Alcohols have lower vaporization temperatures than gasoline, which can cause difficulties during the start of vehicles and degrade the performance in cold weather conditions.

9.2.6 Comparison of Alcohol Types

Alcohol fuels, including ethanol, methanol, butanol, and propanol, have distinguishing features that make them suitable for use in existing IC engines. This section aims to compare the previously mentioned alcohol fuels with a focus on their key parameters such as production methods, energy content, their octane rating, compatibility of the fuels with existing engine structures, the impact of emissions on the environment and the related economic considerations. The inter-relational understanding of the differences that exist between the available alcohol fuels is required for the selection of appropriate alcohol fuel types to obtain optimized performance of IC engines followed by a decrease in emissions. Search and understanding will also help the end user to select the fuel type to meet the specific requirements of various applications.

To compare the alcohol fuel types, the user must learn about properties such as energy content, environmental impact and the octane rating. Ethanol, as such, has a higher-octane rating than the commonly used gasoline; thereby, it is a better option as a fuel additive for improving the overall performance of engines with a significant reduction in knocking. On the contrary, ethanol's energy content per liter is lower, leading to a decrease in fuel efficiency than gasoline. The less-dense nature of methanol than ethanol and gasoline affect the vehicle range. These alcohol fuels are considered more environmentally friendly as they produce fewer greenhouse gases and pollutants than traditional fossil fuels and oils. Methanol produced from CO_2 can also help achieve carbon neutrality goals for nations around the globe (Fig. 9.7).

To compare the alcohol fuel types, one must try to learn about properties such as energy content, environmental impact and the octane rating. In this chain of events ethanol has a higher-octane rating than the commonly used gasoline; thereby, it is a better option as a fuel additive for improving the overall performance of engines with a significant reduction in knocking. On the contrary, ethanol's energy content per liter is lower, leading to a decrease in fuel efficiency than gasoline. The less-dense nature

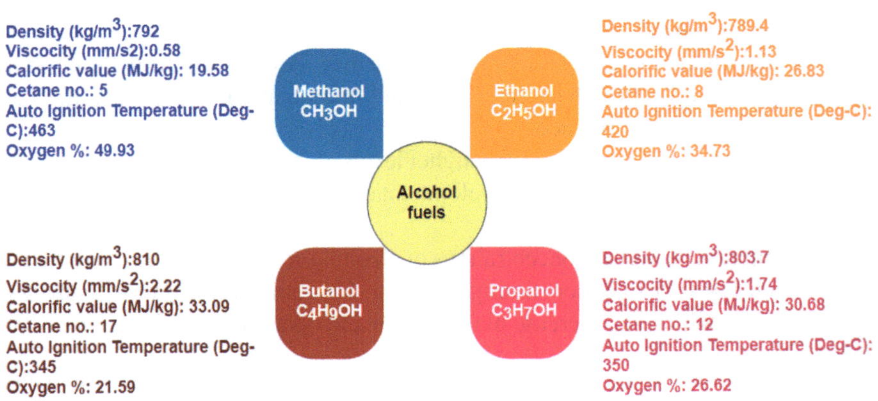

Fig. 9.7 Comparison of alcohol fuel properties (Karthick et al. 2021)

of methanol than ethanol and gasoline affect the vehicle range. These alcohol fuels are considered more environmentally friendly as they produce fewer greenhouse gases and pollutants than traditional fossil fuels and oils. Methanol produced from CO_2 can also help achieve carbon neutrality goals for nations around the globe.

9.3 Production of Alcohol Fuels

Biofuels can be produced in solid, liquid, or gaseous form directly or indirectly from various biomass sources. Liquid biofuels can be produced using both edible and edible oils. Biofuels, such as vegetable oils, are considered renewable and sustainable energy source. Such naturally derived fuels are reported to exhibit properties similar to conventionally available diesel fuel. However, fuels obtained using vegetable oils are prone to viscosity, low boiling point, low cetane number, and reduced volatility (Agarwal 2007). These properties are necessary for the uninterrupted operation of an engine that uses these fuels. These characteristics are even more crucial for CI engines. Any overshooting of these properties can lead to problems such as contamination of engine oil, partial combustion and increased emissions through exhaust, making these fuels less suitable for direct use in CI engines (Rajeswari et al. 2022).

The production processes of alcohol, particularly ethanol and methanol, can be categorized based on the type of feedstock used: non-renewable and renewable. Each category has distinct processes, advantages, and challenges that influence their environmental impact and economic viability.

9.3.1 Alcohol Production Using Non-renewable Feedstock

The primary non-renewable feedstock for alcohol production is fossil fuels, particularly natural gas, coal, and petroleum. Methanol is predominantly produced from natural gas through a process known as steam reforming. In this process, methane (the primary component of natural gas) is reacted with steam at high temperatures (700–1000 °C) in the presence of a catalyst to produce synthesis gas (syngas), which is a mixture of hydrogen (H_2) and carbon monoxide (CO). The reaction can be summarized as follows:

$$CH_4 + H_2O \rightarrow CO + 3H_2 \tag{9.1}$$

This syngas is then converted into methanol using another catalytic process in high-pressure reactors. The chemical reaction for methanol synthesis from syngas can be represented as:

$$CO + 2H_2 \rightarrow CH_3OH \tag{9.2}$$

While this method is efficient and can produce high yields of methanol, it is heavily reliant on fossil fuel inputs, contributing to significant carbon emissions. Furthermore, the energy-intensive nature of the steam reforming process raises concerns regarding sustainability. Ethanol production from non-renewable sources typically involves the fermentation of sugars derived from fossil fuel-based feedstocks such as corn or sugarcane processed with fossil fuels. In this scenario, the sugars are obtained through hydrolysis or enzymatic conversion, followed by fermentation using yeast. However, the carbon footprint associated with these processes remains a significant drawback.

9.3.2 Alcohol Production Using Renewable Feedstock

In contrast, renewable alcohol production primarily utilizes biomass feedstocks such as sugarcane, corn, agricultural residues, or dedicated energy crops like switchgrass. The use of renewable feedstocks offers a more sustainable approach to alcohol production by recycling carbon dioxide through plant growth and reducing overall greenhouse gas emissions.

For ethanol production from biomass, the process begins with the hydrolysis of complex carbohydrates (starch or cellulose) into fermentable sugars. This can be achieved through enzymatic hydrolysis where enzymes such as amylases break down starch into glucose. The hydrolysis reaction can be illustrated as follows:

$$(C_6H_{10}O_5)_n + nH_2O \rightarrow nC_6H_{12}O_6 \quad (9.3)$$

Following hydrolysis, fermentation occurs wherein yeast (e.g., Saccharomyces cerevisiae) converts the sugars into ethanol and carbon dioxide through anaerobic respiration. The fermentation reaction can be represented as:

$$(C_6H_{12}O_6)_n \rightarrow 2C_2H_5OH + 2CO_2 \quad (9.4)$$

After fermentation, the resulting mixture undergoes distillation to separate ethanol from water and other byproducts. This step typically involves multiple distillation stages to achieve a high concentration of ethanol suitable for fuel use. One significant advantage of renewable alcohol production is the ability to integrate renewable energy sources such as solar or wind power into the production process. For example, solar energy can be utilized to power fermentation tanks or distillation units, significantly reducing reliance on fossil fuels and minimizing the carbon footprint associated with alcohol production.

9.3.3 Production from Biomass Sources

The two main techniques for producing biofuel are thermochemical and biochemical (fermentative) processes. Over the past century, methanol manufacturing techniques have evolved, and various feedstocks, including CO_2, biomass, natural gas, and coke oven gas, are now used. For ethanol production, starch-based feedstocks like maize, grain, barley, and grain sorghum, as well as sugar-based feedstocks like sugarcane, fruits, citrus molasses, cane sorghum, and sugar beets, are primarily utilized in the fermentation process. Furthermore, there are a lot of commercial initiatives to make ethanol from cellulosic resources, such as leftover paper and wood (Balat 2005).

Manufacturing methanol and ethanol from biomass, carbon dioxide, water, or fossil fuels is possible. While methanol Comparing methanol and ethanol to more traditional fuels like gasoline and diesel reveals some distinct benefits and drawbacks. Because of their high-octane ratings, these alcohols in IC engines enable higher EGR rates and compression ratios.is often made from synthesis gas, ethanol has historically been derived by the fermentation of carbohydrates. Enzymes are also used in contemporary techniques as a substitute for fermentation. Ethanol can be produced from methanol, which is a simpler molecule. Almost any biomass, including animal dung, can be used to produce methanol industrially by gasifying the biomass into synthesis gas. Additionally, it can be made in labs with enzymes or electrolysis (Thangavelu et al. 2016).

Figure 9.8 shows the steps involved in the production of methanol from biomass and industrial processes. As per the published results, it is found that using alcohol fuels can reduce emissions of NO_x, CO, hydrocarbons, and particulates. For instance, tests on Chevrolet Luminas running on E85 showed a 20–22% reduction in non-methane hydrocarbons (NMHC), a 25–32% reduction in NO_x, and a 12–24% reduction in CO compared to reformulated gasoline. There was also a decrease in toxic emissions such as benzene and 1,3-butadiene, though there was an increase in aldehyde emissions, especially acetaldehyde (Karthick et al. 2021).

Methanol, commonly known as MeOH or methyl alcohol or wood alcohol or wood spirit is the simplest of all the alcohols available in the literature. It is a light, volatile, colorless, flammable liquid with a distinctive odor and is polar at room temperature. Also, this is hygroscopic in nature and has a tendency of attracting water. MeOH mixes well with water, petrol, and many organic compounds. It burns with an almost invisible flame and is biodegradable. It is considered a safe fuel, with toxicity levels (mortality rates) comparable to or better than gasoline. Additionally, methanol biodegrades more quickly than petroleum fuels if spilt.

Ethanol (C_2H_5OH) is mainly produced from biomass through the process of fermentation, and ethanol is widely used as a gasoline additive or alternative fuel. It has a high-octane number, which enhances anti-knock properties and enables more efficient engine performance. The higher oxygen content of ethanol enhances combustion efficiency and leads to higher combustion temperatures. Additionally, ethanol's lower carbon-to-hydrogen ratio and high heat of evaporation contribute to a reduced adiabatic flame temperature. Although ethanol has a lower heating value

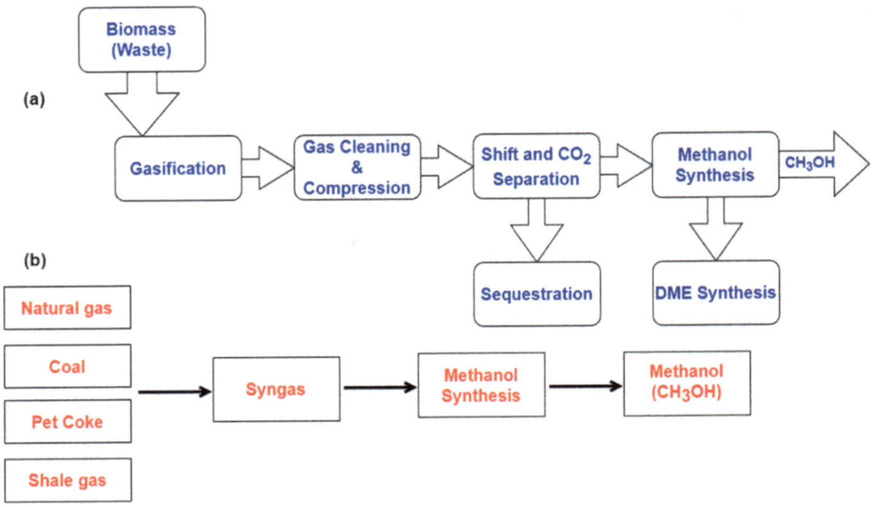

Fig. 9.8 **a** Steps of methanol production from biomass **b** Steps of methanol production from industrial processes

than gasoline, necessitating a greater volume of fuel to achieve the same power output, its higher evaporation heat can increase volumetric efficiency. Furthermore, the high flame speed of ethanol allows for quicker combustion, improving engine performance (Awad et al. 2018).

Figure 9.9 illustrates the ethanol production trends across countries and regions from 2007 to 2022. The United States remains the leading ethanol producer globally, generating over 15 billion gallons in both 2021 and 2022. The U.S. and Brazil together account for 80% of the world's ethanol supply, with the U.S. predominantly using corn and Brazil relying on sugarcane as their primary feedstocks. The United States leads global ethanol production, with an output exceeding 15 billion gallons in both 2021 and 2022. The United States and Brazil are responsible for producing 80% of the world's ethanol supply. In 2022, the worldwide production of methanol was projected to exceed 111 million metric tons, representing a 4% growth compared to the previous year. The global methanol market was valued at more than $37.4 billion in 2021 and is expected to rise to $61.7 billion by 2030 (https://afdc.energy.gov/data).

Ethanol (ethyl alcohol) and methanol (methyl alcohol) are two light alcohols often discussed as alternative fuels. Methanol is considered one of the best fuels for internal combustion (IC) engines, while ethanol is a key component of biodiesel and a promising alternative fuel for IC engines. Both alcohols, as oxygenates, are favored in IC engines for their improved volatility and high latent heat properties. Research focuses on enhancing blends of diesel fuel, biodiesel, and alcohols for CI engines. However, challenges like reduced heating value, phase separation, low pour points, and safety concerns during the storage and transportation of these ternary blends still need to be addressed (Ayas 2018).

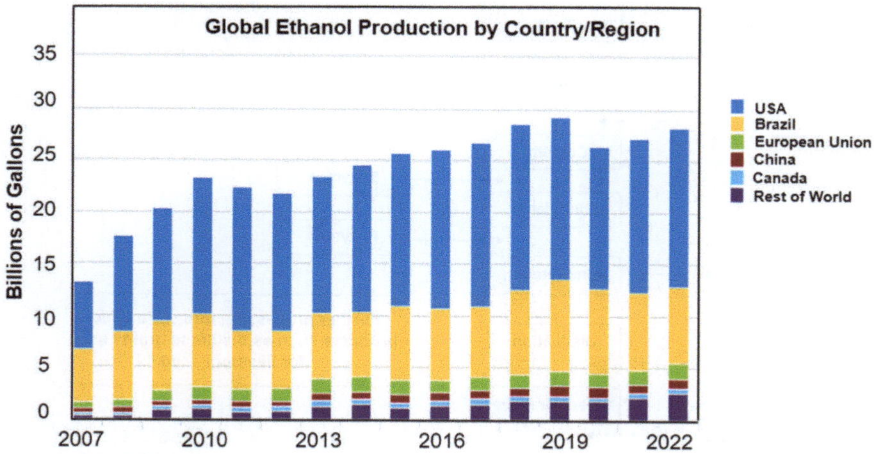

Fig. 9.9 Globally ethanol production from 2007–2022 (https://afdc.energy.gov/data)

Methanol and ethanol contain both soluble and insoluble contaminants, including halide ions, which significantly increase the corrosiveness of these fuels. These ions attack the passivating oxide films on metals, causing pitting corrosion and raising electrical conductivity, which worsens various forms of corrosion within the fuel system.. To mitigate these issues, fuel systems must be made from suitable materials, and the use of high-quality alcohol with low contaminants and added corrosion inhibitors is recommended. Interestingly, water has been shown to inhibit corrosion by ethanol, and experiments with E50 (a more aggressive ethanol blend) demonstrated that increasing water content to 2% in ethanol fuel can stop corrosion (Yusri et al. 2017). Alcohol fuels can also react with certain polymers, causing them to swell and lose tensile strength over time due to the oxidation of the carbon–carbon bonds. However, most modern vehicles are designed to tolerate up to 10% ethanol (E10) without issue. Flex fuel vehicles, in particular, are engineered to handle blends like E85 or M85 and include upgrades such as specialized fuel tanks, pumps, filters, and other components to ensure compatibility with high alcohol content fuels. Ethanol and methanol have different energy contents per liter: ethanol releases 21.1 MJ, methanol 15.8 MJ, and gasoline approximately 32.6 MJ. Although alcohol fuels have lower energy densities, engines can be optimized to use them more efficiently, potentially offsetting the lower energy density (Murray 2024).

Propanol and butanol are much less toxic and volatile compared to methanol. Butanol, especially, has a high flash point of 35 °C, which improves fire safety but can make engine ignition difficult in cold weather. However, during engine operation, the flash point concept is not directly applicable, as the air inside the cylinder is compressed, raising the temperature to several hundred degrees Celsius before ignition. Producing propanol and butanol through the fermentation of cellulose is a complex process (Thangavelu et al. 2016). One can take a look at Fig. 9.10 for the steps involved in production of alcohol from biomass. The Weizmann organism

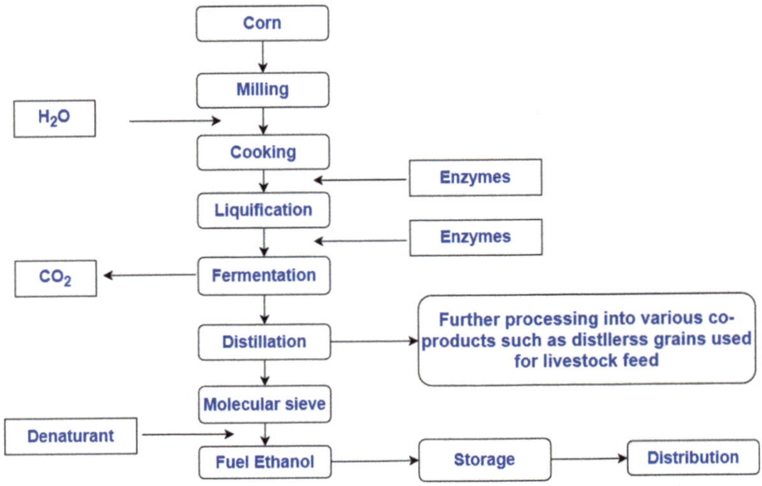

Fig. 9.10 Ethanol production way from biomass

(Clostridium acetobutylicum), mainly used for these conversions, emits a strong, unpleasant odor, which must be considered when planning the location of fermentation plants. Additionally, this organism can only tolerate butanol concentrations up to 2% before it dies. For comparison, yeast can survive in environments with up to 14% ethanol content, and specialized strains like turbo yeast can endure up to 16% ethanol. The development of a strain of the Weizmann organism with greater resistance to butanol, potentially surpassing the natural limit of 7%, is being explored. This advancement would be advantageous, given that butanol possesses a higher energy density than ethanol. Additionally, converting waste fiber from sugar crops into butanol could boost the alcohol yield from these crops without the need for additional planting. (Varma et al. 2021).

9.3.4 Production from Industrial Processes

In industrial processes, alcohol can be synthesized from petrochemical feedstocks. For example, ethylene can be hydrated to produce ethanol. Following fermentation or synthesis, alcohol is purified and separated through distillation. The methods used are chosen based on resource availability, economic feasibility, and environmental considerations.

Fermentation is a biochemical process wherein microorganisms convert organic compounds—primarily sugars into simpler substances, including alcohols, gases, and acids. In alcohol production, fermentation generally refers to anaerobic respiration, where microorganisms like yeast and bacteria produce alcohol and carbon dioxide without oxygen (Wang et al. 2022). Fermentation is crucial in producing alcohols such as ethanol and butanol, especially when using biological feedstocks. This process involves converting sugars into alcohols and other byproducts through the metabolic activities of microorganisms like yeast and bacteria. Here, we delve into the detailed mechanisms, stages, and factors influencing the fermentation process in industrial alcohol production (Erdiwansyah et al. 2019).

Similarly, ethanol is categorized into 1G (first generation) and 2G (second generation) ethanol. 1G ethanol is derived from food crops such as sugarcane, corn, and wheat. While this type of ethanol is widely produced and used, it has faced criticism for competing with food production and contributing to food price volatility. The production process for 1G ethanol typically involves fermentation of sugars extracted from these crops, followed by distillation. On the other hand, 2G ethanol is produced from non-food biomass feedstocks, such as agricultural residues, wood chips, or dedicated energy crops like switchgrass. The production of 2G ethanol involves more complex processes such as enzymatic hydrolysis or thermochemical conversion to break down lignocellulosic materials into fermentable sugars. Although 2G ethanol production is still developing and may have higher costs associated with it, it offers significant advantages in terms of sustainability because it does not compete with food resources and can utilize waste materials. Recent literature emphasizes the importance of these classifications in understanding the sustainability of alcohol fuels. For instance, studies have shown that bio-methanol can significantly reduce greenhouse gas emissions compared to brown or gray methanol by utilizing renewable feedstocks and employing carbon capture technologies. Furthermore, research indicates that while 1G ethanol can lead to increased food prices and land-use changes, 2G ethanol presents a more sustainable alternative to enhance energy security without compromising the food supply.

9.3.5 Chemical and Physical Characteristics of Alcohol Fuels

Table 9.1 presents the physicochemical properties of higher alcohols. As the length of the carbon chain in the alcohol molecule increases, so do the density, kinematic viscosity, low heating value, cetane number (CN), boiling point, and flash point. The density of higher alcohols, excluding aromatic alcohols, closely matches that of diesel fuel. However, alcohols have a much higher viscosity than diesel, which improves lubrication but may adversely affect spray characteristics. In terms of energy content, higher alcohols are comparable to traditional fuels, with the low heating value of n-octanol exceeding that of gasoline and diesel, resulting in a higher energy density than short-chain alcohols. (Karthick and Nanthagopal 2021). The CN, an important measure of fuel's propensity for spontaneous combustion in CI engines, is generally lower for higher alcohols than diesel, which typically ranges from 45 to 55.

Therefore, blending higher alcohols with highly reactive substitutes like ethers, is suggested for them to replace diesel completely. As the carbon chain length increases, the spontaneous combustion properties of higher alcohols improve while anti-knock

Table 9.1 Chemical and physical characteristics of various alcohol fuels used in IC engines (Thiyagarajan et al. 2019; Verma et al. 2016)

Properties	Diesel	Gasoline	Methanol	Ethanol	Butanol	Propanol
Chemical structure	$C_{14}H_{30}$	C_8H_{15}	CH_3OH	C_2H_5OH	C_4H_9OH	C_3H_7OH
Carbon mass (%)	86.13	86	37.48	52.14	64.82	59.96
Hydrogen mass (%)	13.87	14	12.48	13.02	13.49	13.31
Oxygen mass (%)	–	–	49.93	34.73	21.59	26.62
Molecular weight (g/mole)	198.4	111.19	32.04	46.06	74.11	60.09
ΔH Vapor (kJ/kg)	232	351	1168	919.6	707.9	727.88
Vapor pressure (kPa)	0.0533		17	7.33	0.933	2.66
Density (kg/m^3)	849.2	737	792	789.4	810	803.7
Viscosity (mm/s^2)	2.72	0.5–0.6	0.58	1.13	2.22	1.74
Boiling point (°C)	125–400	22–225	64.7	78	118	97.1
Flash point (°C)	> 55	(− 45 to 38)	11	17	3537	11.7
Self-ignition temperature (°C)	254–300	420	463	420	345	350
Octane number	–	8088	89	109	85	
Cetane number	–	15	5	8	17	12
Calorific value (MJ/kg)	42.49	43.5	19.58	26.83	33.09	30.68
Stoichiometric *A*/*F* ratio	14.95	14.58	6.46	8.98	11.17	–

performance decreases, making them more suitable for CI engines, whereas short-chain alcohols are more suited for SI engines (Sarathy et al. 2014). The latent heat of vaporization, an important factor affecting in-cylinder mixture formation and combustion, is only slightly higher for higher alcohols than for diesel, thus not posing significant cold-start issues. Alcohols' boiling and flash points are slightly lower than diesel's. Aromatic alcohols have higher boiling and flash points than linear alcohols. Furthermore, high alcohols exhibit lower hygroscopicity and better blending stability with gasoline and diesel than short-chain alcohols. These advantageous physico-chemical properties make higher alcohols viable substitutes for short-chain alcohols (Kumar et al. 2013).

Adopting alcohol fuels in IC engines offers several environmental and economic benefits: Alcohols burn more cleanly than gasoline, leading to lower emissions of pollutants like CO, PM, and NO_x. Ethanol, in particular, can be produced from biomass through fermentation, making it a renewable fuel. It reduces the carbon footprint and promotes energy sustainability. Diversifying fuel sources with alcohol fuels can enhance energy security by reducing reliance on imported petroleum.

Producing ethanol and methanol can create economic opportunities in agriculture and rural development, particularly in countries with abundant biomass resources (Muthuraman et al. 2022). The choice of alcohol fuel for IC engines depends on various factors, including production methods, energy content, octane rating, engine compatibility, environmental impact, and economic viability. Ethanol and methanol are the most established options with extensive research and application histories. Butanol offers promising advantages, particularly regarding energy density and compatibility with existing infrastructure. Propanol, while less common, also presents potential benefits. The optimal choice depends on specific application needs, local resource availability, and regulatory frameworks. Continued research and technological advancements will play a critical role in overcoming the challenges associated with each type of alcohol fuel, paving the way for more sustainable and efficient energy solutions (Suresh and Porpatham 2024).

9.4 Case Studies of Alcohol Fuels in IC Engines

As a viable solution for reducing greenhouse gases and mitigating energy costs, fuels made from alcohol fuels emerge as the most viable and promising renewable fuels. Among these, ethanol stands out as a potential fuel for spark ignition engines due to several technical advantages. The octane number for ethanol is comparatively higher than popular gasoline fuels, which allows for higher compression ratios and improved thermal efficiency. Higher availability of oxygen in alcohol fuels leads to more complete combustion, thereby reducing carbon monoxide and unburned hydrocarbon emissions. Moreover, ethanol's latent heat of vaporization is significantly higher than that of gasoline, contributing to a cooling effect within the combustion chamber. This cooling effect not only helps in knocking suppression but also enables the engine to operate more efficiently under high load conditions. Additionally, the high flame speed of ethanol ensures more stable combustion, enhancing the overall performance of the engine. However, ethanol also presents some challenges, such as its lower energy density compared to gasoline, which can lead to reduced mileage per gallon. Despite this, the environmental benefits and the ability to blend ethanol with gasoline to create flexible fuel options make it an attractive alternative. The renewable nature of ethanol, derived from biomass and agricultural waste, further underscores its potential to contribute to a sustainable energy future. Overall, the integration of ethanol into fuel systems for spark ignition engines addresses economic and environmental concerns and enhances engine performance. Future research and development in this area are crucial for optimizing engine designs and fuel formulations to exploit the benefits of ethanol and other alcohol fuels fully. A few case studies on ethanol, methanol and butanol are included in following subsections.

Suresh and Porpatham undertook experiments on a spark ignition engine derived from a diesel engine that used a lean mixture of ethanol and gasoline as a fuel. The engine ran at 1500 rpm and had a higher compression ratio valued at 10.5:1. The ethanol blends with gasoline were varied from 5 to 30%. Also, the throttle

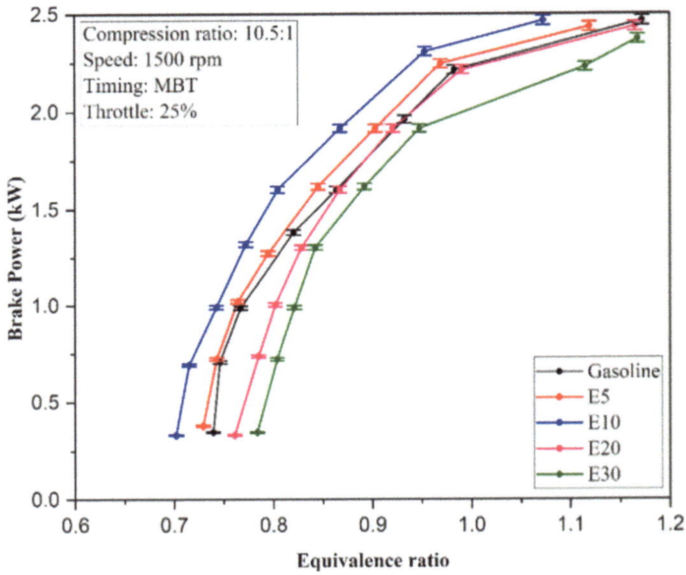

Fig. 9.11 Brake power variations with changes in the equivalence ratio (Meng 2024)

was adjusted to 25% to run the engine at different equivalency ratios and evaluate performance and emissions.

Figure 9.11 provides experimental test results which showcase improvements in brake power at various blends of ethanol. Such improved result warrant a promising future for alcohol based fuels. Also, it was shown by the researchers through their experimental results that a 10% ethanol-gasoline blend hiked the thermal efficiency and power by 7% and 4.5%, respectively, due to the increased compression ratio (Meng 2024).

To increase the efficient and clean use of alcohol fuels, Meng et al. investigated performance improvements by changing the intake valve timing and lift on ethanol, *n*-butanol, and gasoline engines under stoichiometric and lean burn scenarios. Increasing intake valve timing lengthens the combustion process for ethanol and *n*-butanol but shortens it for gasoline. Gasoline improves equivalent brake-specific fuel consumption (ESFC) by 6.9%, but ethanol reduces $BSNO_x$ by 42.07%. Adjusting valve lift has a smaller influence on fuel economy than timing. The combined throttle, valve lift, and timing changes enhanced ESFC by 6.79% for ethanol, 4.35% for *n*-butanol, and 10.98% for gasoline. Furthermore, integrating valve timing and lean burn conditions increases economy and emissions, reducing ESFC by 12.73% for ethanol, 6.87% for *n*-butanol, and 11.96% for gasoline, with $BSNO_x$ reductions of 93.69%, 74.86%, and 61.65%, respectively (Rawat et al. 2023).

Rawat et al. demonstrated the viability of using hydrogen fuel to increase the performance, exhaust emissions, and exergy parameters of a compression ignition engine running on an ethanol-jatropha biodiesel blend. The engine was operated at

25, 50, 75, and 100% load at a constant speed of 1500 rpm (± 5 rpm), with hydrogen injection durations of 1200, 2500, 3700, 4500, and 6000 μs. The 10% ethanol/90% jatropha biodiesel (BE10) blend slightly improved performance, emissions, and exergy measures. Adding hydrogen for a 2500 μs injection interval improved brake thermal efficiency (2.09–2.92%), exergetic efficiency (1.85–2.67%), exergy destruction rate (1.72–7.05%), and entropy generation (4.26–9.47%). Increasing hydrogen reduced unburned hydrocarbons, carbon monoxide, and particulate matter with greater NO_x emissions (Firew et al. 2022).

Firrew et al. investigated the performance and emissions of a 10% ethanol emulsion containing nickel zinc iron oxide ($NiZnFe_2O_4$) nanoparticles compared to diesel fuel. In tests on a single-cylinder diesel engine, E5, E10, and E15 fuels increased fuel consumption by 27%, 34%, and 37%, respectively; reduced braking torque by 15%, 5%, and 15%; and lowered power by 19%, 7%, and 12%. Adding 25, 50, and 100 ppm of nanoparticles to E10 decreased fuel consumption by 16%, 27%, and 27%, respectively; increased torque by 1%, 2%, and 2%; and increased power by 5%, 7%, and 7% (Wai 2022).

Wai et al. aimed to characterize the effects of oxygen surplus biofuels on combustion, thermal efficiency, and emissions by blending ethanol and biodiesel with fossil diesel. Bioethanol at 5 and 10% by weight was added to create various blend ratios for a 3 L four-cylinder common rail diesel engine with various rpm values ranging from 1000 to 2000 with torques for 56–140 Nm. Results showed higher pressure for premixed combustion and improvements in the net amount of heat release, with more than a 50% reduction in particle emissions for ethanol-biodiesel blends (Yue et al. 2024).

Methanol can be use as medium for storing the energy for longer durations in renewable systems, with oxy-fuel combustion offering higher values of carbon recovery and thermal efficiency. In this context, using 1-D thermodynamics simulations, Yue et al. investigated the effects of CO_2 dilution on methanol-based spark ignition (SI) engines. Results showed that net indicated thermal efficiency (ITE_{net}) decreased from 40.5% to 31% by using 0.16 oxygen fraction. The $ITE_{net,r}$ improved to 38.6%, enabling cost-effective CO_2 capture (Shahinsha et al. 2023).

Shahinsha et al., through their ethanol-based combustion experiments, highlighted significant reductions in particulate matter, HC and CO emissions, though NO_x and CO_2 emissions exhibit variations (Torres et al. 2024). On the other hand, Torres et al. analyzed and experimented with a mono-cylinder engine to improve performance and emissions data by using two different compression rates and various blends made using anhydrous ethanol and water. Engine performance parameters fuel mixtures with up to 50% water content were tested for speed values in the range of 2000–4000 rpm. Significant improvements were obtained using blended ethanol as fuel (Domínguez et al. 2024). Transitioning to a carbon-free economy has necessitated advancements in sectors utilizing compression-ignition engines. In this context, Domínguez et al. examined the impact of different renewable short-chain alcohols, namely methanol and ethanol, for improvements in performance, combustion characteristics, emissions, and potency of a dual-fuel engine operating under higher and lower load conditions. Hydrotreated vegetable oil (HVO), a diesel-like fuel with a high cetane number, was employed to assess its effectiveness in dual-fuel mode. The investigation comprised two stages: the first involved optimizing operating parameters such as EGR rate and combustion phasing to achieve an optimum balance between efficiency and NO_x emissions. The second stage focused on maximizing fuel substitution ratios, with the HVO-ethanol combination achieving the highest substitution rate of 84%. Methanol reduced NO_x emissions and enhanced thermal efficiency. Additionally, using HVO and high alcohol ratios effectively reduced large and small particulate emissions compared to diesel or HVO alone (Kalwar et al. 2023).

Methanol usage for light-duty vehicles used for public and goods transport was investigated, focusing on a Bharat Stage-VI (BS-VI) 2020-compliant spark-ignition engine with M15 (15% methanol, 82% gasoline, 3% propanol) by Kalwar et al. Compared to baseline gasoline, M15 reduced unwanted emissions. Catalytic conversion efficiency was higher for CO but lower for HC and NO_x with M15. Unregulated emissions, including ethene, propene, formaldehyde, acetaldehyde, and methane were reduced with M15 (Catapano et al. 2022). Catapano et al. characterized the effect of ethanol blends on sub-23 nm particles and their volatile organic fraction from a turbocharged 1.8 L direct injection spark ignition engine. Tests were conducted using 10%, 25%, and 85% *v*/*v* ethanol blends. Results indicated that ethanol blends increased the fraction of volatile organic matter and sub-23 nm particle numbers, except for E85, which exhibited larger soot particles due to more substantial charge cooling effects (Tipanluisa et al. 2024).

The impact of renewable fuels on vehicle exhaust emissions was investigated by Tipanluisa et al. The effect of butanol on the Selective Catalytic Reduction (SCR) system of a Euro V heavy-duty diesel engine was studied. Various butanol-diesel blends (5/95, 10/90, and 20/80%v) were tested. Results indicated a 4% reduction in exhaust gas temperature with 20% butanol. The SCR system's conversion efficiency increased by 4%, achieving a maximum of 96% with a 10% butanol blend. Reductions in CO and THC emissions were witnessed by the researchers (Musthafa et al. 2023).

Musthafa et al. indicated the need for alternative fuels due to fuel usage and resource depletion increments. Their study explored quinary fuel blends, combining fossil diesel (D) with karanja biodiesel (B), coconut oil, sunflower oil, and alcohols (ethanol, propanol, *n*-butanol) for usage in a diesel engine. The results indicated that the brake thermal efficiency of the blended fuel increased by 9.2%, with decrements in emissions by 37.2% and 53.93%, respectively. Such explorations with various naturally derived additives have kept the quest for IC engine research alive (Ghadikolaei 2016).

Recent studies on alcohol fuels, particularly ethanol and methanol, have demonstrated their potential to enhance engine performance and reduce emissions. Experimental results indicate that a lean mixture of ethanol and gasoline in a spark ignition engine significantly improves thermal efficiency and power output, with a 10% ethanol-gasoline blend increasing thermal efficiency by 7% and power by 4.5%. Research on intake valve timing adjustments revealed that ethanol can substantially reduce nitrogen oxide emissions while improving fuel economy compared to gasoline. Furthermore, the integration of hydrogen with ethanol blends has been shown to enhance brake thermal efficiency and decrease unburned hydrocarbons and carbon monoxide emissions. Recent technological advancements involves use of ethanol emulsions containing nanoparticles, revealing reductions in fuel consumption and increases in torque and power. Additionally, blending ethanol with biodiesel has led to significant reductions in particle emissions while improving thermal efficiency. Methanol's role as an energy storage medium has been highlighted, with studies demonstrating its effectiveness in maintaining thermal efficiency in spark ignition engines. Overall, these findings underscore the advantages of using alcohol fuels in internal combustion engines, showcasing their ability to improve performance while addressing environmental concerns. Continued research into optimizing these fuels and their blends is essential for advancing sustainable transportation solutions.

The carbon footprint of alcohol production processes varies significantly depending on the feedstock and method used, making it a crucial factor in evaluating their environmental impact compared to traditional fossil fuels like diesel and gasoline. Ethanol production, particularly from first-generation (1G) sources such as corn and sugarcane, typically involves significant greenhouse gas emissions associated with agricultural practices, land use changes, and energy-intensive fermentation and distillation processes. However, second-generation (2G) ethanol derived from non-food biomass presents a more sustainable option with a lower carbon footprint, as it utilizes waste materials and does not compete with food production. In contrast, methanol production from fossil fuels—labeled as brown or grey methanol—carries a high carbon footprint due to the extraction and processing of natural gas or coal. Conversely, bio-methanol produced from renewable biomass can substantially reduce emissions by recycling carbon dioxide and utilizing organic waste. When compared to diesel and gasoline production, alcohol fuels generally offer significant advantages in terms of emissions. The higher oxygen content in alcohol fuels leads to more complete combustion, resulting in lower carbon monoxide and unburned hydrocarbon emissions. Additionally, alcohol fuels like ethanol have higher octane ratings, which allow for higher compression ratios in engines, enhancing thermal efficiency and reducing nitrogen oxide emissions. For instance, studies have shown that using ethanol blends in internal combustion engines can lead to substantial reductions in brake-specific nitrogen oxide emissions while improving overall fuel economy (Kumar et al. 2021). Furthermore, the integration of alcohol fuels into existing fuel systems contributes to a reduction in particulate matter emissions compared to traditional fossil fuels. Overall, the adoption of alcohol fuels not only mitigates the carbon footprint associated with fuel production but also enhances the performance of internal combustion engines while addressing critical environmental concerns related to air quality and climate change.

9.5 Challenges and Limitations

Using alcohol fuels in IC engines presents several challenges and limitations. One of the primary issues is the lower energy density of alcohols, such as methanol and ethanol, compared to conventional fossil fuels. This lower energy density results in reduced fuel efficiency and shorter driving ranges. Additionally, alcohol fuels have a high heat of vaporization, which can lead to difficulties in cold-start conditions and extended ignition delays. Another significant challenge is the corrosive nature of alcohols, which can lead to material degradation in engine components and fuel systems. Furthermore, the hygroscopic properties of alcohols, particularly methanol and ethanol, lead to water absorption, potentially causing phase separation and fuel instability. The compatibility of alcohols with existing engine and fuel infrastructure also poses a limitation, often requiring modifications to engines and storage systems. Lastly, the production and distribution of alcohol fuels can be constrained by feedstock availability, economic factors, and regulatory challenges. These issues highlight the need for continued research and development to optimize the use of alcohol fuels in IC engines. Despite their advantages, several challenges must be addressed to optimize the use of alcohols in IC engines:

- Alcohols are more corrosive than conventional fuels and can degrade certain materials used in fuel systems, such as rubber and certain metals. This necessitates the use of compatible materials and coatings.
- Alcohols have lower energy densities than gasoline and diesel, which can result in reduced fuel economy. Engine modifications and optimizations are needed to offset this disadvantage.
- Alcohols have a higher heat of vaporization, which can cause difficulties in cold weather starting. Engine systems may require additional heating elements or other modifications to ensure reliable operation.
- The widespread adoption of alcohol fuels requires significant investments in fuel distribution and storage infrastructure, as well as modifications to existing vehicles and fueling stations.
- Despite the potential benefits, several challenges and limitations must be addressed for the widespread adoption of alcohol fuels in IC engines:
- Large-scale production of alcohol fuels requires significant agricultural and industrial infrastructure. Land use and food security concerns must be balanced with fuel production needs.
- Older engines and certain materials may not be compatible with high concentrations of alcohol fuels, necessitating modifications or developing flexible-fuel vehicles.
- The lower energy density of alcohol fuels necessitates larger fuel tanks or more frequent refueling, which can be a logistical challenge.
- The production and distribution costs of alcohol fuels can be higher than those of conventional fuels, although technological advancements and economies of scale could mitigate this over time (Prajapati et al. 2024).

9.6 Conclusion and Future Scope

9.6.1 Conclusion

In conclusion, the integration of alcohol fuels, particularly ethanol and methanol, into internal combustion (IC) engines presents a compelling opportunity to enhance engine performance while significantly reducing harmful emissions. The emission reduction potential of alcohol fuels is notable, as they consistently demonstrate lower levels of hydrocarbons (HC), carbon monoxide (CO), and particulate matter (PM) compared to conventional fossil fuels. This reduction is attributed to the higher oxygen content in alcohols, which promotes more complete combustion and leads to improved thermal efficiency. Studies indicate that the use of ethanol blends can enhance combustion properties, resulting in greater brake thermal efficiency and reduced specific fuel consumption. Furthermore, the compatibility of alcohol fuels with existing engine designs allows for a smoother transition toward more sustainable fuel options without necessitating extensive modifications to infrastructure.

Sustainability is another critical advantage of alcohol fuels derived from renewable resources such as agricultural biomass and waste materials. This shift not only conserves resources but also aligns with global efforts to minimize reliance on non-renewable energy sources. The ability to blend alcohol fuels with traditional gasoline or diesel enhances their appeal, enabling current vehicles to utilize these cleaner alternatives with minimal adjustments. Overall, the adoption of alcohol fuels in IC engines represents a significant step toward achieving a more sustainable energy future, addressing both economic and environmental challenges while enhancing engine performance and efficiency. Continued research and development in this field will be essential for optimizing fuel formulations and engine technologies, ensuring that the benefits of alcohol fuels are fully realized in the transition to cleaner transportation solutions.

9.6.2 Future Scope

Future research in the realm of alcohol fuels for internal combustion engines should prioritize several key areas to enhance their efficiency and sustainability. One significant focus should be on the development and optimization of advanced fuel blends that combine alcohols with other biofuels and additives. These innovative blends have the potential to improve combustion efficiency, reduce emissions, and enhance overall engine performance. Additionally, continued innovation in engine design specifically tailored for alcohol fuels can lead to greater efficiency gains and emissions reductions. This includes exploring enhanced injection techniques, optimizing combustion chamber designs, and implementing advanced management technologies to mitigate thermal loads.

Comprehensive emission studies are also essential for understanding the long-term impacts of alcohol fuels on both regulated and unregulated emissions across various engine types and operating conditions. Gaining insights into the full spectrum of emissions will be crucial for developing more effective emission control strategies. Furthermore, conducting thorough lifecycle assessments and economic analyses of alcohol fuel production, distribution, and usage will provide valuable information regarding their overall sustainability and feasibility. Such analyses will empower policymakers and industry stakeholders to make informed decisions about the large-scale adoption of alcohol fuels and their integration into the broader energy mix. By addressing these areas, future research can significantly contribute to realizing the full potential of alcohol fuels in promoting cleaner, more sustainable transportation solutions.

References

Agarwal AK (2007) Biofuels (alcohols and biodiesel) applications as fuels for internal combustion engines. Prog Energy Combust Sci. https://doi.org/10.1016/j.pecs.2006.08.003

Alternative fuels data center: maps and data. Accessed 31 July 2024 (Online). Available https://afdc.energy.gov/data

Awad OI et al (2018) Alcohol and ether as alternative fuels in spark ignition engine: a review. Renew Sustain Energy Rev. https://doi.org/10.1016/j.rser.2017.09.074

Ayas N (2018) Solvent materials. In: Comprehensive energy systems: volumes 1–5, vol 2. Elsevier, Amsterdam, pp V2-368–V2-395. https://doi.org/10.1016/B978-0-12-809597-3.00226-1

Bae C, Kim J (2017) Alternative fuels for internal combustion engines. Proc Combust Inst 36(3):3389–3413. https://doi.org/10.1016/j.proci.2016.09.009

Balat M (2005) Current alternative engine fuels. Energy Sour 27(6):569–577. https://doi.org/10.1080/00908310490450458

Catapano F, Di Iorio S, Magno A, Sementa P, Vaglieco BM (2022) Effect of ethanol blends, E10, E25 and E85 on sub-23 nm particle emissions and their volatile fraction at exhaust of a high-performance GDI engine over the WLTC. Fuel 327. https://doi.org/10.1016/j.fuel.2022.125184

Das AK, Sahu SK, Panda AK (2022) Current status and prospects of alternate liquid transportation fuels in compression ignition engines: a critical review. Renew Sustain Energy Rev. https://doi.org/10.1016/j.rser.2022.112358

Domínguez VM, Hernández JJ, Ramos Á, Giménez B, Rodríguez-Fernández J (2024) Exploring the effect of methanol and ethanol on the overall performance and substitution window of a dual-fuel compression-ignition engine fueled with HVO. Fuel 359. https://doi.org/10.1016/j.fuel.2023.130529

Erdiwansyah, Mamat R, Sani MSA, Sudhakar K, Kadarohman A, Sardjono RE (2019) An overview of higher alcohol and biodiesel as alternative fuels in engines. Energy Rep. https://doi.org/10.1016/j.egyr.2019.04.009

Firew D, Nallamothu RB, Alemayehu G, Gopal R (2022) Performance and emission evaluation of CI engine fueled with ethanol diesel emulsion using $NiZnFe_2O_4$ nanoparticle additive. Heliyon 8(11). https://doi.org/10.1016/j.heliyon.2022.e11639

Ghadikolaei MA (2016) Effect of alcohol blend and fumigation on regulated and unregulated emissions of IC engines—a review. Renew Sustain Energy Rev. https://doi.org/10.1016/j.rser.2015.12.128

Göktaş M, Kemal Balki M, Sayin C, Canakci M (2021) An evaluation of the use of alcohol fuels in SI engines in terms of performance, emission and combustion characteristics: a review. Fuel. https://doi.org/10.1016/j.fuel.2020.119425

How to rapidly reduce fossil fuel use—FPIF. Accessed 31 July 2024 (Online). Available https://fpif.org/how-to-rapidly-reduce-fossil-fuel-use/

Kalwar A, Singh RK, Gupta A, Rajak R, Gosakan G, Agarwal AK (2023) Particulates, unregulated and regulated emissions and catalytic converter efficiency evaluation of methanol (M15) fuelled BS-VI compliant light-duty spark-ignition engine. Sci Total Environ 902. https://doi.org/10.1016/j.scitotenv.2023.166047

Karthick C, Nanthagopal K (2021) A comprehensive review on ecological approaches of waste to wealth strategies for production of sustainable biobutanol and its suitability in automotive applications. Energy Convers Manag. https://doi.org/10.1016/j.enconman.2021.114219

Karthick C, Nanthagopal K, Ashok B, Saravanan SV (2021) Influence of alcohol and gaseous fuels on NO_x reduction in IC engines. In: NO_x emission control technologies in stationary and automotive internal combustion engines: approaches toward NO_x free automobiles. Elsevier, Amsterdam, pp 347–385. https://doi.org/10.1016/B978-0-12-823955-1.00012-7

Kumar S, Cho JH, Park J, Moon I (2013) Advances in diesel-alcohol blends and their effects on the performance and emissions of diesel engines. Renew Sustain Energy Rev. https://doi.org/10.1016/j.rser.2013.01.017

Kumar D, Kachhap S, Thokchom SS (2024) Numerically comparative assessment of different biodiesel & its blends on performance, combustion and emission characteristics. Ind Crops Prod 209:117900. ISSN 0926-6690. https://doi.org/10.1016/j.indcrop.2023.117900

Meng X et al (2024) Study on the effects of intake valve timing and lift on the combustion and emission performance of ethanol, N-butanol, and gasoline engine under stoichiometric combustion and lean burn conditions. Energy 300. https://doi.org/10.1016/j.energy.2024.131385

Murray RJ (2024) Methanol and methanol/diesel fuel modes in compression ignition engines. In: Reference module in chemistry, molecular sciences and chemical engineering. Elsevier, Amsterdam. https://doi.org/10.1016/b978-0-443-15740-0.00052-5

Musthafa B, Saravanan B, Asokan MA, Devendiran S, Venkatesan K (2023) Effect of ethanol, propanol and butanol on Karanja biodiesel with vegetable oil fuelled in a single cylinder diesel engine. Egypt J Pet 32(2):35–40. https://doi.org/10.1016/j.ejpe.2023.05.001

Muthuraman VS et al (2022) Progress on compatibility issues of alcohols on automotive materials: kinetics, challenges and future prospects—a comprehensive review. Process Saf Environ Prot. Institution of Chemical Engineers. https://doi.org/10.1016/j.psep.2022.04.022

Percy AJ, Edwin M (2023) Studies on the performance and emission characteristics of a dual fuel VCR engine using producer gas as secondary fuel: an optimization approach using response surface methodology. Energy 263. https://doi.org/10.1016/j.energy.2022.125685

Percy AJ, Edwin M (2024) A comprehensive review on the production and enhancement techniques of gaseous biofuels and their applications in IC engines with special reference to the associated performance and emission characteristics. Sci Total Environ. https://doi.org/10.1016/j.scitotenv.2024.173087

Prajapati AK, Yadav S, Gomey AK, Choubey AK, Kumar R (2024) Assessment of performance and emission characteristics of CI engine using tyre pyrolysis oil and biodiesel blends by nano additives: an experimental study. J Energy Inst 117:101825. ISSN 1743-9671. https://doi.org/10.1016/j.joei.2024.101825

Rajeswari S, Baskaran D, Saravanan P, Rajasimman M, Rajamohan N, Vasseghian Y (2022) Production of ethanol from biomass—recent research, Scientometric review and future perspectives. Fuel 317. https://doi.org/10.1016/j.fuel.2022.123448

Rawat JS, Bhowmik S, Panua R, Madane PA, Triveni MK (2023) Investigation of performance, emission and exergy parameters of a compression ignition engine fuelled with ethanol-jatropha biodiesel blend under varying hydrogen strategies. Int J Hydrogen Energy 48(95):37415–37426. https://doi.org/10.1016/j.ijhydene.2023.06.128

Sarathy SM, Oßwald P, Hansen N, Kohse-Höinghaus K (2014) Alcohol combustion chemistry. Prog Energy Combust Sci. https://doi.org/10.1016/j.pecs.2014.04.003
Shahinsha NM, John JA, Singh K, Kumar N (2023) Reduction in automotive emissions by ethanol blending in SI engine: a review. Mater Today Proc. https://doi.org/10.1016/J.MATPR.2023.03.808
Suresh D, Porpatham E (2024) Influence of high compression ratio on the performance of ethanol-gasoline fuelled lean burn spark ignition engine at part throttle condition. Case Stud Therm Eng 53. https://doi.org/10.1016/j.csite.2023.103832
Thangavelu SK, Ahmed AS, Ani FN (2016) Review on bioethanol as alternative fuel for spark ignition engines. Renew Sustain Energy Rev. https://doi.org/10.1016/j.rser.2015.11.089
Thiyagarajan S et al (2019) NO_x emission reduction using permanent/electromagnet-based fuel reforming system in a compression ignition engine fueled with pine oil. Clean Technol Environ Policy 21(4):815–825. https://doi.org/10.1007/s10098-019-01670-8
Tipanluisa L, Fonseca-González N, Flores N, Ayala A (2024) Butanol and selective catalytic reduction of gaseous emissions in a heavy-duty diesel engine. Fuel 373:132236. https://doi.org/10.1016/j.fuel.2024.132236
Torres DJD, de S Mendes A, Albuquerque C (2024) Performance and emissions data of an internal combustion engine operating with different ethanol/water mixtures and compression ratios. Data Brief 54:110390. https://doi.org/10.1016/J.DIB.2024.110390
Varma JR, Katepalli SKR, Sreeja M, Hadagali B (2021) Comprehensive studies on alcohol using port fuel injection facilitated with spark plug engine. Mater Today Proc 3219–3225. https://doi.org/10.1016/j.matpr.2020.12.379
Verger T, Azimov U, Adeniyi O (2022) Biomass-based fuel blends as an alternative for the future heavy-duty transport: a review. Renew Sustain Energy Rev. https://doi.org/10.1016/j.rser.2022.112391
Verhelst S, Turner JW, Sileghem L, Vancoillie J (2019) Methanol as a fuel for internal combustion engines. Prog Energy Combust Sci. https://doi.org/10.1016/j.pecs.2018.10.001
Verma P, Sharma MP, Dwivedi G (2016) Impact of alcohol on biodiesel production and properties. Renew Sustain Energy Rev. https://doi.org/10.1016/j.rser.2015.11.048
Wai P et al Experimental investigation of the influence of ethanol and biodiesel on common rail direct injection diesel engine's combustion and emission characteristics. Case Stud Therm Eng 39. https://doi.org/10.1016/j.csite.2022.102430
Wang et al (2022) Evaluation of hydrous ethanol as a fuel for internal combustion engines: a review. Renew Energy. https://doi.org/10.1016/j.renene.2022.05.132
Yue Z, Wang X, Liu H, Li B, Yao M (2024) Exploring the application of oxy-fuel combustion to methanol spark ignition engines. Appl Energy 367. https://doi.org/10.1016/j.apenergy.2024.123449
Yusri IM et al (2017) Alcohol based automotive fuels from first four alcohol family in compression and spark ignition engine: a review on engine performance and exhaust emissions. Renew Sustain Energy Rev. https://doi.org/10.1016/j.rser.2017.03.080

Chapter 10
Current Challenges and State-of-the-Art Reviews of Methanol Fueled ICEs

Anbarasan Baluchamy

Abstract Research and development on alternative fuels for internal combustion engines has increased significantly as a result of the search for low-emission and sustainable energy sources. Methanol-fueled internal combustion engines (ICEs) present a promising alternative to conventional fossil fuels, offering significant potential for reducing greenhouse gas emissions and enhancing energy safety. This review synthesizes recent research and technologies advancements in methanol fueled ICEs, addressing critical challenges and highlighting state of the art developments. Significant issues such as fuel properties, combustion characteristics, and emissions control are analyzed. This review work explores advancements of engine modifications to enhance efficiency and emission reduction when using methanol as a fuel. Moreover, this review evaluates the integration of methanol ICEs with energy sources and the role of hybrid systems in optimizing the performance. Despite the benefits, there are still a number of problems to overcome, such as infrastructure development, material compatibility and economic sustainability. In order to overcome these issues and realize the full potential of ICEs powered by methanol in the pursuit of a sustainable energy future, this through review emphasizes the need for ongoing research and collaboration between the sectors.

Keywords Fuel · Methanol · Internal combustion · Engine · Emissions

10.1 Introduction

Methanol-fueled internal combustion engines are emerging as a promising alternative to conventional gasoline and diesel engines due to methanol's potential for reducing greenhouse gas emissions and dependence on fossil fuels. Methanol, a simple alcohol, can be produced from various feedstock's, including natural gas,

A. Baluchamy (✉)
Department of Mechanical Engineering, PSNA College of Engineering and Technology, Dindigul, Tamil Nadu, India
e-mail: anbarasan5mech@gmail.com

A. Bartl et al. (eds.), *Alcohol Production Processes and Their Utilisation in the Transport Sector*, Energy, Environment, and Sustainability,
https://doi.org/10.1007/978-981-96-7384-1_10

coal, and biomass, making it a versatile and sustainable energy carrier (Agarwal et al. 2019a, b). Despite these advantages, the adoption of methanol-fueled ICEs faces several challenges. These include technical issues related to combustion efficiency, material compatibility, and emissions control. Additionally, infrastructure for methanol production, distribution, and refueling remains underdeveloped compared to established fuels (Agarwal et al. 2021a, b). This review aims to provide a comprehensive overview of the current challenges associated with methanol-fueled ICEs, alongside a detailed analysis of the latest technological advancements and state-of-the-art solutions addressing these obstacles. This review seeks to inform future research directions and policy decisions to facilitate the wider adoption of methanol as a sustainable fuel for internal combustion engines.

10.1.1 Background and Significance of Alternative Fuels

Global energy consumption has been steadily rising over the past decades due to population growth, economic development, and technological advancements. Traditionally, fossil fuels such as coal, oil, and natural gas have been the primary sources of energy. These fossil fuels have driven industrialization, transportation, and modern conveniences, leading to substantial economic benefits. However, the environmental and geopolitical drawbacks of relying on fossil fuels have become increasingly apparent (Agarwal et al. 2021a, b). The combustion of fossil fuels is a major source of greenhouse gas emissions, particularly carbon dioxide, which contributes to global warming and climate change. The environmental consequences of these emissions include rising global temperatures, melting polar ice caps, sea-level rise, and increased frequency and severity of extreme weather events. Additionally, fossil fuel extraction and processing can result in significant ecological disruption, habitat destruction, and pollution (Agarwal et al. 2019a, b). The global distribution of fossil fuel reserves is uneven, leading to geopolitical tensions and economic vulnerabilities. Countries that rely heavily on imported fossil fuels are subject to supply disruptions and price volatility, which can impact economic stability and national security. The finite nature of fossil fuel reserves also raises concerns about long-term energy sustainability (Sovacool 2012). In response to these challenges, there has been a growing interest in alternative fuels as a means to diversify energy sources, reduce environmental impact, and enhance energy security. Alternative fuels encompass a broad range of energy sources and technologies, including biofuels, hydrogen, electric batteries, and renewable natural gas. These fuels offer the potential to significantly reduce greenhouse gas emissions and reliance on fossil fuels (Bansal and Meena 2021). One of the primary motivations for the development and adoption of alternative fuels is their potential to mitigate environmental harm. Biofuels, such as ethanol and biodiesel, are derived from renewable biological sources and typically produce lower levels of CO and HC emissions compared to conventional fossil fuels (Bansal and Meena 2021). Alternative fuels can enhance energy security by diversifying the energy supply and reducing dependence on imported fossil fuels. For instance, many countries have the capacity to produce biofuels domestically from agricultural residues, waste products, or dedicated energy crops. Similarly, advancements in renewable energy technologies, such as solar and wind power, can provide clean electricity for EVs and hydrogen production (Sovacool 2012). The transition to alternative fuels can stimulate economic growth and job creation in emerging industries. The development, manufacturing, and maintenance of alternative fuel technologies can create employment opportunities in sectors such as renewable energy, advanced manufacturing, and research and development. Additionally, investing in alternative fuels can reduce long-term energy costs and improve economic resilience (Kumar et al. 2020). The pursuit of alternative fuels drives technological innovation and fosters advancements in various scientific and engineering disciplines. For example, research into biofuel production methods, battery storage technologies, and hydrogen fuel cells has the potential to yield advances with broad applications beyond the energy sector. These innovations can contribute to the overall advancement of technology and industry (Singh et al. 2022). The global demand for energy continues to rise, driven by population growth, economic development, and technological advancements. However, this surge in energy consumption has led to significant environmental challenges, particularly due to the widespread use of fossil fuels. ICEs, which power the majority of vehicles on the road today, are major consumers of these fossil fuels, contributing to air pollution and greenhouse gas emissions. In response to these environmental concerns, the importance of alternative fuels for ICEs has gained considerable attention (Agarwal et al. 2019a, b).

10.2 Objectives and Scope of the Review

The primary objective of this review is to comprehensively examine and synthesize the current challenges and advancements in methanol-fueled internal combustion engines. This encompasses a detailed analysis of the technical, environmental, and economic aspects that influence the adoption and performance of methanol as an alternative fuel. Key areas of focus include the safety factors, performance and emissions characteristics of methanol-fueled ICEs. The modifications required for existing engine technologies to optimize methanol usage, and the integration of methanol fueling setup. Additionally, the review aims to highlight recent and research innovations that address the limitations of methanol fuel, such as its lower energy density and corrosive properties. By providing an in-depth evaluation of both the obstacles and the state-of-the-art solutions, this review seeks to offer valuable insights for researchers, policymakers, and industry stakeholders interested in advancing methanol as a viable and sustainable fuel option for internal combustion engines.

10.3 Methanol Fueled Internal Combustion Engines

10.3.1 Historical Development and Adoption

The historical development and adoption of methanol-fueled internal combustion engines began in the early twentieth century, driven by the search for alternative fuels to gasoline. Methanol, a type of alcohol, gained attention due to its high octane rating and clean-burning properties. During the oil crises of the 1970s, interest in methanol surged as countries sought to reduce dependence on petroleum. Research and development efforts during this period led to the creation of flexible-fuel vehicles capable of running on methanol, gasoline, or a blend of both. Despite its potential, widespread adoption faced challenges, including infrastructure development, production costs, and competition from other alternative fuels such as ethanol (Agarwal et al. 2021a, b). However, advancements in fuel technology and growing environmental concerns have recently renewed interest in methanol as a sustainable fuel, especially in regions with abundant natural gas resources, which can be converted to methanol (Agarwal et al. 2019a, b).

10.3.2 Chemical Properties and Fuel Characteristics

See Table 10.1.

10.3.3 Technological Advancements and State-of-the-Art Reviews

Methanol, a single-carbon alcohol, is increasingly considered a viable alternative fuel for ICE due to its favorable combustion characteristics, availability, and lower emissions compared to conventional fossil fuels. However, the successful utilization of methanol in engines hinges significantly on the design and efficiency of the fuel injection system. This discussion explores the intricacies of fuel injection systems for methanol-fueled engines, covering their types, design considerations, challenges, and advantages (Turner et al. 2011).

Spark Ignition Engines: Methanol can be utilized in SI engines with changes to the fuel injection system owing to its elevated latent heat of vaporization and octane rating (about 109). These qualities provide elevated compression ratios, hence enhancing thermal efficiency.

Challenges: Cold-start difficulties increase due to methanol's lower vapor pressure relative to gasoline. This requires improved fuel atomization technologies and maybe heating methods for the fuel–air mixture.

Table 10.1 Chemical properties and fuel characteristics (Amine and Barakat 2019; Agarwal et al. 2014)

Sl.No.	Property	Methanol	Gasoline
1	Chemical formula	CH_3OH	C_8H_{18}
2	Molecular weight	32.04 g/mol	~ 114–125 g/mol
3	Physical state	Liquid	Liquid
4	Color	Colorless	Colorless
5	Odor	Alcohol-like	Alcohol-like
6	Density	0.7918 g/cm^3 at 20 °C	0.71–0.77 g/mL
7	Melting point	− 97.6 °C	− 40 °C to − 50 °C
8	Boiling point	64.7 °C	25–225 °C
9	Flash point	11 °C (closed cup)	− 43 °C to − 40 °C
10	Auto ignition temperature	464 °C	246 °C to 280 °C
11	Vapor pressure	13.02 kPa at 20 °C	50–75 kPa
12	Solubility in water	Miscible	Miscible
13	Viscosity	0.544 cP at 25 °C	0.4–0.6 cP at 25 °C
14	Heat of vaporization	35.21 kJ/mol	305–380 kJ/kg
15	Heat of combustion	− 726 kJ/mol	43–47 MJ/kg
16	Energy content	19.9 MJ/kg (lower heating value)	31–35 MJ/L
17	Stoichiometric air/fuel ratio	6.4:1	14.7:1
18	Latent heat of vaporization	1101 kJ/kg	305–380 kJ/kg
19	Cetane number	3–5	–

Solutions: Advanced direct injection systems can be designed to accommodate the unique characteristics of methanol, hence enhancing atomization. Moreover, the materials of the cylinder head and the injection timing require tuning to accommodate methanol's elevated flame speed and energy density (Francesconi 2015).

Compression Ignition Engines: Compression ignition engines have greater difficulties with methanol owing to its elevated auto-ignition temperature and inadequate lubricity. The failure of methanol to self-ignition indicates that it does not readily ignite only under compression.

Solutions: Dual-fuel techniques are often used, wherein a minimal quantity of diesel (pilot fuel) ignites combustion, facilitating the effective combustion of methanol. Alternatively, additive techniques or ignition enhancers such as dimethyl ether (DME) may be employed to improve ignition (Zhang et al. 2018).

Adjustments in Technology: Injection systems in CI engines must handle high-pressure injection, requiring opted materials that withstand the corrosive properties of methanol to avert long-term degradation.

Corrosion: Methanol exhibits greater corrosiveness than traditional fuels, especially toward specific metals in fuel systems. Solutions entail utilizing corrosion-resistant materials or implementing protective coatings (Yuan and Zhen 2016). Methanol possesses approximately half the energy content of gasoline and diesel, necessitating larger fuel tanks or increased fuel flow rates in engines to attain comparable power output.

10.3.4 Fuel Injection Systems

Types of Fuel Injection Systems for Methanol fuel.

1. Port Fuel Injection (PFI)

Figure 10.1 shows that port fuel injection system. In PFI systems, methanol is injected into the intake port, where it mixes with air before entering the combustion chamber. This type of injection is simpler and less expensive but may suffer from less precise control of fuel–air mixture compared to direct injection systems (Agarwal et al. 2021a, b). Most commonly used with **spark-ignition (SI) engines**. Methanol's high octane rating makes it suitable for SI engines, which rely on a spark plug to ignite the air–fuel mixture (Francesconi 2015).

2. Direct Fuel Injection (DI)

Figure 10.2 shows that direct fuel injection system. In DI systems, methanol is directly injected into the combustion chamber. This allows for better control over the fuel–air mixture, improved combustion efficiency, and reduced emissions. Utilized in both **spark-ignition (SI) engines** and **compression-ignition (CI) engines** (Zhang et al. 2018).

10.4 Design Considerations

- Methanol is corrosive to certain materials used in conventional fuel injection systems. Components such as fuel lines, injectors, and seals must be made from methanol-resistant materials like stainless steel or certain polymers to ensure longevity and safety (Wei et al. 2020).
- Coatings: Nickel-phosphorus plating or ceramic coatings can protect metal surfaces exposed to methanol (Wang 2023).
- Methanol has lower viscosity and lubricity compared to gasoline and diesel. This necessitates the use of specialized fuel pumps and injectors to ensure proper atomization and to prevent wear and tear of moving parts (Chen et al. 2019).
- Lubricity Additives: Compounds like fatty acid esters (e.g., Methyl oleate) and mono/di-acyl glycerides are often added to methanol to improve lubricity.

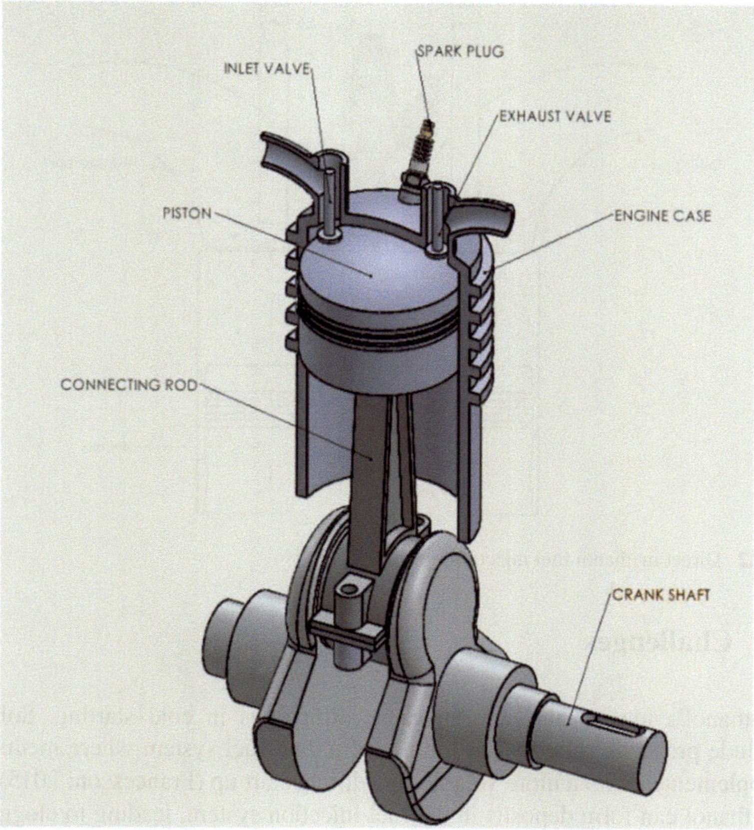

Fig. 10.1 Port fuel injection system

- Injectors for methanol must be designed to handle higher flow rates due to methanol's lower energy density. Additionally, the spray pattern and droplet size must be optimized to ensure complete combustion and minimize emissions (Chen et al. 2019).
- Combustion Additives: Ethanol or dimethyl ether (DME) can be blended with methanol to stabilize combustion, reducing the risks associated with poor vaporization.
- Methanol has a high latent heat of vaporization, which provides a cooling effect when injected. This can be advantageous for charge cooling but may require adjustments in engine calibration to avoid issues like wall wetting and incomplete vaporization.
- Calibration of Injection Timing: Adjusting the injection timing using ECU (Engine Control Unit) mapping ensures that methanol vaporizes completely without causing wetting of cylinder walls (Johnson et al. 2021)

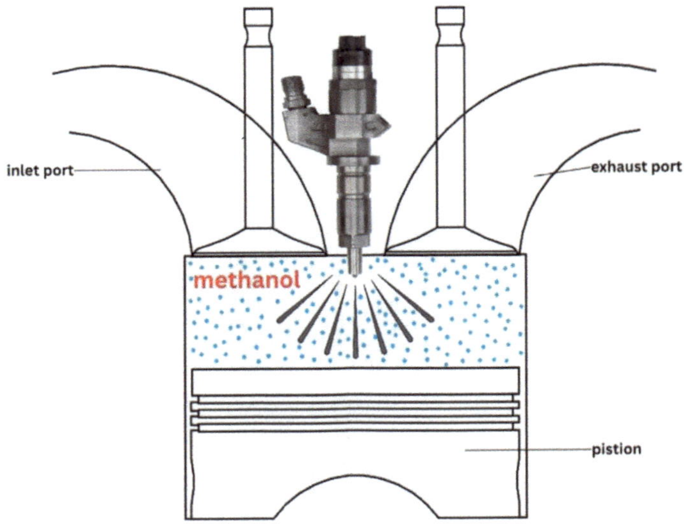

Fig. 10.2 Direct methanol fuel injection system

10.5 Challenges

- Methanol's lower volatility can cause difficulties in cold starting. Solutions include preheating the fuel or the use of a dual-fuel system where methanol is supplemented with a more volatile fuel during start up (Francesconi 2015).
- Methanol can form deposits in the fuel injection system, leading to clogging of injectors. Additives and regular maintenance are required to mitigate this issue (Methanol Institute 2023).
- The need for methanol-compatible materials and the complexity of direct injection systems can drive up costs. This makes the transition to methanol-fueled engines more challenging from an economic perspective (https://www.marketresearchfuture.com/reports/methanol-market-2023).
- Aluminum Alloys: 6061 aluminum alloy (Vergara-Juarez et al. 2017), Stainless Steel: AISI 316 stainless steel (Sugie et al. 2009)

10.6 Combustion and Performance

Turbocharged Methanol Engines

Studies on turbocharged methanol engines have demonstrated significant performance improvements. research by (Wei et al. 2020) showed that a turbocharged methanol engine could achieve a 20% increase in power output and a 15% reduction in fuel consumption compared to a naturally aspirated gasoline engine (Nguyen et al. 2020).

Dual-Fuel Engines

Methanol can also be used in dual-fuel configurations, where it is combined with diesel or gasoline. Experiments by (Zhang et al. 2018) on a dual-fuel methanol-diesel engine reported improved thermal efficiency and reduced NO_x and particulate emissions compared to conventional diesel engines.

Advanced Combustion Strategies

Techniques such as homogeneous charge compression ignition (HCCI) and spark-assisted compression ignition (SACI) have been explored for methanol engines. Research by (Chen et al. 2019) indicated that HCCI operation with methanol can achieve ultra-low NO_x and soot emissions, with thermal efficiencies exceeding 45%.

10.7 Emission Control Technologies

The use of methanol as a fuel in internal combustion engines has been extensively researched for its potential to reduce emissions compared to conventional fuels. Several emission control technologies have been developed and optimized to enhance the environmental benefits of methanol (Gupta et al. 2018).

10.7.1 Catalytic Converters

Catalytic converters have proven effective in reducing harmful emissions from methanol-fueled engines. Methanol combustion typically produces lower levels of NO_x compared to gasoline due to its lower flame temperature. However, the presence of unburned methanol and formaldehyde requires efficient catalytic oxidation. Studies indicate that three-way catalytic converters can effectively reduce CO, HC, and NO_x emissions simultaneously when properly optimized for methanol combustion (Liu et al. 2018).

10.7.2 Exhaust Gas Recirculation (EGR)

EGR is a widely adopted technology that recirculates a portion of the exhaust gases back into the combustion chamber, thereby reducing the oxygen concentration and lowering combustion temperatures. This process significantly reduces NO_x emissions. Research by (Zhao et al. 2020) demonstrated that EGR can reduce NO_x emissions by up to 50% in methanol-fueled engines, highlighting its effectiveness in emission control.

10.7.3 Particulate Filters

Methanol combustion typically produces fewer particulate matters than diesel, the use of particulate filters is still essential to capture any PM that might form. Diesel particulate filters (DPFs) and gasoline particulate filters (GPFs) can be adapted for methanol engines to ensure minimal PM emissions. (Smith et al. 2019) found that the integration of GPFs in methanol engines reduced PM emissions by over 90%.

10.7.4 Lean NO_x Traps (LNT)

Lean NO_x traps are effective in capturing and reducing NO_x emissions in lean-burn engines. Methanol-fueled engines, which can operate under lean conditions, benefit from the application of LNTs. Recent advancements have shown that LNTs, combined with appropriate regeneration strategies, can achieve NO_x conversion efficiencies of up to 80% (Johnson et al. 2021).

10.7.5 Emission Reductions

Methanol's high oxygen content contributes to a more complete combustion process, reducing CO and HC emissions. Methanol also produces fewer NO_x emissions due to its lower combustion temperature. According to a study by (Shen et al. 2017) methanol engines exhibited a reduction of CO and HC emissions by 30–50% compared to gasoline engines.

10.7.6 Cold Start Issues

Methanol has a higher latent heat of vaporization, which can lead to difficulties in cold starting. Advanced fuel injection systems and pre-heating technologies have been developed to address this issue. Studies by (Kim et al. 2019) demonstrated that with proper engine modifications, cold start performance of methanol engines can be significantly improved. The use of methanol as a fuel in IC engines, combined with advanced emission control technologies, offers a promising pathway to reduce harmful emissions and enhance engine performance. Catalytic converters, EGR, particulate filters, and lean NO_x traps are critical in optimizing methanol combustion and minimizing emissions. While challenges such as formaldehyde formation and cold start issues exist, ongoing research and technological advancements continue to address these obstacles, making methanol a viable alternative fuel for cleaner and more efficient IC engines.

10.8 Methanol Fuel

CH_3OH a simple alcohol, has gained attention as an alternative fuel for internal combustion engines. Its favorable properties, such as high octane rating and clean combustion characteristics, make it a potential substitute for conventional gasoline and diesel fuels. This discussion explores the availability and production of methanol, highlighting its potential as a sustainable fuel option for IC engines (Yang 2024).

10.8.1 *Production of Methanol*

Methanol can be produced from a variety of feedstock's including natural gas, coal, biomass, and even carbon dioxide. The most common production method involves the steam reforming of natural gas, which accounts for approximately 65% of global methanol production. This process involves the following steps:

1. Steam Reforming: Natural gas (primarily methane) reacts with steam to produce synthesis gas (syngas), a mixture of hydrogen, carbon monoxide, and carbon dioxide (Palo et al. 2007).

$$CH_4 + H_2O \rightarrow CO + 3H_2 \tag{10.1}$$

2. Methanol Synthesis: The syngas is then converted to methanol over a catalyst at high pressure and temperature.

$$CO + 2H_2 \rightarrow CH_3OH \tag{10.2}$$

Alternative methods include coal gasification, where coal is converted into syngas, and biomass gasification, which utilizes agricultural residues, wood, and other biomass sources. Recently, the conversion of CO_2 to methanol using renewable hydrogen (produced via electrolysis using renewable electricity) has emerged as a promising sustainable route (Guil-López et al. 2019).

10.8.2 *Availability of Methanol*

The global methanol market is substantial, with production exceeding 100 million metric tons per year. Major methanol-producing regions include China, the Middle East, and the United States. China alone accounts for over half of the global production, driven by its large coal reserves and strategic focus on alternative fuels.

Methanol distribution infrastructure is well-established, with extensive networks of pipelines, storage facilities, and transport vessels. This infrastructure facilitates the availability of methanol for various applications, including fuel for IC engines, chemical feedstock, and energy carrier (Olah et al. 2018).

1. **High Octane Rating**: Methanol has a high octane rating (around 110), which allows for higher compression ratios and improved engine efficiency.
2. **Clean Combustion**: Combustion of methanol produces fewer emissions of particulate matter, nitrogen oxides, and carbon monoxide compared to gasoline and diesel.
3. **Renewable Potential**: Methanol can be produced from renewable sources, contributing to a reduction in greenhouse gas emissions.

However, there are challenges to its widespread adoption, including its lower energy density compared to gasoline (about half), which necessitates larger fuel tanks or more frequent refueling. Additionally, methanol is corrosive to certain materials, requiring modifications to engine components and fuel systems (Olah et al. 2011).

Methanol presents a viable alternative fuel for IC engines, with established production processes and availability. Its clean combustion properties and potential for renewable production make it an attractive option for reducing environmental impact. Continued advancements in production technologies, particularly those leveraging renewable resources, will further enhance methanol's role in future energy landscape (International Energy Agency 2019).

10.8.3 Infrastructure and Distribution

Methanol can be produced from various feedstock, including natural gas, coal, biomass, and even carbon dioxide. The choice of feedstock significantly impacts the location and type of production facilities required:

Natural Gas: Methanol production from natural gas typically occurs in regions with abundant gas supplies. These facilities are often integrated with existing petrochemical complexes.

Coal: Coal-to-methanol plants are more common in coal-rich regions, such as China. These plants require additional infrastructure for coal handling and gasification.

Biomass: Biomass-based methanol production requires facilities close to agricultural or forestry operations to minimize transportation costs of the raw material.

Carbon Dioxide: Emerging technologies convert CO_2 to methanol, which necessitates proximity to CO_2 sources, such as industrial plants or power stations (Guil-López et al. 2019).

10.8.4 Storage Infrastructure

Methanol storage infrastructure must accommodate its physical and chemical properties. Methanol is corrosive to certain metals and absorbs water from the atmosphere, necessitating specialized storage solutions:

Storage Tanks: Tanks made from compatible materials, such as stainless steel or lined carbon steel, are essential to prevent corrosion. These tanks should also have vapor recovery systems to handle methanol's volatility.

Blending Facilities: For methanol to be used as a fuel, it is often blended with Diesel-Methanol Blend. Four blends were prepared by mixing diesel with methanol 5, 10, 15 and 20% by mass The results indicate that methanol blend is an encouraging alternative for lower smoke at the cost of CO and HC emissions. Altogether, it is concluded that diesel-methanol blends can be suitably used in CRDI diesel engines after making a good trade-off between performance and emission (Sahoo et al. 2019). Blending facilities must ensure precise mixing and maintain the stability of the fuel mixture (https://www.osha.gov/methanol).

10.8.5 Transportation Infrastructure

The transportation of methanol involves several modes, each requiring specific infrastructure adjustments:

Pipelines: Existing pipelines can be adapted for methanol transport with modifications to prevent corrosion. However, the extensive pipeline network needed for methanol is not yet in place.

Rail and Road: Rail tank cars and tanker trucks designed for methanol transport must comply with safety regulations due to methanol's flammability and toxicity. Dedicated fleets and trained personnel are essential.

Shipping: For international transport, methanol can be shipped in bulk using chemical tankers. Ports need appropriate loading and unloading facilities, including storage tanks and transfer equipment (Smith et al. 2024).

10.8.6 Distribution Network

The distribution of methanol as a fuel involves a complex network ensuring its availability and accessibility:

Fuel Stations: To facilitate the use of methanol as a fuel, existing fuel stations must be retrofitted or new ones built to accommodate methanol dispensing:

Retrofit Existing Stations: Modifying existing gasoline stations to dispense methanol involves installing methanol-compatible storage tanks, pumps, and dispensing nozzles. This approach leverages the existing network but requires significant investment (Gao 2022).

Challenges and Considerations

Cost: The high initial investment in production facilities, storage, transportation, and distribution infrastructure is a significant barrier.

Compatibility: Ensuring the compatibility of existing infrastructure with methanol's chemical properties requires careful material selection and design adjustments (Nitsche and Kulkarni 2021).

Safety and Regulations: Strict safety protocols and compliance with environmental regulations are crucial to prevent accidents and minimize environmental impact (Gao et al. 2024).

Public Acceptance: Educating the public and stakeholders about the benefits and safety of methanol as a fuel is essential for widespread adoption. Developing the infrastructure and distribution network for methanol as a fuel involves a multifaceted approach, addressing production, storage, transportation, and dispensing challenges. While significant investments and strategic planning are required, the potential environmental and economic benefits of methanol as a clean fuel make it a promising alternative in the transition to sustainable energy systems (https://www.phmsa.dot.gov/hazmat/erg/emergency-response-guidebook-erg).

10.9 Safety and Handling

- Methanol is a colorless, volatile liquid that serves as a versatile chemical in various industries, including as a fuel. Despite its utility, methanol poses significant health and safety risks, necessitating stringent safety measures and handling protocols (Johnson 2024)
- Methanol is toxic if ingested, inhaled, or absorbed through the skin. Acute exposure can result in symptoms such as headaches, dizziness, nausea, and in severe cases, central nervous system depression, metabolic acidosis, blindness, and death. Chronic exposure can lead to cumulative health issues affecting the liver, kidneys, and nervous system (https://www.niehs.nih.gov/health/topics/agents/methanol/index.cfm.)
- Methanol is highly flammable, with a flashpoint of 11 °C. It can form explosive mixtures with air, making it a significant fire and explosion hazard. Methanol fires are difficult to detect and extinguish due to their almost invisible flame. Proper fire safety measures, including the use of appropriate fire extinguishers (alcohol-resistant foam, dry chemical powder, CO_2), are critical in environments where methanol is used or stored (Kumar et al. 2024a, b).

- Safe handling of methanol requires the use of personal protective equipment such as gloves, safety goggles, and protective clothing to minimize skin and eye contact. Inhalation hazards necessitate adequate ventilation or the use of respiratory protection in enclosed spaces. Storage of methanol should be in tightly closed containers in a cool, well-ventilated area away from heat sources, open flames, and incompatible substances like strong oxidizers (Zhang et al. 2021)
- In the event of a methanol spill, it is crucial to remove all sources of ignition and ventilate the area. Spills should be contained and cleaned up using non-sparking tools and absorbent materials designed for chemical spills. Proper disposal methods for contaminated materials must be followed in accordance with local regulations to prevent environmental contamination (Wang et al. 2022).
- Emergency protocols should be in place for incidents involving methanol exposure. This includes first aid measures such as flushing exposed skin or eyes with water for at least 15 min and seeking immediate medical attention. For inhalation exposure, the affected individual should be moved to fresh air and provided with oxygen if breathing is difficult, while methanol is a valuable industrial chemical and fuel, its handling and storage require rigorous safety measures to protect health and prevent accidents. Adherence to established safety guidelines and protocols is essential to mitigate the risks associated with methanol use (Chen 2022).

10.10 Future Prospects and Research Directions

10.10.1 Future Prospects

- Methanol is considered a cleaner-burning fuel compared to gasoline and diesel, producing fewer pollutants such as nitrogen oxides and particulate matter. Its use in internal combustion engines and as a feedstock for fuel cells can significantly reduce the carbon footprint of transportation and power generation sectors (Zhang et al. 2021). Additionally, methanol can be produced from renewable sources such as biomass and municipal waste, further enhancing its environmental benefits (Gao et al. 2022).
- The economic feasibility of methanol fuel is influenced by production costs, infrastructure, and market demand. Recent advancements in methanol synthesis, such as the development of more efficient catalysts and processes, have the potential to lower production costs (Li et al. 2023). Furthermore, the establishment of methanol refueling infrastructure and incentives for its use in vehicles could stimulate market growth and drive down costs through economies of scale (Smith et al. 2024).
- The integration of methanol fuel into existing energy systems presents both opportunities and challenges. Methanol can be used in modified ICEs and in methanol fuel cells, offering flexibility in application (Wang et al. 2022). Moreover, methanol's compatibility with existing infrastructure, such as pipelines and

storage tanks, facilitates its adoption as a transitional fuel in the shift toward more sustainable energy systems (Chen et al. 2023).

10.10.2 Research Directions

- Future research should focus on enhancing methanol production technologies to make them more efficient and cost-effective. This includes developing novel catalysts for methanol synthesis from both traditional fossil fuels and renewable sources (Kumar et al. 2024a, b). Research into alternative feedstock's and innovative production methods, such as electrochemical reduction of CO_2, could also play a crucial role in advancing methanol as a sustainable fuel (Johnson et al. 2023).
- Improving the performance of methanol-fueled engines is essential for its wider adoption. Research should focus on optimizing engine design and combustion processes to maximize efficiency and reduce emissions (Zhang et al. 2021). This includes exploring advanced fuel injection systems, combustion chamber designs, and engine control strategies.
- Developing infrastructure to support methanol fuel is critical for its commercial viability. This includes designing and constructing refueling stations, storage facilities, and distribution networks. Research into the economic and logistical aspects of infrastructure development will help address barriers to widespread methanol adoption (Smith et al. 2024).
- Comprehensive lifecycle assessments are needed to evaluate the environmental and economic impacts of methanol fuel compared to other alternatives. Research should focus on assessing the full lifecycle emissions, resource use, and economic implications of methanol production, distribution, and consumption (Gao et al. 2022).
- Policymaking plays a crucial role in shaping the future of methanol fuel. Research should explore the impact of different policy measures, such as subsidies, tax incentives, and emission regulations, on the development and adoption of methanol fuel (Chen et al. 2023). Understanding the interplay between policy and market dynamics will be essential for creating a supportive environment for methanol fuel.

10.10.3 Fuel Blending and Additives

Blending methanol with other fuels or additives can further enhance engine performance and efficiency:

Methanol-Gasoline Blends: Blends such as M85 (85% methanol, 15% gasoline) can leverage the high octane number of methanol while maintaining the energy density of gasoline, resulting in improved overall engine efficiency.

Performance Additives: Additives such as anti-corrosion agents and lubricants can mitigate some of the challenges associated with methanol use, such as fuel system corrosion and lubrication issues, thus enhancing long-term engine performance.

The utilization of methanol as a fuel for IC engines presents numerous opportunities for efficiency improvements. Engine design modifications, advanced combustion control techniques, effective after treatment technologies, and optimized fuel blends contribute to the enhanced performance, reduced emissions, and overall sustainability of methanol-fueled engines. Continued research and development in these areas will further solidify methanol's role as a key alternative fuel in the transition to cleaner and more efficient transportation solutions (Kalghatgi 2014).

10.11 Practical Implementations and Case Studies

Methanol combustion produces significantly lower emissions of CO_2, NO_x, and particulates, making it a cleaner alternative to traditional fossil fuels. Studies have demonstrated that methanol can reduce greenhouse gas emissions by up to 15% compared to gasoline (Kalghatgi 2014).

10.11.1 Practical Implementations

10.11.1.1 Dual-Fuel Systems

Dual-fuel systems, where methanol is used in combination with diesel or gasoline, have been a focal point of recent research. This approach leverages methanol's clean-burning properties while maintaining the high energy density of traditional fuels. For instance, research conducted by (Dong et al. 2020) demonstrated that a dual-fuel engine using methanol and diesel could achieve a 10–15% increase in thermal efficiency and a 30% reduction in NO_x emissions.

10.11.1.2 Dedicated Methanol Engines

Dedicated methanol engines, designed specifically to run on methanol, are being developed to maximize the benefits of this fuel. These engines incorporate advanced fuel injection systems and optimized combustion chambers to improve efficiency and reduce emissions. For example, a study by (Wang and Wright 2021) showed that a dedicated methanol engine achieved a thermal efficiency of 42%, which is significantly higher than conventional gasoline engines.

10.11.1.3 Retrofitting Existing Engines

This practical approach involves retrofitting existing gasoline or diesel engines to run on methanol. This method is cost-effective and can be quickly implemented in existing vehicle fleets. A case study in China, where over 10,000 buses were retrofitted to run on methanol, showed a substantial decrease in operating costs and emissions. (https://www.man-es.com/company/press-releases/press-details/2023/07/19/methanol-retrofit-solutions-to-be-developed-for-medium-speed-marine-engines).

10.11.2 Case Studies

10.11.2.1 China's Methanol Fuel Program

China has been at the forefront of methanol fuel implementation, driven by its abundant coal resources and the need to reduce air pollution. The country's methanol fuel program includes the use of methanol-gasoline blends (M15, M85) and pure methanol (M100) in various applications. A comprehensive study of this program revealed significant improvements in air quality and a reduction in fossil fuel dependence (Chinese Ministry of Industry and Information Technology (MIIT): Methanol Vehicle Policies 2022).

10.11.2.2 Methanol in Racing Cars

Methanol has also been used in racing due to its high octane rating and excellent combustion properties. The use of methanol in IndyCar racing is a well-documented example. Methanol-fueled race cars have shown enhanced performance and safety, as methanol fires can be extinguished with water, unlike gasoline fires (McKinsey & Company 2021 "Methanol in Internal Combustion Engines").

10.12 Comparative Analysis of Methanol, Ammonia, and Hydrogen Fueled ICEs

In the ongoing mission for sustainable and efficient fuel alternatives, methanol, ammonia, and hydrogen have emerged as significant contenders. This section provides a comparative overview of their fuel characteristics, which are crucial for understanding their potential and limitations as fuels for internal combustion engines (Ogden et al. 1999).

10.12.1 Methanol Fuel

Methanol is a simple alcohol with a relatively high hydrogen content, making it an attractive alternative fuel. It is liquid at ambient conditions, which simplifies its storage and handling compared to gaseous fuels.

Energy Density: Methanol has a lower energy density than gasoline, with a value of approximately 19.7 MJ/kg compared to gasoline's 44.4 MJ/kg. This means more methanol is needed to achieve the same energy output.

Combustion Characteristics: Methanol burns cleanly, producing fewer NO_x and particulate emissions than gasoline. It has a high octane rating, typically around 109, which allows for higher compression ratios and better engine efficiency.

Safety and Handling: Methanol is toxic and corrosive, posing significant risks during handling and storage. It is also hygroscopic, meaning it absorbs water from the atmosphere, which can lead to contamination issues (Global Methanol Institute: Methanol Market Trends Report 2023).

10.12.2 Ammonia Fuel

Ammonia is a nitrogen–hydrogen compound that offers a carbon-free combustion process, making it appealing from an environmental perspective.

Energy Density: Ammonia has an energy density of about 18.6 MJ/kg, which is lower than methanol and significantly lower than traditional fossil fuels (Zamfirescu and Dincer 2009).

Combustion Characteristics: Ammonia has a narrow flammability range and a high ignition temperature, which presents challenges for efficient combustion. It produces no carbon dioxide when burned, but can result in NO_x emissions if not properly controlled (Schüth et al. 2012).

Safety and Handling: Ammonia is highly toxic and has a pungent odor, which can act as a deterrent for leaks. It is also stored as a liquid under moderate pressure, which simplifies storage compared to hydrogen (Valera-Medina et al. 2018).

10.12.3 Hydrogen Fuel

Hydrogen is the lightest and most abundant element, and its use as a fuel offers the potential for zero-emission vehicles (Hydrogen Council 2021). Table 10.2 shows that comparison analysis of three different fuels.

Table 10.2 Comparison analysis of different fuel

Sl.No.	Fuel	Energy density (MJ/kg)	Combustion emissions	Storage and handling characteristics	Safety considerations
1	Methanol	19.7	Low NO_x, low particulates	Liquid at ambient conditions	Toxic, corrosive, hygroscopic
2	Ammonia	18.6	Zero CO_2, potential NO_x	Liquid under moderate pressure	Highly toxic, pungent odor
3	Hydrogen	120 (by mass)	Water vapor	High pressure or cryogenic storage	Highly flammable, requires robust systems

Each fuel presents unique advantages and challenges. Methanol's compatibility with existing infrastructure and clean combustion makes it a practical choice, while ammonia offers a carbon-free combustion pathway. Hydrogen, with its high energy density and clean combustion, holds the promise of zero emissions but requires significant advancements in storage and safety systems (Rivard et al. 2019).

10.13 Efficiency and Performance Metrics

The efficiency and performance metrics of internal combustion engines fueled by methanol, ammonia, and hydrogen are critical in evaluating their viability as alternative fuels. This discussion examines the thermal efficiency, power output, and emissions of each fuel type, supported by relevant studies and data.

10.13.1 Methanol-Fueled ICEs

Methanol is a promising alternative fuel due to its high octane number and ability to reduce emissions. Methanol-fueled ICEs typically exhibit improved thermal efficiency compared to conventional gasoline engines. Studies indicate that methanol can achieve a thermal efficiency of around 42% under optimal conditions. Methanol also produces fewer NO_x emissions due to its lower flame temperature and higher latent heat of vaporization, which helps in cooling the combustion chamber. However, methanol has a lower energy density compared to gasoline, which can affect the overall power output of the engine. Despite this, methanol engines can still deliver comparable performance due to their higher compression ratios and better knock resistance (Nguyen et al. 2020).

10.13.2 Ammonia-Fueled ICEs

Ammonia is another alternative fuel that has gained attention for its zero-carbon content. Ammonia can be used in ICEs with some modifications, including changes to the fuel injection system and combustion chamber design to handle its lower flame speed and higher ignition temperature. The thermal efficiency of ammonia-fueled ICEs is generally lower than that of methanol and hydrogen. Studies report thermal efficiencies in the range of 30–35%. The primary challenge with ammonia is its relatively low energy density and the need for catalytic support to achieve complete combustion, which can complicate engine design and operation. However, ammonia's advantages lie in its clean combustion, producing only water and nitrogen as by-products, significantly reducing NO_x and particulate emissions (Reiter and Kong 2011).

10.13.3 Hydrogen-Fueled ICEs

Hydrogen is considered the ultimate clean fuel due to its high energy content per unit mass and zero-carbon emissions. Hydrogen-fueled ICEs can achieve thermal efficiencies of up to 45–50%. The high flame speed of hydrogen allows for more complete combustion, contributing to higher efficiency and power output. Additionally, hydrogen combustion produces only water vapor, eliminating CO_2 and significantly reducing NO_x emissions when lean burn strategies or EGR are employed. However, hydrogen's low energy density requires high-pressure storage systems, and its propensity for pre-ignition and knock poses engineering challenges. Advanced ignition systems and precise fuel management are essential to harness hydrogen's full potential in ICEs (Das 1992).

Comparing the efficiency and performance metrics of methanol, ammonia, and hydrogen as fuels for ICEs reveals distinct advantages and challenges for each. Methanol offers a balance of high efficiency and reduced emissions, while ammonia provides a zero-carbon alternative with clean combustion, albeit at lower efficiencies (Singh et al. 2024). Hydrogen stands out with the highest efficiency and zero-carbon emissions, but requires sophisticated engine technologies to mitigate its inherent challenges. The choice of fuel ultimately depends on the specific application, infrastructure availability, and environmental considerations (White et al. 2006).

10.14 Emission Profiles and Environmental Impact

The comparative analysis of methanol, ammonia, and hydrogen as fuels for ICEs highlights their distinct advantages and challenges in terms of emissions and environmental impact. Ammonia offers a carbon-free combustion process but requires effective NO_x control and sustainable production methods to minimize its environmental footprint. Hydrogen, with its zero-carbon emissions, presents a promising alternative, though its high NO_x emissions and the environmental impact of its production need addressing. Transitioning to these alternative fuels requires a comprehensive approach, including advancements in production technologies, emission control systems, and infrastructure development. Policy support and investments in renewable energy will play crucial roles in making these fuels viable and sustainable options for the future (Bromberg et al. 2008) (Table 10.3).

Table 10.3 Emission comparison analysis for different fuels

Sl. No.	Fuel	CO_2 emissions	NO_x emissions	Particulate matter (PM)	Other pollutants
1	Methanol	Carbon-based fuel, produces CO_2 upon combustion. However, it generates about 25% less CO_2 compared to gasoline	Combustion results in lower NO_x emissions due to its lower combustion temperature	Combustion produces significantly fewer particulates than gasoline or diesel engines	Combustion can produce formaldehyde and other aldehydes, which are toxic pollutants
2	Ammonia	It does not produce CO_2 during combustion as it does not contain carbon	Combustion can produce significant NO_x emissions unless properly managed with catalytic converters or other technologies	Combustion typically results in very low particulate emissions	Unburned NH_3 can be emitted, which is harmful to the environment and human health
3	Hydrogen	Combustion produces no CO_2 as it is a carbon-free fuel	Combustion can result in high NO_x emissions due to the high flame temperature, but this can be mitigated with appropriate technologies	Combustion produces negligible particulate emissions	Combustion typically does not produce other significant pollutants

10.15 Economic and Practical Viability

The economic and practical viability of methanol, ammonia, and hydrogen as fuels for internal combustion engines is a crucial factor determining their adoption in the automotive and industrial sectors. This analysis will compare these three alternative fuels based on their production costs, infrastructure requirements, fuel efficiency, and overall feasibility (Klerke et al. 2008). Table 10.4 shows that economic and practical viability comparison analysis for three different fuels.

Table 10.4 Economic and practical viability comparison analysis for different fuels

Sl.No.	Fuel	Production cost	Infrastructure	Fuel efficiency
1	Methanol	Methanol production costs are relatively low compared to hydrogen and ammonia It can be produced from a variety of feedstock's, making it a versatile option. Current production costs range from Rs. 34–51 per L, depending on the feedstock and production technology used	Existing fuel infrastructure can be adapted for methanol with minor modifications. This includes storage tanks, pipelines, and refueling stations, which can be upgraded to handle methanol with minimal investment	Its high octane rating improves engine efficiency and performance, potentially offsetting the higher consumption rate
2	Ammonia	It's is higher than methanol but lower than hydrogen The production cost ranges from Rs. 43–85 per kg	It's requires specialized handling and storage due to its toxicity and corrosive nature This necessitates significant investment in infrastructure upgrades, including corrosion-resistant storage tanks and safety systems at refueling stations	Lower energy density than methanol, leading to even higher fuel consumption Lower flame speed and stability
3	Hydrogen	Electrolysis and natural gas reforming are the primary production methods, with costs ranging from Rs. 85–255 per kg, depending on the technology and scale	Entirely new infrastructure for storage and distribution High-pressure tanks and specialized refueling stations	Highest energy density by weight but a very low density by volume This necessitates high-pressure storage or liquefaction, both of which are energy-intensive processes

The comparative analysis of methanol, ammonia, and hydrogen reveals distinct economic and practical advantages and challenges for each fuel. Methanol stands out for its low production costs and adaptable infrastructure, making it a viable short-term solution (Kumar et al. 2024a, b). Ammonia offers a moderate balance between production cost and infrastructure requirements but faces challenges in fuel efficiency and handling. Hydrogen, while presenting the most significant infrastructure challenges and highest production costs, holds the potential for superior fuel efficiency and environmental benefits in the long term (Zamfirescu and Dincer 2009; Turner et al. 2011).

10.16 Policy, Regulation, and Market Trends

Methanol is gaining attention as an alternative fuel due to its potential to reduce greenhouse gas emissions and its capability to be produced from renewable resources. The push toward cleaner energy sources has made methanol an attractive option for policymakers, regulators, and market players. This discussion delves into the current policy frameworks, regulatory measures, and market trends influencing the adoption of methanol as a fuel (Zhang et al. 2021).

10.16.1 Policy Frameworks

Several countries have recognized the potential of methanol and have formulated policies to promote its use. These policies often aim to diversify energy sources, enhance energy security, and mitigate environmental impacts.

China: China has been a frontrunner in adopting methanol as a transportation fuel. The government has implemented policies supporting methanol vehicles, such as subsidies for methanol-fueled taxis and trucks. The Chinese Ministry of Industry and Information Technology (MIIT) has also developed standards for methanol fuel and methanol vehicles (Chinese Ministry of Industry and Information Technology (MIIT): Methanol Vehicle Policies 2022).

European Union: The EU's Renewable Energy Directive encourages the use of renewable fuels, including methanol, to achieve its climate targets. The directive sets binding targets for the share of renewable energy in the transportation sector (European Union Renewable Energy Directive (RED II): Directive (EU) 2018/2001).

United States: In the US, methanol is part of the broader strategy to reduce dependence on oil imports and lower emissions. The Department of Energy has funded research and development projects to improve methanol production and utilization technologies (US Department of Energy (DOE): Methanol Fuel Research 2021).

10.16.2 Regulatory Measures

Regulatory frameworks play a crucial role in the adoption of methanol as a fuel. These regulations ensure the safe production, distribution, and use of methanol, and also establish standards for emissions and fuel quality.

Emissions Standards: Countries like China and Europe have stringent emissions standards that methanol-powered vehicles must meet. These standards drive technological advancements in engine design and after-treatment systems to reduce pollutants.

Fuel Quality Standards: Regulatory bodies have established specifications for methanol fuel to ensure its compatibility with existing infrastructure and vehicles. For example, ASTM International provides standards for methanol fuel blends in the US.

10.17 Market Trends

The market for methanol as a fuel is influenced by several factors, including production costs, availability of feedstock's, and advancements in technology.

Production and Supply: Methanol can be produced from various feedstock's, including natural gas, coal, and biomass. The choice of feedstock affects production costs and environmental impact. The development of cost-effective and sustainable production methods is crucial for market growth.

Technological Innovations: Innovations in methanol production and engine technologies are driving the market. For instance, direct methanol fuel cells and methanol-to-gasoline processes are emerging as promising technologies.

Market Adoption: The adoption of methanol as a fuel varies across regions. China leads in methanol vehicle deployment, while Europe focuses on methanol as a marine fuel. In the US, methanol is being explored as a blending component in gasoline and diesel (Agarwal et al. 2019a, b). Figure 10.3 shows the Global Methanol Fuel Market Trends.

Methanol holds significant promise as a cleaner alternative fuel, supported by favorable policies, stringent regulations, and evolving market dynamics. Continued collaboration between governments, industry stakeholders, and researchers is essential to overcome challenges and fully realize the potential of methanol in the global energy landscape (International Renewable Energy Agency (IRENA) (2020) "Innovation Outlook: Renewable Methanol").

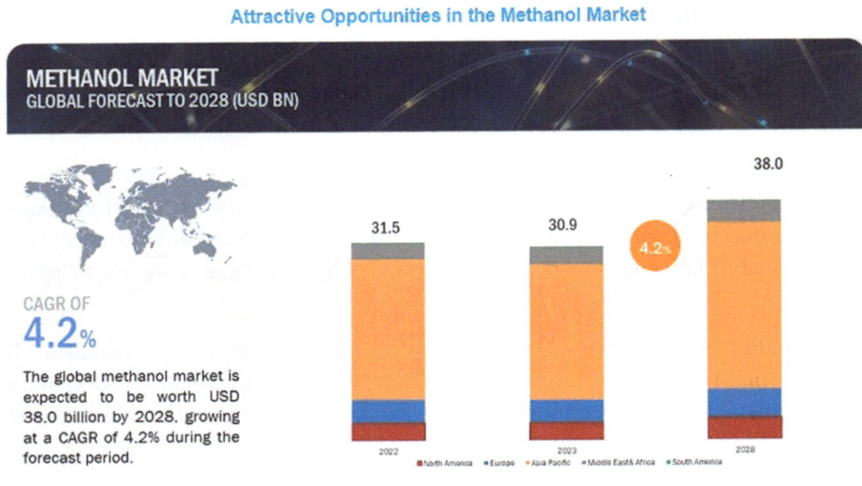

Fig. 10.3 Global methanol fuel market trends (https://www.marketsandmarkets.com/Market-Reports/methanol-market-425.html)

10.18 Summary and Way-Forward

In conclusion, the utilization of methanol as a fuel for internal combustion engines presents a promising yet complex alternative to conventional fossil fuels. The primary challenges revolve around methanol's lower energy density compared to gasoline, which necessitates larger fuel tanks or more frequent refueling. Additionally, the corrosive nature of methanol requires specialized materials for engine components and fuel delivery systems to prevent degradation. State-of-the-art reviews highlight significant advancements in overcoming these issues, such as the development of methanol-compatible materials and fuel additives that enhance energy efficiency and reduce corrosion, Hence Methanol is an acceptable replacement fuel for ICEs as compared to hydrogen and ammonia, mainly due to its economic efficiency, compatibility with current infrastructure, and less maintenance. In contrast to hydrogen, which faces storage and distribution difficulties. The ammonia presents toxicity issues, methanol may be sustainably synthesized from biomass or captured CO_2, resulting in reduced carbon emissions and enhanced compatibility with internal combustion engine systems. Despite methanol's inferior energy density compared to gasoline, it is nevertheless viable for upgrades to traditional engines, facilitating a smoother transition. Methanol and diesel blend provides better performance and lower emission in HCCI engine. Furthermore, advancements in engine tuning and fuel injection technologies have optimized combustion processes, improving power output and emissions profiles. Despite these advancements, widespread adoption is delayed by the lack of infrastructure for methanol distribution and the economic feasibility compared to other alternative fuels. Steam methane reforming is obsolete and renewable hydrogen is used is not emphasized. This is because energy-intensive process and carbon dioxide emission due to the process. Future research is essential to address these infrastructure challenges and to further enhance engine modifications and fuel formulations, paving the way for methanol-fueled ICEs to become a viable and sustainable option in the quest for reducing carbon emissions and dependence on traditional fossil fuels.

Nomenclature

NH_3	Ammonia
H_2	Hydrogen
CH_3OH	Methanol
CO	Carbon monoxide
HC	Hydro carbon
NO_x	Nitrogen oxides
ICE	Internal Combustion Engines
CI	Compressed Ignition
SI	Spark Ignition
DMFC	Direct Methanol Fuel Cells
MTG	Methanol-To-Gasoline
EGR	Exhaust Gas Recirculation

References

Agarwal AK, Karare H, Dhar A (2014) Combustion, performance, emissions and particulate characterization of a methanol–gasoline blend (gasohol) fueled medium duty spark ignition transportation engine. Fuel Process Technol 121:16–24

Agarwal AK, Gautam A, Sharma N, Singh AP (2019a) Introduction of methanol and alternate fuel economy. In: Methanol and the alternate fuel economy, pp 3–6

Agarwal AK, Gautam A, Sharma N, Singh AP (eds) (2019b) Methanol and the alternate fuel economy. Springer, Singapore

Agarwal AK, Valera H, Pexa M, Cedik J (2021a) Methanol

Agarwal AK, Valera H, Pexa M, Čedík J (eds) (2021b) Methanol: a sustainable transport fuel for CI engines. Springer, Singapore

Amine M, Barakat Y (2019) Properties of gasoline-ethanol-methanol ternary fuel blend compared with ethanol-gasoline and methanol-gasoline fuel blends. Egypt J Pet 28(4):371–376

Bansal P, Meena R (2021) Methanol as an alternative fuel in internal combustion engine: scope, production, and limitations. In: Methanol: a sustainable transport fuel for SI engines. Springer, Singapore, pp 11–36

Bromberg L, Cohn DR, Heywood JB, Hadidi K (2008) Emissions analysis of methanol, ammonia, and hydrogen fueled internal combustion engines. MIT Laboratory for Energy and the Environment

Chen J (2022) Direct injection methanol engines: a review of recent advances. Energy Rep 8:1543–1560

Chen X, Liu H, Jiang D (2019) Investigation of HCCI combustion with methanol fuel: combustion characteristics, performance, and emissions. Energy Convers Manage 196:437–447
Chen Y, Zhang L, Li X (2023) Methanol as an alternative fuel: current status and future perspectives. J Energy Res Technol 145(2):023203
Chinese ministry of industry and information technology (MIIT): methanol vehicle policies (2022)
Das LM (1992) Hydrogen engines: a view of the past and a look into the future. Int J Hydrogen Energy 17(7):445–474
Dong Y, Kaario O, Hassan G, Rant O, Larmi M, Johansson B (2020) High-pressure direct injection of methanol and pilot diesel: a non-premixed dual-fuel engine concept. Fuel 277:117932
Emergency response guidebook 2020—US Department of Transportation. https://www.phmsa.dot.gov/hazmat/erg/emergency-response-guidebook-erg
European Union renewable energy directive (RED II): directive (EU) 2018/2001
Francesconi A (2015) Challenges and benefits of using methanol as a fuel in spark ignition engines. SAE Technical Paper 2015-01-9040
Gao J, Wang X, Zhang Y (2022) Renewable methanol production: recent advances and future trends. Renew Sustain Energy Rev 148:111329
Gao Y (2024) Hybrid powertrains with methanol: opportunities and challenges. J Clean Prod 367:1325–1340
Global methanol institute: methanol market trends report (2023)
Global methanol market report (2023)
Guil-López R, Mota N, Llorente J, Millán E, Pawelec B, Fierro JLG, Navarro RM (2019) Methanol synthesis from CO_2: a review of the latest developments in heterogeneous catalysis. Materials 12(23):3902
Gupta A, Smith L, Kumar R (2018) Optimization of three-way catalytic converters for methanol-fueled engines. J Emission Control Res 15(3):210–225
https://www.marketresearchfuture.com/reports/methanol-market-2023
Hydrogen Council (2021) Hydrogen insights report (2021)
International Energy Agency (2019) The future of hydrogen. https://www.iea.org/reports/the-future-of-hydrogen
International Renewable Energy Agency (IRENA) (2020) Innovation Outlook: renewable methanol
Johnson M (2024) Renewable methanol production: technologies and economic viability. Renew Energy 185:1234–1248
Johnson P, Zhao J, Nguyen T (2021) Advances in lean NO_x traps for methanol-fueled engines. Catal Today 342:150–162
Johnson R, Smith A, Liu M (2023) Electrochemical CO_2 reduction to methanol: a review of recent developments and future directions. Chem Eng J 460:142389
Kalghatgi G (2014) Fuel/engine interactions. SAE International
Kim S, Park J, Lee D (2019) Improvements in cold start performance of methanol engines. J Engine Technol 31(4):335–350
Klerke A, Christensen CH, Nørskov JK, Vegge T (2008) Ammonia for hydrogen storage: challenges and opportunities. J Mater Chem 18(20):2304–2310
Kumar N, Tomar M, Sonthalia A, Sidharth Kumar P, Pali HS, Mishra D (2020) Methanol-based economy: a way forward to hydrogen. In: Sustainable development for energy, power, and propulsion. Springer, Singapore, pp 563–585
Kumar S, Agarwal AK, Khandelwal B, Singh P (2024a) Ammonia and hydrogen for green energy transition. Springer, Singapore (Imprint)
Kumar S, Sharma P, Gupta R (2024b) Advanced catalysts for methanol synthesis: current progress and future outlook. Catal Sci Technol 14(3):565–583
Li H, Wang L, Zhou Y (2023) Cost reduction strategies for methanol production: a comprehensive review. Energy Convers Manage 270:116929
Liu Q, Wang X, Tang C (2018) Reduction of formaldehyde emissions from methanol combustion using catalytic converters. Environ Sci Technol 52(6):3420–3428
McKinsey & Company (2021) Methanol in internal combustion engines

Methanol Institute (2023) Methanol as an engine fuel. https://www.methanol.org/methanol-engine-fuel/

Methanol spill response—National institute of environmental health sciences (2018). https://www.niehs.nih.gov/health/topics/agents/methanol/index.cfm

Nguyen L, Chen Y, Zhao J (2020) Energy efficiency of methanol-fueled internal combustion engines. Energy Convers Manage 206:112–123

Nitsche M, Kulkarni S (2021) Advantages and challenges of methanol as a fuel for IC engines. J Clean Prod 279:123658

Ogden JM, Steinbugler MM, Kreutz TG (1999) A comparison of hydrogen, methanol and gasoline as fuels for fuel cell vehicles: implications for vehicle design and infrastructure development. J Power Sour 79(2):143–168

Olah GA, Goeppert A, Prakash GKS (2011) Beyond oil and gas: the methanol economy, 2nd edn. Wiley-VCH.

Olah GA, Goeppert A, Prakash GKS (2018) Beyond oil and gas: the methanol economy. Wiley-VCH

OSHA safety and health topics: methanol (2019). https://www.osha.gov/methanol

Palo DR, Dagle RA, Holladay JD (2007) Methanol steam reforming for hydrogen production. Chem Rev 107(10):3992–4021

Reiter AJ, Kong SC (2011) Combustion and emissions characteristics of compression-ignition engine using dual ammonia-diesel fuel. Fuel 90(1):87–97

Rivard E, Trudeau M, Zaghib K (2019) Hydrogen storage for mobility: a review. Materials 12(12):1973

Sahoo S, Nayak C, Tripathy S, Srivastava D (2019) Experimental investigation on performance and emission characteristics of a single cylinder CRDI engine fueled with diesel-methanol blend (No. 2019-28-2380). SAE Technical Paper

Schüth F, Palkovits R, Schlögl R, Su DS (2012) Ammonia as a possible element in an energy infrastructure: catalysts for ammonia decomposition. Energy Environ Sci 5(4):6278–6289

Shen G, Luo H, Zhang Y (2017) Comparative study on emissions of methanol and gasoline engines. Fuel 198:137–145

Singh J, Dhar A, Kumar P (2022) Methanol fuel in compression ignition engines. In: Application of clean fuels in combustion engines. Springer, Singapore, pp 71–101

Singh P, Agarwal AK, Thakur A, Sinha RK (2024) Challenges and opportunities in green hydrogen production. Springer

Smith J, Wang H, Liu Y (2019) Performance of gasoline particulate filters in methanol engines. Fuel Process Technol 188:87–95

Smith B, Allen R, Thompson P (2024) Infrastructure Needs for Methanol Fuel: Economic and Logistical Considerations. Trans Res Part D Trans Environ 120:103694

Sovacool BK (2012) Energy security: challenges and needs. Wiley Interdisc Rev Energy Environ 1(1):51–59

Sugie Y, Fukutsuka T, Matsuo Y, Shintani D (2009) Effect of water on corrosion behavior of stainless steel under high-temperature and high-pressure methanol solution containing chloride ions. In: NACE corrosion. NACE, pp NACE-09253

Turner JWG, Pearson RJ, Holland B, Dekker E (2011) Alcohol-based fuels in high-performance engines. Proceedings of the Institution of Mechanical Engineers, Part D: J Automob Eng 225(11):1519–1534

US Department of Energy (DOE) (2021) Methanol fuel research

Valera-Medina A, Xiao H, Owen-Jones M, David WI, Bowen PJ (2018) Ammonia for power. Prog Energy Combust Sci 69:63–102

Vergara-Juarez F, Rosales-Cadena I, Salinas-Bravo VM, Lucio-Garcia MA, Porcayo-Calderon J, Gonzalez-Rodriguez JG, Martinez-Gomez L (2017) Effect of methanol on the corrosion behaviour of Al–Cu alloys in sulphuric acid. Corros Eng Sci Technol 52(6):476–483

Wang L (2023) Novel combustion chamber designs for methanol engines: emission and efficiency benefits. J Power Sour 552:230–245

Wang Y, Wright LA (2021) A comparative review of alternative fuels for the maritime sector: economic, technology, and policy challenges for clean energy implementation. World 2(4):456–481

Wang Z, Zhao J, Li (2022) Methanol fuel cells: recent advances and future prospects. J Power Sour 526:231208

Wei H, Yang W, Zhang S, Hu E (2020) Performance and emission characteristics of a turbocharged methanol engine. Appl Therm Eng 180:115868

White CM, Steeper RR, Lutz AE (2006) The hydrogen-fueled internal combustion engine: a technical review. Int J Hydrogen Energy 31(10):292–1305

Yang P (2024) Blended methanol fuels: enhancements in combustion stability and performance. Fuel Process Technol 260:107–120

Yuan W, Zhen X (2016) Experimental and numerical study on the combustion and emission characteristics of a methanol/diesel dual-fuel engine. Appl Energy 161:199–210

Zamfirescu C, Dincer I (2009) Ammonia as a green fuel and hydrogen source for vehicular applications. Fuel Process Technol 90(5):729–737

Zhang L, Yao M, Zheng Z (2018) Combustion and emission characteristics of a dual-fuel methanol-diesel engine. Fuel 215:595–605

Zhang X, Liu J, Zhang Q (2021) Methanol as a transportation fuel: opportunities and challenges. Energy Policy 154:112289

Zhao X, Li Y, Chen Z (2020) Effect of exhaust gas recirculation on NO_x emissions in methanol-fueled engines. Int J Engine Res 21(5):412–428

Chapter 11
Comprehensive Review of Regulated, Unregulated, and Particulate Emissions in Methanol-Powered Engines

Md Shadab Reza, Jai Gopal Gupta, and S. Anil Lal

Abstract Some forecasts predict that conventional automotive fuels will run out in the next three to four decades. Furthermore, traditional fuel burning generates a substantial amount of gaseous and particulate pollutants. Because of these issues, the research community has to look for alternative solutions to conventional fuels. Alcohol-based fuels are the most extensively researched alternative fuel due to their numerous beneficial features that make them suited as an internal combustion engine fuel. These fuels are considered environmentally benign since they lower the quantity of hazardous gaseous, particulate, and greenhouse gas emissions. Additionally, blending these fuels with conventional fuels can extend the run-out period of conventional fuels. Methanol is among the most often utilized fuels in the alcohol family due to its several advantageous properties, including a high-octane number, high heat of vaporization, high flame speed, and a low stoichiometric air–fuel ratio, etc., making it a suitable alternative fuel for IC engines. Having a high oxygen-to-carbon and hydrogen to carbon ratio, methanol generally produces less amount of CO_2, CO, HC, Soot, and smoke emissions. The emission of NO_X depends upon the operating conditions and quantity of blending with conventional automotive fuels. The main challenge in the application of methanol as a fuel is the unregulated emissions such as unburned methanol, formaldehyde, benzene, toluene, soluble organic fraction, etc. Studies indicate that an equal blending ratio of methanol and gasoline lowers the HC, CO and NO_X emissions between 30 and 40%. The blending of 10% methanol with diesel lowers the CO and THC emissions by 30 and 37%, respectively, but the NO_X emission was increased by 23% as compared to diesel alone. With increasing the amount of methanol content in the fuel blend, the emissions of formaldehyde, unburned methanol, benzene, toluene, and soluble organic fraction also increase. However, the emissions of ethylene, 1,3-butadiene, isobutene, and propylene show a slight decrement. Xylene, acetaldehyde, and acetone emissions remain almost the same. This chapter includes an in-depth review of the regulated, unregulated, and particulate characteristics of methanol-powered engines.

M. S. Reza · J. G. Gupta (✉) · S. Anil Lal
Department of Mechanical Engineering, National Institute of Technology Sikkim, Ravangla, Sikkim, India
e-mail: jaig@nitsikkim.ac.in

A. Bartl et al. (eds.), *Alcohol Production Processes and Their Utilisation in the Transport Sector*, Energy, Environment, and Sustainability,
https://doi.org/10.1007/978-981-96-7384-1_11

Keyword Methanol · Internal combustion engines · Regulated emissions · Unregulated emissions

Abbreviations

CO_2	Carbon dioxide
CO	Carbon monoxide
THC	Total Hydrocarbon
NO_X	Nitrogen oxides
SO_X	Sulfur oxides
PM	Particulate matter
H_2	Hydrogen
ICEs	Internal combustion engines
SI	Spark Ignition
CI	Compression ignition
BTX	Benzene, Toluene, Xylene
BTEX	Benzene, Toluene, Ethylbenzene, Xylene
MSR	Methanol substitution ratio
DI	Direct injection
CRDI	Common rail direct injection
PFI	Port fuel injection
TWC	Three way catalyst
DOC	Diesel oxidation catalyst
OCC	Oxidation catalytic converter
SOF	Soluble organic fraction
MDI	Methanol direct injection
GPI	Gasoline port injection
MPI	Methanol port injection
GDI	Gasoline direct injection

11.1 Introduction

At the current consumption rate of conventional automobile fuels, it will run out in the next three to four decades, according to some forecasts. The International Energy Agency articulates that the demand for oil, gas, and coal will reach its peak by 2030 (https://www.weforum.org/agenda/2023/10/iea-energy-peak-fossil-fuel-demand-by-2030/). Gasoline and diesel are still the most widely used automobile fuels for internal combustion engines (ICEs) in the world. The combustion of these automobile fuels into the ICEs produces a large amount of air pollutants like Carbon monoxide (CO), Carbon dioxide (CO_2), Nitrogen oxides (NOx),

Hydrocarbons (HC), Particulate Matter (PM), Sulfur oxides (SO_X), Volatile organic compounds (VOCs) and others that cause acid rain, the greenhouse effect, photochemical smog, and a host of other diseases in flora and fauna. Internal combustion engines are responsible for emissions of about 23% of carbon dioxide (Osman et al. 2021). The use of alternative automobile fuels such as Alcohol, Ammonia, and Hydrogen can solve these issues to a great extent because these fuels produce lower quantities of hazardous gaseous, particulate, and greenhouse gas emissions. Also, blending these alternative fuels with conventional fuels can extend the run-out time of conventional fuels. To accomplish the target of net-zero carbon by 2050, the investment in research and production of alternative fuels has increased significantly in recent years across the globe. As per the World Economic Forum's Fostering Effective Energy Transition report for 2023, the investment in alternative energy sources in 2022 was $1.3 trillion, which is 19% more than the previous year and 70% more than pre-COVID-19 levels (https://www.weforum.org/publications/fostering-effective-energy-transition-2023/). The utilization of ammonia and hydrogen as automobile fuels is still in the experimental stage, but the application of alcohol-based fuels has already been commercialized in different countries of the world. The alcohol fuels, such as methanol, ethanol, propanol, and butanol are becoming popular due to a large number of advantages such as (Verhelst et al. 2019):

- It improves the thermal efficiency of the ICEs due to the high speed of laminar flame propagation.
- It produces less amount of gaseous and particulate emissions due to its high stoichiometric fuel–air ratio, rich oxygen concentration, high hydrogen-to-carbon ratio, and less amount of sulfur.
- Its injection, atomization, and mixing with air are easier as compared to diesel, due to its lesser viscosity than diesel.
- It enhances the volumetric efficiency of the ICEs due to more evaporative cooling and less amount of work input during compression stroke.

Methanol (methyl alcohol) and ethanol (ethyl alcohol) from the alcohol family are the most frequently experimented and used IC engine fuels due to their ease of production and relatively low costs of manufacturing (Jamrozik 2017). Methanol is a great replacement for conventional automobile fuels due to its numerous beneficial properties, including a highoctane number, high heat of vaporization, high laminar flame speed, and a low stoichiometric air–fuel ratio, etc. It has less flammability than gasoline making it safer to use in ICEs; Also, the storage and distribution of methanol is comparatively easier because of its property to remain in a liquid state at ambient temperatures (Pu et al. 2024). The utilization of methanol as an ICE fuel can decrease the regulated emissions of carbon dioxide by 60–95%, nitrogen oxide by 60–80%, and nearly eliminate the emissions of sulfur oxides and particulate matter (https://www.methanol.org/wp-content/uploads/2020/03/Future-Fuel-Strategies-Methanol-Automotive-Fuel-Primer.pdf). With the increment of methanol content in the fuel blend, the unregulated emissions of ethylene, 1,3-butadiene, isobutene, and propylene demonstrated a slight decrement. Xylene, acetaldehyde, and acetone emissions remain almost the same, however, the emissions of formaldehyde, unburned methanol, benzene, toluene, and soluble organic fraction increased with the methanol percentage (Zhang et al. 2010).

Due to the various advantageous properties of methanol, a lot of countries in the world are working to adopt methanol as a blending substitute for gasoline and diesel. China is the largest producer and consumer of methanol in the world. At present, methanol utilization is around 9% of the total transportation fuel in China. China already implemented nationwide standards for the utilization of M85 and M100 in 2009. They converted more than 180,000 automobiles to run on 100% methanol (Li et al. 2023). Countries such as Israel and Italy also adopted the 15% blending of methanol with gasoline. Australia also approved Gasoline, Ethanol, and Methanol (GEM) blending with up to 56% blending of methanol (https://www.methanol.org/wp-content/uploads/2019/02/4.-Greg-Dolan-Overview-of-Global-Methanol-Fuel-Blending.pdf). In India, the present annual demand for gasoline is around 29,000 million liters and 90,000 million liters of diesel. Which will be two-fold by 2030 and India will become the 3rd largest consumer of petroleum products (https://vikaspedia.in/energy/energy-basics/methanol-economy-in-india). Therefore, India is also looking for alternate options for the reduction in the import of conventional fossil fuels. The aim of the Indian government through its methanol economy is to cut down the import of gasoline through blending of methanol 15% by volume with it. That will also reduce the greenhouse gas emissions by 20% and the fuel costs by 30%. The installed production capacity of methanol production per annum in India was 4.74 Lakh metric tons in the financial year 2020–21. The government planned to increase the installed production capacity of methanol up to 20 million metric tons per annum by 2025 (https://www.niti.gov.in/methanol-economy). In 2023 the global methanol production capacity was around 170 million tons per annum. The valuation of the global methanol market size in 2023 was USD 31.26 billion. The projected market size by 2032 is USD 46.32 billion with a compound annual growth rate of 4.5%. The Methanex corporation is the leading methanol producer in the world with 66.42 million tons of production in 2023 (https://www.fortunebusinessinsights.com/industry-reports/methanol-market-101552).

Methanol is a kind of toxic alcohol; therefore, precautions should be taken while dealing with it. It can enter the human body through inhalation, skin or eye contact, and accidental ingestion. It can be fatal if the ingested amount is more than 30 mL or 1 g per kg of body weight. The initial symptoms after ingestion are nausea, dizziness, headache, vomiting, weakness, blurred vision, abdominal pain etc. If the patient is left untreated it may cause permanent blindness, also the patient may go into a coma and even death is possible (Ashurst and Nappe 2018). Methanol-fueled engines can emit a large number of unregulated carcinogenic emission species such as Formaldehyde, Benzene, Acetaldehyde, etc. Also, some other species like Formic acid, Acetylene, and Ethylene, etc., these species are harmful to the respiratory system, heart, and brain functioning (Benjamin 2013).

As far as storage and fire hazards are concerned, methanol is safer to store as compared to gasoline due to its high flash and fire point temperature (Machiele 1990).

Table 11.1 Comparison of different properties between Gasoline, Methanol, and Diesel (Pu et al. 2024)

Property	Gasoline	Methanol	Diesel
Density[#] (kg/m^3)	740	790	845
Lower heating value (MJ/kg)	42.90	20.09	43.2
Heat of vaporization (kJ/kg)	180–350	1100	225–280
Adiabatic flame temperature (K)	~ 2275	2143	~ 2200
Volumetric energy density (MJ/L)	31.75	15.87	36.5
Stoichiometric air–fuel ratio (kg/kg)	14.7	6.5	14.7
Flammability limit (vol%)	1.3–7.6	6.7–36	0.6–6.5
Laminar burning velocity[##] (m/s)	~ 0.3	0.37	n.a
Autoignition temperature (K)	465–743	738	483–503
Research octane number	85–95	92–109	n.a
Cetane number (Derived)	8–14	3	> 51

[#]At 273.15 K & 1 bar; [##]At 298 K & 1 atm

It is generally not ignitable below 10 °C and has a higher autoignition temperature and heat of vaporization as compared to gasoline and diesel (see Table 11.1) (Pu et al. 2024). The only concern with pure methanol flame is its invisibility in sunlight, but it is not an issue for blended fuel. Also, pure methanol fire can be extinguished with the help of water (Verhelst et al. 2019).

In this chapter the various production techniques of methanol have been described, after that the methods for utilizing the methanol in SI and CI engines are discussed in detail. Finally the Regulated, Unregulated, and Particulate Characteristics of methanol-powered SI and CI engines are provided in tabular form with details of the engine type, fuel used and results of the experimental studies.

11.2 Methanol Production Techniques

Methanol was conventionally produced from wood; therefore, it is commonly known as 'wood alcohol' also. Methanol can also be produced from a large number of feedstocks, which include both non-renewable (e.g., natural gas, coal, coke oven gas, etc.) and renewable (e.g., H_2, CO_2, biomass, and municipal waste, etc.) sources (Chen et al. 2018). About 110 million tons of methanol is produced globally per annum (Tabibian and Sharifzadeh 2023), of which around 30–35% is generated by coal gasification (coded as brown methanol) and around 55–65% from natural gas by steam methane reforming (coded as gray methanol). The color codes of methanol are related to its methods of production. Color coding such as green, brown, blue,

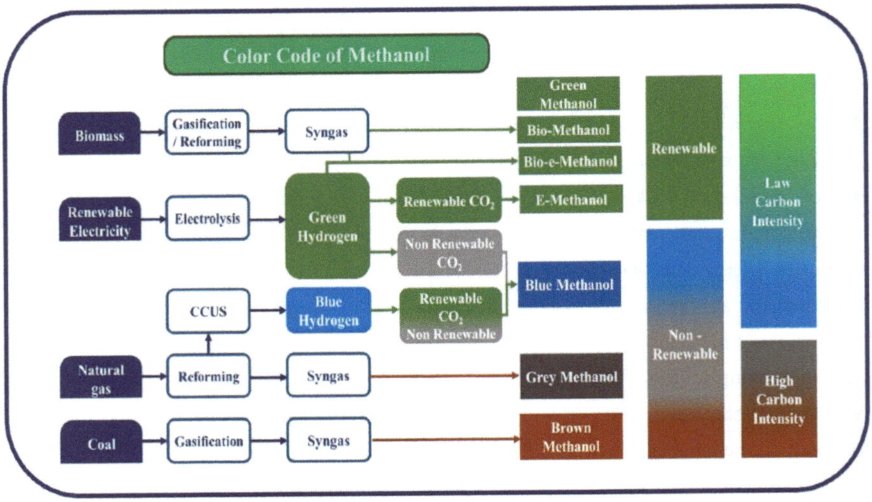

Fig. 11.1 Color codes for methanol production (Reddy et al. 2023)

gray, etc. (see Fig. 11.1) (Reddy et al. 2023), is generally used to indicate the sustainability and the environmental impact of specific production methods. (Tabibian and Sharifzadeh 2023; Sollai et al. 2023).

At present, methanol is produced by synthesis gas (syngas) in most industrial procedures, which is a combination of carbon monoxide (CO), carbon dioxide (CO_2) and hydrogen (H_2), used as feedstock to convert them into methanol (Verhelst et al. 2019). The reason for that is due availability of low-cost and ample carbon resources like natural gas and coal. Apart from this, other feedstocks such as biomass, biogas, and biological municipal waste, etc., can also be used for the production of bio-methanol (Kasteren 2016).

11.2.1 Methanol Production by Gasification

The various steps for the production of methanol by fossil fuel-based carbon resources are as follows: (a) gasification of natural gas or coal, (b) gas refinement—to remove some impurities such as nitrogen and sulfur contents, but this process reduces the CO and H_2 availability (c) reforming of hydrocarbon, (d) water–gas shift reaction, to recover the lost CO and H_2 during the refinement process, (e) carbon dioxide removal and/or hydrogen addition, and (f) synthesis of methanol and filtration (Galindo and Badr 2007). Bioethanol production mainly from biomass sources requires pre-treatment of raw material. After that, the treated biomass is gasified for syngas production.

For optimizing the formation of H_2 and CO and reducing excess water and CO_2, a limited supply of oxygen is provided during feedstock heating at temperatures above

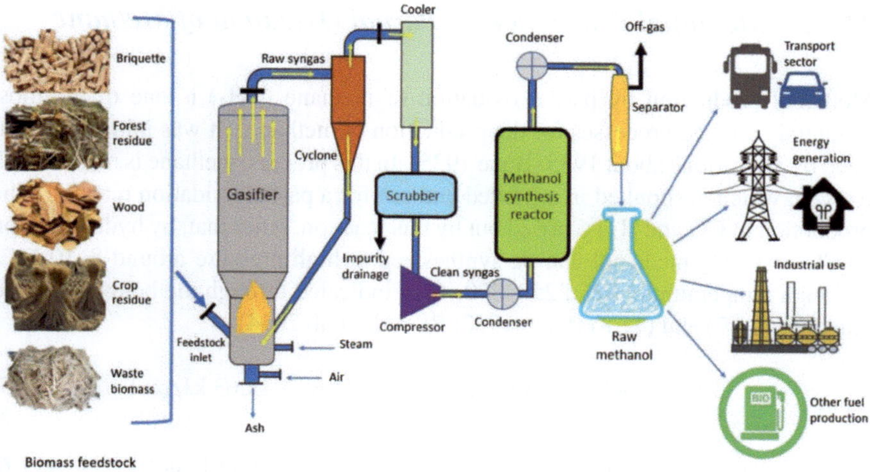

Fig. 11.2 Various steps of methanol production through gasification (Deka et al. 2022)

700 °C. Elimination of the impurities is done before the syngas passes through a lot of conditioning phases for the production of syngas with double H_2 molecules compared to CO molecules (see Fig. 11.2) (Deka et al. 2022). The following reactions are involved in methanol production through gasification (Bansal and Meena 2021).

Syngas Production

$$3C + O_2 + H_2O \rightarrow 3CO + H_2 \qquad \Delta H = -142.22\ \text{kJ/mol} \tag{11.1}$$

Water–gas shift reaction

$$CO + H_2O \leftrightarrow CO_2 + H_2 \qquad \Delta H = -41.20\ \text{kJ/mol} \tag{11.2}$$

Methanol synthesis reaction

$$CO + 2H_2 \leftrightarrow CH_3OH \qquad \Delta H = -90.7\ \text{kJ/mol} \tag{11.3}$$

$$CO_2 + 3H_2 \leftrightarrow CH_3OH + H_2O \qquad \Delta H = -49.5\ \text{kJ/mol} \tag{11.4}$$

Methanol production through this route is very energy-intensive and produces very high carbon emissions. The carbon footprint of methanol production through gasification of coal is about 300 g CO_2 eq/MJ, this amount is nearly 3 times higher as compared to the natural gas-based methanol (https://www.methanol.org/wp-content/uploads/2022/01/CARBONFOOTPRINT-OF-METHANOL-PAPER_1-31-22.pdf).

11.2.2 Methanol Production by Partial Oxidation of Methane

Methanol production by partial oxidation of methane (CH_4) is one of the most frequently applied processes for the production of methanol. It was initially formulated by W.A. Bone about 1935 (Bone 1935). In this process, methane is reacted with oxygen, which is supplied in a limited amount for a partial oxidation reaction. The production of CO and H_2 is carried out by this reaction. After that, by hydrogenation of CO and CO_2 methanol can be synthesized at high pressure around 8–10 MPa and high temperature around 250–300 °C as indicated through the below-specified reactions (11.7) and (11.8) (Klier 1982; Brynolf et al. 2014).

$$CH_4 + {}^1\!/_2 O_2 \rightarrow CO + 2H_2 \qquad \Delta H = -22.63\ \mathrm{kJ/mol} \tag{11.5}$$

$$CH_4 + 2O_2 \rightarrow CO_2 + 2H_2O \qquad \Delta H = -891.64\ \mathrm{kJ/mol} \tag{11.6}$$

$$CO + 2H_2 \leftrightarrow CH_3OH \qquad \Delta H = -90.7\ \mathrm{kJ/mol} \tag{11.7}$$

$$CO_2 + 3H_2 \leftrightarrow CH_3OH + H_2O \qquad \Delta H = -49.5\ \mathrm{kJ/mol} \tag{11.8}$$

The conversion of methane to methanol can help for reducing the emissions of methane into the atmosphere. Methane is the second most abundant greenhouse gas after CO_2; it has a very high global warming potential. The carbon footprint of methanol production from methane (Generated through municipal solid waste) may vary in the range of 10–55 g CO_2 eq/MJ (https://www.methanol.org/wp-content/uploads/2022/01/CARBONFOOTPRINT-OF-METHANOL-PAPER_1-31-22.pdf).

11.2.3 Methanol Production Through Biological Conversion of Methane

Methanol production through the biological conversion of methane into biomethanol is one more attractive method. Methanol production through this route involves two different groups of bacteria, ammonia-oxidizing bacteria, and methanotrophic bacteria are utilized for the activation of methane's carbon-hydrogen bond. The ammonia-oxidizing bacteria use ammonia as an energy source and produce biomethanol by partial oxidation of methane (Taher and Chandran 2013). For methanotrophic bacteria, methane is the only source of carbon and energy. All methanotrophic bacteria produce methanol by activating methane with the application of the methane monooxygenase enzyme (Han 2013) (Fig. 11.3) (Srivastava et al. 2022).

The biological conversion of methane into methanol is an energy-efficient and eco-friendly process. The carbon footprint of methanol production from biomethane

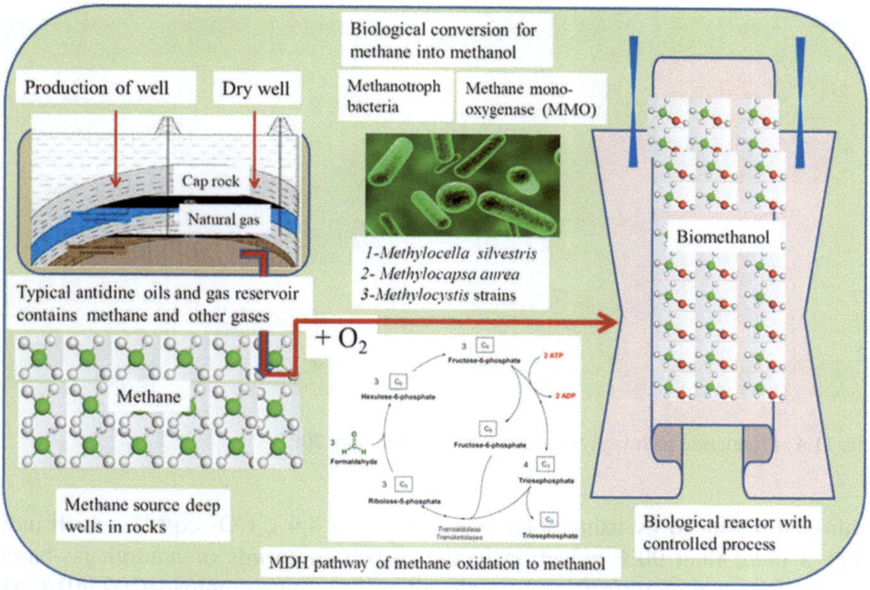

Fig. 11.3 Production of biomethanol through biological conversion of methane (Srivastava et al. 2022)

can vary in the range of − 103 to + 38 g CO_2 eq/MJ (https://www.methanol.org/wp-content/uploads/2022/01/CARBONFOOTPRINT-OF-METHANOL-PAPER_1-31-22.pdf). A negative carbon footprint is an indication of the removal of more CO_2 from the atmosphere compared to CO_2 emissions during production. However, there are some challenges such as the limitation of gas to liquid mass transfer, over oxidation of methanol, toxicity of methanol, etc. which require proper strategies to enhance the process efficiency and methanol yield (Charlotte et al. 2018).

11.2.4 Production of Green Methanol

The attention of the researchers at the current time is to reduce the carbon emissions from the methanol production methods, i.e., the production of green or e-methanol. When methanol is produced through biomass or by using captured carbon and hydrogen produced through renewable sources of energy, it is considered green methanol or e-methanol (Osman 2022b). For the production of green methanol, the CO_2 is obtained through carbon capture and utilization techniques either directly from the atmosphere or industrial waste gases. This method alleviates the excess CO_2 emissions from the atmosphere, which is beneficial for our environment and atmosphere. The lifecycle carbon footprint of green

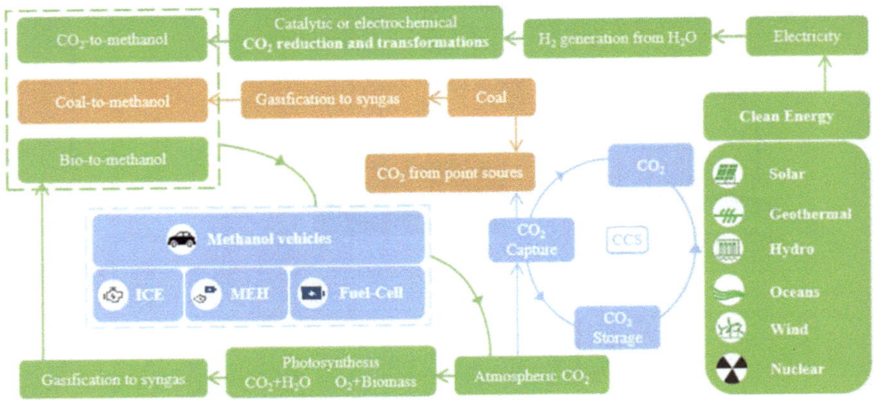

Fig. 11.4 Production pathways for green methanol (Li et al. 2023)

hydrogen produced by using solar energy is about 4.4 g CO_2 eq/MJ, which indicates a more than 90% reduction in the carbon footprints of natural gas-based methanol production (https://www.methanol.org/wp-content/uploads/2022/01/CARBONFOOTPRINT-OF-METHANOL-PAPER_1–31–22.pdf).

The production of green methanol by employing CO_2 collected through the exhaust gases of biomass-based industries and green hydrogen (produced by using renewable resources), can create a carbon–neutral methanol manufacturing system (Hillestad 2023) (Fig. 11.4) (Li et al. 2023).

One of the major hurdles in the way of the production and application of green methanol is the high production cost associated with it. For the production of green methanol in a 1000 kg amount, around 1400 kg of CO_2, 200 kg of H_2 and 1700 kg of water are required. Apart from these, renewable electricity of about 10–11 MWh is required for the production of 1000 kg of green methanol (Irena and Methanol Institute 2021). According to the Methanol Institute and the International Renewable Energy Agency, the estimated cost of production for bio-methanol is in the range of $ 700 to 900 per metric ton, whereas the estimated production cost for green methanol is in the range of $ 1200 to 1600 per metric ton (Dalena et al. 2018).

11.3 Methods of Utilizing Methanol as SI Engine Fuel

Methanol is primarily considered as a spark ignition engine fuel, due to its favorable properties such as high-octane number, heat of vaporization, autoignition resistance, and less difference in its density as compared to gasoline (see Table 11.1). The SI engine does not require a lot of modifications for its operation with low blending amount of methanol (Verhelst et al. 2019). In SI engines methanol can be utilized either as a neat fuel or in a blended form with gasoline. Neat methanol can be a good option as an SI engine fuel, particularly in warm climate regions, where the issue of cold start is less in warm atmospheres. The low and medium blends of up to 30% by volume methanol have a calorific value almost equal to gasoline, making them suitable options as SI engine fuels. However, the calorific value decreases for higher than 30% blends of methanol, making it less efficient SI engine fuel.

Methanol and gasoline have different polarities; methanol is polar but gasoline is non-polar. Methanol is miscible with water and easily forms hydrogen bonds with it. This difference in polarities makes the methanol-gasoline blends unstable and often results in phase separation (Waluyo et al. 2021). Therefore, to avoid this situation, some co-solvents are required to make the blend stable. Ethanol, propanol, butanol, etc. are generally used as co-solvents to make the blends stable (Schifter et al. 2019). In the United States., ASTM D 4814-10a standard limits methanol blending up to 0.3% by volume normally and up to 2.75% by volume with an equal amount of cosolvent such as butanol, or alcohol of higher molecular weight (https://www.iea-amf.org/content/fuel_information/methanol).

Generally, light alcohols have a more corrosive nature than gasoline to ferrous and non-ferrous metals. Due to their polarity methanol and ethanol induce dry corrosion. Dry corrosion can also take place due to the presence of ionic impurities like chloride ions in the fuel. Methanol is also electrically conductive in nature, which increases the chances of galvanic and electrochemical corrosion (Yuen et al. 2010). To reduce the corrosion formation in the existing engines, fuel additives, corrosion inhibitors, and correctly formulated engine oils should be used with methanol-blended fuels (Bromberg and Cheng 2010). The utilization of methanol in SI engines reduces the amount of pollutant gases due to its rich oxygen concentration, faster flame propagation, and absence of a C–C bond. HC, CO, and PM emissions are generally lower as compared with gasoline; however, the NO_X emissions do not always reduce and it also increases the amount of some unregulated gas emissions (Yanju et al. 2008).

11.4 Methods of Utilizing Methanol as CI Engine Fuel

The methods for the application of methanol as CI engine fuel are blending, dual-fuel injection, fumigation, and emulsification (Agarwal 2007). Among all these methods, blending and methanol-diesel dual-fuel injection are the most frequently used methods. The blending of methanol with diesel is generally carried out in low to medium blends up to 30% by volume, due to methanol's poor miscibility with mineral diesel. Whereas by the methanol-diesel dual-fuel injection method, high methanol concentrations up to 80% by volume can be used (Sayina et al. 2009). The poor miscibility issue of methanol can be resolved by adding an appropriate additive or co-solvent such as 1-dodecanol (Agarwal et al. 2019). The blended and emulsified fuel can be directly injected into the engine cylinder; it is the easiest and cost-effective method. Whereas in the Port fuel injection (fumigation) technique, methanol is injected via an inlet manifold throughout the intake stroke, and diesel is injected directly near the top dead center. To overcome the issue of solubility and phase separation, direct fuel injection is used. In this technique, diesel and methanol are both directly injected into the engine cylinder through separate fuel injection systems (Li et al. 2020). Also, CI engines require more significant modifications to utilize methanol in high volume. The very low cetane number of methanol and the large density difference as compared to diesel (see Table 11.1) require different measures to use it as a CI engine fuel (Verhelst et al. 2019). The utilization of methanol in CI engines reduces the NO_X and PM emissions, however; it increases the unregulated emissions of unburned methanol, formaldehyde, and BTX emissions (Zhang et al. 2010).

11.5 Regulated, Unregulated, and Particulate Characteristics of Methanol- Powered SI Engines

11.5.1 Regulated Emissions

The utilization of methanol in SI engines generally reduces the amount of CO and HC emissions. An experimental study by Rifal et al. (Rifal and Sinaga 2016) demonstrated the reduction in CO and HC emissions by 33% and 45%, respectively. NO_X emissions can be reduced significantly by using a high percentage (75–85%) of methanol blending with traditional gasoline fuel (Yanju et al. 2008). However, the CO_2 emissions had been seen increasing in most of the studies. A comparative summary of regulated emission characteristics of different methanol-powered SI engines is provided in Table 11.2.

11.5.2 Unregulated Emissions

The emissions of unburned methanol and formaldehyde from methanol-fueled SI engines are a big challenge to resolve. Generally, the amount of these unregulated emissions increases with increasing methanol content (Mishra et al. 2020). However, the BTEX emissions decrease by increasing the methanol percentage (Yanju et al. 2008). A comparative summary of unregulated emission characteristics of different methanol-powered SI engines is provided in Table 11.3.

11.5.3 Particulate Emissions

The particulate emissions can be reduced by up to 40% as compared to baseline gasoline by utilizing methanol in SI engines (Wang et al. 2015). However, the particulate emissions can be increased at very low ambient temperature conditions as compared

Table 11.2 Experimental studies for regulated emissions from methanol-powered SI engines

Year	Researcher	Engine specifications	Fuel used	Methanol utilization technique	Test results for emissions
2008	Yanju et al. (2008)	3-cylinder, 4-stroke, PFI, electronic control engine	M0 M10 M20 M85	Blending	The emissions of NO_X and CO were reduced by 80% and 25%, respectively, by using the M85 blended fuel compared to baseline gasoline
2015	Wang et al. (2015)	4-cylinder, Turbocharged GDI engine	M0 M15 M25 M40	Blending	CO and HC emissions decreased in the range of 9.4–33.2% and 9.7–36.9%, respectively, while NO_X emissions increased by increasing the methanol content
2016	Rifal et al. (2016)	4-cylinder 4-stroke Variable Valve Timing Injection (VVTi) engine	M0 M15 M30 M50	Blending	HC and CO emissions decreased by 45% and 33%, but CO_2 emissions increased by 10% by using the blended fuels
2017	Alexandru et al. (2017)	Single cylinder, 4-stroke, air cooled engine	M0 M5 M10 M15 M20 M25	Blending	CO and HC emissions decreased, but CO_2 increased by using the blended fuels
2020	Mishra et al. (2020)	Single cylinder, 4-stroke, air-cooled engine	M0 M5 M10 M15	Blending	HC, CO_2, and CO emissions decreased with the application of 10% methanol blending, while NO_X emissions increased from all blended fuels
2021	Zhu et al. 2021)	(i) 4-cylinder, Turbocharged GDI engine with Three Way Catalyst (TWC) (ii) 4-cylinder, PFI naturally aspirated engine with TWC	M0 M15	Blending	**In the GDI engine:** CO and THC emissions decreased at 30 °C but increased at -7 °C ambient temperature, NO_X emissions increased at 30 °C but decreased at -7 °C **In PFI engine**- CO and THC emissions increased significantly at -7 °C, and NO_X emissions were almost doubled at 30 °C but showed a slight increase at -7 °C

(continued)

Table 11.2 (continued)

Year	Researcher	Engine specifications	Fuel used	Methanol utilization technique	Test results for emissions
2021	Malik et al. (2021)	4-stroke, 163 cc, air cooled engine	M0 M12	Blending	CO and HC emissions decreased, but CO_2 and NO_X emissions increased as compared to the baseline gasoline
2022	Zhang et al. (2022)	4-cylinder, Turbocharged, gasoline direct injection (GDI) engine pc	M0 M10 M20 M30 M50 M75	Blending	CO emission increased and the emissions of NO_X, CO_2 and THC decreased by increasing the methanol blending ratio
2023	Yang et al. (2023)	4-cylinder, 4 stroke, water cooled naturally aspirated engine	M0 M20 M40 M60 M80 M100	**MDI + GPI mode:** Methanol was directly injected into the combustion chamber and Gasoline was injected through the intake port **MPI + GDI mode:** Methanol was injected through the intake port and Gasoline was directly injected into the combustion chamber	CO and HC emissions reported to be decreased significantly, whereas a slight reduction in NO_X emissions reported with increasing methanol ratio for both fuel injection modes
2024	Catapano et al. (2024)	4-cylinder, 4-stroke, Turbocharged, naturally aspirated engine	M0 M100	Methanol and Gasoline were injected through port fuel injectors	CO, CO_2 and NO_X emissions reduced significantly, but HC emissions increased significantly as compared to the baseline gasoline

to normal ambient conditions (Zhu et al. 2021). A comparative summary of particulate emission characteristics of different methanol- powered SI engines is provided in Table 11.4.

11.6 Regulated, Unregulated, and Particulate Characteristics of Methanol-Powered CI Engines

11.6.1 Regulated Emissions

In most of the studies the NO_X emissions are reported to be decreased by increasing the methanol concentration due to reduced combustion temperatures. However, in several studies, the HC and CO emissions are reported to be increased by increasing

Table 11.3 Experimental studies for unregulated emissions from methanol-powered SI engines

Year	Researcher	Engine Specifications	Fuel used	Methanol utilization technique	Test results for emissions
2008	Yanju et al. (2008)	3-cylinder, 4-stroke, Port fuel injection	M0 M10 M20 M85	Blending	Unburned methanol emissions increased by increasing the methanol content in the fuel blend
2009	Fan at el. (2009)	4-cylinder Multi-point PFI gasoline engine	M0 M10 M15 M20 M30	Blending	Methanol and formaldehyde emissions increased linearly, whereas the acetaldehyde and ethanol emissions were slightly changed by increasing the methanol content
2015	Wang et al. (2015)	4-cylinder Turbocharged GDI engine	M0 M15 M25 M40	Blending	Formaldehyde emissions increased by 115%, whereas BTEX emissions decreased by increasing the methanol content

the methanol concentration, due to incomplete combustion as a result of lower combustion temperature. According to the experimental study of Agarwal et al. (2019), blending of 15% (v/v) methanol with conventional diesel results in simultaneous reduction of HC, CO, and NOx emissions. A comparative summary of regulated emission characteristics of different methanol- powered CI engines is provided in Table 11.5.

11.6.2 Unregulated Emissions

The emissions of unburned methanol, formaldehyde, 1,3-butadiene, ethyne, and ethene in methanol-fueled CI engines are reported to be increased by increasing the methanol concentration in most of the studies. BTX emissions were low at lower methanol concentrations, but increased significantly by increasing the methanol fraction (Zhang et al. 2010). A comparative summary of unregulated emission characteristics of different methanol- powered CI engines is provided in Table 11.6.

11.6.3 Particulate Emissions

The smoke opacity as well as soot emissions are reported to be decreased by increasing the methanol energy fraction in most of the studies as compared to baseline diesel. Particle mass and particle number also reduced but the soluble organic fraction was increased by increasing the methanol concentration (Qiu 2018). A comparative

Table 11.4 Experimental studies for particulate emissions from methanol-powered SI engines

Year	Researcher	Engine specifications	Fuel used	Methanol utilization technique	Test results for emissions
2015	Wang et al. (2015)	4-cylinder, Turbocharged GDI engine	M0 M15 M25 M40	Blending	PM emission reduced in the range of 33.2–40.25% as compared to the baseline gasoline
2021	Zhu et al. (2021)	(i) 4-cylinder, Turbocharged GDI engine with TWC (ii) 4-cylinder, PFI naturally aspirated engine with TWC	M0 M15	Blending	PM emissions increased by up to 70% at -7 °C as compared to 30 °C ambient temperature
2022	Zhang et al. (2022)	4-cylinder, turbocharged, gasoline direct injection (GDI) engine	M0 M10 M20 M30 M50 M75	Blending	PM emissions were reduced by increasing the methanol blending ratio
2023	Yang et al. 2023)	4-cylinder, 4-stroke, water-cooled, naturally aspirated engine	M0 M20 M40 M60 M80 M100	**MDI + GPI mode:** Methanol was directly injected into the combustion chamber and Gasoline was injected through the intake port **MPI + GDI mode:** Methanol was injected through the intake port and Gasoline was directly injected into the combustion chamber	For a less than 60% methanol ratio, Particle Number emission was less in MDI + GPI mode whereas, for a more than 60% methanol ratio, Particle Number emission was less in MPI + GDI mode
2024	Catapano et al. (2024)	4-cylinder, 4-stroke, turbocharged, naturally aspirated engine	M0 M100	Methanol and Gasoline were injected through port fuel injectors	Particle Number reduced significantly as compared to baseline gasoline at higher speeds

Table 11.5 Experimental studies for regulated emissions from methanol-powered CI engines

Year	Researcher	Engine specifications	Fuel used	Methanol utilization technique	Test results for emissions
2008	Yao et al. (2008)	4-cylinder, naturally aspirated, modified DI engine with Oxidation Catalytic Converter (OCC)	Diesel was used to ignite the mixture of methanol and air. Methanol was injected into each cylinder to produce a homogeneous mixture with the intake air	Methanol was introduced through Port Fuel Injection (PFI) and Pilot Diesel is introduced directly into the combustion chamber	**Without OCC** The emissions of NO_X were reduced; however the CO and HC emissions increased by using the diesel methanol compound combustion system **With OCC** The emissions of NO_X, CO and HC reduced as compared to the baseline diesel fuel
2009	Cheung et al. (2009)	4-cylinder, naturally aspirated, water-cooled, DI engine	M0 M5 M10 M15	Blending (Methanol + Biodiesel)	The NO_X emissions reduced with increasing mass fraction of methanol, CO emissions from M10 and M15 increased as compared to biodiesel, The HC emissions from M5 and M10 are lower compared to biodiesel
2010	Cenk Sayin (2010)	Single cylinder, naturally aspirated DI engine	M0 M5 M10 with 1% (v/v) dodecanol	Emulsion	CO and HC emissions decreased, but NO_X emissions increased by increasing the methanol content in the blend

(continued)

Table 11.5 (continued)

Year	Researcher	Engine specifications	Fuel used	Methanol utilization technique	Test results for emissions
2010	Zhang et al. (2010)	4-cylinder, Water-cooled DI engine with and without Diesel Oxidation Catalyst (DOC)	10%, 20%, 30% Fumigation methanol	Methanol was introduced through PFI and Pilot Diesel was introduced directly into the combustion chamber	**Without DOC** NO_X emissions reduced by 11.3%, but HC and CO emissions increased by 76% & 69% at 30% fumigation methanol **With DOC** NO_X emissions were reduced further by about 6%, and CO and HC emissions also reduced by more than 90% in fumigation modes
2011	Ciniviz et al. (2011)	4-cylinder, Turbocharged DI engine	M0 M5 M10 M15	Blending	NO_X emissions increased, whereas, CO and HC emissions decreased by increasing the methanol mass fraction
2014	Yasin et al. (2014)	4-cylinder, Water-cooled DI engine	M0 B20M5 (B20-Blend of 80% by volume mineral diesel and 20% by volume biodiesel)	Blending	CO and CO_2 emissions decreased significantly, but NO_X and NO emissions increased as compared to the baseline diesel
2016	Li et al 2016)	6-cylinder, heavy-duty (HD) CRDi engine	20.93% to 37.16% Methanol co-combustion ratio	Methanol was introduced through PFI and Pilot Diesel was introduced directly into the combustion chamber	CO and HC emissions increased, whereas NO_X emissions decreased in the methanol-diesel dual fuel mode compared to baseline diesel

(continued)

Table 11.5 (continued)

Year	Researcher	Engine specifications	Fuel used	Methanol utilization technique	Test results for emissions
2018	Wu et al. (2018)	6-cylinder, Turbocharged CRDI engine	0%, 15%, 25%, 35% 50% MSR	Methanol was introduced through PFI and Pilot Diesel was introduced directly into combustion chamber	NO_X emissions decreased by increasing the MSR and the late injection quantity of diesel
2018	Saccullo et al. (2018)	Single-cylinder, Volvo D13 CRDI experimental engine	Diesel was used to start the engine, while Methanol was used as the main fuel	Direct injection of Methanol and Diesel	Methanol produced slightly lower NO_X emissions as compared to conventional diesel fuel
2018	Zhou and Qiu (2018)	6-cylinder, Turbocharged, intercooled, CRDI engine with DOC	0% to 40% Methanol energy fraction in total fuel	Methanol was introduced through PFI and Pilot Diesel was introduced directly into the combustion chamber	NO_2 emission from the dual fuel mode increased significantly as compared to the baseline diesel fuel
2019	Agarwal et al. (2019)	Single-cylinder, 4-stroke, constant-speed DI engine	M0 M10 M15 with 1% (v/v) dodecanol	Emulsion	Blending of 15% methanol with diesel reduced the HC, CO, and NOx emissions
2021	Panda and Ramesh (2021)	Single-cylinder, 4-stroke, naturally aspirated, water-cooled, CRDI engine	Methanol-Diesel energy share of around 50%	Methanol was introduced through PFI with Pilot, main and post injection of Diesel directly into the combustion chamber	The NO_X emissions reduced by increasing the amount of post-injected diesel, and the HC and CO emissions were also reduced

(continued)

Table 11.5 (continued)

Year	Researcher	Engine specifications	Fuel used	Methanol utilization technique	Test results for emissions
2021	Zhang et al. (2021)	4-cylinder, 4-stroke engine	M0 M10 M20 M30 M40	Blending	The emissions of NOx, and CO were reduced, but HC Emissions increased by increasing the methanol content in the blend
2022	El-seesy et al. (2022)	Single-cylinder, 661 cc engine	M0 M5 M10 M15 with 20% n-decanol as cosolvent	Emulsion	NO_X emissions reduced by about 30%, but UHC and CO concentrations increased by around 60% and 50%, respectively, from the combustion of blended fuel as compared to diesel
2023	Hassan et al. (2023)	Single-cylinder, 4-stroke pneumatic cooling engine	M0 M7 M14 M21 with 1% (v/v) dodecanol	Emulsion	CO, CO_2 and HC emissions reported to be decreased by 69–83%, 60–69% and 80–83%, but NO_X emissions increased by 135–346% as methanol blending ratio increased from minimum to maximum
2024	Cung et al. (2024)	Single-cylinder, 4-valve research engine	Port fuel injection of Methanol substitution rates up to 83.4% with renewable diesel	Methanol is introduced through PFI and Pilot renewable Diesel is introduced directly into the combustion chamber	NO_X emissions decreased, but CO and HC emissions increased by increasing the methanol substitution rates

Table 11.6 Experimental studies for unregulated emissions from methanol-powered CI engines

Year	Researcher	Engine specifications	Fuel used	Methanol utilization technique	Test results for emissions
2009	Cheung et al. (2009)	4-cylinder, naturally aspirated, water-cooled, DI engine	M5 M10 M15	Blending (Methanol + Biodiesel)	The emissions of unburned methanol, formaldehyde, 1,3-butadiene and acetaldehyde increased, but Benzene emissions reduced by increasing the methanol fraction. Toluene emissions were equal to diesel fuel. An equal amount of xylene is produced by all test fuels
2010	Zhang et al. (2010)	4-cylinder, Water-cooled DI engine with and without Diesel Oxidation Catalyst (DOC)	10%, 20%, 30% fumigation methanol	Methanol was introduced through PFI and Pilot Diesel was introduced directly into the combustion chamber	**Without DOC** The emissions of ethyne, ethene and 1,3-butadiene were reduced but the emissions of unburned methanol, formaldehyde and BTX increased significantly with increasing fumigation methanol percentage **With DOC** Almost all unregulated emissions were reduced with the utilization of DOC
2018	Zhou and Qiu (2018)	6-cylinder, turbocharged, intercooled, CRDI engine with DOC	0% to 40% Methanol energy fraction in total fuel	Methanol was introduced through the PFI and Pilot Diesel was introduced directly into the combustion chamber	Formaldehyde emission from the diesel-methanol mode was higher as compared to pure diesel and diesel-ethanol mode, however the acetaldehyde emission from the diesel-methanol combustion was almost nil as compared to other fuels

summary of particulate emission characteristics of different methanol-powered CI engines is provided in Table 11.7.

Table 11.7 Experimental studies for particulate emissions from methanol-powered CI engines

Year	Researcher	Engine specifications	Fuel used	Methanol utilization technique	Test results for emissions
2008	Yao et al. (2008)	4-cylinder, naturally aspirated, modified DI engine with oxidation catalytic converter (OCC)	Diesel was used to ignite the mixture of methanol and air, methanol was injected into each cylinder to produce a homogeneous mixture with the intake air	Methanol was introduced through PFI and Pilot Diesel was introduced directly into the combustion chamber	**Without OCC** Smoke opacity was reduced, however, the PM emissions increased **With OCC** Smoke opacity and PM emissions were reduced as compared to the baseline diesel fuel
2009	Cheung et al. (2009)	4-cylinder, naturally aspirated, water-cooled, DI engine	M5 M10 M15	Blending (Methanol + Biodiesel)	PM emissions decreased by 32%, while a 26% decrease was reported by using pure biodiesel
2010	Cenk Sayin (2010)	Single-cylinder, naturally aspirated DI engine	M5 M10 with 1% (v/v) dodecanol	Emulsion	Smoke opacity decreases by increasing the methanol fraction in the fuel
2010	Zhang et al. (2010)	4-cylinder, Water-cooled DI engine with and without Diesel Oxidation Catalyst (DOC)	10% 20% 30% fumigation methanol	Methanol is introduced through PFI and Pilot Diesel is introduced directly into the combustion chamber	**Without DOC** Particulate mass and particulate number reduced but SOF increased with increasing methanol content **With DOC** Particulate mass and number were reduced further with the simultaneous reduction of SOF

(continued)

Table 11.7 (continued)

Year	Researcher	Engine specifications	Fuel used	Methanol utilization technique	Test results for emissions
2018	Wu et al. (2018)	6-cylinder, turbocharged CRDI engine	0%, 15%, 25%, 35% 50% MSR	Methanol is introduced through PFI and Pilot Diesel is introduced directly into the combustion chamber	PM emissions in accumulation mode decreased by increasing the MSR and late injection quantity of diesel
2018	Saccullo et al. (2018)	Single-cylinder, Volvo D13 CRDI experimental engine	Diesel was used to start the engine, while methanol was used as a main fuel	Direct injection of Methanol and Diesel	Methanol reduces the soot emissions significantly
2019	Agarwal et al. (2019)	Single-cylinder, 4-stroke, constant-speed DI engine	M0 M10 M15 with 1% (v/v) dodecanol	Emulsion	M15 blend produced lower particulate mass and lower particulate number as compared with the baseline diesel fuel
2021	Li et al. (2021)	Single-cylinder, 4-st roke, air-cooled, natural aspiration DI engine	M0 M10 M20	Blending (Methanol + Biodiesel)	Soot content, particle size and carbon mass in the particulates reduced with increasing the methanol blending ratio, whereas the SOF contents increased
2021	Panda and Ramesh (2021)	Single-cylinder, 4-stroke, naturally aspirated, water-cooled, CRDI engine	Methanol-diesel energy share of around 50%	Methanol is introduced through PFI with Pilot, main, and post injection of Diesel directly into the combustion chamber	Post-injection of diesel in small quantities reduced the soot emissions

(continued)

Table 11.7 (continued)

Year	Researcher	Engine specifications	Fuel used	Methanol utilization technique	Test results for emissions
2021	Zhang et al. (2021)	Four-cylinder, 4-stroke engine	M0 M10 M20 M30 M40	Blending	The soot emissions were reduced by the addition of methanol
2022	El-seesy et al. (2022)	Single-Cylinder, 661 cc engine	M0 M5 M10 M15 with 20% n-decanol as co-solvent	Emulsion	Smoke opacity intensity was reduced by about 50% as compared to diesel fuel
2023	Hassan et al. (2023)	Single-cylinder, 4-stroke pneumatic cooling engine	M0 M7 M14 M21 with 1% (v/v) dodecanol	Emulsion	PM emission reduced in the range of 25–30% as the methanol blending ratio increased from minimum to maximum

11.7 Conclusions

Methanol has become an important alternative Ices fuel due to the uncertain future of conventional fossil fuels and issues of regulated, unregulated, and particulate emissions from their combustion. The utilization of methanol in Ices can decrease the regulated emissions of carbon dioxide by 60–95%, nitrogen oxides by 60–80%, unregulated emissions, and nearly complete elimination of the emissions of SO_X and PM. At present time methanol is produced mainly from coal and natural gas. Producing methanol from fossil fuels is a very energy-intensive process and generates a lot of CO_2 emissions. Therefore, to reduce CO_2 emissions, methanol production from biomass and captured carbon is becoming popular nowadays. The generated methanol from these routes is known as green methanol which produces a negligible amount of carbon footprint. Methanol is primarily used as SI engine fuel due to some similar properties to gasoline, but it can also be utilized in CI engines with some alteration in the engine and fueling systems. In SI engines, methanol can be used as a neat fuel, in a blended form with gasoline, and through direct or port fuel injection. Whereas in the CI engine, methanol is used by blending, dual-fuel injection, fumigation, and emulsification methods. Low volumes of methanol can be utilized in both SI and CI engines by blending techniques with very few modifications in the engine and fueling systems. Due to the poor miscibility of methanol with gasoline and diesel, some co-solvents are required to avoid the chances of phase separation of the blend. By using methanol as SI engine fuel HC, CO, and PM emissions are generally less as compared with baseline gasoline; however, the NO_X emissions do not always reduce and it also increases the amount of some unregulated gas emissions. The utilization of methanol in CI engines reduces the NO_X and PM emissions, howeve; it increases the unregulated emissions of unburned methanol, formaldehyde, and BTX emissions. The NO_X and unregulated emissions from methanol utilization can be reduced significantly by the utilization of proper exhaust after treatment devices/processes such as TWC, DOC, Selective catalytic reduction, etc. Overall, methanol shows a good characteristic for a reduction in pollutant emissions and it is an important alternative to fossil fuels.

References

Agarwal AK (2007) Biofuels (Alcohols and Biodiesel) applications as fuels in internal combustion engines. Prog Energy Combust Sci 32(3):233–271

Agarwal AK, Sharma N, Singh AP, Kumar V, Satsangi DP, Patel C (2019) Adaptation of methanol–dodecanol–diesel blend in diesel Genset engine ASME. J Energy Resour Technol 141(10):102203. https://doi.org/10.1115/1.4043390

Alexandru D, Dragos T (2017). Evaluation of performance and emissions characteristics of methanol blend (gasohol) in a naturally aspirated spark ignition engine. In: IOP conference series: materials science and engineering, vol 252, No 1, p. 012086. IOP Publishing

Ashurst JV, Nappe TM (2018) Methanol Toxicity. [Updated 2023 Jun 12]. In: StatPearls [Internet]. Treasure Island (FL): StatPearls Publishing; 2024 Jan-. Available from: https://www.ncbi.nlm.nih.gov/books/NBK482121/

Bansal P, Meena R (2021) Methanol as an alternative fuel in internal combustion engine: scope, production, and limitations. In: Agarwal AK, Valera H, Pexa M, Čedík J (eds) Methanol. energy, environment, and sustainability. Springer, Singapore. https://doi.org/10.1007/978-981-16-1224-4_2

Benjamin YH (2013) Chapter 47—Toxic and metabolic brain disease. Imaging of the Brain, pp 951–971. https://doi.org/10.1016/B978-1-4160-5009-4.50055-8

Bone WA (1935) Initial formation of methyl alcohol in the oxidation of methane. Nature 136:910–910. https://doi.org/10.1038/136910a0

Bromberg L, Cheng W (2010) Methanol as an alternative transportation fuel in the US: Options for sustainable and/or energy-secure transportation. Tech. Rep. Massachusetts Institute of Technology

Brynolf S, Fridell E, Andersson K (2014) Environmental assessment of marine fuels: liquefied natural gas, liquefied biogas, methanol and bio-methanol. J Clean Prod 74:86–95

Catapano F, Di Iorio S, Magno A, Sementa P, Vaglieco BM (2024) A complete assessment of the emission performance of a SI engine fueled with methanol, methane and hydrogen. Energies 17:1026. https://doi.org/10.3390/en17051026

Charlotte EB, Paul DD, Jagroop P (2018) Biotechnological conversion of methane to methanol: evaluation of progress and potential. AIMS Bioeng 5(1):1–38. https://doi.org/10.3934/bioeng.2018.1.1

Chen Y, Ma J, Han B, Zhang P, Hua H, Chen H, Su X (2018) Emissions of automobiles fueled with alternative fuels based on engine technology: a review. J Traffic Transp Eng 5(4):318–334

Cheung CS, Zhu L, Huang Z (2009) Regulated and unregulated emissions from a diesel engine fueled with biodiesel and biodiesel blended with methanol, Atmos spheric Environment 43:4865-4872, ISSN 1352-2310, https://doi.org/10.1016/j.atmosenv.2009.07.021

Ciniviz M, Köse H, Canli E, Solmaz O (2011) An experimental investigation on effects of methanol blended diesel fuels to engine performance and emissions of a diesel engine. Sci Res Essays 6(15):3189–3199

Cung KD, Wallace J, Kalaskar V, Smith EM, Briggs T, Christopher B (2024) Experimental study on engine and emissions performance of renewable diesel methanol dual fuel (RMDF) combustion, Fuel 357(B):129664, ISSN 0016-2361, https://doi.org/10.1016/j.fuel.2023.129664

Qiu CD (2018) Experimental study on unregulated emissions characteristics of alcohol-diesel dual-fuel combustion with diesel oxidation catalyst. J Energy Eng 145(2). https://doi.org/10.1061/(ASCE)EY.1943-7897.0000596

Dalena F, Senatore A, Marino A, Gordano A, Basile M, BasileA (2018) Methanol production and applications: an overview. In: Methanol: science and engineering; Amsterdam, The Netherlands, Elsevier, pp 3–28

Deka TJ, Osman AI, Baruah DC et al (2022) Methanol fuel production, utilization, and techno-economy: a review. Environ Chem Lett 20:3525–3554. https://doi.org/10.1007/s10311-022-01485-y

El-Seesy AI, Waly MS, Nasser A, El-Zoheiry RM (2022) Improvement of the combustion, emission, and stability features of diesel-methanol blends using n-decanol as cosolvent. Sci Rep 12(1):18963. https://doi.org/10.1038/s41598-022-20326

Fan Z, Xia Z, Shuai S, Jianhua X, Jianxin W (2009) Unregulated emissions and combustion characteristics of low-content Methanol−Gasoline blended fuels. Energy and Fuels - Energy Fuel 24:10:1021/ef900974p

Galindo CP, Badr O (2007) Renewable hydrogen utilisation for the production of methanol. Energy Convers Manage 48:519–527. https://doi.org/10.1016/j.enconman.2006.06.011

Han JS (2013) Partial oxidative conversion of methane to methanol through selective inhibition of methanol dehydrogenase in methanotrophic consortium from landfill cover soil. Appl Biochem Biotechnol 171:1487–1499. https://doi.org/10.1007/s12010-013-0410-0

Hassan H, Alalwan H, Mohammed M, Mohammed M (2023) Identifying the impact of methanol-diesel fuel on the environment using a four-stroke CI engine. J Appl Eng Sci 21:1–6. https://doi.org/10.5937/jaes0-39809

Hillestad M (2023) Direct hydrogenation of carbon dioxide to methanol: systematic generation of multi-stage designs. J CO_2 Util 74:102535

Ijaz MMA, Usman M, Hayat N, Zubair SWH, Bashir R, Ahmed E (2021) Experimental evaluation of methanol-gasoline fuel blend on performance, emissions and lubricant oil deterioration in SI engine. Adv Mech Eng 13(6). https://doi.org/10.1177/16878140211025213

Irena and Methanol Institute (2021) Innovation outlook: renewable methanol; international renewable energy agency: Masdar City, United Arab Emirates

Jamrozik A (2017) The effect of the alcohol content in the fuel mixture on the performance and emissions of a direct injection diesel engine fueled with diesel-methanol and diesel-ethanol blends. Energy Convers Manage 148:461–476. https://doi.org/10.1016/j.enconman.2017.06.030

Kasteren JMN (2016) 16 - Production of bioalcohols via gasification. In: Luque R et al (Eds), Handbook of biofuels production (Second Edition). Woodhead Publishing, pp 495–507. https://doi.org/10.1016/B978-0-08-100455-5.00016-3

Klier K (1982) Methanol synthesis. In: Advances in catalysis, vol 31. Academic Press, pp 243–313

Li G, Zhang C, Li Y (2016) Effects of diesel injection parameters on the rapid combustion and emissions of an HD common-rail diesel engine fueled with diesel-methanol dual-fuel. Appl Therm Eng 108:1214–1225. https://doi.org/10.1016/j.applthermaleng.2016.08.029

Li Z, Wang Y, Geng H, Zhen X, Liu M, Shuai X (2020) Investigation of injection strategy for a diesel engine with directly injected methanol and pilot diesel at medium load. Fuel 266:116958. https://doi.org/10.1016/j.fuel.2019:116958

Li R, Wang Z, Tang C, Meng X (2021) Study on particulate structure characteristics of diesel engines fueled with a methanol/biodiesel blend. ACS Omega 6(9):6081–6087. https://doi.org/10.1021/acsomega.0c04371

Li C, Jia T, Wang S, Wang X, Negnevitsky M, Wang H, Hu Y, Xu W, Zhou N, Zhao G (2023) Methanol vehicles in China: A review from a policy perspective. sustainability 15:9201. https://doi.org/10.3390/su15129201

Machiele P (1990) Summary of the fire safety impacts of methanol as a transportation fuel. SAE technical paper 901113

Mishra PC, Gupta A, Kumar A, Bose A (2020) Methanol and petrol blended alternate fuel for future sustainable engine: a performance and emission analysis. Measurement 155:107519

Osman AI, Hefny M, Abdel Maksoud MIA (2021) Recent advances in carbon capture storage and utilisation technologies: a review. Environ Chem Lett 19:797–849. https://doi.org/10.1007/s10311-020-01133-3

Osman AI (2022b) Hydrogen production, storage, utilisation and environmental impacts: a review. Environ Chem Lett 20:153–188. https://doi.org/10.1007/s10311-021-01322-8

Panda K, Ramesh A (2021) Diesel injection strategies for reducing emissions and enhancing the performance of a methanol based dual fuel stationary engine. Fuel 289:119809, ISSN 0016-2361

Pu YH, Dejaegere Q, Svensson M, Verhelst S (2024) Renewable methanol as a fuel for heavy-duty engines: a review of technologies enabling single-fuel solutions. Energies 17:1719. https://doi.org/10.3390/en17071719

Reddy VJ, Hariram NP, Maity R, Ghazali MF, K MS (2023) Sustainable E-fuels: green hydrogen, methanol and ammonia for carbon-neutral transportation. World Electr Veh J 14:349. https://doi.org/10.3390/wevj14120349

Rifal M, Sinaga N (2016) Impact of methanol-gasoline fuel blend on the fuel consumption and exhaust emission of a SI engine. In: AIP conference proceedings 1725(1)

Saccullo M, Benham T, Denbratt I (2018) Dual fuel methanol and diesel direct injection HD single cylinder engine tests. SAE Technical Paper 2018-01-0259. https://doi.org/10.4271/2018-01-0259

Sayin C (2010) Engine performance and exhaust gas emissions of methanol and ethanol–diesel blends. Fuel 89:3410–3415. https://doi.org/10.1016/j.fuel.2010.02.017

Sayina C, Ilhan M, Canakci M, Gumus M (2009) Effect of injection timing on the exhaust emissions of a diesel engine using diesel–methanol blends. Renew Energy 34:1261–1269

Schifter I, Diaz L, Gonzalez U, Gonzalez-Macias C, Mejía-Centeno I (2019) The effects of addition of co-solvents on the physicochemical properties of gasoline–methanol blended fuels. Int J Engine Res 20(5):501–509. https://doi.org/10.1177/1468087418757855

Sollai S, Porcu A, Tola V, Ferrara F, Pettinau A (2023) Renewable methanol production from green hydrogen and captured CO_2: a techno-economic assessment. J CO_2 Util 68:102345

Srivastava R, Sarangi P, Bhatia L, Singh A, Shadangi K (2022) Conversion of methane to methanol: technologies and future challenges. Biomass Convers Biorefinery. https://doi.org/10.1007/s13399-021-01872-5

Tabibian SS, Sharifzadeh M (2023) Statistical and analytical investigation of methanol applications, production technologies, value chain and economy with a special focus on renewable methanol. Renew Sustain Energy Rev 179:113281

Taher E, Chandran K (2013) High-rate, high-yield production of methanol by ammonia-oxidizing bacteria. Environ Sci Technol 47:3167–3173. https://doi.org/10.1021/es3042912

Verhelst S, Turner JW, Sileghem L, Vancoillie J (2019) Methanol as a fuel for internal combustion engines. Prog Energy Combust Sci 70:43–88

Waluyo B, Setiyo M, St MES, Wardana ING (2021) Fuel performance for stable homogeneous gasoline-methanol-ethanol blends. Fuel 294:120565. https://doi.org/10.1016/j.fuel.2021.120565

Wang X, Ge Y, Liu L, Peng Z, Hao L, Yin H, Wang J (2015) Evaluation on toxic reduction and fuel economy of a gasoline direct injection-(GDI) powered passenger car fueled with methanol–gasoline blends with various substitution ratios. Appl Energy 157:134–143

Wu T, Yao A, Yao C, Pan W, Wei H, Chao C, Gao J (2018) Effect of diesel late-injection on combustion and emissions characteristics of diesel/methanol dual fuel engine. Fuel 233:317–327. https://doi.org/10.1016/j.fuel.2018.06.063

Yang S, Feng J, Sun P, Wang Y, Dong W, Yu X, Li W (2023) Combustion and emissions characteristics of methanol/gasoline CISI engines under different injection modes. Fuel 333:126506. https://doi.org/10.1016/j.fuel.2022.126506

Yanju W, Shenghua L, Hongsong L, Rui Y, Jie L, Ying W (2008) Effects of methanol/gasoline blends on a spark ignition engine performance and emissions. Energy Fuels 22(2):1254–1259

Yao C, Cheung C, Cheng C, Wang Y, Chan T, Lee S (2008) Effect of diesel/methanol compound combustion on diesel engine combustion and emissions. Energy Convers Manage Energ Conv Manage 49:1696–1704. https://doi.org/10.1016/j.enconman.2007.11.007

Yasin MM, Yusaf T, Mamat R, Yusop AF (2014) Characterization of a diesel engine operating with a small proportion of methanol as a fuel additive in biodiesel blend. Appl Energy 114:865–873

Yuen PKP, Villaire W, Beckett J (2010) Automotive materials engineering challenges and solutions for the use of ethanol and methanol blended fuels. SAE paper no. 2010-01-0729

Zhang Z, Tian J, Li J, Ji H, Tan D, Luo J, Jiang Y, Yang D, Cui S (2021) Effects of different mixture ratios of methanol-diesel on the performance enhancement and emission reduction for a diesel engine. Processes 9:1366. https://doi.org/10.3390/pr9081366

Zhang Z, Wen M, Cui Y, Ming Z, Wang T, Zhang C, Ampah JD, Jin C, Huang H, Liu H (2022) Effects of methanol application on carbon emissions and pollutant emissions using a passenger vehicle. Processes 10:525. https://doi.org/10.3390/pr10030525

Zhang ZH, Cheung CS, Chan TL, Yao CD (2010) Experimental investigation on regulated and unregulated emissions of a diesel/methanol compound combustion engine with and without diesel oxidation catalyst. Sci Total Environ 15:408(4):865–872

Zhu R, Hu J, He L et al (2021) Effects of ambient temperature on regulated gaseous and particulate emissions from gasoline-, E10- and M15-fueled vehicles. Front Environ Sci Eng 15:14. https://doi.org/10.1007/s11783-020-1306-1

Chapter 12
Innovative Fuel Injection Strategies for Enhancing Methanol Utilization in Internal Combustion Engines

Reetu Raj, S. Prabakaran, and Neeraj Kumbhakarna

Abstract The chapter provides a comprehensive examination of the different fuel injection methodologies employed to optimize the use of methanol in Internal Combustion (IC) engines. As the global focus intensifies on sustainable energy sources, methanol emerges as a promising alternative to conventional fossil fuels, necessitating refined injection techniques to enhance performance, efficiency, and emissions characteristics of the IC engines. This chapter delves into multiple strategies, including blending, Port Fuel Injection (PFI), high-pressure direct injection (HPDI), and emulsions, each tailored to maximize the advantages of methanol while addressing their inherent challenges. In this chapter, blending involves the mixture of methanol with conventional fuels, facilitating a seamless transition and compatibility with existing engine technologies. PFI, a widely used technique, introduces methanol into the intake port, promoting homogeneous mixing and improved combustion stability. HPDI, on the other hand, directly injects methanol into the combustion chamber under high pressure, offering precise control over the injection process, thereby enhancing combustion efficiency and reducing emissions. The chapter also explores the use of emulsions, where methanol is combined with water or other fluids to form a stable mixture, resulting in a more complete combustion process and significant emission reductions. This chapter also aims to thoroughly review different fuel injection strategies, such as adjusting injection pressure, altering injection timing, and implementing multiple injections, to enhance engine performance and control emissions. Each section provides an in-depth analysis of the operational principles, advantages, and limitations of these techniques, supported by recent research findings and case studies. In conclusion, the chapter underscores the critical role of innovative fuel injection technologies in the broader adoption of methanol, highlighting their potential to transform the energy landscape by providing cleaner, more efficient, and sustainable fuel options for IC engines.

R. Raj (✉) · S. Prabakaran · N. Kumbhakarna
Department of Mechanical Engineering, Indian Institute of Technology Bombay, Maharashtra 400076, India
e-mail: rturaj.2013@gmail.com

A. Bartl et al. (eds.), *Alcohol Production Processes and Their Utilisation in the Transport Sector*, Energy, Environment, and Sustainability,
https://doi.org/10.1007/978-981-96-7384-1_12

Keywords Methanol · Port fuel injection · Direct injection · Injection pressure · Engine performance

12.1 Introduction

The Internal Combustion Engine (ICE) has long been the cornerstone of automotive propulsion, continually evolving to meet stringent emission standards and efficiency demands. Fuel injection systems, a critical component of ICEs, play a pivotal role in determining engine performance and emissions. Traditional fuel injection techniques, including Port Fuel Injection (PFI) and Direct Injection (DI), have been optimized primarily for conventional fossil fuels like petrol and diesel. However, the growing interest in alternative fuels, particularly methanol, necessitates innovative fuel injection techniques to fully exploit their potential. Methanol, a versatile alcohol-based fuel, offers several advantages over conventional fuels. It burns more cleanly, reducing harmful emissions such as NOx, CO, and Particulate Matter (PM) (Agarwal et al. 2021). Moreover, methanol can be generated from renewable resources, helping to lower the carbon footprint.

Fuel injection systems in ICEs can be categorized into three primary types: PFI, DI, and dual-fuel injection systems. In PFI systems, fuel is introduced into the intake port, where it blends with air prior to entering the combustion chamber. This method is relatively simple and cost-effective but often results in lower combustion efficiency and higher emissions compared to direct injection. Research has shown that PFI can be effectively adapted for methanol utilization by modifying injector materials and adjusting fuel flow rates (Zhou et al. 2023, 2022). In contrast, DI systems deliver fuel directly into the combustion chamber, enabling more precise control over the air–fuel mixture and the combustion process. This leads to improved fuel efficiency and lower emissions. DI is particularly advantageous for methanol and methanol-blended fuels due to its high precision in fuel delivery, which is essential given methanol's unique properties (Agarwal et al. 2021; Verhelst et al. 2019). Moreover, dual-fuel injection systems combine PFI and DI techniques to leverage the advantages of both. Dual-fuel systems offer greater flexibility in controlling the combustion process, which can be crucial for optimizing performance and emissions when using methanol. Studies has shown that dual-fuel systems can attain higher thermal efficiency and reduced emissions when using methanol blends (Li et al. 2022a). The choice of fuel injection techniques for IC engines using methanol as fuel can significantly influence engine performance, emissions, and overall efficiency. PFI, DI, and dual injection systems each offer distinct advantages and disadvantages in the context of methanol utilization. Port Fuel Injection is a cost-effective and simpler technology that injects fuel into the intake port, allowing methanol to mix with air before entering the combustion chamber (Lius et al. 2022). This method provides uniform fuel distribution and is easier to implement in existing engines. Methanol's ability to mix well with air can result in a homogeneous air–fuel mixture, reducing NOx emissions due to lower peak combustion temperatures (Zhou et al. 2022). However, PFI tends to suffer from reduced combustion efficiency because methanol may not fully vaporize before entering the combustion chamber (Tornatore et al. 2019). Additionally, the power output is typically lower compared to DI systems, as PFI does not fully capitalize on methanol's high octane rating and potential for higher compression ratios. Moreover, methanol's high latent heat of vaporization can cause cold start issues, where the fuel fails to vaporize adequately at lower temperatures, leading to ignition difficulties (Zhu et al. 2013). Direct Injection (DI), on the other hand, injects fuel directly into the combustion chamber, enabling greater control over the combustion process and more precise fuel delivery (Feng et al. 2024; Zhao et al. 2024). This technique enhances fuel efficiency and power output, particularly when utilizing methanol, as it allows for higher compression ratios without the risk of knocking. Methanol's cooling effect during DI further improves engine performance, allowing for more aggressive tuning (Pu et al. 2024a). Additionally, DI systems can reduce CO_2 emissions due to leaner air–fuel mixtures (Golzari et al. 2021; Wang et al. 2024). However, DI comes with higher costs and complexity due to the need for high-pressure pumps and advanced control systems. Furthermore, the higher combustion temperatures in DI systems can lead to increased NOx emissions, and methanol's corrosive properties may cause injector fouling, requiring more frequent maintenance (Ladommatos et al. 1996; Yao et al. 2009; Ye and Boehman 2010). Dual Injection systems, which combine PFI and DI, aim to leverage the strengths of both techniques. This hybrid approach optimizes performance across various engine conditions, switching between PFI for low-load scenarios, which aids fuel atomization and cold starts, and DI for high-load conditions, enhancing power and efficiency (Yu et al. 2024a; Mohamed et al. 2024). Dual injection offers the flexibility to reduce both NOx and particulate emissions by adjusting the air–fuel ratio based on load demands (Chen 2000). However, this method is the most complex and costly, requiring sophisticated control systems and posing potential space constraints in engine design (Huang et al. 2021). The combined use of PFI and DI also introduces durability concerns, especially with methanol's tendency to cause component wear due to its corrosive nature. Thus, the choice of fuel injection system should align with the specific performance goals, emissions targets, and cost considerations for the methanol-powered engine. The fuel injection technique plays a crucial role in determining the efficiency, emissions, and overall performance of methanol-fueled internal combustion engines. Methanol, with its high octane rating and oxygen content, can enhance combustion efficiency, but its physical properties, such as low energy density and high latent heat of vaporization, require precise control over fuel delivery (Verhelst et al. 2019; Wang et al. 2019). An appropriate fuel injection system ensures the correct amount of methanol is atomized and mixed with air, optimizing combustion and reducing the risk of pre-ignition or incomplete combustion. Selecting the right fuel injection system is crucial because it directly affects engine stability, power output, and emission characteristics, particularly in advanced combustion modes like homogeneous charge compression ignition (HCCI) or reactivity controlled compression ignition (RCCI), where precise fuel–air mixture control is essential for achieving the desired combustion dynamics and emission reduction targets (Li et al. 2003; Egler 2010). Moreover, it helps maintain an ideal air–fuel ratio under varying operating conditions, which is essential for minimizing engine knock and ensuring reliable engine performance (Heywood 1988). The precise control offered by modern fuel injection systems is particularly important in meeting stringent emissions standards while ensuring the longevity and reliability of the engine (Mohan et al. 2013). Thus, the choice of injection technique directly impacts the engine's ability to harness methanol's advantages while addressing its inherent challenges. However, adapting PFI, DI, and dual-fuel systems for methanol utilization involves addressing its distinct characteristics:

- **PFI with Methanol**: Modifying PFI systems for methanol involves using materials resistant to methanol's corrosive effects and adjusting injector sizes to handle increased flow rates. The cooling effect can be beneficial for preventing knock but must be managed to avoid intake manifold condensation (Zhou et al. 2023, 2022).
- **DI with Methanol**: DI systems can offer precise control over the fuel–air mixture, making them well-suited for methanol. Innovations in injector design, such as using advanced alloys and coatings, can mitigate corrosion issues. Controlling injection timing and pressure is critical to optimize methanol's combustion characteristics (Verhelst et al. 2019).
- **Dual-Fuel Injection with Methanol**: Dual-fuel systems can optimize methanol's advantages by using PFI during low-load conditions and DI during high-load conditions. This approach improves efficiency and reduces emissions but requires sophisticated control systems and robust hardware to handle methanol's properties (Li et al. 2022a)

The dire need of alternative fuel presents methanol with several advantages including lower emissions and the potential for renewable production. Methanol's octane rating of 109 allows for higher compression ratios and improved thermal efficiency (Stone 1999). Despite these benefits, methanol's physical and chemical properties present unique challenges for fuel injection systems designed for traditional fuels. However, its adoption in ICEs poses specific challenges that necessitate innovative fuel injection techniques:

- **Lower Energy Density**: Methanol has an energy density of 19.9 MJ/kg, considerably lower than gasoline's 44 MJ/kg (Heywood 1988). This requires higher fuel flow rates to maintain engine performance, which traditional injection systems may not adequately support.
- **High Latent Heat of Vaporization**: Methanol's latent heat of vaporization is 1100 kJ/kg, which is almost three times higher than gasoline's 370 kJ/kg (Stone 1999). This property cools the intake air, reducing the risk of knock but complicating the fuel–air mixture preparation.
- **Corrosiveness and Lubricity**: Methanol is corrosive to some materials found in conventional fuel systems and has lower lubricity, which increases wear and tear on fuel injection components (Jin et al. 2024; Olah et al. 2018).

Methanol-fueled engines are gaining increasing attention in the commercial sector due to their potential to reduce greenhouse gas emissions and reliance on traditional fossil fuels (Verhelst et al. 2019; Zhen and Wang 2015). Methanol, which can be produced from a variety of feedstocks including natural gas, coal, and renewable sources like biomass, offers a cleaner-burning alternative with lower emissions of nitrogen oxides (NO_x) and particulate matter compared to diesel engines (Patil 2024). Commercial adoption of methanol as a fuel is particularly evident in the marine industry, where it is being used to comply with stringent emission regulations. Additionally, there is growing interest in using methanol in automotive and power generation sectors, supported by advancements in engine technology that enhance fuel efficiency and performance. However, the widespread commercial deployment of methanol-fueled engines still faces challenges, including the need for an extensive refueling infrastructure and overcoming the higher costs associated with fuel production and distribution (Srivastava et al. 2022). Nevertheless, with increasing regulatory pressure to reduce emissions and the drive toward sustainable energy solutions, methanol-fueled engines are poised to play a significant role in the future energy landscape (Zhen and Wang 2015).

In conclusion, this chapter explores how integrating methanol into ICEs through advanced fuel injection systems can significantly enhance engine performance. Additionally, it highlights the potential for these systems to contribute to reduced emissions, making methanol a promising alternative fuel. Traditional injection systems such as PFI and DI have laid the groundwork for this evolution, but adapting them for methanol requires addressing the unique properties and challenges associated with this alternative fuel. Methanol's clean-burning characteristics, higher octane rating, and potential for renewable production make it an appealing option for sustainable automotive propulsion. However, the successful implementation of methanol in ICEs hinges on overcoming several technical hurdles. Modifying PFI systems involves enhancing material resistance to methanol's corrosiveness and adjusting for its higher flow rates. DI systems, with their precision and control, are well-suited for methanol but require advancements in injector design and combustion management. Dual-fuel systems, combining the strengths of PFI and DI, offer a flexible solution but necessitate sophisticated control mechanisms to handle methanol's distinct attributes.

The challenges posed by methanol's lower energy density, high latent heat of vaporization, and corrosiveness underscore the need for innovative engineering solutions. By addressing these issues, methanol can be effectively utilized to achieve higher thermal efficiency, reduced emissions, and a reduced carbon footprint. As the automotive industry searches for sustainable alternatives to traditional fossil fuels, methanol's potential as a cleaner and renewable fuel source highlights its importance as a key area for future research and development in fuel injection technologies.

12.2 Different Fuel Injection Techniques

12.2.1 Port Fuel Injection

Port Fuel Injection (PFI) is a widely utilized technique in internal combustion engines where fuel is injected into the intake manifold, upstream of the intake valves. In this approach, fuel is injected into the airstream via separate fuel injectors located in the intake ports, where it blends with the incoming air before reaching the combustion chamber. Figure 12.1 shows the schematic diagram of the PFI configuration. The primary advantages of PFI include improved fuel atomization and mixing, which promote more efficient combustion and better throttle response. Additionally, PFI helps to reduce particulate emissions as the fuel's vaporization process cools the intake air, leading to enhanced charge density and reduced knocking tendencies. However, PFI can be less effective in controlling emissions under low-load conditions compared to direct injection (DI) systems, as the fuel may not fully combust at lower engine loads and speeds (Alkidas 2007; Zhao and Ladommatos 2001). This might be due to that in PFI systems, fuel is injected into the intake manifold before being drawn into the combustion chamber. At low-load and low-speed conditions, the air–fuel mixture can struggle to atomize fully, leading to incomplete combustion. This can cause an increase in unburned hydrocarbons (HC) and carbon monoxide (CO) emissions due to inefficient burning of the fuel. Whereas DI systems inject fuel directly into the combustion chamber, allowing for more accurate control of the fuel quantity, timing, and spray pattern. This leads to better atomization of the fuel and more efficient combustion, even at low engine speeds and loads. This method is often employed in conjunction with DI systems in modern engines to combine the benefits of both techniques, optimizing performance, fuel efficiency, and emissions across various operating conditions.

- **Advantages of Port Fuel Injection (PFI)**: PFI systems are simpler and cheaper to manufacture and maintain compared to DI systems (Pal 2023). PFI systems generally produce fewer particulate emissions than DI systems, making them a cleaner option in terms of PM (Wu et al. 2019). PFI tends to offer smoother engine operation with less noise compared to DI systems.
- **Limitations of Port Fuel Injection (PFI)**: PFI systems are generally less efficient than DI systems because the fuel is injected into the intake port rather than directly into the combustion chamber, leading to potential fuel losses (Golzari et al. 2021). PFI has a slower throttle response compared to DI, as the fuel has to travel further before reaching the combustion chamber (Cowart and Cheng 2000).

12.2.2 High-Pressure Direct Injection

High-Pressure Direct Injection (HPDI) is a modern technology used in diesel engines to enhance performance, improve fuel efficiency, and reduce emissions. It involves

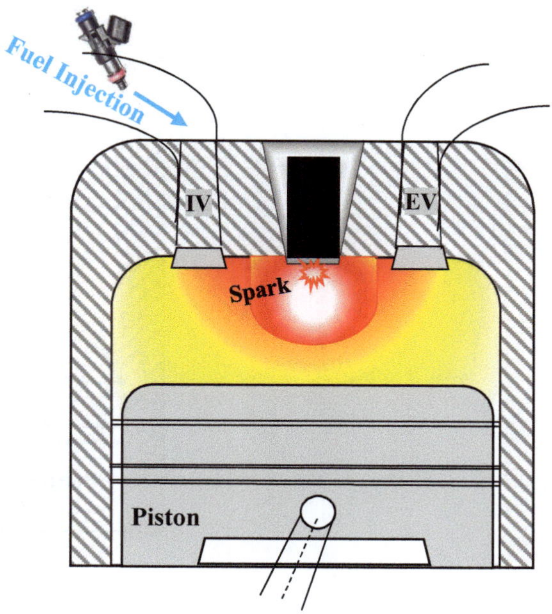

Fig. 12.1 Schematic diagram of PFI configuration

the direct injection of fuel at very high pressures directly into the combustion chamber, which allows for better fuel atomization and more efficient combustion. This system injects fuel directly into the combustion chamber, offering better control of air–fuel ratios for heavy-duty diesel engines. HPDI systems were originally developed for Compression Ignition (CI) engines. While, they are now used in both Spark Ignition (SI) and CI engines. In SI engines, HPDI systems are often used in Gasoline Direct Injection (GDI) setups. However, in gasoline engines, lower pressure direct injection systems are more common because gasoline has different combustion characteristics than diesel. Figure 12.2 shows the schematic representation of a conventional direct injection system being used in an IC engine. Typically, heavy vehicles are fitted with two injectors: one for injecting diesel directly into the combustion chamber at high pressure, and another for injecting methanol into the port at relatively lower pressure (Pu et al. 2024b). This allows for the effective use of both fuels, optimizing performance and efficiency. In a diesel engine, fuel is drawn from the tank by a low-pressure pump and compressed to a high pressure by a high-pressure pump. This fuel is stored in a common rail that distributes it to the injectors. The Electronic Control Unit (ECU) regulates the injectors to atomize the fuel for efficient combustion in the chamber, where it auto-ignites to produce power. After combustion, exhaust gases are released, and modern emission control systems are also used like diesel particulate filters and selective catalytic reduction to support global emissions standards. Direct Injection in diesel engines improves fuel efficiency, power, and emissions control. By precisely managing fuel–air mixtures, DI systems enhance thermal efficiency, power output, and fuel economy while reducing emissions to meet

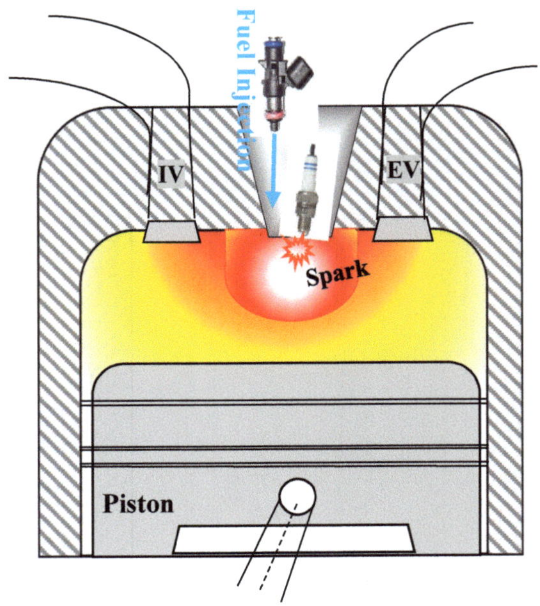

Fig. 12.2 Schematic diagram of direct injection fuel system

environmental regulations. Diesel fuel has been crucial for heavy-duty machinery due to its high energy density and reliability. However, with fossil fuel resources depleting and environmental concerns rising, researchers are exploring alternative fuels for IC engines. Methanol has appeared as a viable option, and current studies focus on its potential to enhance engine performance and sustainability. Table 12.1 reveals the key differences between the effect of methanol substitution in SI and CI engines.

- **Advantages of High-Pressure Direct Injection**: High-Pressure Direct Injection (HPDI) is a fuel delivery technique that enhances the efficiency and performance of internal combustion engines. By injecting fuel directly into the combustion chamber at very high pressures, typically exceeding 1000 bar, HPDI allows for better atomization of the fuel, leading to more complete and efficient combustion (Jin et al. 2021). This results in improved fuel efficiency, higher power output, and reduced emissions of pollutants such as unburnt hydrocarbons and carbon monoxide (Wang et al. 2024a). Additionally, HPDI systems offer greater control over the combustion process, enabling advanced combustion strategies like stratified charge and lean-burn operation, which can further enhance engine efficiency and reduce fuel consumption.
- **Limitations of High-Pressure Direct Injection**: Despite its advantages, high-pressure direct injection has some limitations. One of the primary challenges is the increased complexity and cost associated with HPDI systems (Mumford et al. 2017). The high-pressure fuel pumps, injectors, and control systems required for HPDI are more expensive and intricate than those used in conventional fuel injection systems. Additionally, the higher injection pressures can lead to increased

Table 12.1 Differences between effect of methanol substitution in SI and CI engines

Aspect	Effect of methanol substitution in SI engine	Effect of methanol substitution in CI engine
Fuel injection method	PFI or DI	PFI or DI
Ignition method	Spark plug ignition	Compression ignition with pilot fuel
Methanol role	Primary fuel (Göktaş et al. 2021) or blended with gasoline (Yanju et al. 2008)	Secondary fuel, blended with diesel or pilot fuel (Sayin et al. 2010)
Combustion Process	Homogeneous charge	Stratified charge, dual-fuel combustion (Li et al. 2020)
Effect on engine efficiency	Potential increase due to higher octane rating	Improved thermal efficiency and combustion
Emissions impact	Lower CO and NOx, but potential increase in formaldehyde emissions (Yanju et al. 2008)	Lower emissions and smoke opacity, but potential increase in NOx emissions (Sayin et al. 2010)
Knock resistance	Improved due to higher octane number (Göktaş et al. 2021)	Higher resistance to knock and misfire (Li et al. 2020)
Engine modifications required	Minor adjustments in fuel delivery system	Significant changes in injection system and control
Cold start performance	Challenging due to methanol's lower vapor pressure	Improved with pilot fuel-aiding ignition
Fuel economy	Slight decrease due to lower energy density (Göktaş et al. 2021)	Potential improvement due to efficient combustion
Thermal efficiency	Can improve with optimized blending ratios (Yanju et al. 2008)	Higher due to methanol's high latent heat
Durability and wear	Requires corrosion-resistant materials	Similar concerns, with focus on injector durability
Engine power output	Potential increase with higher methanol content (Yanju et al. 2008)	Improves with optimized blending ratios
Combustion temperature	Lower peak temperatures	Lower peak temperatures, reducing NOx formation
Fuel handling and storage	Requires special handling due to methanol's corrosive nature	Similar concerns, with diesel/methanol compatibility
Modifications	Dual-fuel DI, blended fuels, corrosion resistance	Diesel-methanol compound combustion engine, dual-fuel DI, blended fuels, direct dual-fuel stratification integration (Li et al. 2020)

wear and tear on engine components, potentially reducing the longevity of the engine (Krieger et al. 2000). HPDI systems also demand precise calibration and maintenance to ensure optimal performance, and any malfunctions can result in severe engine damage or performance issues. Furthermore, while HPDI can reduce some emissions, it may increase the production of particulate matter and nitrogen oxides (NO_x), which require additional after treatment systems to mitigate (Hanson et al. 2009).

Cenk et al. (2010) found, increasing injection pressure from 180 to 220 bar in a DI diesel engine using methanol-blended diesel improved the mixing process and reduced smoke opacity, but this also led to higher NOx levels due to increased combustion temperature. However, retarded injection timing from 25° to 15° CA BTDC tends to lower NOx emissions as peak combustion temperatures are reduced, but it can also lead to incomplete combustion, decreasing brake thermal efficiency (BTE) and increasing fuel consumption (BSFC) and PM emissions. Chunde et al.(2008) did an experimental analysis of the Diesel Methanol Compression Combustion (DMCC) engine demonstrating reduced diesel consumption (approximately 30%), gas temperatures (around 10–15%), and smoke opacity (50%) but increased emissions except NOx. The addition of an Oxidation Catalytic Converter (OCC) further mitigated the increased HC and CO emissions, reducing them by up to 80%, while improving overall engine efficiency by approximately 5–10%. Zhiyong et al. (2021a) analyzed injection strategies and found that methanol/diesel/methanol (M/D/M) mode offers the best overall fuel economy and emissions. M/D mode is optimal for low-load applications and diesel/methanol mode is best for high-load applications. The most suitable methanol injection timings are obtained in M/D mode, namely 20° CA BTDC at low load and 30° CA BTDC at high load. Xiaojun et al. (2023) studied M/D dual-direct injection in a single-cylinder diesel engine and found that, at full load, the engine could substitute 65.1% of its energy with methanol, but its peak thermal efficiency (43.4%) was observed at a slightly lower load (79.5% load), when methanol contributed 52.4% of the energy. This demonstrates that the engine was most thermally efficient at moderate loads rather than full load. Li et al. (2020) investigated that methanol-Diesel Dual Direct Fuel Stratification (DDFS) provides higher thermal efficiency, lower HC, CO and soot emissions and better energy recovery than reactivity controlled compression ignition at low loads. DDFS operates at lower initial temperatures, uses higher EGR rates and features a two-stage combustion process. Optimizing methanol fraction, injection timings, and spray angles enhances combustion efficiency and lowers NOx emissions. Yabin et al. (2020) observed a HPDI M/D dual-fuel engine with an innovative cylinder head design and found that high methanol substitution ratios (>90%) provided stable combustion and low HC emissions at medium to high loads (>8 bar IMEP). Advancing fuel injection timing further enhanced stability. In conclusion, from the extensive literature review conducted, Bin et al. (2020), Cenk et al. (2010) for methanol-diesel dual-fuel direct injection configurations, there is a drop in emissions and soot formation as the methanol substitution percent increases. Cenk et al. (2010) showed that optimum BTE of 26.7% was achieved at around 10–15% methanol blend by volume with reduced CO (40–50%) and HC (35–48%) emissions as compared to diesel fuel. Ahmad et al. (2021) reported that the highest BTE was achieved at higher engine loads with a 10% methanol blend in the fuel. This study also observed that the BSFC increased when higher percentages of methanol were blended with diesel, particularly at various engine load conditions. Methanol concentrations of 30% and 40% led to an increase in BSFC by 28% and 55%, respectively. Valera et al. (2019) also showed the modifications of ignition improvers and the use of glow plugs in diesel engines for better performances of methanol blended direct injection fuel system.

12.2.3 Emulsion

The emulsion of fuels is an unconventional method of immersing droplets of a liquid into another liquid with the help of an emulsifying agent or surfactant, this process of two or more immiscible fluids mixing is called emulsification. These mixtures are called emulsions, often of polar and non-polar fluids, which are thermodynamically unstable but can be kinetically stabilized with emulsifying agents, making them useful in diesel engines (Debnath et al. 2015; Ghosh and Ravikrishna 2021). In IC engines, emulsified fuels work by dispersing one liquid phase into another, stabilized by emulsifying agents. In CI engines, emulsion fuels create finer droplets for better air mixing and combustion under high compression, leading to efficient burning and reduced emissions. In SI engines, the emulsion improves fuel–air mixing, ignited by a spark plug, enhancing combustion efficiency. However, maintaining emulsion stability and precise ignition timing is challenging. One of the less-used emulsion blends is made by mixing methanol and diesel with a surfactant and emulsifying agent. In methanol-diesel emulsion blends, commonly used surfactants include Span 80 (Sorbitan Monooleate) and Tween 80 (Polysorbate 80), which help stabilize the emulsion. Emulsifying agents like oleic acid and lecithin are also used to enhance the stability of the mixture. This mixture is then agitated to disperse methanol into small droplets within the diesel. The surfactant stabilizes the emulsion by reducing surface tension and preventing droplet coalescence to maximize surface area, resulting in a stable, homogeneous fuel blend. This type of emulsion is referred to as a methanol-in-diesel (M/D) emulsion (Ghosh and Ravikrishna 2021). Figure 12.3 illustrates a schematic of the phases involved in the methanol-in-diesel emulsification process.

In a diesel engine, diesel-methanol emulsified fuel blends enhance Brake-Specific Fuel Consumption (BSFC) and Brake Thermal Efficiency (BTE) compared to conventional diesel, reflecting improved fuel economy. These blends notably decrease emissions of CO, HC, and Particulate Matter (PM) due to methanol's oxygen content, which facilitates more complete combustion. However, they lead to increased NOx emissions because of the higher combustion temperatures. Overall, while diesel-methanol emulsions improve fuel efficiency and reduce CO HC, and PM emissions, they result in higher NOx emissions (Kumar et al. 2021a).

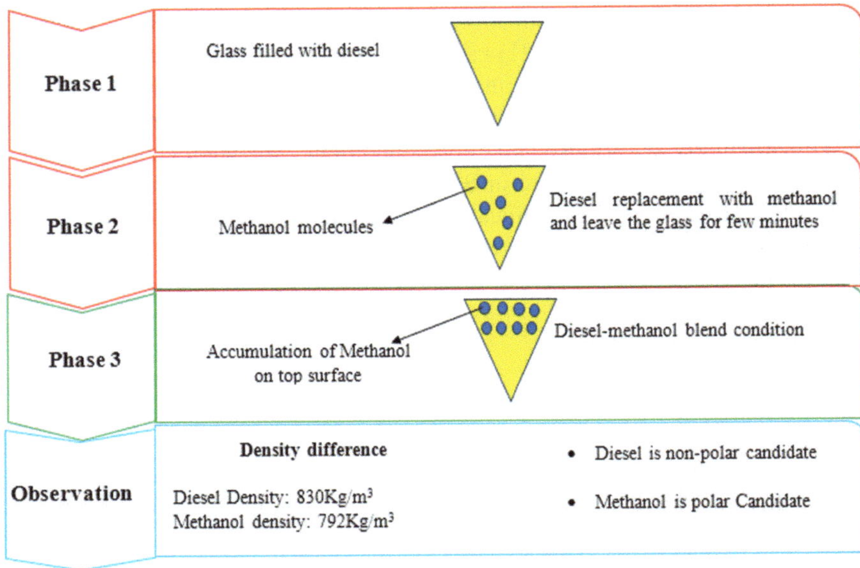

Fig. 12.3 Diesel–methanol emulsion process

Methanol-diesel emulsions face challenges like phase separation, increased NOx emissions, and high maintenance costs. Existing diesel engines and infrastructure also are not designed for these emulsions, which limits their widespread adoption. Debnath et al. (2015) stated that methanol improves peak pressure in diesel-animal fat emulsion and enhances both premixed and diffusion combustion in methanol diesel emulsions, leading to better combustion efficiency. Additionally, methanol helps reduce NO_X and CO emissions. Ghosh et al. (2021) investigated methanol-in-diesel emulsions under high temperature and pressure, examined spray characteristics, the impact of surfactant boiling points on liquid length, and the stability of emulsions at 1500 bar. A novel 1-dodecanol-based microemulsion allows 25% methanol in diesel, showing potential for direct engine use without modifications. Agarwal et al. (2016) studied that methanol-based diesel emulsions, such as M5 (methanol 5% by volume), result in higher methane emissions compared to mineral diesel. The study highlights that methanol increases methane emissions, which poses significant environmental concerns. Valera et al. (2019; Huang et al. 2004; Agarwal et al. 2019) studied a diesel-methanol emulsion is prepared with oleic acid, Isobutanol, n-butanol, and dodecanol. However, more emulsifier altered the chemical composition and combustion process (Huang et al. 2004). It reduced HC, CO, and NOx emissions due to higher oxygen content, hydrogen to carbon ratios, and methanol's latent heat, though engine conditions may affect overall results. In conclusion, from the extensive literature review conducted, it is noticed that this fuel type is not widely used or recommended in CI engines due to challenges regarding phase separation and instability (Ghosh and Ravikrishna 2021), increased engine wear high global warming potential (Agarwal et al. 2016), and the need for significant modifications to existing engine designs.

Table 12.2 provides an overview of the key differences and considerations for using an emulsion-based fuel injection system with methanol in SI and CI engines.

Table 12.2 Emulsion-based fuel injection system with methanol in SI and CI engines

Feature	SI engine (emulsion-based fuel injection)	CI engine (emulsion-based fuel injection)
Fuel type	Gasoline-methanol emulsion	Diesel-methanol emulsion
Ignition method	Spark ignition	Compression ignition
Fuel injection system	Port Fuel Injection / Direct Injection	Common Rail / Direct Injection
Methanol content	Up to 20–30%	Up to 10–20%
Emulsifier requirement	Yes	Yes
Combustion efficiency	Improved with homogeneous mixture	Improved due to better atomization
Engine modifications	Minor adjustments in fuel system	Significant modifications to fuel system
Power output	Comparable to pure gasoline	Slight reduction compared to pure diesel
Emission characteristics	Lower NOx and CO emissions	Lower particulate matter and NOx emissions
Knock resistance	Higher due to methanol's high-octane number	Improved due to methanol's cooling effect
Lubrication	Requires lubrication additives	Potentially more demanding on lubrication
Cold start	More challenging in colder climates	More challenging without proper additives
Fuel storage	Requires anti-corrosive measures	Requires anti-corrosive measures
Engine durability	Potential for corrosion	Higher risk of corrosion and wear
Operational cost	Potentially lower with cheaper methanol	Higher due to maintenance and fuel system
Applications	Light-duty vehicles, motorcycles	Heavy-duty vehicles, marine engine
Market adoption	Limited, mainly in research and niche markets	Very limited, primarily in experimental stages

12.3 Impact of Fuel Injection System on Performance of Methanol-Fueled Engines

12.3.1 Port Fuel Injection of Methanol Blends

In the PFI method for methanol and DI of pilot diesel approach, methanol is added to the intake air stream, while diesel is injected directly into the combustion chamber. The methanol, injected through the port, contributes to the engine's power output. This process can be categorized into major and minor fraction PFI: major fraction refers to using more than 50% methanol, while minor fraction involves less than 50% methanol. This approach provides a quick way to incorporate methanol into diesel engines, leading to smoother operation and allowing for a larger quantity of methanol to be used. However, it requires a control system to regulate the quantities and timing of methanol and diesel injections, and it also adds to the system's overall weight due to the extra fuel injection equipment required for methanol.

12.3.1.1 Impact of Injection Pressure

Injection pressure is a crucial parameter influencing the performance of methanol-fueled engines. It directly affects combustion efficiency, fuel consumption, brake power, and emissions by controlling the atomization of methanol fuel. In particular, injection pressure is critical for the atomization of methanol fuel. Adequate injection pressure is essential for effective atomization of the fuel, ensuring a uniform and well-mixed air–fuel mixture. High injection pressures significantly improve fuel atomization, resulting in smaller droplets that mix more uniformly with air. This enhanced mixing of methanol with air creates a more consistent air–fuel mixture. In contrast, lower injection pressures produce larger fuel droplets, which may not mix as thoroughly with air. This can lead to incomplete combustion and higher engine emissions. Studies have examined a wide range of injection pressures to optimize engine performance. For example, in a Port Fuel Injection system, researchers have explored pressures ranging from 3 to 10 MPa for effective atomization of methanol and optimal air–fuel mixing. Wei et al. (2023) reported that increasing injection pressure from 5 to 15 MPa in a direct injection system improved BTE to 42.97% and reduced BSFC by enhancing combustion efficiency. Similarly, investigations by Liu et al. (2015) revealed that BSFC of diesel/ methanol dual-fuel (DMDF) mode decreases with the increase of injection pressure, and is lower than that of diesel mode for injection pressure over 115 MPa. Both of NO_X and smoke emissions are reduced in DMDF mode. Moreover, an increase in injection pressure from 70 to 130 MPa, leads to decrease in HC emission of DMDF mode by 49.11%. These findings highlight that injection pressure should be carefully optimized, as higher pressures generally lead to reduced ignition delay, shorter combustion duration, and increased power output. Higher injection pressures generally lead to increased in-cylinder pressures and temperatures, influencing both the ignition delay and the combustion duration. Increased injection pressure reduces ignition delay by creating a more readily combustible mixture. Additionally, higher injection pressures shorten the combustion duration, leading to a more rapid release of energy and higher peak pressures (Khalid and Manshoor 2012). This accelerated rate of heat release improves engine efficiency and increases power output. Higher injection pressures can boost engine power by enhancing the combustion process. This leads to more efficient methanol combustion, increased thermal efficiency, and more effective energy conversion. Enhanced atomization and combustion can also boost torque output, particularly at lower engine speeds. Fuel efficiency is another critical parameter affected by injection pressure. Higher pressures contribute to better fuel efficiency by ensuring more complete combustion of methanol (Yao et al. 2017). This optimization lowers BSFC by reducing fuel wastage and enhancing thermal efficiency, thereby converting more of the fuel's chemical energy to useful work.

12.3.1.2 Impact of Injection Timing

Injection timing, or the precise moment fuel is introduced into the combustion chamber, significantly impacts combustion efficiency, power output, fuel consumption, and emissions. It can be categorized into three primary phases: early injection, middle injection, and late injection. Injection timing is typically classified into three phases: early injection (e.g., 30°–20° BTDC), optimal or mid-range injection (typically around 10°–5° BTDC), and late injection (after 5° BTDC). Methanol combustion, due to the fuel's high volatility and high latent heat of vaporization, responds notably to variations in these timings.

When injection occurs early in the compression stroke, fuel–air mixing has more time to occur. This can potentially enhance combustion efficiency by improving homogeneity, but the downside includes increased heat losses to the cylinder walls due to prolonged exposure, and the risk of pre-ignition due to methanol's low auto-ignition temperature. Studies, such as those by Li et al. (2022b), show that early injection timing can lead to lower thermal efficiency and power output due to these heat losses and the potential for premature combustion. In addition, the cooling effect of methanol vaporization can increase charge density, providing a power boost if pre-ignition is managed properly (Sayin et al. 2009). Injecting fuel closer to top dead center (TDC), typically around 8° BTDC, achieves a balance between thorough fuel–air mixing and an optimal combustion start. This timing allows combustion to occur during the expansion stroke, maximizing the pressure exerted on the piston and improving engine efficiency. A study by Xie et al. (2013) suggests that injection at 8° BTDC for methanol engines provides the highest combustion efficiency and power output, as it avoids early combustion-related heat losses while still achieving near-complete combustion. Late injection results in less time for charge mixing, which can lead to incomplete combustion. As reported by Canakci et al. (2008), injecting methanol at 2° BTDC or later often results in higher levels of HC and CO emissions, due to incomplete combustion. Though late injection can mitigate knocking and pre-ignition issues, it sacrifices efficiency and power output. This is especially relevant for methanol, which requires sufficient mixing time to vaporize fully. Regarding power output, early injection can lead to power losses due to increased heat losses and potential knocking. However, methanol's high latent heat of vaporization can cool the charge, making the mixture denser and potentially increasing power output if managed properly. Mid injection generally maximizes power output by ensuring peak pressure occurs shortly after TDC, optimizing leverage for the piston during the power stroke (Li et al. 2021a). In contrast, late injection typically results in lower power output due to less efficient combustion and reduced peak pressure. In terms of fuel consumption, early injection can lead to higher fuel consumption due to increased heat losses and the necessity for richer mixtures to avoid pre-ignition (Canakci et al. 2008). Mid injection optimizes fuel consumption by ensuring efficient combustion and maximizing energy extraction from the fuel. Conversely, late injection often leads to higher fuel consumption because of incomplete combustion and the requirement for extra fuel to reach the desired power output (Canakci et al. 2008).

12.3.1.3 Impact of Multiple Injections

Multiple injection strategies have emerged as a key approach for optimizing methanol-fueled engines, as they allow better control of the combustion process. Rather than injecting fuel in a single event, multiple injections distribute fuel into the combustion chamber over several stages. This technique involves three primary phases: pilot injection, main injection, and post injection. Each phase plays a distinct role in optimizing combustion performance, improving emissions, and ensuring engine reliability. Pilot injection introduces a small amount of fuel before the main injection event (Kim et al. 2024). Its primary objective is to initiate combustion in a controlled manner, promoting more homogeneous mixing and reducing the pressure rise rate. This has been shown to lower peak cylinder temperatures, which directly impacts the formation of NOx (Kim et al. 2024). Studies by Wei et al. (2017) demonstrate that pilot injection timings between 15° and 10° BTDC help to stabilize combustion in methanol engines by reducing ignition delay and promoting smoother pressure development. For methanol, which has a high latent heat of vaporization, pilot injection plays an essential role in enhancing charge cooling, increasing air density, and improving the vaporization of methanol fuel. This results in a denser fuel–air mixture, which helps optimize power output while maintaining controlled combustion (Tao et al. 2022). The main injection provides the bulk of the fuel for combustion and is critical for determining engine performance and emissions. Optimal timing of this phase is crucial for achieving peak cylinder pressure shortly after top dead center. According to Ruan et al. (2021), main injection timing around 5°–8° BTDC in methanol-fueled engines leads to improved combustion efficiency and reduced HC emissions. This is because it allows sufficient mixing time for the fuel to vaporize and mix with air, while also aligning the combustion event with the piston's downstroke to maximize power extraction.

Methanol, being a low-carbon fuel, tends to produce fewer particulate emissions. However, multiple injections can further reduce PM formation, as the finer atomization of fuel during both pilot and main injections ensures a more thorough and complete combustion process (Panda et al. (2021). Methanol's lower carbon content compared to diesel also minimizes soot production, a common issue in conventional diesel engines (Svensson et al. 2016). Post injection involves injecting a small quantity of fuel after the main combustion event, typically during the expansion stroke (Gaur et al. 2024). This phase is primarily used for emissions control, specifically targeting soot oxidation and reducing particulate matter formation. In methanol engines, post injection can also serve to raise exhaust gas temperatures, which is beneficial for after treatment systems such as selective catalytic reduction (SCR) or diesel particulate filters (DPFs) (Chen et al. 2023). A study by Shubham et al. (2014) suggests that post injection at around 10°–15° ATDC helps optimize the trade-off between NOx and PM emissions, particularly in lean-burn conditions.

Multiple injection strategies offer significant advantages in controlling combustion efficiency and emissions. By breaking down the fuel delivery process into smaller stages, multiple injections provide better control over air–fuel mixing and combustion phasing, allowing for finer adjustment of combustion pressure and temperature (Yin et al. 2019). This approach leads to more complete combustion, as demonstrated by Zhen et al. (2015), who found that multiple injection strategies improved thermal efficiency in methanol-fueled engines by 5–10% compared to single injections. From an emissions standpoint, multiple injections play a crucial role in managing both NOx and particulate emissions (Han 1996). Pilot injection reduces the combustion rate, preventing excessive NOx formation, while post injection enhances soot oxidation, reducing particulate emissions (Gambhir et al. 2018). Peterson et al. (2012) showed that using a dual-pilot injection system with methanol engines reduced NOx emissions by up to 20% compared to single injections, without compromising fuel efficiency. Multiple injections contribute positively to engine durability by reducing peak cylinder pressures and distributing combustion more evenly across the power stroke. This reduces mechanical stress on engine components, such as pistons and cylinder heads, thereby extending their lifespan (Gallas et al. 2018). For methanol-fueled engines, which are prone to higher wear due to methanol's lower lubricity compared to diesel, the staged combustion process helps to maintain controlled combustion temperatures, mitigating excessive wear on critical components (Ijaz et al. 2021).

Furthermore, multiple injections improve injector durability by limiting deposit formation and maintaining optimal temperatures during combustion (Roth 2005). This is especially critical in methanol engines, where material compatibility is a concern due to methanol's corrosive nature. Compatible materials for the fuel system, including injectors and seals, must be selected to prevent degradation over time, as reported by Kumar et al. (2020, 2019). While the benefits of multiple injection strategies are significant, they require advanced control systems and precise injector design to function effectively. The timing and quantity of each injection phase must be accurately managed to avoid unwanted effects, such as increased emissions or combustion instability. For methanol-fueled engines, cold start performance is another challenge

(Hu et al. 2007). Methanol's lower volatility makes it harder to vaporize in cold conditions, which can lead to incomplete combustion or misfires during engine startup. However, research by Gong et al. (2021) indicates that optimized pilot injection timing can mitigate cold start issues by improving fuel atomization and vaporization.

12.3.2 Direct Injection of Methanol Blends

Direct injection (DI) of methanol blends into internal combustion engines has become a focal point for optimizing performance, efficiency, and emissions. This section explores the impact of various direct injection parameters on methanol-fueled engines, with a focus on injection pressure, injection timing, and multiple injections.

12.3.2.1 Impact of Injection Pressure

Injection pressure plays a crucial role in the performance of methanol-fueled engines, affecting both atomization and combustion efficiency. Increased injection pressures result in finer atomization of the methanol blend, improving vaporization and creating a more uniform air–fuel mixture. It has been reported that M15 (methanol 15% by volume) has been adopted commercially in many countries (Agarwal et al. 2021). Hence, the analysis is done on M15 blended fuel. Figure 12.4 shows the variations of BSFC and BTE with injection pressure of different authors for direct injection M15 methanol blends (Sayin et al. 2010, 2009; Liu et al. 2015; Ruan et al. 2021; Sumanlal et al. 2019; Gumus et al. 2010; Tian et al. 2022). This improved atomization reduces the likelihood of incomplete combustion, thereby increasing engine power output and BTE. Moreover, improved atomization stabilizes the combustion process, reducing problems like knock or pre-ignition and enabling more aggressive adjustments to parameters such as boost pressure and ignition timing. However, higher injection pressures might also result in increased specific fuel consumption because methanol has a lower energy density than gasoline. Balancing injection pressure with fuel consumption is crucial for optimizing engine performance.

12.3.2.2 Impact of Injection Timing

Injection timing is another crucial factor influencing methanol-fueled engines. The timing of fuel injection relative to the engine's intake and compression strokes has a significant effect on combustion characteristics and overall engine performance. Figure 12.5 shows the variations of BSFC and BTE with injection timing of different authors for direct injection M15 methanol blends (Sayin et al. 2010, 2009; Ruan, et al. 2021; Sumanlal, et al. 2019; Gumus et al. 2010; Tian et al. 2022). It can be observed that advancing injection timing, by injecting fuel earlier in the intake

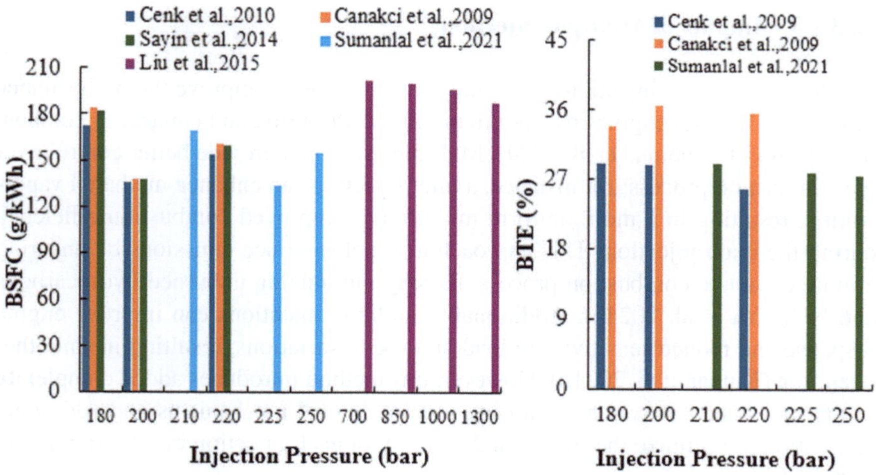

Fig. 12.4 BSFC, and BTE vs Injection pressure for direct injection of methanol blends

stroke, can enhance the mixing of methanol with air, leading to improved combustion. However, excessively early injection may result in increased heat loss and diminished BTE. The optimal injection timing typically involves injecting methanol just before or during the intake stroke to ensure sufficient vaporization and mixing. Conversely, delaying injection timing may enhance power output by allowing for a more complete combustion process but can increase the risk of knocking and higher exhaust temperatures, necessitating careful calibration to prevent adverse effects on engine longevity.

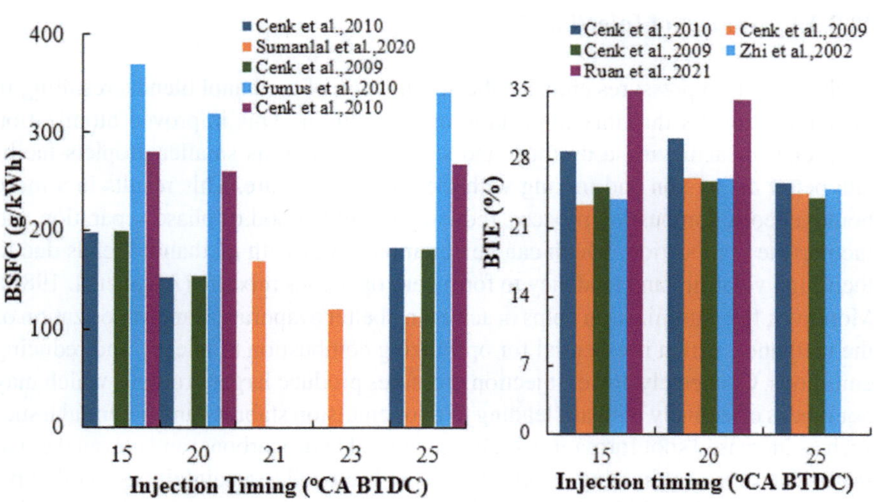

Fig. 12.5 BSFC, and BTE vs Injection timing for direct injection of methanol blends

12.3.2.3 Impact of Multiple Injections

Employing multiple injections per cycle can significantly improve the performance of methanol-fueled engines by optimizing fuel distribution and enhancing combustion dynamics (Ahamad et al. 2024). Multiple injections provide better control over the combustion process; for instance, a pilot injection can enhance methanol vaporization, resulting in a more uniform mixture and improved combustion efficiency during the main injection. This approach also helps reduce emissions by ensuring a more complete combustion process, thereby minimizing unburned hydrocarbons and NO_x (Yu et al. 2024b). Additionally, multiple injections can improve engine response and reduce sensitivity to load and speed variations, resulting in smoother operation (Kumar et al. 2021b). However, this method introduces added complexity to the fuel injection system, requiring precise control mechanisms and additional calibration to optimize the timing and quantity of each injection event (Kumar and Ashok 2024).

12.3.3 Emulsion of Methanol Blends

Emulsions of methanol blends have emerged as a promising approach for enhancing the performance, efficiency, and emissions characteristics of internal combustion engines. This section delves into the impact of various emulsion-related parameters on methanol-fueled engines, emphasizing the roles of water content, surfactant concentration, and fuel preparation techniques.

12.3.3.1 Impact of Injection Pressure

Higher injection pressures enhance the atomization of methanol blends, resulting in finer fuel droplets that mix more thoroughly with air. This improved atomization is crucial for achieving a uniform and stable emulsion, as smaller droplets facilitate better dispersion and mixing with the air–fuel mixture. This results in a more homogeneous combustion process, reducing the likelihood of phase separation and incomplete combustion, which can be common issues with methanol blends due to their high volatility and tendency to form heterogeneous mixture (Aigal et al. 1986). Moreover, finer atomization helps in achieving better evaporation and vaporization of the methanol, which is essential for optimizing combustion efficiency and reducing emissions. Conversely, lower injection pressures produce larger droplets, which may not mix as effectively with air, leading to poor emulsion stability and potential issues such as increased soot formation, higher unburned hydrocarbons, and elevated emissions. Therefore, optimizing injection pressure is crucial for maintaining an effective emulsion of methanol blends, enhancing combustion performance, and minimizing environmental impact (Xu et al. 2022).

12.3.3.2 Impact of Injection Timing

Injection timing significantly impacts the emulsion of methanol blends, influencing their combustion behavior and efficiency. Early injection, which involves introducing fuel into the combustion chamber before the compression stroke, provides more time for mixing methanol with air. This can lead to better atomization and a more homogeneous emulsion, improving the vaporization of methanol and reducing the risk of phase separation. However, early injection might also result in increased heat losses and a potential risk of pre-ignition due to methanol's high volatility (Wang et al. 2024b). Conversely, middle injection, which occurs around the top dead center (TDC) or just before the combustion phase, optimizes the balance between fuel–air mixing and the start of combustion. This timing typically results in a more stable emulsion, enhanced fuel atomization, and better combustion efficiency, as it allows for effective energy release during the power stroke. Late injection, where fuel is introduced after the optimal combustion phase, can lead to poorer emulsion stability. The reduced time for mixing means that larger fuel droplets may persist, resulting in incomplete combustion, higher emissions, and reduced thermal efficiency. Late injection may also reduce the risk of pre-ignition and knock but often at the expense of emulsion quality and overall engine performance. Therefore, precise control of injection timing is essential to maintain an effective emulsion of methanol blends, ensuring optimal combustion characteristics and reducing environmental impact (Sayin et al. 2009). Wang et al. (2024b) discusses the impact of adding 20% water to methanol on engine performances in Diesel-Methanol Diffusion Combustion (DMDC) and Diesel-Hydrous Methanol Diffusion Combustion (DHMDC) modes. Figure 12.6 shows the impact of water addition and Methanol Injection Timing (MIT) on Indicated Thermal Efficiency (ITE). The DHMDC mode exhibits lower ITE compared to the DMDC mode. ITE increases initially with advanced MIT due to higher heat release but decreases with further advancement due to inhibited diesel combustion and methanol ignition. Water addition suppresses temperature rise, dilutes oxygen, and accelerates combustion, affecting efficiency. At 100% load, better water atomization and evaporation dominate, slightly increasing Diesel Replacement (DR) indicated thermal efficiency. Overall, the effect of water in methanol is very sensitive to in-cylinder temperature (Wang et al. 2024b).

12.3.3.3 Impact of Multiple Injections

Multiple injection strategies in methanol-fueled engines can significantly influence the emulsion of methanol blends, affecting both performance and emissions. In a multi-injection system, methanol is injected into the combustion chamber at multiple intervals during the intake or combustion stroke, rather than a single, continuous injection. This approach provides improved control over the air–fuel mixture and enhances methanol atomization, leading to better dispersion in the mixture and

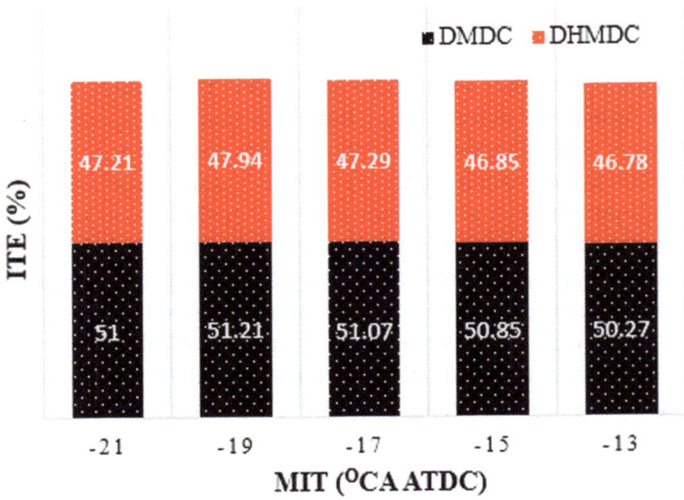

Fig. 12.6 Impact of water addition and MIT on ITE

a more uniform emulsion. Improved atomization leads to more efficient combustion and better thermal management, reducing incomplete combustion and minimizing particulate emissions. Furthermore, multiple injections can assist in handling methanol's high latent heat of vaporization, which might otherwise cause problems such as cylinder wall wetting and inadequate vaporization. By strategically timing the injections, the engine can maintain optimal combustion temperatures and pressures, which helps to lower emissions and enhance overall efficiency of the engine. However, the complexity of multiple injection strategies requires precise control and calibration to avoid potential drawbacks such as increased fuel consumption or engine knocking. Properly implemented, multiple injection can therefore significantly enhance the performance and environmental benefits of methanol blends in internal combustion engines (Yin et al. 2023). Figure 12.7 shows the different injection positioning and its corresponding change in thermal efficiency under different methanol substitution rate. Injecting methanol at the pre-intercooler position under high load reduces intake air charge due to higher turbocharged temperatures, leading to incomplete combustion and lowered ITE (Xu et al. 2022). At a 30% methanol substitution rate, oxygen content increases, enhancing combustion. Post-intercooler injection under high load also promotes combustion due to oxygen-rich methanol but achieves lower ITE compared to intake manifold injection. Under high-load conditions at the intake manifold, methanol evaporation and mixing are promoted, which improves combustion and ITE (Xu et al. 2022).

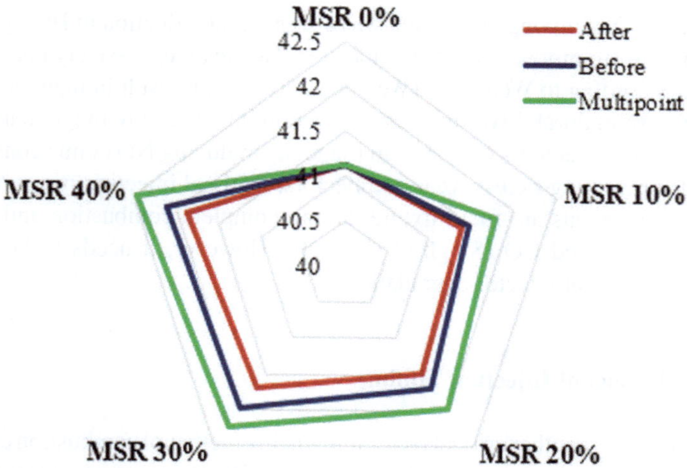

Fig. 12.7 Different injection positions indicate the change of thermal efficiency

12.4 Impact of Fuel Injection System on Emission of Methanol-Fueled Engines

The fuel injection system plays a pivotal role in controlling the emission profile of methanol-fueled engines. The influence of various injection parameters—such as injection pressure, timing, and strategies like multiple injections—on emissions needs to be understood to optimize performance while minimizing pollutants like HC, CO, and NOx. This section elaborates on these influences with supporting literature and key findings.

12.4.1 Port Fuel Injection of Methanol Blends

Port Fuel Injection (PFI) systems are critical in optimizing the performance and emissions of methanol-fueled engines. This section examines how different PFI parameters, such as injection pressure, injection timing, and multiple injections, affect the emissions of methanol blends.

12.4.1.1 Impact of Injection Pressure

The injection pressure significantly affects atomization and vaporization, which are crucial for methanol due to its high latent heat of vaporization. A study by Radzali et al. (Radzali et al. 2020) indicated that increasing injection pressure from 0.1 to 0.3 MPa enhanced atomization quality, producing smaller fuel droplets and improving air–fuel mixing. The result was a significant reduction in HC by 20% and CO by 15% due to more complete combustion. However, excessively high injection pressures, according to Wen et al. (Wen et al. 2023), can result in injector wear and affect fuel system durability. They reported an optimal pressure range that balanced atomization improvements with system longevity, reducing NOx emissions by up to 10% while preventing excess wear on injectors. Optimal injection pressure ensures a more homogeneous air–fuel mixture, aiding complete combustion and reducing emissions of HC and CO (Singh et al. 2020). However, it needs to be carefully controlled to prevent injector degradation.

12.4.1.2 Impact of Injection Timing

Injection timing is another critical factor influencing methanol combustion emissions. Early injection timing, as suggested by Park et al. (2013), improves vaporization and air–fuel mixing. In their study, early injection timings (15° before the intake valve closing) reduced HC emissions by 18% and CO emissions by 12%, as more fuel had time to vaporize and mix before combustion. However, excessively early injection led to higher intake manifold wall wetting, causing a rise in HC emissions. Late injection timing, on the other hand, can help control combustion temperatures and NOx emissions. According to Panda et al. (2021), injecting methanol closer to the combustion phase reduced NOx by 30% due to cooler combustion temperatures while maintaining adequate combustion efficiency. The optimal injection timing must balance early fuel atomization with the risk of higher NOx emissions (Chen et al. 2024). Proper control strategies can ensure that both early and late injection timings enhance the overall emission profile of methanol-fueled engines.

12.4.1.3 Impact of Multiple Injections

Multiple injection strategies involve injecting methanol at several points throughout the intake or combustion cycle. This technique enhances the distribution and atomization of the fuel, promoting a more homogeneous air–fuel mixture (Panda and Ramesh 2021). The primary advantage of multiple injections is the improved control over the combustion process, which can lead to more efficient burning of methanol and reduced emissions (Yu et al. 2024b). By splitting the fuel into multiple pulses, the engine can better manage the high latent heat of vaporization associated with methanol, reducing the risk of incomplete combustion and minimizing HC and CO emissions (Dong et al. 2020; Li et al. 2022b; Aziz et al. 2020). Li et al. (2021b) conducted experiments using double injection strategies, reporting a 25% reduction in HC and a 20% reduction in CO emissions compared to single injections, as the split injections allowed more time for methanol to vaporize fully. Additionally, multiple injections can help in managing engine knock and optimizing combustion temperatures, further reducing NO_x emissions (Panda and Ramesh 2021). However, implementing multiple injections requires precise control mechanisms and sophisticated engine management systems to ensure that the benefits outweigh the complexities and potential drawbacks, such as increased fuel consumption or system cost (Badami et al. 2003). In summary, parameters of the Port Fuel Injection system, including injection pressure, timing, and multiple injections, are essential in shaping the emission characteristics of methanol-fueled engines. Properly optimized, these factors can significantly enhance combustion efficiency and emission control, making methanol blends a more viable alternative fuel option.

12.4.2 Direct Injection of Methanol Blends

Direct injection (DI) of methanol blends into ICEs has gained considerable interest because of its potential to enhance efficiency and lower emissions. This section examines how various parameters related to direct injection systems, such as injection pressure, injection timing, and multiple injections, affect the emissions of methanol-fueled engines.

12.4.2.1 Impact of Injection Pressure

Injection pressure is pivotal in determining both the performance and emission characteristics of methanol-fueled engines. Figure 12.8 shows the variations of engine emissions with injection pressure of different authors for direct injection M15 methanol blends (Sayin et al. 2010, 2009; Ruan et al. 2021; Sumanlal et al. 2019; Gumus et al. 2010; Tian et al. 2022). Higher injection pressures usually result in a finer atomization of the fuel, which promotes more thorough combustion. This improved atomization enhances the mixing of methanol with air, leading to more efficient combustion and lower emissions of HC and CO. Additionally, higher injection pressures can help mitigate issues related to incomplete combustion and carbon deposition, which are common in methanol engines. Conversely, excessively high injection pressures can result in higher NOx emissions because of the increased combustion temperatures (McTaggart-Cowan et al. 2005). Therefore, optimizing the injection pressure is essential to balance improved combustion efficiency and minimized NO_x emissions. The effect of injection pressure on emissions varies significantly based on the particular engine design and the conditions under which it operates.

12.4.2.2 Impact of Injection Timing

Injection timing, which refers to the point in the engine cycle when the fuel is introduced, significantly influences combustion characteristics and emissions. Figure 12.9 shows the variations of engine emissions with injection timing of different authors

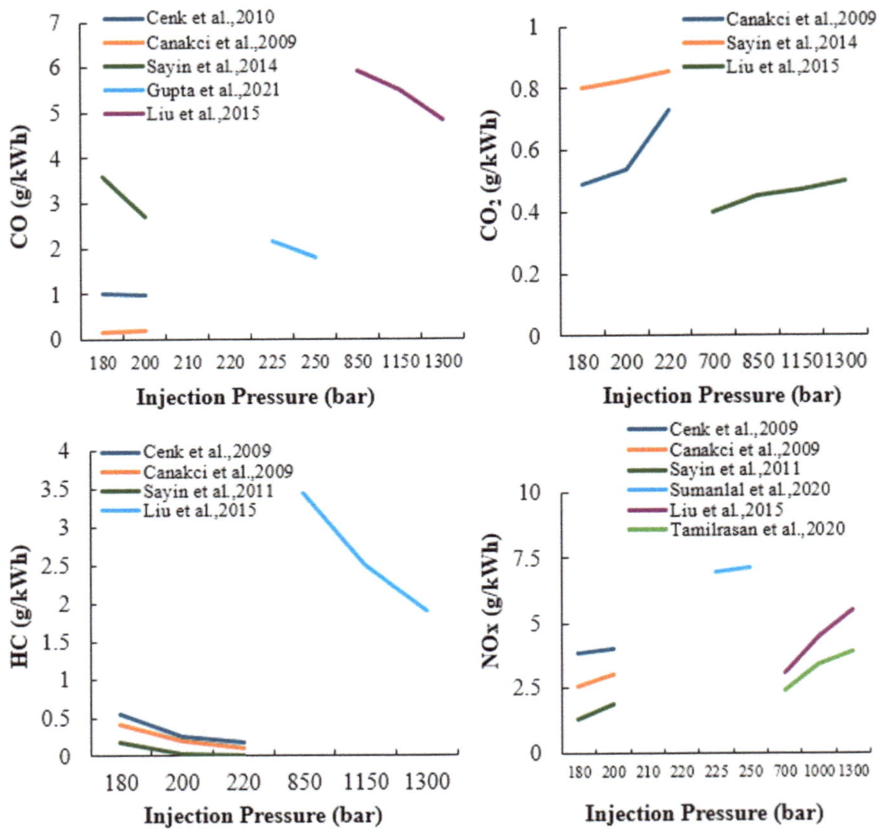

Fig. 12.8 Impact of injection pressure on engine emissions for DI of methanol blends

for direct injection M15 methanol blends (Sayin et al. 2010, 2009; Ruan, et al. 2021; Sumanlal, et al. 2019; Gumus et al. 2010; Tian et al. 2022). In methanol-fueled engines, advancing the injection timing generally leads to improved combustion efficiency. Injecting methanol early improves its mixing with intake air, creating a more uniform air–fuel mixture and better combustion. This process can help decrease CO and HC emissions. However, advancing the injection timing too much can lead to increased NO_x emissions due to higher combustion temperatures. This may also negatively impact engine performance, leading to lower power output and higher fuel consumption. Consequently, it is crucial to carefully calibrate the injection timing to strike a balance between lowering HC and CO emissions while managing NOx emissions effectively (Canakci et al. 2008).

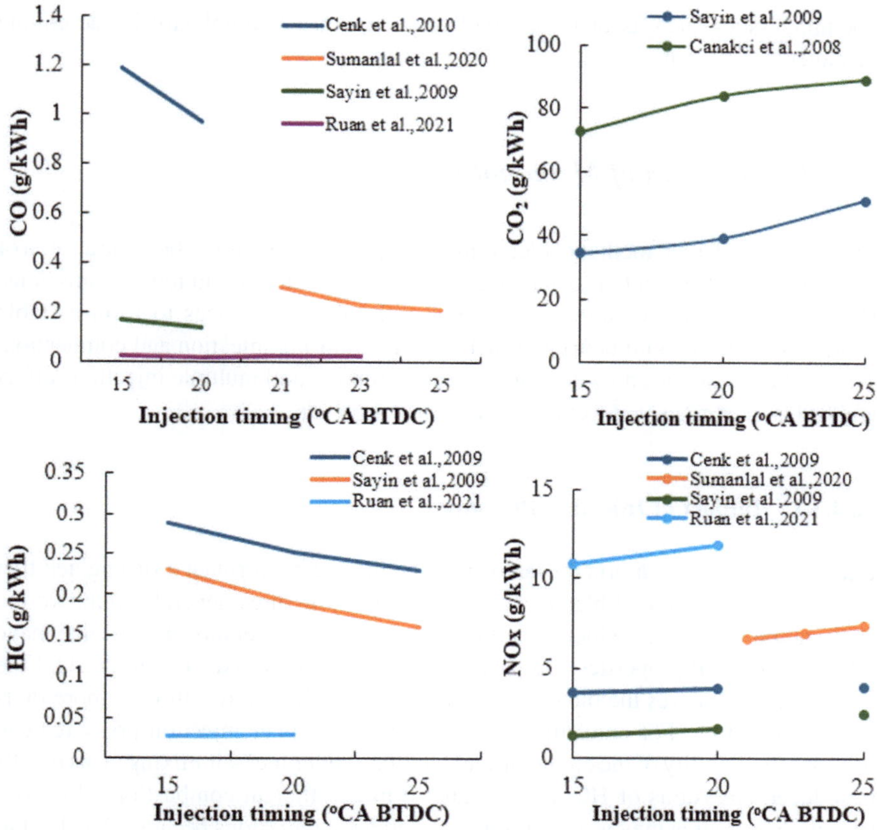

Fig. 12.9 Impact of injection timing on engine emissions for DI of methanol blends

12.4.2.3 Impact of Multiple Injections

The strategy of employing multiple injections, where the fuel is injected at several points during the engine cycle, can significantly influence emissions in methanol-fueled engines. Multiple injections can improve the overall combustion process by enhancing the mixing of methanol with air, which leads to more complete combustion and reduced emissions of HC and CO (Han et al. 2020). One of the main advantages of multiple injections is the ability to control combustion phasing more precisely. For instance, splitting the fuel injection into a main injection and a pilot injection can help achieve a more stable and efficient combustion process. The pilot injection assists in lowering both peak pressure and temperature, which in turn helps decrease NO_x emissions, while the main injection ensures complete combustion of the methanol (Aziz et al. 2020). Additionally, multiple injections can help manage engine knocking and improve engine performance under various operating conditions. However, this approach requires sophisticated control strategies and careful calibration to avoid potential issues such as increased fuel consumption or complexity in the engine management system.

12.4.3 Emulsion of Methanol Blends

The use of emulsified methanol blends in fuel injection systems has become an area of interest due to its potential to enhance combustion efficiency and reduce emissions. Emulsification involves mixing methanol with other substances to form a stable emulsion, which can influence the fuel's behavior during injection and combustion. This section examines how injection pressure, timing, and multiple injections affect emissions in engines using emulsified methanol blends.

12.4.3.1 Impact of Injection Pressure

Injection pressure is a crucial factor influencing the performance of engines that use emulsified methanol blends. Higher injection pressures generally improve the atomization of the fuel, which is particularly important for emulsions that may have different physical properties compared to pure methanol (Saxena et al. 2021). Fine atomization improves the blending of the emulsion with air, resulting in more thorough combustion. For emulsified methanol blends, higher injection pressures can improve the stability of the emulsion and facilitate better fuel–air mixing. This results in reduced emissions of HC and CO due to more efficient combustion. However, very high pressures may also result in elevated NOx emissions because they lead to higher combustion temperatures. Therefore, optimizing injection pressure is essential to balance improved atomization with controlled NO_x emissions (Sayin et al. 2009).

12.4.3.2 Impact of Injection Timing

Injection timing affects how the emulsified methanol blend interacts with the combustion process. For emulsions, the timing of the fuel injection must be carefully controlled to ensure that the emulsion is adequately mixed with the air before combustion begins. Early injection timing generally allows for better mixing and more complete combustion, which can lead to lower HC and CO emissions. However, advancing injection timing too much can result in increased combustion temperatures, leading to higher NO_x emissions (Khan and Gollahalli 1981). The unique properties of emulsified methanol blends can alter the combustion characteristics, making it crucial to calibrate the injection timing to optimize both emissions and performance (Wu et al. 2015). Proper adjustment ensures that the emulsion is effectively utilized, reducing Total Hydrocarbon (THC) and CO while keeping NO_x emissions in check (Vigneswaran et al. 2015). Figure 12.10 shows the impacts of water addition and change in methanol injection timing on CO, THC, and NO_x emissions. The advanced injection results in an increased trend of NOx emissions and reduced HC and CO emissions due to better combustion. Adding water increases CO emissions due to incomplete combustion from local anoxic regions, which decrease with higher engine load and initially with advancing Methanol Injection Timing (MIT) before increasing (Wang et al. 2024b). DHMDC mode has higher THC emissions due to local low temperatures and oxygen dilution, which decrease with higher engine load and initially with advancing MIT before increasing. Water addition reduces NOx emissions by up to 40% by diluting oxygen and accelerating combustion. Higher CO emissions in DHMDC also help reduce NOx. Advancing MIT increases NOx emissions due to higher temperatures but aids oxygen dilution. Additionally, under full engine load, the DHMDC mode enhances both efficiency and emissions relative to traditional diesel combustion. It boosts the ITE by 0.55% and decreases CO and NOx emissions by 23.19% and 67.54%, respectively, as illustrated in Fig. 12.11.

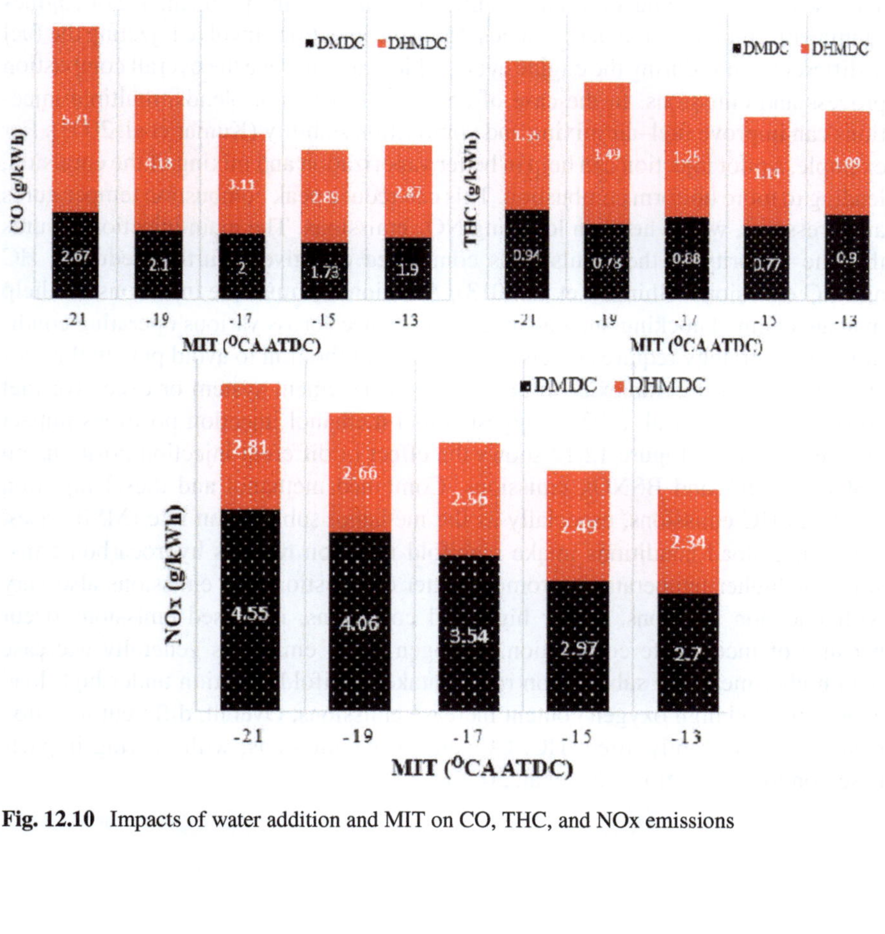

Fig. 12.10 Impacts of water addition and MIT on CO, THC, and NOx emissions

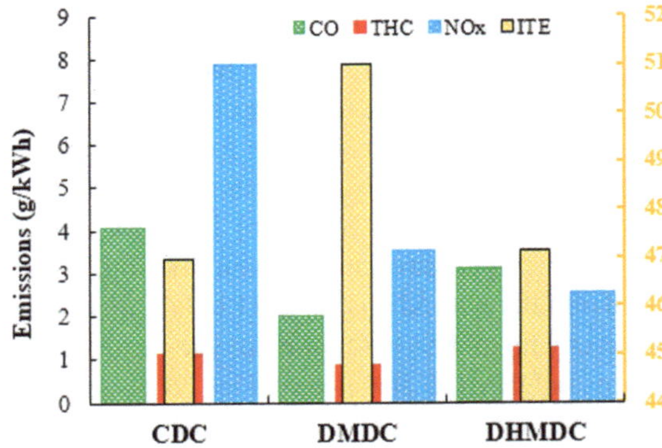

Fig. 12.11 Comparison of three combustion modes at 100% load

12.4.3.3 Impact of Multiple Injections

The use of multiple injections can significantly enhance the performance of engines running on emulsified methanol blends. Multiple injections involve injecting the fuel at different points during the engine cycle, which can improve the overall combustion process and emissions. In the case of emulsified methanol blends, multiple injections can improve fuel–air mixing and combustion stability (Kumar et al. 2005). For example, a pilot injection can help in better vaporization and mixing of the emulsion, leading to more uniform combustion. This can reduce peak combustion temperatures and pressures, which helps in lowering NO_x emissions. The main injection ensures that the majority of the emulsion is combusted effectively, further reducing HC and CO emissions (Bhimani et al. 2013). Additionally, multiple injections can help manage engine knocking and enhance performance across various operating conditions. This strategy requires precise control and calibration to avoid potential issues such as increased complexity in the engine management system or excessive fuel consumption. Xu et al. (2022) suggested that methanol injection positions impact engine emissions. Figure 12.12 shows the effect of different injection positions on BSCO, BSHC, and BSNOx emissions. Combined methanol and diesel injection increases HC emissions, especially as the methanol substitution rate (MSR) rises. Under high-load conditions intake manifold injection reduces hydrocarbon emissions, as higher temperatures promote better combustion. CO emissions also vary with injection positions. Under high-load conditions, increased emissions occur because of incomplete combustion. Nitrogen oxide emissions generally decrease with higher methanol substitution rates. Intake manifold injection under high-load conditions and high oxygen content increase emissions. Overall, different injection positions significantly affect HC, CO, and NO_x emissions, with varying impacts based on load conditions (Xu et al. 2022).

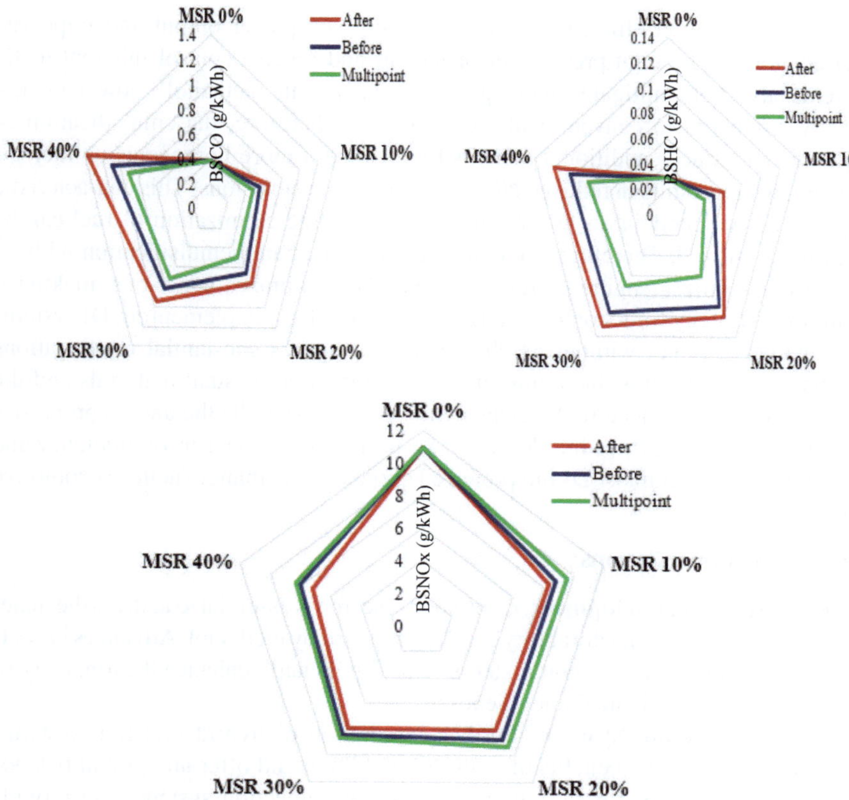

Fig. 12.12 Impact of different injection positioning on BSCO, BSHC, and $BSNO_x$

12.5 Conclusion and Future Recommendations

This chapter delves into the exploration of innovative fuel injection techniques designed to enhance the utilization of methanol in internal combustion engines. As a promising alternative fuel, methanol offers several notable advantages, such as a higher octane number, cleaner combustion profiles, and potential cost-effectiveness. However, the successful integration of methanol into existing engine technologies necessitates a thorough examination of fuel injection strategies to fully harness its benefits. To achieve this objective, three primary fuel injection methods were investigated: Port Fuel Injection (PFI), Direct Injection (DI), and dual injection systems, assessing their respective performance and emissions impacts when using methanol blends and emulsified methanol. Among these, Direct Injection (DI) has emerged as the most advantageous for locomotive engines. The DI systems stand out due to their ability to deliver superior atomization and precise fuel delivery, which are critical for maximizing the performance and efficiency of the large engines typically used in locomotives. When methanol blends are employed, DI systems significantly enhance combustion efficiency, leading to increased power output and improved fuel economy. The high-pressure environment and precise control inherent to DI systems also help mitigate challenges associated with methanol's lower energy density compared to conventional diesel fuel. Furthermore, the emulsification of methanol introduces additional benefits by creating a more homogeneous fuel–air mixture, which results in reduced NOx and particulate emissions. This characteristic is particularly beneficial in DI systems, where the fine atomization of fuel can be effectively utilized. Therefore, the combination of methanol emulsification with DI not only maximizes engine performance but also minimizes emissions, making it an optimal strategy for locomotive engines. Nevertheless, implementing DI systems with methanol is not without its challenges. It requires substantial modifications to existing fuel systems, including the use of corrosion-resistant materials and the development of advanced fuel management systems to handle the unique properties of methanol. Despite these hurdles, the long-term benefits in terms of efficiency and emissions reduction make DI the preferred choice for methanol-fuelled locomotive engines.

Future Recommendations

1. **Research and Development**: Continued research is essential to address the material compatibility and durability challenges posed by methanol. Advances in coatings and corrosion-resistant materials can significantly enhance the longevity of engines running on methanol blends.
2. **Hybrid Injection Systems**: Further exploration of hybrid injection systems, which combine the benefits of both PFI and DI, could offer an optimal balance of performance and emissions reduction. Dual injection systems can provide flexibility in fuel management and optimize combustion under varying load conditions.
3. **Fuel Formulation**: Development of advanced methanol emulsions and blends tailored specifically for DI systems can enhance fuel properties and combustion characteristics. Additives that improve lubricity and stability can also be a focus of future studies.
4. **Engine Control Systems**: Innovations in Engine Control Units (ECUs) that can dynamically adjust injection parameters based on real-time feedback can further optimize the use of methanol in ICEs. Intelligent control systems can enhance performance, efficiency, and emissions control.
5. **Pilot Programs**: Implementing pilot programs for methanol-fuelled locomotives can provide valuable real-world data and drive further innovation. Collaboration between industry stakeholders, including fuel suppliers, engine manufacturers, and transportation companies, will be key to the successful adoption of methanol as a mainstream alternative fuel.

In conclusion, the adoption of direct injection systems for methanol utilization in locomotive engines represents a promising pathway toward enhanced performance and reduced emissions. Continued innovation and collaborative efforts will be crucial in overcoming technical challenges and realizing the full potential of methanol as a sustainable fuel for the future.

References

Agarwal AK et al (2016) Unregulated emissions and health risk potential from biodiesel (KB5, KB20) and methanol blend (M5) fuelled transportation diesel engines. Renewable Energy 98:283–291

Agarwal AK et al (2019) Adaptation of methanol–dodecanol–diesel blend in diesel genset engine. J Energy Res Technol 141(10):102203

Agarwal AK et al (2021) Methanol: a sustainable transport fuel for CI engines. Springer Nature

Ahamad J, Kumar P, Dhar A (2024) Effect of multi-injection strategy on characteristics of methanol-fueled direct injection spark ignition engine. Phys Fluids, 36(4)

Aigal A, Pundir B, Khatchian A (1986) High pressure injection and atomization characteristics of methanol. SAE Trans 691–708

Alkidas AC (2007) Combustion advancements in gasoline engines. Energy Convers Manage 48(11):2751–2761

Aziz A et al (2020) Impact of multiple injection strategies on performance and emissions of methanol PPC under low load operation. SAE Technical Paper

Badami M et al (2003) Experimental investigation on the effect of multiple injection strategies on emissions, noise and brake specific fuel consumption of an automotive direct injection common-rail diesel engine. Int J Engine Res 4(4):299–314

Bhimani S et al (2013) Emission characteristics of methanol-in-canola oil emulsions in a combustion chamber. Fuel 113:97–106

Canakci M, Sayin C, Gumus M (2008) Exhaust emissions and combustion characteristics of a direct injection (DI) diesel engine fueled with methanol– diesel fuel blends at different injection timings. Energy Fuels 22(6):3709–3723

Chen SK (2000) Simultaneous reduction of NOx and particulate emissions by using multiple injections in a small diesel engine. SAE transactions, pp 2127–2136

Chen Y-J et al (2023) Temperature, particulate emission characteristics, and emission reduction performance for SCR coated on DPF under drop to idle regeneration. Energy 268:126764

Chen Z et al (2024) An optical study on the cross-spray characteristics and combustion flames of automobile engine fueled with diesel/methanol under various injection timings. Energy 290:130286

Cowart JS, Cheng WK (2000) Throttle movement rate effects on transient fuel compensation. In: A port-fuel-Injected SI engine. SAE transactions, pp 1632–1640

Debnath BK, Saha UK, Sahoo N (2015) A comprehensive review on the application of emulsions as an alternative fuel for diesel engines. Renew Sustain Energy Rev 42:196–211

Dong Y et al (2020) High-pressure direct injection of methanol and pilot diesel: a non-premixed dual-fuel engine concept. Fuel 277:117932

Egler W et al (2010) Fuel injection systems. Handbook of diesel engines, pp 127–174

Feng H et al (2024) Effects of methanol direct injection and high compression ratio on improving the performances of a spark-ignition passenger car engine. Fuel 357:130052

Gallas D et al (2018) Construction changes for vehicle engines powered with LPG made to extend the engine lifespan based on 1.4 gasoline engine. J Mech Transp Eng 70(3):5–12

Gambhir H, Barman J, Patchappalam K (2018) Experimental investigation on the effect of pilot and post injection on engine performance and emissions. SAE Technical Paper

Gaur A, Saha K, Ghai DP (2024) Experimental characterization of early, quasi-steady and post-injection evolution of single-hole direct injection sprays for spark-ignited engines. Thermal Sci Eng Prog 47:102270

Ghosh A, Ravikrishna R (2021) Evaporating spray characteristics of methanol-in-diesel emulsions. Fuel 290:119730
Göktaş M et al (2021) An evaluation of the use of alcohol fuels in SI engines in terms of performance, emission and combustion characteristics: a review. Fuel 286:119425
Golzari R et al (2021) Impact of intake port injection of water on boosted downsized gasoline direct injection engine combustion, efficiency and emissions. Int J Engine Res 22(1):295–315
Gong C et al (2021) Numerical study of cold-start performances of a medium compression ratio direct-injection twin-spark plug synchronous ignition engine fueled with methanol. Fuel 285:119235
Gumus M, Sayin C, Canakci M (2010) Effect of fuel injection timing on the injection, combustion, and performance characteristics of a direct-injection (DI) diesel engine fueled with canola oil methyl ester– diesel fuel blends. Energy Fuels 24(5):3199–3213
Han Z et al (1996) Mechanism of soot and NOx emission reduction using multiple-injection in a diesel engine. SAE Trans 837–852
Han T et al (2020) Multiple injection for improving knock, gaseous and particulate matter emissions in direct injection SI engines. Appl Energy 262:114578
Hanson R, Splitter D, Reitz RD (2009) Operating a heavy-duty direct-injection compression-ignition engine with gasoline for low emissions. SAE Technical Paper
Hasan AO et al (2021) An experimental study of engine characteristics and tailpipe emissions from modern DI diesel engine fuelled with methanol/diesel blends. Fuel Process Technol 220:106901
Heywood JB (1988) Internal combustion engine fundamentals. (No Title)
Hu T et al (2007) Improvement of spark-ignition (SI) engine combustion and emission during cold start, fueled with methanol/gasoline blends. Energy Fuels 21(1):171–175
Huang Z et al (2004) Engine performance and emissions of a compression ignition engine operating on the diesel-methanol blends. Proc Inst Mech Eng, Part d: J Automob Eng 218(4):435–447
Huang Y et al (2021) Dual injection: an effective and efficient technology to use renewable fuels in spark ignition engines. Renew Sustain Energy Rev 143:110921
Ijaz Malik MA et al (2021) Experimental evaluation of methanol-gasoline fuel blend on performance, emissions and lubricant oil deterioration in SI engine. Adv Mech Eng 13(6):16878140211025212
Jin S et al (2021) Effect of the HPDI and PPCI combustion modes of direct-injection natural gas engine on combustion and emissions. Energies 14(7):1957
Jin C et al (2024) Progress in corrosion research on alternative liquid fuels. Energies 17(12):2803
Kalwar A, Agarwal AK (2020) Overview, advancements and challenges in gasoline direct injection engine technology. In: Advanced combustion techniques and engine technologies for the automotive sector, pp 111–147
Khalid A, Manshoor B (2012) Effect of high injection pressure on mixture formation, burning process and combustion characteristics in diesel combustion. In: Proceedings of world academy of science, engineering and technology. World Academy of Science, Engineering and Technology (WASET)
Khan ZA et al (2016) Biodiesel performance within internal combustion engine fuel system: a review. Tribol Ind 38(2):197–213
Khan N, Gollahalli S (1981) Performance and Emission Characteristics of a Diesel Engine Burning Unstabilized Emulsions of Diesel Fuel with Water, Methanol, and Ethanol. SAE Technical Paper
Kim J et al (2024) Effects of pilot injection on ignition performance for F-24/Gasoline fuel blends. Proc Combust Inst 40(1–4):105218
Krieger K, Hummel H-G, Naik L (2000) Diesel Fuel Injection Technology-An essential contribution towards an environment friendly powerful diesel engine. SAE Technical Paper
Kumar TS, Ashok B (2024) Development of combustion control map for flex fuel operation in methanol powered direct injection SI engine. Energy 288:129695
Kumar C, Rana KB, Tripathi B (2021a) Combustion, performance and emission analysis of diesel-methanol fuel blend in CI engine. In: Agarwal AK et al (eds) Methanol: a sustainable transport fuel for CI engines. Springer Singapore, Singapore, pp 229–246

Kumar D et al (2021b) Simulations of methanol fueled locomotive engine using high pressure co-axial direct injection system. Fuel 295:120231

Kumar V, Agarwal AK (2019) Mater compatibility aspects and development of methanol-fueled. Adv Combustion Tech Eng Technol Automot Sect 37

Kumar V, Agarwal AK (2020) Material compatibility aspects and development of methanol-fueled engines. Adv Combust Tech Eng Tech Automotive Sector 37–51

Kumar MS, Bellettre J, Tazerout M (2005) Investigations on a CI engine using animal fat and its emulsions with water and methanol as fuel. SAE Technical Paper

Ladommatos N et al (1996) The effect of exhaust gas recirculation on combustion and NOx emissions in a high-speed direct-injection diesel engine. SAE Technical Paper

Li Y, Liu G, Zhou X (2003) Fuel-injection control system design and experiments of a diesel engine. IEEE Trans Control Syst Technol 11(4):565–570

Li Y et al (2020) Multiple-objective optimization of methanol/diesel dual-fuel engine at low loads: a comparison of reactivity controlled compression ignition (RCCI) and direct dual fuel stratification (DDFS) strategies. Fuel 262:116673

Li Z et al (2021a) Effect of injection strategy on a diesel/methanol dual-fuel direct-injection engine. Appl Therm Eng 189:116691

Li Y et al (2021b) Effects of diesel pre-injection on the combustion and emission characteristics of a common-rail diesel engine fueled with diesel-methanol dual-fuel. Fuel 290:119824

Li Z et al (2022a) Effects of fuel injection timings and methanol split ratio in M/D/M strategy on a diesel/methanol dual-fuel direct injection engine. Fuel 325:124970

Li J et al (2022b) Effects of pilot injection strategy on in-cylinder combustion and emission characteristics of PODE/methanol blends. Fuel Process Technol 228:107168

Liu J, Yao A, Yao C (2015) Effects of diesel injection pressure on the performance and emissions of a HD common-rail diesel engine fueled with diesel/methanol dual fuel. Fuel 140:192–200

Lius A, Cronhjort A, Stenlaas O (2022) Evaluation of cylinder state estimator using fuel evaporation assessment in a PFI methanol HD SI engine. SAE Technical Paper

McTaggart-Cowan G et al (2005) The effects of high-pressure injection on a compression-ignition, direct injection of natural gas engine. In: Internal combustion engine division fall technical conference

Mohamed M et al (2024) Effects of dual injection operations on combustion performances and particulate matter emissions in a spark ignition engine fuelled with second-generation biogasoline. Int J Engine Res p 14680874241261128

Mohan B, Yang W, Kiang Chou S (2013) Fuel injection strategies for performance improvement and emissions reduction in compression ignition engines—A review. Renew Sustainable Energy Rev 28:664–676

Mumford D, Goudie D, Saunders J (2017) Potential and Challenges of HPDI. SAE Technical Paper

Olah GA, Goeppert A, Prakash GS (2018) Beyond oil and gas: the methanol economy. John Wiley & Sons

Pal MK et al (2023) Development and study of a low-cost mass flow characterization technique for port fuel injector. In: International conference on advances in mechanical engineering and material science. Springer

Panda K, Ramesh A (2021) Diesel injection strategies for reducing emissions and enhancing the performance of a methanol based dual fuel stationary engine. Fuel 289:119809

Park SH, Yoon SH, Lee CS (2013) HC and CO emissions reduction by early injection strategy in a bioethanol blended diesel-fueled engine with a narrow angle injection system. Appl Energy 107:81–88

Patil T et al (2024) Sustainable methanol production from carbon dioxide: advances, challenges, and future prospects. Environ Sci Pollut Res 1–41

Peterson A (2012) The effect of biodiesel blends on a multi-cylinder engine and the impact on after treatment

Pu Y-H, Dierickx J, Verhelst S (2024a) Modelling the evaporative cooling effect from methanol injection in the intake of internal combustion engines. Fuel 372:132131

Pu Y-H et al (2024b) Renewable methanol as a fuel for heavy-duty engines: a review of technologies enabling single-fuel solutions. Energies 17(7):1719
Radzali MH, Khalid A, Jacob DW (2020) Effect of methanol-gasoline blend and ambient pressure on flame propagation and exhaust emission of spark ignition (SI) engine. Fuel, Mixture Formation Combust Process 2(2)
Roth H et al (2005) Effect of multi-injection strategy on cavitation development in diesel injector nozzle holes. SAE transactions, 1029–1045
Ruan J et al (2021) Effects of injection timing on combustion performance and emissions in a diesel engine burning biodiesel blended with methanol. Thermal Sci 25(4 Part A):2819–2829
Saxena MR, Maurya RK, Mishra P (2021) Assessment of performance, combustion and emissions characteristics of methanol-diesel dual-fuel compression ignition engine: a review. J Traffic and Transp Eng (English Edition) 8(5):638–680
Sayin C et al (2009) Effect of injection timing on the exhaust emissions of a diesel engine using diesel–methanol blends. Renewable Energy 34(5):1261–1269
Sayin C, Ozsezen AN, Canakci M (2010) The influence of operating parameters on the performance and emissions of a DI diesel engine using methanol-blended-diesel fuel. Fuel 89(7):1407–1414
Shubham K et al (2014) NOx and PM control using multiple injections strategies in diesel engine: A Review
Singh AP et al (2020) Effect of fuel injection pressure and premixed ratio on mineral diesel-methanol fueled reactivity controlled compression ignition mode combustion engine. J Energy Res Technol 142(12):122301
Srivastava RK et al (2022) Conversion of methane to methanol: technologies and future challenges. Biomass Convers Biorefinery 12(5):1851–1875
Stone R (1999) Introduction to internal combustion engines, vol 3. Springer
Sumanlal M et al (2021) Effect of operating parameters on the performance and emission of a diesel engine fuelled with diesel–methanol blend. In: Advances in IC engines and combustion technology: select proceedings of NCICEC 2019.. Springer
Svensson E et al (2016) Potential levels of soot, NO x, HC and CO for methanol combustion. SAE Technical Paper
Tao W et al (2022) The effect of diesel pilot injection strategy on combustion and emission characteristic of diesel/methanol dual fuel engine. Fuel 324:124653
Tian Z et al (2022) The effect of methanol production and application in internal combustion engines on emissions in the context of carbon neutrality: a review. Fuel 320:123902
Tornatore C et al (2019) Experimental comparative study on performance and emissions of e85 adopting different injection approaches in a turbocharged pfi si Engine. Energies 12(8):1555
Valera H, Agarwal AK (2019) Methanol as an alternative fuel for diesel engines. Methanol Alternate Fuel Econ 9–33
Verhelst S et al (2019) Methanol as a fuel for internal combustion engines. Prog Energy Combust Sci 70:43–88
Vigneswaran R et al (2015) Experimental investigation of a diesel engine with diesel/water emulsion at various injection timing. J Chem Pharm Sci www.jchps.com. 974:2115
Wang C et al (2019) Methanol as an octane booster for gasoline fuels. Fuel 248:76–84
Wang B et al (2020) In-depth comparison between pure diesel and diesel methanol dual fuel combustion mode. Appl Energy 278:115664
Wang W et al (2024a) Towards optimized excess air ratio and substitution rate for a dual fuel HPDI engine. Appl Therm Eng 253:123797
Wang Q et al (2024b) Experimental study on the effects of water addition in methanol on the performance of diesel-methanol diffusion combustion on a high-speed engine. J Clean Prod 446:141436
Wang Y et al (2024) Numerical study on the potential of stratified mixture to improve thermal efficiency and reduce carbon emissions in high-speed gasoline direct injection engine. J Eng Gas Turb Power 146(12)

Wei H et al (2017) Experimental investigations of the effects of pilot injection on combustion and gaseous emission characteristics of diesel/methanol dual fuel engine. Fuel 188:427–441

Wei H et al (2023) Experimental study of thermal efficiency and NOx emission of turbocharged direct injection hydrogen engine based on a high injection pressure. Int J Hydrogen Energy 48(34):12905–12916

Wen H et al (2023) Numerical investigation on the influence of injection location and injection strategy on a high-pressure direct injection diesel/methanol dual-fuel engine. Energies 16(11):4518

Wu T et al (2019) A reduced PM index for evaluating the effect of fuel properties on the particulate matter emissions from gasoline vehicles. Fuel 253:691–702

Wu D et al (2015) Experimental investigation of spray characteristics of diesel-methanol-water emulsion. Atomization and Sprays 25(8)

Xie F-X et al (2013) Research on using EGR and ignition timing to control load of a spark-ignition engine fueled with methanol. Appl Therm Eng 50(1):1084–1091

Xu C et al (2022) Effect on the performance and emissions of methanol/diesel dual-fuel engine with different methanol injection positions. Fuel 307:121868

Yanju W et al (2008) Effects of methanol/gasoline blends on a spark ignition engine performance and emissions. Energy Fuels 22(2):1254–1259

Yao C et al (2008) Effect of diesel/methanol compound combustion on diesel engine combustion and emissions. Energy Convers Manage 49(6):1696–1704

Yao M, Zheng Z, Liu H (2009) Progress and recent trends in homogeneous charge compression ignition (HCCI) engines. Prog Energy Combust Sci 35(5):398–437

Yao C et al (2017) Effects of injection pressure on ignition and combustion characteristics of diesel in a premixed methanol/air mixture atmosphere in a constant volume combustion chamber. Fuel 206:593–602

Ye P, Boehman AL (2010) Investigation of the impact of engine injection strategy on the biodiesel NOx effect with a common-rail turbocharged direct injection diesel engine. Energy Fuels 24(8):4215–4225

Yin L et al (2019) High efficient internal combustion engine using partially premixed combustion with multiple injections. Appl Energy 233:516–523

Yin X et al (2023) Investigation into the operating range of a dual-direct injection engine fueled with methanol and diesel. Energy 267:126625

Yu J et al (2024a) Effects of dual injection strategies on combustion, heat-work conversion process and emissions of a spark ignition engine fuelled with E20 blends. J Energy Inst 112:101498

Yu J et al (2024b) Research on the influence of injection strategies on the in-cylinder combustion process and emissions of methanol. Biomass Bioenerg 188:107340

Zhao Y, Liu X, Kook S (2024) Effect of three-hole nozzle orientations on sprays and combustion in methanol-diesel dual direct injection engines. Appl Therm Eng 254:123953

Zhao H, Ladommatos N (2001) Engine combustion instrumentation and diagnostics. SAE international

Zhen X, Wang Y (2015) An overview of methanol as an internal combustion engine fuel. Renew Sustain Energy Rev 52:477–493

Zhou Y et al (2022) Potential of compression ratio and exhaust gas dilution on improving combustion and nitrogen oxides emission performance on a PFI engine fueled with methanol. Fuel 323:124197

Zhou F et al (2023) The application prospect and challenge of the alternative methanol fuel in the internal combustion engine. Science of The Total Environment, p 169708

Zhu H et al (2013) Droplet vaporization characteristics of multicomponent mixtures of methanol and gasoline surrogate in opposed stagnation flows. Proc Combust Inst 34(1):1545–1552

Chapter 13
Bioethanol: Production Processes and Strategies for Utilization in Diesel Engines

Krishnangshu Guha and Avinash Kumar Agarwal

Abstract The interest in ethanol as a potential alternative to fossil fuels has gained momentum since the mid-twentieth century. Ethanol, being an oxygenated fuel, promotes more complete combustion, which reduces particulate emission from Compression Ignition Engines. The cooling effect of ethanol, due to its higher latent heat of vaporization, also reduces the emission of oxides of nitrogen. The chapter starts with a discussion on the physicochemical properties of ethanol relevant to engine application. It is followed by a description of various processes to produce ethanol, and its classification based on the feedstock used. There are four main strategies for using ethanol as a fuel in CI engines: fumigation, blending by adding cosolvents, blending by emulsification, and dual-fuel mode. The chapter addresses the technical challenges associated with blending ethanol with diesel and discusses the potential solutions. The safety aspects of using ethanol as a fuel have also been brought into the scope of discussion.

Keywords Ethanol · Alcohol fuels · Biofuels · Compression ignition engines · Alternate fuels

Glossary

AFE	Ammonia fibre explosion
ARP	Ammonia recycling percolation
APL-EA	Aganonerion polymorphum leaf-ethyl acetate
ASTM	American society for testing and materials
atdc	After top dead centre
btdc	Before top dead centre
bsfc	Brake specific fuel consumption

K. Guha · A. K. Agarwal (✉)
Engine Research Laboratory, Department of Mechanical Engineering, Indian Institute of Technology, Kanpur, India
e-mail: akag@iitk.ac.in

A. Bartl et al. (eds.), *Alcohol Production Processes and Their Utilisation in the Transport Sector*, Energy, Environment, and Sustainability,
https://doi.org/10.1007/978-981-96-7384-1_13

CBP	Consolidated bioprocessing
CI	Compression ignition
CVCC	Constant volume combustion chamber
DI	Direct injection
DDGS	Dried distillers grain solubles
DLC	Diamond-like carbon
EGR	Exhaust gas recirculation
FIP	Fuel injection pressure
GM	Genetically modified
HC	Hydrocarbons
HDPE	High-density polyethylene
HLB	Hydrophilic lipophilic balance
HMF	Hydroxymethyl furfural
IMEP	Indicated mean effective pressure
LA	Levulinic acid
LGT	Lateral gene transfer
LHW	Liquid hot water
NO_x	Oxides of nitrogen
PFI	Port fuel injection
PM	Particulate matters
PN	Particulate numbers
PODE	Polyoxymethylene dimethyl ethers
RCCI	Reactivity controlled compression ignition
SAA	Soaking aqueous ammonia
SE	Steam explosion
SCF	Supercritical fluid
SI	Spark ignition
SO_x	Oxides of sulphur
SSF	Simultaneous saccharification and fermentation
SSCF	Simultaneous saccharification and co-fermentation
TAN	Total acid number
TBHQ	Tert-butylhydroquinone
TDC	Top dead centre
THF	Tetrahydrofuran
UK	United Kingdom
USA	United States of America

13.1 Introduction

Since the invention of automobiles in the late-nineteenth century, the transport sector has been heavily dependent on the use of non-renewable fossil fuels like diesel and gasoline. However, the growing concern regarding the environmental implications of burning fossil fuels has driven the research on the feasibility of using various alternative fuels. Studies have shown that the emission of harmful gases from engines can be successfully mitigated to comply with stringent emission norms, by using fuels like alcohols, natural gas, hydrogen, and ammonia. The potential of different alcohol fuels like ethanol and methanol has been a major research focus due to several emission and performance benefits including simultaneous reduction of soot and NO_x with increased brake thermal efficiency. Ethanol, being a renewable fuel, decreases the dependency on fossil fuels which is beneficial from the perspective of energy security. Ethanol has a very high octane rating, and it has been widely used as a fuel for SI engines by blending with gasoline. As shown in Fig. 13.1, many countries including the USA, the UK, Canada, Brazil, India, China, and Australia have successfully substituted gasoline with ethanol at varying percentages (https://grains.org/bioethanol/). In India, ethanol blending with gasoline has been underway with the National Ethanol Blended Petrol Programme since 2003 (Aatmanirbhar et al. nd). In 2024, India successfully implemented 15% ethanol blending with gasoline throughout the nation (Larsen et al. 2024). The country is aiming to increase the level of ethanol blending to 20% by 2025. It is noteworthy that ethanol has a very low cetane rating, which makes it an unsuitable CI engine fuel. Mixing ethanol with diesel to prepare a stable blend is also problematic. However, when used as a CI engine fuel, ethanol shows immense potential to decrease emissions. The use of ethanol–diesel blends in diesel engines simultaneously reduces the emission of soot and NO_x by overcoming the dreaded trade-off between them. Ethanol, being an oxygenated fuel, also promotes cleaner combustion and decreases carbon emissions. Extensive research since the 1970s has led to the development of several technical solutions to use diesel blended with ethanol in varying proportions. These efforts have led to the emergence of ethanol as a clean alternative fuel for CI engines.

13.1.1 Properties of Ethanol Influencing Its Use in CI Engines

The chemical formula for ethyl alcohol or ethanol is $CH_3{-}CH_2{-}OH$. It is a colourless, volatile, flammable liquid with a pleasant odour and pungent taste. It is considered a mildly toxic substance. The molecular structure of ethanol consists of a two-carbon straight chain attached to a Hydroxyl (−OH) group. A comparison of the properties of ethanol with that of diesel and gasoline is presented in Table 13.1.

Molecular Weight

The molecular weight or molar mass of ethanol (46.07 g/mol) is significantly lower than that of diesel (~150–250 g/mol) (Source: https://www.iea-amf.org/content/fuel_information/diesel_gasoline). Graham's law of diffusion states that the rate of diffusion is inversely proportional to a compound's molecular weight. So, ethanol, being

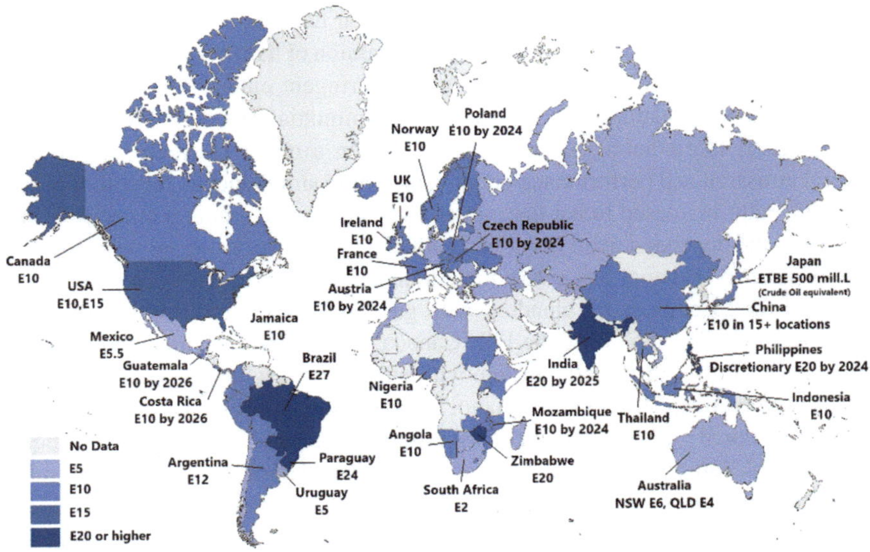

Fig. 13.1 Global ethanol blending levels in gasoline (https://grains.org/bioethanol/)

a lighter fuel, has a higher rate of diffusion compared to diesel. This, in turn, causes lower emissions.

Combustion Properties

Pure ethanol produces a smokeless blue flame which is difficult to see under typical lighting conditions. Within the combustion chamber of a CI engine, ethanol blends burn faster than diesel due to its higher flame speed. In air, the adiabatic flame temperature of ethanol is 2355 K, while that of diesel is more than 2600 K. The presence of ethanol in the fuel blend lowers the adiabatic flame temperature as ethanol has a higher latent heat of vaporization than diesel. The auto-ignition temperature of ethanol is higher than that of diesel. In the air, ethanol autoignites at 423 °C while diesel autoignites at 315 °C (ref.: Table 13.1). This leads to an increase in ignition delay with an increasing amount of ethanol in ethanol–diesel blends. Ethanol's high volatility increases the amount of combustible mixture available for premixed combustion as it evaporates quicker than conventional diesel. The duration and rate of heat release of the premixed combustion phase increase with increasing ignition delay. The amount of fuel available for premixed combustion also increases with ignition delay. An increase in the amount of fuel burnt in the premixed phase reduces the amount of fuel burned in the diffusion phase, thereby shortening its duration. An increased proportion of premixed combustion increases NO_x emissions.

Latent Heat of Vaporization

The latent heat of vaporization of ethanol is 910 kJ/kg and that of diesel is 230 kJ/Kg (ref.: Table 13.1). As the latent heat of vaporization of ethanol is significantly higher

Table 13.1 Comparison between the properties of ethanol, diesel, and gasoline Larsen et al. (2009)

Property	Ethanol	Diesel	Gasoline
Chemical formula	C_2H_5OH	C_3 to C_{25}	C_4 to C_{12}
Molar mass (g/mol)	46.07	~ 200	100–105
Carbon (mass %)	52.2	84–87	85–88
Hydrogen (mass %)	13.1	13–16	12–15
Oxygen (mass %)	34.7	0	0
Liquid density, 20 °C (kg/l)	0.792	0.81–0.88 (37 °C)	0.72–0.78
Viscosity (cSt)	1.52	2–4 (37 °C)	0.4–0.5 (20 °C)
Boiling temperature, 1 atm (°C)	78.4	288–340	27–225
Reid vapour pressure (kPa)	16	0.1–0.55	50–100
Flammability limit, 20 °C (vol %)	3.3–19	0.6–6.5	1.4–7.6
Stoichiometric air/fuel ratio	9	14.6–15	14.5–14.7
Flash point temperature, closed cup, atmospheric conditions (°C)	12	74	– 42
Auto-ignition temperature (°C)	423	~ 315	257
Heat of vaporization (kJ/kg)	910	230	330–400
Heat of combustion (lower heating value) (kJ/kg)	26,900	42,800–45,300	42,000–44,000
Heat of combustion (lower heating value) (kJ/litre)	21,300	~ 37,200	~ 32,000
Research octane number	92	N/A	90–98
Motor octane number	80	N/A	81–90
Cetane number	5–15	40–55	5–20
Water tolerance, volume (%)	Completely miscible	Negligible	Negligible

than that of diesel, this hinders complete in-cylinder combustion and increases HC and CO emissions, especially at low engine speed and load. However, it also lowers the adiabatic flame temperature and decreases NO_x emission by compensating for the impacts of an increased proportion of premixed combustion. When ethanol is injected in the inlet manifold, the cooling effect due to its higher latent heat of vaporization decreases the temperature and increases the density of charge air. This allows a greater amount of air to enter the combustion chamber, leading to an increase in combustion efficiency and power output. However, there can be cold start issues due to the presence of ethanol in the fuel blend, owing to its higher latent heat of vaporization.

Cetane Number

Ethanol's cetane number has been estimated to be between 5 and 15. According to Hardenberg and Ehnert, it is unreliable to use cetane numbers to describe ignition characteristics of ethanol–diesel blends due to discrepancies in the determination of cetane numbers below 30 (Hardenberg and Ehnert 1981). The cetane number

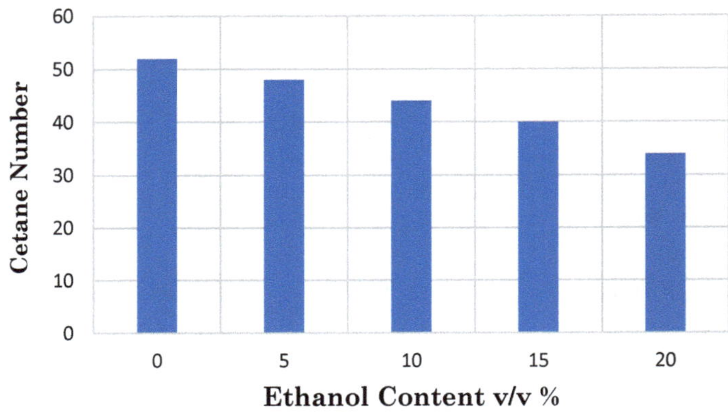

Fig. 13.2 Variation of cetane number of ethanol–diesel blends with increasing ethanol content (Hajba et al. 2011)

of ethanol–diesel blends decreases with increasing ethanol content as shown in Fig. 13.2.

The ignition delay increases as the cetane number decreases. Higher ignition delay results in a greater fraction of premixed combustion, which leads to higher NO_x emission. So, to reduce the cetane number penalty with increasing ethanol content, various ignition improvers are added to ethanol–diesel blends. Alkyl nitrates like amyl, hexyl, and octyl nitrates and alkyl peroxides are some commonly used ignition improvers that raise the cetane number of ethanol–diesel blends (Yahuza and Dandakouta 2015). Alkyl nitrates have high efficiency, but they are toxic and corrosive. Alkyl peroxides remain stable in storage and are not decomposed by heat or contamination with water. The decrease of cetane number in the ethanol–diesel blend made by emulsification is less than that of the stable blends of aqueous ethanol and diesel made by adding cosolvents. This occurs due to a shielding effect of the emulsion structure which delays the evaporation of the alcohol from the fuel droplets, whereas the ethanol molecules in the blends prepared by adding cosolvents are free to evaporate quickly (Yahuza and Dandakouta 2015).

Blend Stability

Ethanol is polar like water, while hydrocarbons like diesel are non-polar. This makes ethanol difficult to blend with diesel. There are two ways to produce homogenous ethanol–diesel blends without any phase separation: emulsification and 'splash blending' by adding cosolvents. The stability of ethanol–diesel blends is primarily influenced by factors like ambient temperature, the ethanol's water content, and the composition of the diesel fuel. At temperatures above 30 °C, up to 15% v/v anhydrous ethanol can be blended with diesel without any additives, and no phase separation is observed. However, at temperatures below 10 °C, considerable phase separation is noticed. Furthermore, anhydrous ethanol is hygroscopic and absorbs water when stored, which contributes to the blend instability. The blend loses its homogeneity when phase separation occurs. Homogenous blends are important for controlling the precise amount of fuel injected. Otherwise, this would lead to combustion issues. Another important factor is the aromatic content of the diesel fuel used. Although the aromatics are primarily non-polar, the delocalized electrons in their ring structure make them polar to some extent. Hence, they act as a bridging agent between the polar ethanol and non-polar diesel molecules, thereby enhancing their solubility.

Heating Value

The lower heating value of ethanol is significantly lower than that of diesel. This implies that less energy is obtained from ethanol when equal masses of ethanol and diesel are combusted. The CI engines fuelled by ethanol–diesel blends have a lower power output for the same mass flow of fuel than those fuelled by diesel, due to the lower energy content of ethanol. Therefore, an increase in *bsfc*, almost proportional to the decrease in energy content, is noticed while using ethanol–diesel blends. Researchers have observed that the energy content of ethanol–diesel blends decreases approximately by 5% for each 2% v/v ethanol added (Lee et al. 2019). However, the use of ethanol–diesel blends with up to 10% v/v ethanol has a negligible impact on vehicle performance.

Corrosive Nature and Material Compatibility

Corrosion transforms the material into refined forms like oxides, sulphides, and hydrides by continuous electrochemical and/or chemical reactions in ambient conditions. This transformation is mostly detrimental. All alcohols, including ethanol, are corrosive. Corrosion due to the presence of ethanol in fuel blends can be classified into the following categories based on the causative agent:

1. **Wet Corrosion**: Ethanol, being inherently hygroscopic, absorbs atmospheric water while in storage. The presence of this azeotropic water causes corrosive reactions between alcohols and metals/alloys. An increase in the ethanol content of the blend also increases its capacity to absorb moisture. The corrosive nature of the fuel blends increases with increasing ethanol content. The water absorbed along with oxygen causes an electrochemical reaction in which the material is transformed into metal ions in the presence of alcoholic electrolyte media. Many tiny galvanic cells are developed in the metal, creating anode and cathode regions. The metal undergoes oxidation at the anode by providing electrons to the cathode to form metal ions. So, corrosion occurs at the anode. Reduction occurs at the cathode where oxygen and water react to form hydroxyl ions. The rate of corrosion can be measured by chemical kinetics which measures the weight loss of the component due to corrosion. Finally, the metal ions react with the hydroxyl ions to form an insoluble hydrolytic thin film on the anode zone which prevents further dissolution of metal ions. The reactions involved in the process are shown below (Muthuraman et al. 2022).

$$O_2 + 2H_2O + 4e^- \rightarrow 4OH^- \text{(Reduction at Cathode)}, \tag{13.1}$$

$$M \rightarrow M^{n+} + \text{ne}^{-}(\text{Oxidation at Anode}), \quad (13.2)$$

$$M^{n+} + n\text{H}_2\text{O} \rightarrow M\,(\text{OH})_n + n\text{H}^{+}. \quad (13.3)$$

However, in the presence of water, acetic acid and formic acid are formed in the diesel-ethanol blend which dissolves the metal hydroxide film on the surface and promotes corrosion. The acidity of the fuel blend can be quantified by the *Total Acid Number* (TAN). TAN measures the quantity of by-products of oxidation in the fuel blends. An increase in the TAN of fuel suggests the presence of acids which accelerate corrosion.

2. **Dry Corrosion**: It occurs due to the polarity of ethanol molecules which directly attacks metals like magnesium, lead, aluminium, etc. Ethanol facilitates higher dissolution of gases like oxygen due to its polarity. Hence, the presence of ethanol promotes the oxidative corrosion process. Unlike wet corrosion, dry corrosion occurs in the absence of water and no anode or cathode region is created in the metal. Corrosion occurs uniformly on the exposed metallic surface. The oxidation reactions involved in the dry corrosion of aluminium are shown below (Muthuraman et al. 2022):

$$3\,\text{C}_2\text{H}_5\text{OH} + \text{Al} \rightarrow \text{Al}\,(\text{C}_2\text{H}_5\text{O})_3 + 3/2\,\text{H}_2, \quad (13.4)$$

$$\text{Al}\,(\text{C}_2\text{H}_5\text{O})_3 + 3\text{H}_2\text{O} \rightarrow \text{Al}\,(\text{OH})_3 + 3\,\text{C}_2\text{H}_5\text{OH}, \quad (13.5)$$

$$2\,\text{Al}\,(\text{C}_2\text{H}_5\text{O})_3 \rightarrow \text{Al}_2\text{O}_3 + 6\,\text{C}_2\text{H}_4 + 3\,\text{H}_2\text{O}. \quad (13.6)$$

In the first reaction, ethanol gets converted to alcoholates along with the formation of hydrogen gas. By measuring the mass of released hydrogen, the corrosion rate can be evaluated. The alcoholates formed undergo hydrolysis and decomposition in the second and third reactions. The oxide layer formed on the metal surface by this reaction protects the surface from further corrosion. However, in a water-deficient environment, this passive layer of metal oxide is destroyed due to the formation of several alkaline compounds in the oxidation reaction. This phenomenon accelerates corrosion. Inhibitors, like sodium silicate in alkaline media, can reduce the dissolution rate of the protective film (Linder nd).

Wet corrosion is generally more severe than dry corrosion since it occurs in the presence of water as an electrolyte, which speeds up the reaction. Moreover, it is difficult to keep ethanol completely dry due to its hygroscopic nature. The presence of absorbed moisture in the fuel blend also contributes to making wet corrosion more severe than dry corrosion. Ethanol also has detrimental effects on elastomers which lose their tensile strength upon exposure to ethanol. Non-metallic materials like rubber, leather, polyurethane, polyvinyl chloride, methyl-methacrylate plastics, nylon 6/6 and some thermoplastic and thermoset polymers degrade when in contact with fuel ethanol (Safety and Procedural Requirements … 2020). It causes degradation, e.g. softening and swelling of the seals and O-rings of the engine and fuel systems. The presence of ethanol might make the elastomers brittle enough to fail. To address the problems due to the corrosive nature of ethanol–diesel blends, several preventive measures have been taken by the engine designers. Some of them are as follows:

- Metals with good corrosion resistance should be selected for design purposes. Metals like zinc, brass, lead, aluminium, etc., should be avoided consciously as they tend to oxidize easily when brought in contact with ethanol (Safety and Procedural Requirements … 2020). Alloys are preferred over pure metal as they contain certain second-phase elements that inhibit corrosion. For example, chromium is added as a corrosion inhibitor in stainless steel. Steel, magnesium, and aluminium alloys are commonly used in ethanol-fuelled engines to avoid corrosion. The design should not have any voids, surface porosity, irregularities, moisture traps, or sharp corners as these imperfections lead to corrosion. Old engines with aluminium cylinder heads are more susceptible to corrosion due to ethanol-blended diesel fuel. However, modern engines are designed to be unaffected by corrosion. Pitting corrosion in stainless steel can be lowered if the concentration of water, chloride, and acetic acid is kept in check (Abel and Virtanen 2015). Alloys like *stainless steel 304* and *aluminium 6061* possess good corrosion resistance.
- A protective coating of high-density polyethylene (HDPE) is applied to the fuel tank and various other engine and fuel system components to prevent corrosion. Metal surfaces that are exposed to ethanol undergo various surface treatment processes like anodization, keronite coating, diamond-like carbon (DLC) coating, and nickel coating. Moreover, barrier technologies like fluorination, sulphonation, coextrusion, etc., can cause a significant reduction in alcohol permeation rate.
- Rubber parts of engine and fuel systems can be replaced by materials like HDPE, nylon, and fluorinated plastics like Teflon.

Fuel storage and supply units can be designed to be at low temperatures to prevent contamination of fuel blends by moisture oxygen, chloride, and acids. Corrosion inhibitors like *tert-butylhydroquinone* or TBHQ and *Aganonerion Polymorphum Leaf-Ethyl Acetate extract* or APL-EA can also be added to fuel blends.

Lubricity and Viscosity

When fuel viscosity drops below a certain threshold, it will affect its lubricity. Several injection system components including rotary distribution pumps and high-pressure pumps delivering fuel to the rail in common rail systems rely on the fuel's lubricity to prevent wear. The sliding surfaces of fuel injection pumps are protected from wear by hydrodynamic and boundary lubrication. Parts of the fuel injector like the needle-guide nozzle-body interface and the plunger-barrel interface also depend on fuel for lubrication to some degree. The viscosity of ethanol is lower than that of diesel. So, the presence of ethanol lowers the viscosity of the fuel blends. Lower fuel viscosity leads to greater pump and injector leakage, fuel pump failure and injector degradation. According to ASTM standards, the minimum viscosity of 2.0 mm^2/s at 37.8 °C should be maintained for diesel fuel. To improve the lubricity of ethanol–diesel blends, various lubricity additives can be added to them. Lubricity additives are mainly various surface-active agents having an affinity for metallic surfaces. They form a protective film over the surface. Mono and diacylglycerol, glycerol, ester and fatty acids are used as lubricity additives to ethanol–diesel blends. Biodiesel is another very widely used lubricity additive for ethanol–diesel blends.

13.1.2 Opportunities

The main opportunities for using ethanol as a fuel in CI engines can be listed as follows:

- Ethanol, especially bioethanol is a renewable fuel, obtained from various natural sources like edible, non-edible, and algal biomass. Bioethanol can also be produced from agricultural, industrial, and municipal waste, thus ensuring the conversion of waste to wealth.
- Replacing a fraction of conventional fuels like diesel with indigenously produced ethanol decreases the dependency on fossil fuels, which are very often imported. An oil importing nation, like India, can have substantial benefits in terms of energy security by using ethanol as an alternative fuel.
- Using ethanol as a fuel in CI engines has certain emission benefits. Ethanol, being an oxygenated fuel, decreases the PM emissions from diesel engines as it causes rapid and more complete combustion. Using ethanol in fumigation or dual-fuel mode decreases NO_x emissions as well, thereby beating the dreaded soot-NO_x trade-off.
- Substituting a fraction of diesel fuel with ethanol increases the thermal efficiency of the engine.

13.1.3 Challenges and Potential Solutions

- The calorific value of ethanol is 26.9 MJ/kg which is substantially lower than that of diesel (Larsen et al. 2009). So, compared to diesel, more ethanol is needed to reach a similar fuel energy input. This can be done by increasing the nozzle diameter of the injectors.
- Ethanol has a low cetane number compared to diesel, which causes ignition delay and increases knocking tendency. This issue can be addressed by using various commercially available cetane improver additives for diesel fuel.
- Ethanol has low viscosity which affects the lubrication of injection system components. Lubricity can be increased by adding various lubricity additives to ethanol–diesel blends. The corrosive nature of ethanol is another problem that is mitigated by avoiding plastic and rubber parts in fuel injection system components.

- Ethanol has poor miscibility with diesel fuel. This causes phase separation of the blend. However, the stability of diesel-ethanol blends can be increased substantially by using a small amount of cosolvents or emulsifiers in the blends.
- From an emission point of view, using ethanol–diesel blends increases HC and CO emissions from the engine. Although ethanol is an oxygenated fuel promoting combustion, it has a high latent heat of vaporization, which hinders complete combustion. As a result, the unburnt hydrocarbons in the exhaust increase. CO emission is generally higher in the mid-load range due to low combustion temperature and high ignition delay, causing partial oxidation. At higher loads, complete oxidation of CO occurs, producing CO_2 instead. Increasing the compression ratio decreases HC and CO emissions due to the higher temperature of the combustion chamber at the end of the compression stroke. Moreover, by using suitable after-treatment technologies like Diesel Oxidation Catalyst (DOC), HC and CO in the exhaust are oxidized to form harmless H_2O and CO_2, respectively. However, using ethanol in CI engines increases the emission of aldehydes.
- Higher NO_x emissions can be an issue during DI of ethanol–diesel blends (made by emulsification or by adding cosolvents). However, the NO_x emissions decrease by fumigation or port fuel injection of ethanol in a dual-fuel engine, due to the cooling effect of ethanol caused by high latent heat of vaporization of ethanol.

13.1.4 Ethanol Production from Renewable Sources

There are multiple ways to produce ethanol on an industrial basis. Some methods are renewable while others are not. For example, ethanol produced by hydrolysis of ethylene, a petrochemical, is non-renewable. However, the sustainability associated with using ethanol as an alternate fuel in engines lies in the fact that the ethanol used is produced strictly from renewable sources. Ethanol is produced renewably from four main sources: sugar, starch, lignocellulosic biomass, and algae. Ethanol produced from renewable sources is called bioethanol (Dr. Murry Tamers. 2006). Thus, for this section, we will be using the term bioethanol, instead of ethanol. Based on the feedstock used, bioethanol can be classified into four generations. While the first (1G) and second (2G)-generation bioethanol production technologies are used to produce a major fraction of global bioethanol, the third (3G) and fourth (4G)-generation technologies are still an emerging field. The following paragraphs contain details about the processes involved in the production of different generations of bioethanol (Fig. 13.3).

13.1.4.1 First-Generation (1G) Bioethanol

It is produced by processing various edible feedstocks, rich in carbohydrates like soluble sugars and starch. Carbohydrate is the main fermentable component that produces bioethanol. Sugar-containing feedstocks like sugarcane, sugar beet, sweet

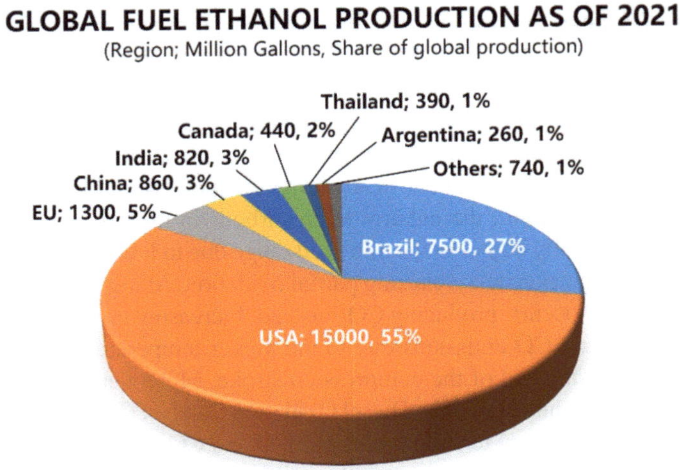

Fig. 13.3 Global fuel ethanol production as of 2021. (*Source* https://energytracker.asia/pros-and-cons-of-ethanol/) (Article: Eric Koons 2023)

sorghum, fruits like dates, grapes, watermelon, etc., and starch-containing feedstocks like corn, barley, rye, legumes, wheat, cassava, tubers, sweet potatoes, etc., have been successfully used across the globe to produce first-generation bioethanol (Fig. 13.4).

Biomass containing soluble sugar in glucose, sucrose, and fructose form can be readily fermented by microorganisms like yeast (saccharomyces) to produce bioethanol. However, the glucose present in starch exists in long-chain polymeric form and it needs to be converted to glucose syrup before fermentation, as yeast cannot ferment polymeric substrates.

Hydrolysis

The process implemented for depolymerization of starch is called hydrolysis. It uses amylolytic enzymes like α-amylase and glucoamylase to convert amylose and amylopectin molecules to soluble sugar. Hydrolysis of starch mainly comprises two steps: (1) liquefaction and (2) saccharification. During liquefaction, the starch substrate is first converted into a gelatinous form using α-amylase. The subsequent process of liquefaction converts glucose polymers present in starch to shorter chains of dextrins, maltotriose, and maltose. In the next step, dextrin is converted to fermentable glucose by glucoamylase. This preprocessing step of hydrolysis is not involved in producing bioethanol from sugar-containing biomass, which makes it comparatively cheaper.

Fermentation

The glucose solution produced after hydrolysis undergoes a metabolic process called fermentation where it is converted to bioethanol by various microorganisms like bacteria, fungi, and yeast. The chemical reaction involved in the fermentation of glucose into ethanol can be expressed as:

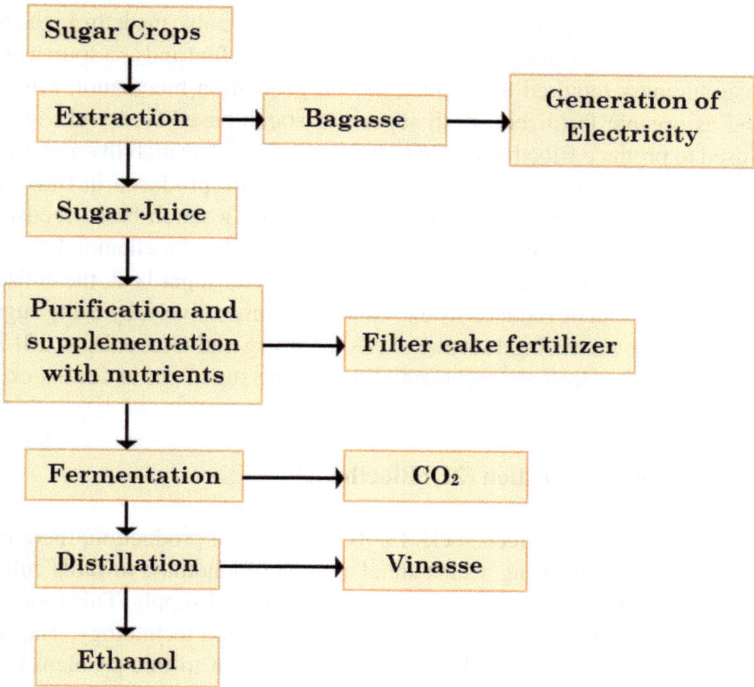

Fig. 13.4 Schematic of production of bioethanol from sugar-based feedstocks (Zabed et al. 2017)

$$C_6H_{12}O_6 + H_2O + \text{Yeast} \rightarrow 2CO_2 + 2C_2H_5OH + H_2O + \text{Heat} \quad (13.7)$$

The alcohol produced by fermentation undergoes distillation and dehydration to reach its final form of ~ 96% v/v concentration (Fig. 13.5).

A highly nutritious livestock feed called Dried Distillers Grains with Solubles (DDGS) is produced from the residue of the distillation process, as a by-product.

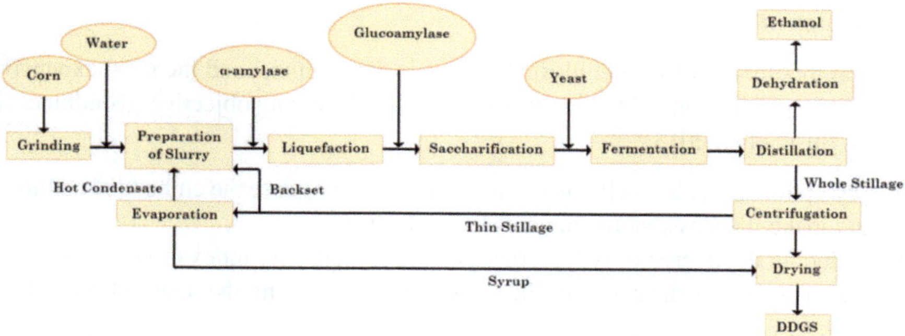

Fig. 13.5 Schematic of production of bioethanol from starch-based feedstocks (Zabed et al. 2017)

Other useful by-products like Bagasse and filter cakes are also made from the residue of sugar-containing feedstocks. Bagasse finds its use as a fuel and as a source of lignocellulosic biomass required to produce second-generation bioethanol. Filter cakes are used as organic fertilizers. Both sugar and starch-based feedstocks are extensively used to produce Bioethanol due to their widespread availability and economic advantages. For example, a major portion of bioethanol produced in Brazil comes from sugarcane, while the USA relies mostly on corn, a starch-based feedstock. In Europe, sugar beet is primarily used as feedstock to produce bioethanol. In a comparative study, it has been found that compared to corn and sugar beet, the emission of greenhouse gases can be reduced by 30% and 40%, respectively, by using sugarcane as a feedstock for producing bioethanol (Goldemberg and Guardabassi 2010). The production cost also decreases substantially by using sugarcane as feedstock.

13.1.4.2 Second-Generation (2G) Bioethanol

As food crops are used as feedstocks for 1G Bioethanol production, there exists a dilemma of whether diverting a substantial amount of landmass towards cultivating crops for producing biofuels would affect the global food supply. This food vs. fuel debate is the main drawback of 1G bioethanol production technology. The second-generation (2G) bioethanol production offers a solution to this problem by using non-food lignocellulosic feedstocks. Agricultural and forestry residues like switch grass, corn stover, straw, wood, rice hulls, sugarcane bagasse; industrial wastes like wastepaper, paper mill waste; food industry wastes; and various municipal solid wastes are used as feedstock to produce 2G bioethanol. Using lignocellulosic biomass for bioethanol production has certain advantages like sustainability, easy availability, and low procurement cost (Fig. 13.6).

However, the high processing cost involved harms the economy of the process. Moreover, the physicochemical property of lignocellulosic biomass varies widely due to the versatility of sources. This makes it difficult to standardize and optimize the conversion process. Lignocellulosic feedstock majorly contains lignin, cellulose, hemicellulose, and protein. Converting it to bioethanol involves the following steps:

Pretreatment

The pretreatment of lignocellulosic biomass is the foremost and the most expensive step in the production of 2G bioethanol. It has the following objectives (Sindhu et al. 2019; Bušić et al. 2018):

- Disruption of the lignin hemicellulose structure to release the embedded cellulose and make it accessible to enzymes for hydrolysis.
- Reducing the degree of polymerization and crystallinity index of cellulose.
- Increasing the surface area of the substrate by increasing the material porosity.

It has been observed that toxins like furan derivatives, carboxylic acid, and phenolic compounds are synthesized during pretreatment due to the prevailing harsh conditions. These compounds are yeast cell growth inhibitors, and they have adverse

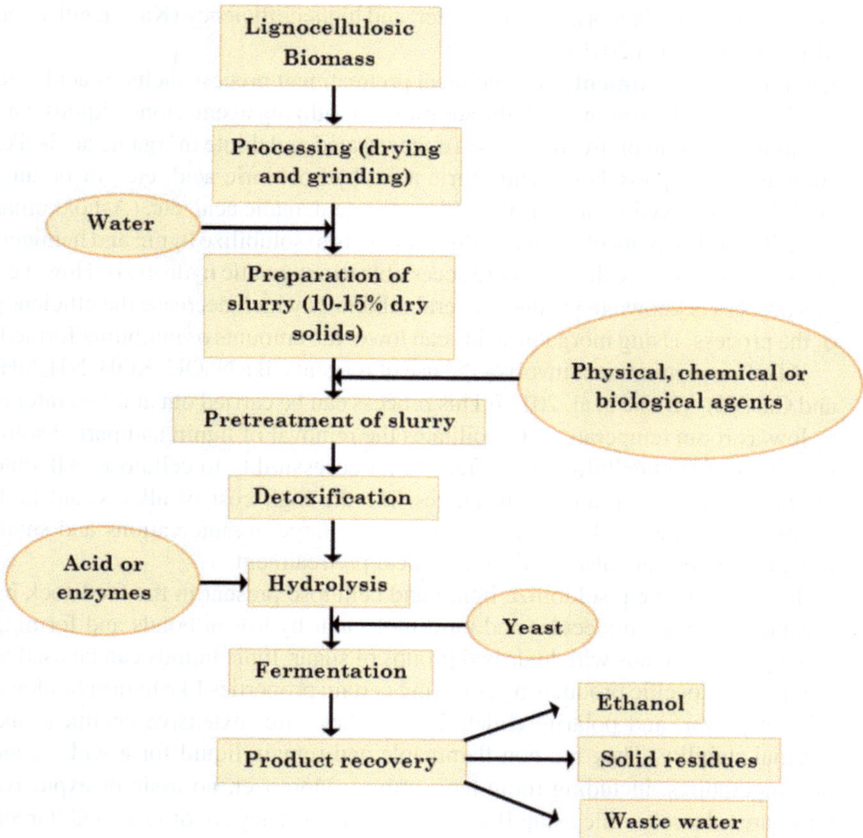

Fig. 13.6 Schematic of steps involved in the production of 2G bioethanol (Zabed et al. 2017)

effects on the efficiency of the bioconversion process. The type and amount of the inhibitors formed depend on the nature of the biomass and the pretreatment process involved. Minimizing the formation of these inhibitors should be taken care of. The pretreatment process can be classified into four types, and they are as follows:

a. **Physical Pretreatment**: The physical pretreatment process includes extensive mechanical grinding, thermolysis, ultrasound, and irradiation by gamma rays, electron beams, or microwaves. These processes reduce the size of the feedstock and increase its porosity to facilitate better enzymatic hydrolysis. However, the very low size of the substrate is unwanted as it increases energy consumption. Lignin and cellulose are depolymerized to some extent by these processes. However, they are energy-intensive and mostly non-eco-friendly. Chemical processes are often combined with physical pretreatment processes to increase their efficiency. In recent times, extrusion is gaining popularity as a physical pretreatment process, owing to its capacity of superior mixing, higher shear rate, increased heat transfer, and higher efficiency (Karunanithy and Muthukumarappan 2010).

b. **Chemical Pretreatment**: The chemical pretreatment process includes acid, and alkali, pretreatment along with the addition of oxidizing agents, ionic liquids, and organosolvs. Acid pretreatment uses concentrated and dilute inorganic acids like sulphuric acid, phosphoric acid, nitric acid, hydrochloric acid, etc., or organic acids like carboxylic acid, formic acid, oxalic acid, malic acid, etc. (Ashokkumar et al. 2022). The main objective of this process is to solubilize lignin and hemicellulose and make the cellulose more accessible to enzymatic hydrolysis. However, this process is known to produce several inhibitors which decrease the efficiency of the process. Using inorganic acids can lower the amounts of inhibitors formed.

   Alkaline pretreatment involves the use of reagents like NaOH, KOH, NH_4OH, and $Ca(OH)_2$ (Bušić et al. 2018). This process can be carried out at temperatures as low as room temperature. It facilitates the removal of lignin and partial solubilization of hemicellulose and offers better accessibility to cellulose. Alkaline pretreatment has certain disadvantages like the high cost of alkalis and high energy requirements. Ionic liquids, containing large organic cations and small inorganic anions are also used for chemical pretreatment.

   Ionic liquids help solubilize lignin and cellulose present in the feedstock by breaking the inter-molecular and intra-molecular hydrogen bonds and forming new hydrogen bonds with hydroxyl groups of sugar. Ionic liquids can be used to obtain very specific products by adjusting certain properties like hydrophobicity, solvent power, and polarity (Edeh 2020). They offer extensive chemical and thermal stability. They are non-flammable and remain liquid for a wide range of temperatures, including room temperature. Moreover, no toxic or explosive gases are released while using ILs for pretreatment. They are often called 'Green Solvents' for being eco-friendly (Tomás-Pejó et al. 2011).

   Organosolvs contain organic solvents like ethanol, methanol, phenol, or ethylene glycol along with small quantities of acids like hydrochloric, sulphuric, oxalic, or salicylic acid to increase the efficiency by breaking hemicellulose bonds. They are gaining popularity as a means of chemical pretreatment in recent years. However, the need to recycle the solvents used makes the process quite expensive.

c. **Physicochemical Pretreatment**: Both physical and chemical pretreatment processes are combined here to maximize efficiency. The commonly used physicochemical pretreatment processes are steam explosion (SE), ammonia fibre explosion (AFE), ammonia recycling percolation (ARP), soaking aqueous ammonia (SAA), liquid hot water (LHW), supercritical fluid (SCF), etc. Steam explosion technique involves the use of high-pressure saturated steam causing explosive decompression. SE is the most frequently used pretreatment process for producing 2G bioethanol. It has several advantages low cost, high energy efficiency, less environmental impact, and complete sugar recovery (Alvira et al. 2010). Ammonia is used along with high temperature and pressure in AFE to achieve higher surface accessibility and decrystallization of cellulose, with partial depolymerization of hemicellulose. In SCF, the explosive release of supercritical carbon dioxide increases the surface area of the substrate and disrupts cellulose and hemicellulose. SCF is an inexpensive, eco-friendly process, which does not form any inhibitors.

d. **Biological Pretreatment**: The biological pretreatment process is an eco-friendly, inexpensive, and energy-efficient pretreatment process. This process produces the least amount of inhibitors. Microorganisms like fungi are used in this process for the pretreatment of lignocellulosic biomass. The fungi used include white rot, brown rot, and soft rot fungi. Brown rot fungi act on cellulose while the white rot and soft rot varieties act on both lignin and cellulose (Zabed et al. 2017; Ceballos 2017). Microorganisms are allowed to grow directly on the biomass in this process. However, this process lowers the rate of hydrolysis and has an incubation time that is quite longer than the other pretreatment processes (Fig. 13.7).

Hydrolysis

The product of the pretreatment process consists of a solid fraction, containing lignin and cellulose along with a liquid fraction, containing hemicellulose and sugars. After pretreatment, the hemicellulose can either be completely converted to monomeric sugar or can remain partially polymerized, in oligosaccharide form, requiring further depolymerization. This depends on the process implemented for pretreatment. The cellulose remains almost unconverted. These cellulosic and hemicellulosic products of pretreatment are converted to fermentable monomeric sugar by hydrolysis. The process of hydrolysis can be catalysed by acids and enzymes. Acid-catalysed hydrolysis involves the use of sulphuric, hydrochloric, nitric, or phosphoric acid both in concentrated and dilute forms. Concentrated acid-catalysed hydrolysis is carried out at lower temperatures. It can hydrolyse both cellulose and hemicellulose. The process is fast, but it is associated with certain disadvantages like high cost, corrosion, difficult acid recovery or neutralization, and degradation of sugar monomers due to harsh acidic environment. Dilute acids are used as catalysts for hydrolysis of hemicellulose and pretreatment of cellulose. The process is carried out in two stages. The first stage has a lower temperature and facilitates depolymerization of hemicellulose. The second stage converts cellulose to glucose at an elevated temperature. Dilute acid-catalysed hydrolysis is often associated with the production of inhibitors.

Both cellulose and hemicellulose can undergo hydrolysis in the presence of enzymes as catalysts. The process occurs at low temperatures. Unlike acid-catalysed hydrolysis, this process does not cause any corrosion. Reduced production of inhibitors and less degradation of sugar also make this process advantageous.

Fermentation

It is the bioconversion process that converts monomeric sugar obtained after hydrolysis to ethanol. Microorganisms like yeast, bacteria, or fungi facilitate this process. Saccharomyces cerevisiae is the most used yeast species for fermentation. However, it can only cause fermentation of hexose sugar. So, other species like *Pichia stipitis* and *Pichia angophorae* are used for the fermentation of pentose sugar. Simultaneous

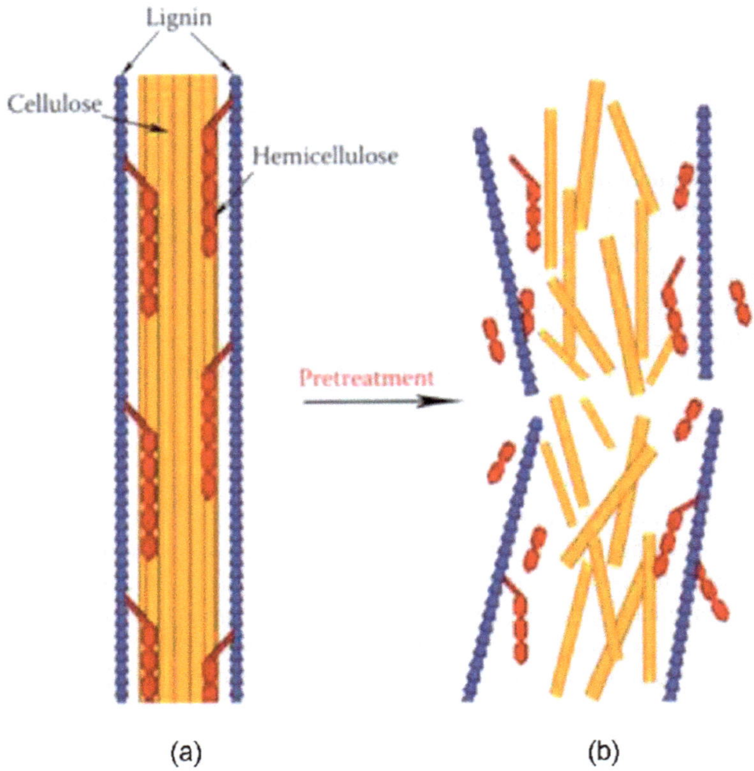

Fig. 13.7 Effect of pretreatment on lignocellulosic biomass **a** before pretreatment, **b** after pretreatment (Edeh 2020)

saccharification and fermentation (SSF) and simultaneous saccharification and co-fermentation (SSCF) are the most frequently used fermentation processes. In SSF, saccharification of cellulose and fermentation of monomeric sugar occurs simultaneously. The product of hydrolysis is readily available for fermentation in this process. It also mitigates the formation of inhibitors. In SSCF, saccharification and hydrolysis are carried out simultaneously with the co-fermentation of pentose sugar (Edeh 2020). Other fermentation processes include batch fermentation, continuous fermentation, non-isothermal simultaneous saccharification and fermentation, consolidated bioprocessing (CBP), etc. The fermentation process is followed by the recovery and dehydration processes to obtain anhydrous bioethanol.

13.1.4.3 Third-Generation (3G) Bioethanol

Several technical challenges are associated with the production of 2G bioethanol. High production costs, the requirement of advanced technology, and reduced accessibility to cellulose due to the presence of lignin are some of them. To overcome these issues, the third-generation (3G) bioethanol uses marine microorganisms, like algae and seaweeds as feedstock. Using algal biomass for ethanol production has certain advantages. They are rich in lipids and carbohydrates. They can be cultivated in a wide variety of environments like freshwater, wastewater, or seawater, without using any nutrients. Algae contain a lower amount of lignin and hemicellulose compared to 2G feedstocks. Besides, the process is carbon neutral as algae absorb the CO_2 released in the process during photosynthesis. Algae can be classified into two categories: (i) microalgae and (ii) macroalgae. Previously, only microalgae were used for bioethanol production. However, in recent years, macroalgae have also been brought to the research focus as a third-generation feedstock.Microalgae are unicellular and can float on water. They can be grown in water with a wide range of pH values, temperatures, and nutrient content, and lighting conditions. Microalgae are rich in carbohydrates and lipids, which makes them an ideal feedstock. Some common examples of microalgae, being used for bioethanol production are *Chlorococcum infusionum*, *Chlamydomonas reinhardtii*, and *Chlorella vulgaris* (Daroch et al. 2013).

Macroalgae are usually known as seaweeds. They are multicellular and are mostly attached to underwater anchorage structures like rocks. They can be classified into three categories: brown seaweed (Phaeophyceae), red seaweed (Rhodophyceae), and green seaweed (Chlorophyceae). The carbohydrate, lipid, and protein content vary with the type. Seaweed does not contain lignin, hence, allowing its easier conversion to simple sugars. Red seaweed has the maximum carbohydrate among all three categories. Some common examples of seaweeds used to produce bioethanol are *Ulva lactuca* (green), *Gelidium amansii* (red), and *Laminaria hyperborea* (brown).The bioconversion of macro and microalgal biomass to ethanol involves three steps: (i) pretreatment, (ii) hydrolysis, and (iii) fermentation (Fig. 13.8).

Pretreatment

The first step in the pretreatment of algal feedstock is drying them, and to prevent the formation of a gelatinous substance. Next, various physical processes like cutting, crushing, milling, and homogenisation are employed to reduce the size and increase the surface area of the dried feedstock. A technique called hydrothermal fractionation is often used to liquefy the dried feedstock using distilled water to facilitate the separation of various cellular components.

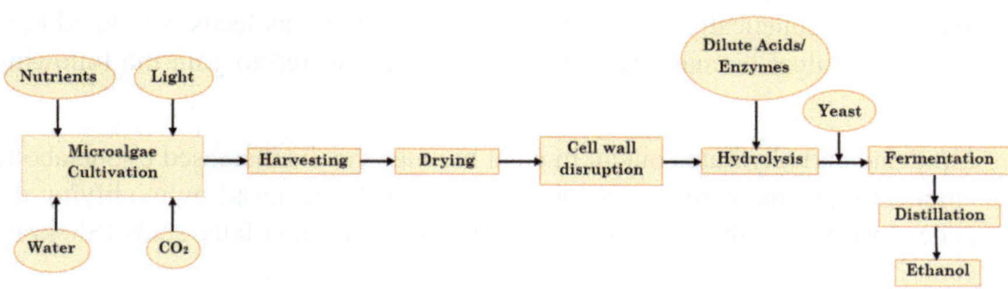

Fig. 13.8 Schematic of steps involved in the production of 3G bioethanol

Hydrolysis

The cell walls of algal biomass need to be depolymerized to release the polysaccharide contents like alginates, laminaran, fucans, carrageenans, ulvans, etc. (Jambo et al. 2016). The polysaccharides can be readily fermented to form bioethanol. The two methods implemented for hydrolysis of algal biomass are: (i) acid hydrolysis and (ii) enzymatic Hydrolysis.

Acid hydrolysis uses dilute mineral acids, like sulphuric and hydrochloric acid at 120–130 °C to facilitate the release of polysaccharides by depolymerizing the cell wall. It also helps in hydrolysis of the polysaccharides to fermentable sugar. The hydrogen bonds, making long chains of polysaccharides, get ruptured in the presence of acids. As a result, the polysaccharides become amorphous. Acid also acts as a catalyst in the subsequent conversion of the polysaccharides to monomeric sugar. However, acid hydrolysis is known to form inhibitors like levulinic acid (LA) and 5-hydroxymethyl furfural (HMF), which causes the degradation of fermentable sugar (Handbook of Biofuels nd).

Enzymatic hydrolysis or Enzymatic Saccharification, on the other hand, does not form any inhibitors. The cell wall of algal biomass is mostly made up of cellulose. Enzymes are used to rupture this cellulosic structure and depolymerize the polysaccharides. An enzyme called cellulase is commonly used for this purpose. The enzyme used mainly depends on the nature of the feedstock. This process is considered more eco-friendly compared to acid hydrolysis. However, the high cost of enzymes may impose an economic constraint on the production process.

Fermentation

The sugar monomers obtained as a product of hydrolysis are subjected to fermentation in the presence of microorganisms like bacteria, yeast, fungi, etc. to form ethanol. The choice of the microorganism used is a crucial step in the process of fermentation as it directly affects the yield of alcohol. The product of the fermentation process undergoes extensive purification and recovery to reach its final form.

13.1.4.4 Fourth-Generation (4G) Bioethanol

The main drawback of producing third-generation bioethanol is its low yield, which decreases its viability from an economic point of view. The fourth-generation (4G) bioethanol uses genetically modified (GM) algal biomass as feedstock, to address this issue. The algal biomass can be genetically engineered to gain the following advantages:

1. Lipid and carbohydrate content in algal biomass can be increased by metabolic engineering to increase the ethanol yield. It can be achieved by modifying the genes involved in the biosynthesis of triacylglycerol and fatty acids (Shokravi et al. 2021).
2. Increased photosynthetic efficiency can be achieved by expanding the spectrum range of light absorbed for photosynthesis.

3. Light penetration into densely cultivated algal biomass can be improved by truncating the chlorophyll antenna of the chloroplast (Abdullah et al. 2019).
4. The presence of lignin makes the cellulose and hemicellulose inaccessible for bioconversion. Lignin content can be reduced by modifying the structure of the cell wall and it can help to bypass the expensive and time-consuming pretreatment processes.
5. The high cost of enzymes for enzymatic hydrolysis can be reduced by genetically modifying the algal feedstock to express the enzymes. Yeast can also express cellulase enzyme upon genetic modification to facilitate simultaneous saccharification and fermentation (Devi et al. 2023).

Fermenting microbes can be genetically modified to use pentose sugar for fermentation, thereby increasing ethanol yield.

13.2 Strategies for Using Ethanol in a CI Engine

Several strategies have been developed by researchers to replace varying fractions of diesel energy shared by ethanol. Ethanol can be used to fuel a CI engine in four different ways. One way is to use ethanol by fumigation, which involves injecting ethanol in the stream of intake air followed by in-cylinder injection of diesel fuel. However, the most popular way to fuel CI engines with ethanol is to mix different proportions of ethanol with diesel. There are two ways to blend ethanol with diesel fuel: Emulsification and the addition of cosolvents. The fourth strategy is to use ethanol and diesel in dual-fuel mode in a CI engine. The description of each of these four methods has been discussed in this part along with some previous research works relevant to each technique.

13.2.1 Fumigation

In this process, ethanol or ethanol–water solution is injected into the stream of intake air by using an injector, vaporizer, or carburettor. Diesel fuel is injected into the cylinder as usual. The injected alcohol undergoes compression and then ignites along with air and diesel. In this way, ethanol is utilized to produce a certain fraction of engine power. The alcohol injection can be both single-point and multipoint. In a turbocharged engine, the injector can be placed either upstream of the turbocharger or in between the turbocharger and the intake manifold

Fumigation of ethanol has certain advantages. The first one is the increased energy contribution from ethanol. When ethanol is blended with diesel, the maximum energy share of ethanol has been observed to be 25% due to its limited miscibility with diesel. However, extensive research has shown that ethanol, when used in fumigation mode can contribute up to 50% of the total energy input by fuel (Broukhiyan and Lestz

1981). Some studies have also increased the energy share of ethanol well beyond this limit. In such cases, a small pilot injection of diesel is used just to ignite the ethanol fuel. Moreover, diesel changes its properties when blended with alcohol, which affects engine performance. Decrease of viscosity, drop of cetane number, and heating value are some of them. Using ethanol in fumigation mode is a solution to these issues. Besides, ethanol has a higher latent heat of evaporation than diesel. The temperature of the intake air reduces when ethanol is vaporized after being injected into the stream of intake air. This increases the density of air and allows the delivery of more air to the cylinder. This increases the power produced by the engine with the addition of the right amount of fuel. This cooling effect of ethanol also helps to reduce NO_x in the exhaust, especially at low loads. However, the presence of ethanol promotes the formation of NO, thereby increasing its emission in the exhaust. Using ethanol fumigation, the peak in-cylinder pressure is found to be lower than that of baseline diesel at low loads and equivalent to it at medium loads. This is due to the lower combustion temperature, and longer ignition delay at lower load ranges, which allows the start of combustion to occur further away from TDC. At higher loads, the peak in-cylinder pressure is higher than diesel. This is attributed to higher combustion temperature and shorter ignition delays at high load ranges. (Ghadikolaei et al. 2019) (Fig. 13.9).

For implementing ethanol fumigation, the engine needs to be modified to incorporate a separate fuel supply system to deliver the ethanol fuel. This adds to the design complexities and increases the weight of the engine. On the bright side, having a separate fuel system for ethanol allows flexible switching between single-fuel and dual-fuel modes. Some important experimental results involving ethanol fumigation are included in Table 13.2.

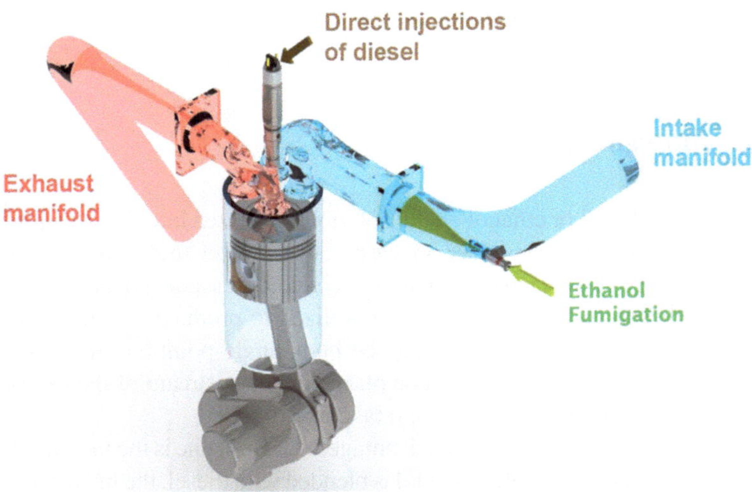

Fig. 13.9 Fumigation of ethanol into the intake air in a diesel engine (Reproduced from: Pedrozo et al. (2018))

Table 13.2 Important studies on ethanol fumigation

Year	Researcher	Test fuel	Remarks
1981	Broukhiyan and Lestz (1981)	Up to 50% of energy is supplied by ethanol. Remaining by diesel	Ethanol was fumigated in the intake manifold using air-atomizing nozzle. Thermal efficiency increased due to ethanol fumigation at higher rack conditions, i.e. higher load. Ethanol fumigation decreased NO_x for all test conditions. It is noted that at lower rack conditions ethanol fumigation produced more NO_2 than NO. Ethanol fumigation increased CO and HC emissions for all test conditions. It also decreased the mass of particulates but increased their biological activities
1986	Savage et al. (1986)	Ethanol (amount fumigated was adjusted until the misfire stops and knocking is within acceptable limits) + diesel	Ethanol was fumigated using three different injection strategies. First, half the ethanol was fumigated during the revolution with intake valve open in the intake stroke and the other half with the intake valve closed in the power stroke. The other two strategies were fumigating the entire volume in the intake stroke and the power stroke, respectively. The presence of ethanol increased thermal efficiency slightly at high loads. It decreased NO in exhaust at low and medium loads. However, no significant improvement was observed in high loads. HC and CO emissions increased under all conditions
1990	Van Gerpen and Van Meter (1990)	10%, 20%, and 30% torque replaced by ethanol. Remaining from diesel. Up to 170-proof ethanol was used	The effect of ethanol fumigation during the transient operation of the engine was studied. NO_x decreased, HC and CO increased in exhaust as the percentage of engine torque replaced by ethanol increased from 10 to 30%. As proof of ethanol increased, the NO_x emissions increased with little effect on HC and CO. 80 proof ethanol was said to be optimum. (Proof is the measure of alcohol content. It is defined as twice the percentage of alcohol by volume)
1999	Abu-Qudais et al. (nd)	D100, E20D80 (used in both fumigation and blending mode)	Diesel injection timing was maintained at 20° *btdc* with a compression ratio of 18 throughout the experiment. Brake thermal efficiency increased more in fumigation than blending. HC and CO emissions increased more in blending than in fumigation. Soot decreased more by fumigation compared to blending
2010	Tsang et al. (2010)	D100 (Euro V), E5D95, E10D90, E15D85, E20D80	Diesel was injected at 8° *btdc*. Ethanol is injected in the inlet manifold at a pressure of 0.35 MPa. A decrease in NO_x is observed by ethanol fumigation. The reduction was lower at higher loads. NO2 emission increased at lower loads and with increasing ethanol content. HC and CO emissions increased while PM, PN, and smoke decreased at all loads

(continued)

Table 13.2 (continued)

Year	Researcher	Test fuel	Remarks
2011	Zhang et al. (2011)	D100, E10D90, E20D80	Ethanol was injected in the inlet manifold at a pressure of 0.35 MPa. NO_x decreased with increasing ethanol percentage at all loads. NO_2 emission increased with increasing ethanol percentage at medium loads. HC and CO emissions increased while PM and PN decreased with increasing ethanol percentage at all loads
2015	Şahin et al. (2015)	D100, E2D98, E4D96, E6D94, E8D92, E10D90, E12D88	Tests were done for half, three quarter, and full rack conditions at three different engine speeds: 2000 rpm, 2500 rpm, and 3000 rpm. Ethanol is fumigated in the inlet manifold using a carburettor. NO_x decreased with increasing ethanol content at all load ranges. This effect was more pronounced for higher loads
2015	Morsy (2015)	Neat diesel, diesel + 25% v/v, 50% v/v, 75%v/v, and 100% v/v ethanol/ water solution	Ethanol/water was injected in the intake manifold using a BoschEV6 port fuel injector at a constant injection pressure of 1 MPa. NO emission was lower than neat diesel values while using 25% v/v, 50% v/ v, and 75% v/v ethanol/Water solution at all engine speeds. However, it was more than that of neat diesel at all speeds when 100% ethanol + diesel was used. HC and CO emissions were more than neat diesel at all speeds, but they decreased with increasing fraction of ethanol in the ethanol/water solution
2019	Ghadikolaei et al. (2019)	D100, D80B5E15 in fumigation, fumigation + blending, and blending mode	Diesel was injected at 8° *btdc*. Ethanol was injected into the intake manifold at a pressure of 0.35 MPa through four injectors, one for each cylinder. Fumigation increased CO and HC more than the blending mode. NO_x decreased but NO_2 increased more by fumigation compared to blending

13.2.2 Blending by Addition of Cosolvents

Diesel-ethanol solution tends to undergo phase separation while in storage. The soluble temperature (the temperature beyond which phase separation is observed) increases to reach a maximum of about 40 °C when the ethanol content of the blend is nearly 50% (v/v) and then decreases. The phase stability also worsens with increasing water content of the blend when exposed to the moisture in air (Figs. 13.10 and 13.11).

The phase stability of the diesel-ethanol solution can be improved by adding cosolvents. The cosolvents possess polar extremities. The hydrocarbon chain of cosolvents is non-polar which can interact with the non-polar diesel molecules with Van der Waals forces. These forces are generated due to transitory dipoles formed in the molecules due to electron movement. Attached to the hydrocarbon chain there are

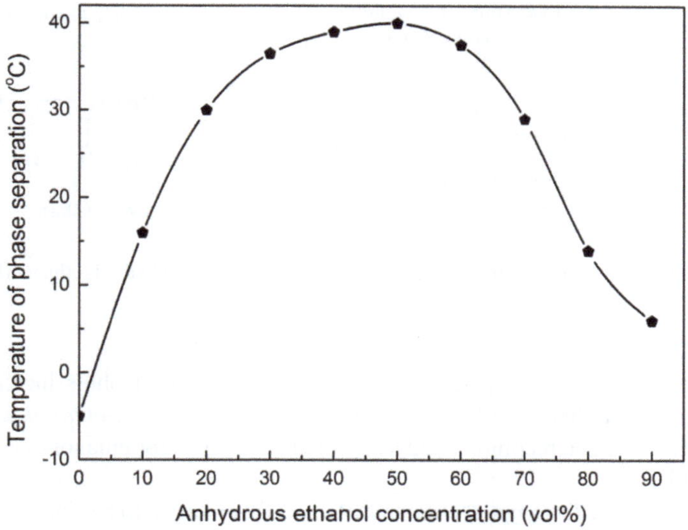

Fig. 13.10 Temperature of phase separation of anhydrous-ethanol/diesel blends (Liu et al. 2016)

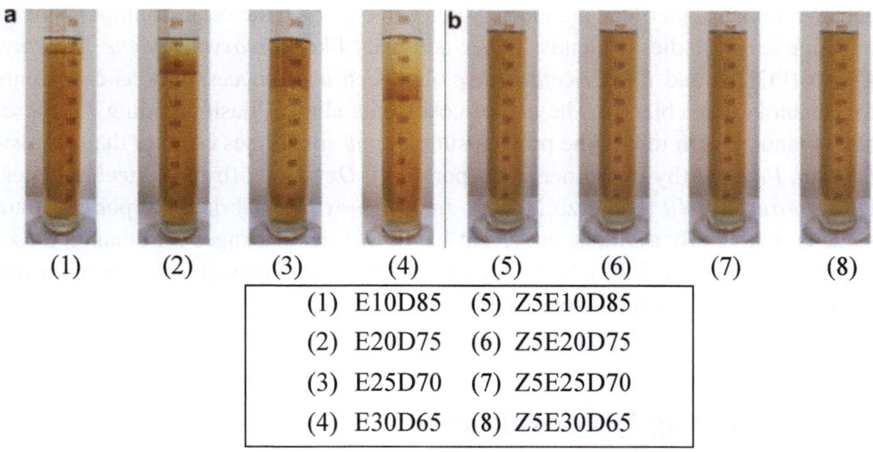

Fig. 13.11 **a** Phase separation in diesel-ethanol solutions of varying proportions without cosolvents, **b** homogenous ethanol–diesel blends produced after adding 5% (v/v) n-butanol (Z) as cosolvent and stirring (Huang et al. 2009)

various hydrogen donor and/or acceptor groups that interact with the polar ethanol and water molecules by making hydrogen bridges. Alcohols with higher carbon numbers like *butanol*, *hexanol*, *octanol*, and *dodecanol* when used as cosolvents can enhance the miscibility of hydrous ethanol-in-diesel. Studies have inferred that straight-chain alcohols are better as cosolvents compared to cyclic and branched-chain isomers. However, *dodecanol* when used as a cosolvent leads to gel formation

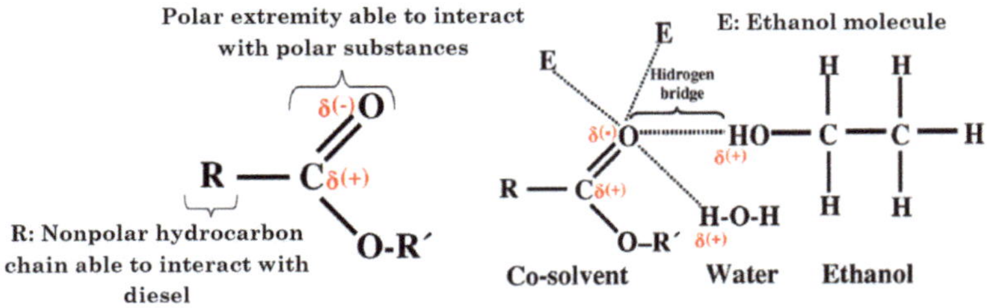

Fig. 13.12 Polarity of cosolvents and their interaction with diesel and hydrous ethanol (Reyes et al. 2009)

in the blends at lower temperatures, as the pour point of alcohols increases with the increasing number of carbon atoms. Keeping this in mind, *n-butanol, n-hexanol*, and *n-octanol* can be recommended as optimal cosolvents for ethanol–diesel blends (Fig. 13.12).

In recent times, *Tetrahydrofuran* or THF has also emerged to be a better cosolvent than higher alcohols, offering better solubility of ethanol–diesel blends. THF has higher density and higher oxygen content than alcohols, which adds to the emission and performance benefits of using ethanol-blended diesel. Moreover, THF also has a cetane number higher than alcohols, which can cut down the cetane number penalty of using ethanol–diesel blends. Other additives like *Polyoxymethylene Dimethyl Ethers* (PODE) and *Ethyl Acetate* have also been used successfully as cosolvents for ethanol–diesel blends. The use of cosolvents allows 'flash-blending' of diesel and ethanol, which makes the process simpler and sometimes cheaper than emulsification. *Puranol* (by Pure Energy Corporation), *OxyDiesel* (by AAE technologies, UK), *Beraid ED-10* (by Akzo Nobel), and *Purinox* by (Lubrizol Corporation) are some commercially available cosolvent-based additive packages for ethanol–diesel blends (Ribeiro et al. 2007). Some important experimental results relevant to the use of cosolvents are listed in Table 13.3.

13.2.3 Blending by Emulsification

Emulsification is a process often used to produce thermodynamically stable homogenous ethanol–diesel blends. An emulsion is an opaque system containing two immiscible liquids, one dispersed in the other in the form of small droplets, with or without the use of an emulsifier. Generally, one of these liquids is polar and the other is non-polar. The droplets of the dispersed phase are usually greater than 400 nm in size. Phase separation is observed in an emulsion after a certain amount of time. Emulsions can be of two types: oil-in-water and water-in-oil. Diesel-ethanol emulsion is a water-in-oil type emulsion with ethanol as water and diesel as oil phase. However, ethanol–diesel blends are not exactly emulsions; rather they can be more specifically categorized as microemulsions. Microemulsions are thermodynamically

Table 13.3 Important studies on ethanol blending by adding cosolvents

Year	Researcher	Test fuel	Remarks
2009	Huang et al. (2009)	E10D90, E20D80, E25D75, E30D70,Z Z5E10D85, Z5E20D75, Z5E25D70, Z5E30D65 (Z represents n-butanol as additive)	The blended fuel was injected at 18° *btdc* at 16 MPa FIP. Thermal efficiencies tested at different loads and speeds remain almost unchanged by using ethanol–diesel blends. CO emissions were less compared to baseline diesel beyond half load. HC emissions were higher but decreased with increasing load. No stable trend was obtained for NO_x. Smoke emissions were reduced under all tested conditions
2015	Gnanamoorthi and Devaradjane (2015)	D100, E10D88, E20D78, E30D68, E40D58 with 1% (v/v) ethyl acetate and 1% (v/v) diethyl carbonate as additive	Combustion, performance, and emission analysis were performed at three different compression ratios of 17.5:1, 18.5:1, and 19.5:1. The FIP was 255 bar. CO and HC emissions increased with increasing ethanol content for all loads (more increment at lower load). NO_x decreased with increasing ethanol content compared to diesel. Smoke emission was lower than diesel at all loads at an elevated compression ratio
2019	Lee et al. (2019)	D100, IBE20D80, IBE40D60 (IBE represents isopropanol-butanol-ethanol additive blended)	Tested at CVCC. Heat release rate, pressure rise, and ignition delay of IBE blended diesel were similar to baseline diesel. Although, at decreased ambient conditions ignition delay increased with higher IBE content. Shorter spray penetration length of IBE blends caused by increased evaporation. Decreased flame luminosity at higher IBE content indicated lower soot emissions

(continued)

Table 13.3 (continued)

Year	Researcher	Test fuel	Remarks
2019	Nour et al. (2019)	D100, Pe10E10D80, Oc10E10D80 (Pe and Oc represent pentanol and octanol additives, respectively.)	The blended fuel was injected at 24° *btdc* at 17.5 MPa FIP. Brake thermal efficiency increased with Pe10E10D80 but decreased with Oc10E10D80 compared to D100. NO_x and smoke opacity decreased more with octanol than with pentanol additive (both less than that of D100). Both CO and CO_2 emission decreased with Pe10E10D80 and Oc10E10D80 than D100
2019	Liu et al. (2019)	D100, D80P20, D75P20E5, D65P20E15 (P represents PODE additive)	The blended fuel was injected at 600 bar, 750 bar, and 1000 bar FIP for low, medium, and high load conditions, respectively. The injection timing was 2° *atdc* for low load and 3° *atdc* for medium and high load. High-pressure common rail direct injection system was used. Brake thermal efficiency using D100, and blended fuels was almost similar. CO decreased using blended fuels at all loads while CO_2 increased at higher loads using blends. HC emissions decreased more at lower loads using blends NO_x increased and soot decreased at all loads

(continued)

Table 13.3 (continued)

Year	Researcher	Test fuel	Remarks
2020	Wu et al. (2020)	D100, D95E0T5, D85E10T5, D65E30T5 (T represents THF additive)	The blended fuel was injected by a common rail direct injection system at 750 bar for low load and at 1020 bar for high load conditions. Injection timing was 2° *atdc*, without any pre-injection, for all conditions tested. Increased brake thermal efficiency was observed by using D85E10T5 and D65E30T5. Compared to D100, CO emissions decreased by using D95E0T5 and D85E10T5 but increased with D65E30T5. HC and NO_x increased while soot decreased compared to D100 by using D95E0T5, D85E10T5, and D65E30T5, respectively
2021	Liang et al. (2021)	D100, E20T0D80, E20T10D70, E30T10D60 (T represents THF additive)	The effect of varying EGR rates on emissions by using the blends specified was studied. The indicated thermal efficiency of the ethanol and THF blended fuel was higher at a lower EGR rate and less at a higher EGR percentage compared to neat diesel. CO emissions increased with increasing EGR rate and decreased with increased ethanol content. HC increased with increasing EGR rate and with increasing ethanol content. NO_x decreased with increasing EGR rate. At lower EGR rates, NO_x increased with increasing ethanol content, but the trend is reversed at higher EGR rates. Soot emission increased and then decreased with increasing EGR rate for all blends except for E30T10D60, where no substantial change was observed. At all tested conditions soot was less than D100

and chemically stable colloidal dispersions of a liquid/dispersed phase in a host/continuous phase. The micelles or droplets of the dispersed phase of a microemulsion are 100 nm to 400 nm in size and are smaller than that of an emulsion. This makes the microemulsions semi-opaque and bluish white (Fig. 13.13).

Diesel-ethanol microemulsions are produced by adding various emulsifiers, with gradual heating and mechanical mixing processes. The emulsifiers have a characteristic called *amphiphilicity*. This means they have both '*hydrophilic*' (water-loving or

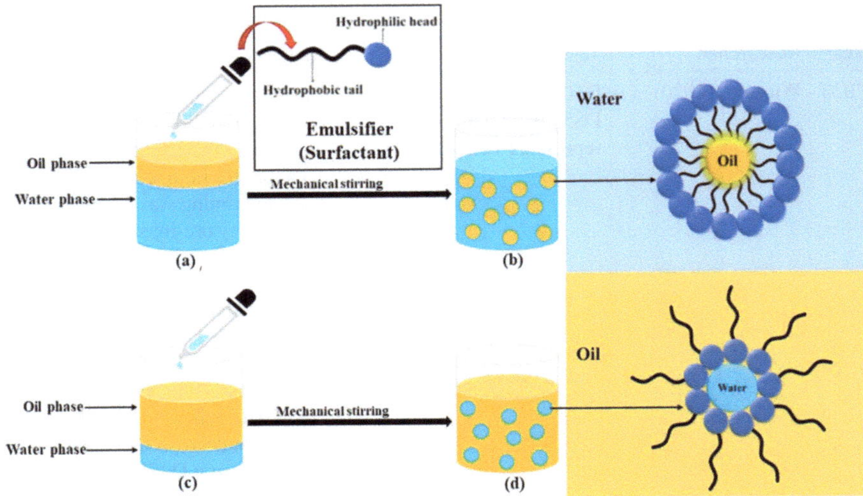

Fig. 13.13 Working of emulsifiers to produce oil-in-water and water-in-oil emulsions (Singh and Pulikkal 2022)

polar) and '*lipophilic*' (fat-loving or non-polar) properties. This property gives them affinity towards both polar and non-polar molecules. HLB or '*Hydrophilic Lipophilic Balance*' gives every amphiphile a value between 0 and 20. Mathematically HLB can be expressed as:

$$\text{HLB} = 20\frac{M_h}{M},$$

where M_h is the mass of the hydrophilic part of the molecule and M is the total mass of the molecule. HLB value of 0 means completely lipophilic and 20 means completely hydrophilic. Surfactants or surface-acting agents are one such widely used amphiphilic emulsifier used to make diesel-ethanol emulsion. Surfactants, by their affinity towards both polar ethanol/water molecules and non-polar diesel molecules, decrease the interfacial surface tension between the two liquids allowing the formation of diesel-ethanol emulsion containing small diesel droplets dispersed in ethanol. Biodiesel, which is essentially methyl or ethyl esters, is another majorly used amphiphilic emulsifier for diesel-ethanol emulsion. Studies show that when diesel concentration is higher the ratio of biodiesel to ethanol should be more than 1:1, to prepare a stable microemulsion (Ribeiro et al. 2007). Various other emulsifiers like *SPAN80* or *TWEEN80* (sorbitan ester), *AEP-102* (Soybean methyl ester), castor oil, fluorine oil, tea oil, acid glyceride, tung oil are also used. Adding emulsifiers to ethanol–diesel blends also enhances their lubricity and low-temperature properties. Important research works involving the blending of ethanol and diesel by emulsification are presented in Table 13.4.

Table 13.4 Important studies on ethanol blending by emulsification

Year	Researcher	Test fuel	Remarks
2003	He et al. (2003)	D100, E10D90, E30D70, E10D88AI2, E30D68AI2 (AI represents an additive prepared based on principles of both emulsification and cosolvents)	Fuel was injected at 21° *btdc* for all test conditions. 0.1% v/v and 1% v/v isooctyl nitrate were added as ignition improvers to the E10D88AI2 and E30D68AI2 blends, respectively. At high loads, NO_x, smoke, and CO_2 emissions decreased with increasing ethanol content. However, an increased fraction of ethanol caused an increase in CO, acetaldehyde, and HC emissions at all loads. By adding an ignition improver and additive, HC emissions decreased below baseline diesel values. The effect of ethanol blending was less prominent at lower loads
2006	Kwanchareon et al. (2007)	D100, B100, D90B10E0, D90B5E5, D90B0E10, D85B15E0, D85B10E5, D85B5E10, D85B0E15, D80B15E5, D80B10E10, D80B5E15 (B represents biodiesel)	Biodiesel derived from palm oil was effective in maintaining the stability of the ethanol–diesel blends. CO in exhaust decreased at full load. D80B15E5 had the least CO emission at full load. Blends having higher ethanol content had higher HC emissions. However, the presence of biodiesel helps to lower HC. At low load, no significant improvement in HC and CO emission was observed. All blended fuels reduced NO_x emission at no-load conditions. At all other load ranges, especially at high loads, NO_x emissions increased by blending ethanol and biodiesel with diesel
2008	Rakopoulos et al. (2008)	D100, D95E5, D90E10. An emulsifier by 'Betz GE' was used without any ignition improver	Fuel was injected at 21° *btdc* with 250 bar FIP. Smoke emission was significantly reduced by using ethanol-blended diesel. CO emissions were slightly reduced by using ethanol–diesel blends. The reduction increased with the ethanol content of the blend. NO_x emission decreased slightly. HC emissions increased with increasing ethanol content

(continued)

Table 13.4 (continued)

Year	Researcher	Test fuel	Remarks
2011	Park et al. (2011)	D100, D90E10, D80E20. Soybean methyl ester is added as an emulsifier	The effect of ethanol blending on combustion, performance, emission, and overall spray characteristics of a four-cylinder diesel engine was studied at different engine loads and injection timings. The four injection timings studied were: *tdc*, 3° *btdc*, 6° *btdc*, and 9° *btdc*. The ethanol-blended fuel showed shorter spray tip penetration length than conventional diesel. Unstable ignition characteristics were observed when the fuel is injected at the top dead centre. An increase in combustion pressure was observed by advancing the injection timing due to sufficient premixed duration. An increase in ethanol content decreased NO_X emission. NO_X increased with increasing engine load and injection timing advance. HC and CO emissions were more from ethanol-blended diesel. With increasing engine load and injection timing advance, the HC and CO decreased in the exhaust
2012	Lei et al. (2012)	D100, D95E5, D90E10, D85E15. A novel emulsifier called *CLZ* was used	Biofuel, castor oil, and some other single emulsifiers were mixed to develop a novel emulsifier called '*CLZ*' with HLB values 4.1–4.4. *CLZ* could maintain the stability of blends containing 10% and 15% ethanol at a wide range of temperatures and blends containing 20% and 25% ethanol under normal temperatures. Smoke in exhaust decreased, especially at high loads by using ethanol–diesel blends. CO emissions increased at low loads but decreased with increasing load. NO_X emission decreased and emission of HC increased
2019	Boggavarapu and Ravikrishna (2019)	D100, D60E20B20 (B represents *Jatropha Methyl Ester* or biodiesel used as an emulsifier)	The evaporative spray characteristics of the fuel blend were studied in engine-like conditions in terms of vapour penetration, liquid penetration, liquid length, and spray cone angle. A striking similarity was observed between the spray characteristics of the blended fuel and regular diesel fuel. The study inferred the feasibility of using the tested blend in a conventional diesel engine without any modifications

(continued)

Table 13.4 (continued)

Year	Researcher	Test fuel	Remarks
2021	Srikanth et al. (2021)	D100, D95E5, D90E10, D85E15, D85E5B10, D80E10B10, D75E15B10, D75E5B20, D70E10B20, D65E15B20. (B represents biodiesel) Niger seed oil Methyl Ester was added as an emulsifier	Niger seed oil methyl ester was synthesized as an emulsifier for the ethanol–diesel blends and was added from 5 to 20% v/v to maintain blend stability. Fuel was injected at 23° *btdc* with 220 bar FIP. The brake thermal efficiency was observed to be highest for the D75E15B10 blend. CO_2 emissions increased with increasing ethanol content with the maximum being for D75E15B10. The highest NO_x emission was observed for D65E15B20. The smoke density was also the lowest for D65E15B20
2023	Nageswara Rao and Sreenivasulu Reddy (2023)	D100, D46B50E4, D42B50E8, D38B50E12	With increasing ethanol content, the brake thermal efficiency decreased slightly, while HC and CO emissions increased. NO_x emissions are reduced, especially at higher loads by using blended fuels

13.2.4 Dual-Fuel Mode

Another technique used to utilize ethanol in CI engines is to develop diesel-ethanol dual-fuel engines. Two approaches are commonly used to use ethanol in dual-fuel mode. The first one is the diesel-ethanol RCCI or Reactivity-Controlled Compression Ignition combustion strategy. In this technique, the low-reactivity ethanol fuel is injected in the intake port at comparatively low FIP, and the high reactivity diesel fuel is injected directly into the cylinder. This establishes a reactivity gradient in the combustion chamber (Fig. 13.14).

By adjusting the proportion of high and low-reactivity fuels, along with optimizations of the injection timing, and other means to control combustion like EGR rate and turbocharger boost, better control in combustion phasing can be achieved. Although this method might appear to have similar arrangements to the fumigation method, one major difference between the two is that in RCCI, the ethanol is injected in the intake port while in fumigation ethanol is injected further upstream in the intake manifold. Studies show that RCCI helps to beat the soot-NO_x trade-off by decreasing soot and NO_x simultaneously. It also increases the thermal efficiency of the engine. The second method is diesel-ethanol direct injection where both diesel and ethanol fuel are injected directly into the combustion chamber by using two separate fuel delivery systems and injectors. By injecting both the high and low-reactivity fuel directly into the combustion chamber, an in-cylinder reactivity gradient higher than RCCI can be achieved which gives more control over combustion phasing. By following this strategy, researchers have used both options: injection of ethanol before the diesel pilot injection and injection of diesel before the ethanol injection.

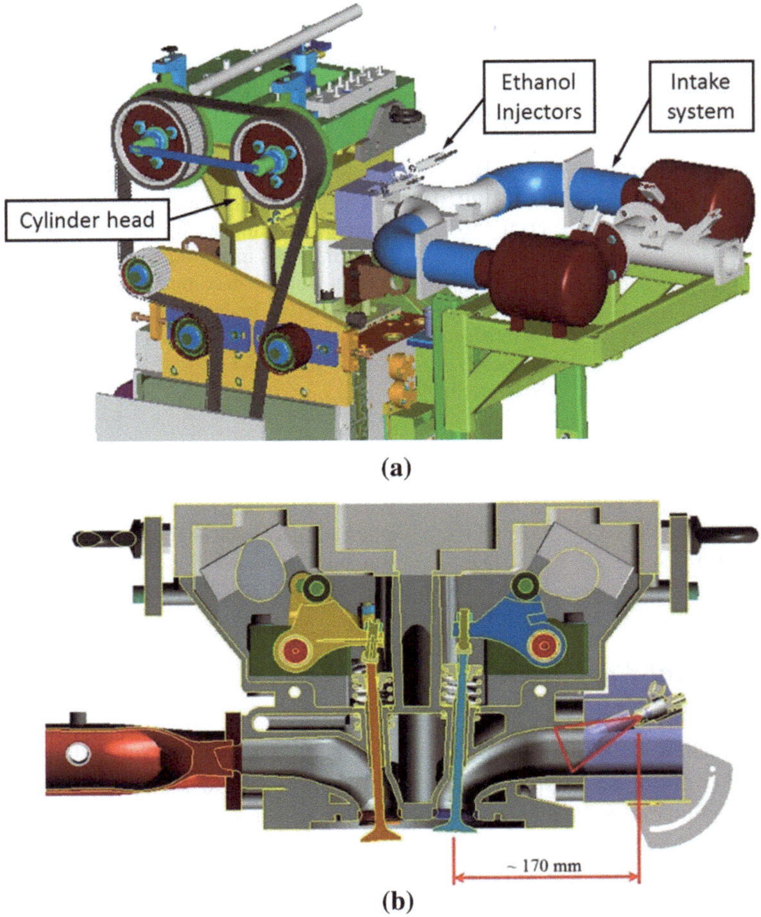

Fig. 13.14 Location (**a**) and spray cone (**b**) of ethanol injectors (2015)

Split injection of the diesel fuel has also been investigated and is beneficial beyond 50% energy share of ethanol. PFI of ethanol and DI of diesel in RCCI mode allow gasoline-like homogenous and premixed combustion. On the contrary, the DI of both diesel and ethanol allows diffusion combustion to achieve better fuel conversion efficiency. Moreover, the RCCI strategy shows increased HC and CO emissions along with the elevated rate of pressure rises at higher loads due to squish and crevices. Direct injection of ethanol and diesel better performance is achieved at an extended load range. However, the DI of both diesel and ethanol calls for added complexities to the design of the cylinder head to accommodate separate injectors for diesel and ethanol. Some notable research using both RCCI and direct injection strategies are listed in Table 13.5.

Table 13.5 Important studies on ethanol blending in dual-fuel mode

Year	Researcher	Test fuel	Remarks
2012	Boretti (2012)	–	In-cylinder combustion using DI of both diesel and ethanol has been modelled using WAVE Diesel Wiebe Combustion object. Reduced CO_2, smoke, and particulate emissions have been pointed out to be some of the advantages
2013	Sarjovaara et al. (2014)	40 and 80% of diesel substitution by ethanol at lower load and higher load, respectively	The effect of split injection on the PFI of ethanol and DI of diesel was studied with alterations of the injection timings. The percentage of ethanol substitution was limited in both low and high loads without split injection. It was observed that it is beneficial to divide the diesel injection into more than one portion if more than 50% ethanol energy substitution is desired. With the implementation of split injection and adjustments of pre and main-injection timings, a substantial increase has been noticed in the ethanol energy share. 80% of the diesel energy was substituted by ethanol at 75% and 100% loads
2014	Britto and Martins (Britto and Martins 2015)	Diesel substitution by ethanol ranging from 10% to approx. 50% using low-flow injectors and 25–65% using high-flow injectors	DI of diesel and PFI of ethanol has been studied under varying compression ratios (14:1, 16:1, and 17:1), diesel injection flow (35 g/s ad 45 g/s), diesel injection pressure (800 bar,1000 bar,1200 bar, and 1400 bar) and different in-cylinder turbulence type (quiescent and high swirl) has been studied. Diesel injection timing and energetic substitution rate of diesel by ethanol were the control variables for engine calibration at each test condition. The highest substitution of diesel by ethanol was observed to be 65% at 17:1 compression ratio using a high-flow injector having flow rate of 45 g/s, in a high swirl combustion chamber. The maximum diesel injection pulse width window for this condition was 19° *btdc* to 14° *atdc*
2015	Britto and Martins (2016)	Up to 65% diesel substitution by ethanol	The engine used in the previous study was run at a 17:1 compression ratio using a high-flow injector (45 g/s) at a high swirl combustion chamber for emission measurements. NO_x emission was found to be reduced by up to 60% with an increase in in emission of HC, CO and aldehydes
2016	Pedrozo et al. (2013)	Up to 69% diesel substitution by ethanol	PFI of ethanol and DI of diesel were studied at 0.615 MPa IMEP and 1200 rpm. The aim was improvement of combustion efficiency, maximization of ethanol substitution and minimization of Soot and NO_x. Effect of different diesel injection strategies on combustion, emission and efficiency was studied. The optimum injection strategy was found to be an ethanol substitution ratio of 54% on energy basis, with diesel pre-injection timing around 36.5° *atdc* with a split ratio corresponding to 53–55% of total energizing time. Soot and NO_x emissions were reduced by 23% and 65% respectively

(continued)

Table 13.5 (continued)

Year	Researcher	Test fuel	Remarks
2018	Lee et al. (2018)	Ethanol substitution by 0%, 36%, 51% at 0.2 MPa IMEP; 0%,17%, 28%, 36%, 48% at 0.4 MPa IMEP, 0%, 11%, 33%, 53%, 63%, 71% at 0.6 MPa IMEP and 0%, 14%, 27%, 42%, 56% at 0.8MPA IMEP	Both ethanol and diesel were injected directly into the cylinder using two separate injectors. The FIP for diesel was 120 MPa and for ethanol was 20 MPa. The split injection timings for diesel at different IMEP (MPa) were as follows: 0.2 MPa—Pilot 20° *btdc*, Main 10° *btdc*; 0.4 MPa—Pilot 17° *btdc*, Main 7° *btdc*; 0.6 MPa—Pilot 15° *btdc*, Main 5° *btdc*; 0.8 MPa—Pilot 15° *btdc*, Main 5° *btdc*. Ethanol injection timings for different IMEP (MPa) were as follows: 0.2 MPa—30° *btdc*, 0.4 MPa—60° *btdc*, 0.6 MPa—120° *btdc*, 0.8 MPa—180° *btdc*. Reduction in NO_X and PM was observed with increasing ethanol substitution percentage. High CO and HC emissions were observed at 0.2 and 0.4 MPa IMEP. Indicated thermal efficiency was observed to be lower for dual-fuel mode, compared to using neat diesel. Efficiency decreased at higher ethanol fractions without optimization of injection strategy. The study suggests suitable ethanol substitution fractions at different load conditions
2022	Frat et al. (2022)	–	Dual direct injection of both diesel and ethanol, RCCI (DI of diesel and PFI of ethanol), and conventional diesel injection were compared. Dual direct injection caused lower emissions of CO and HC compared to the RCCI strategy. For conventional diesel operation, the FIP was 300 bar and injection timing was 21° *btdc*. For dual direct injection mode, the FIP for diesel injection was 300 bar with injection timing at 21° *btdc*. In this mode, ethanol was injected at 15° *btdc* with FIP of 50 bar. In RCCI mode, diesel was injected at 21° *btdc* with 300 bar FIP, while ethanol was injected at 15° *btdc* with 5 bar FIP. Compared to port injection of ethanol, CO emission decreased by 41% at 60% load by using direct injection of ethanol. Port injection of ethanol decreased the NO_X emissions more significantly than direct injection of ethanol. NO_X reduced by 60% at 20% load by using direct injection of ethanol. Port injection of ethanol was also more successful in reducing smoke opacity

13.2.5 Surface Ignition of Ethanol

Substituting a fraction of the energy of the diesel fuel with ethanol causes ignition delay due to the higher auto-ignition temperature of ethanol. To overcome this issue, the ethanol fuel is ignited by injecting it directly onto a hot surface, whose temperature exceeds the auto-ignition temperature of ethanol. The surface ignition plug or glow plug is an insulating material with coils of heating wire wound around it. It is heated with electricity and placed in the combustion chamber so that the ethanol, just after injection, comes in contact with the surface ignition plug. This initiates the ignition. A pre-combustion chamber can also be used for the initiation of ignition by this process. Since the fuel burns on the heated surface and it is insulated, soon this ignition plug does not need any external supply of electricity to maintain its temperature. Surface ignition also has significant cold start benefits.

13.3 Safety Aspects of Ethanol Fuelled Engines

Ethanol is a highly flammable fuel. So, safety against fire hazards is an important concern when using ethanol as a fuel in a CI engine. Flammability range, volatility and self-ignition temperature are the most important properties of ethanol from the fire safety perspective. Besides, the density of fuel vapour, diffusivity, and source of injection also play an important role (Valera and Agarwal 2019). The potential fire risk of using ethanol fuel can be studied by considering two situations: (1) Fuel enclosed in a tank and (2) Fuel exposed to the atmosphere. Both these situations have been discussed in the following paragraphs.

13.3.1 Fuel Enclosed in a Tank

The flammability range of ethanol is much more than that of diesel as shown in Table 13.1. In storage fuel volatility, flammability limit and source of ignition are the key factors influencing fire risk, when stored in an enclosed space like a tank, the vapour of neat diesel does not form a combustible mixture below 64 °C as they are too lean. However, in the case of ethanol–diesel blends, the fuel vapour forms a combustible mixture under normal ambient temperature conditions, in an enclosed space when the ethanol fraction is more than 10% and it can cause an explosion when exposed to any source of ignition. Below 4% ethanol content, the properties of ethanol–diesel blends, like vapour pressure, and flammability range resemble that of diesel, but, beyond 10% the properties are like that of ethanol. So, E-diesel storage facilities like vehicular fuel tanks, underground storage tanks, and delivery trucks transporting ethanol–diesel blends have a combustible mixture of fuel vapours. Precautions should be taken so that no ignition source comes into the vicinity of these storage spaces. The most important sources of ignition are static discharge, sparks from welding and operator negligence.

The safety recommendations to avoid fire hazards from ethanol–diesel blends in an enclosed tank can be listed as follows (Waterland et al. 2003):

1. All the fuel storage tanks should have vents to remove the fuel vapour and the openings should be equipped with flame arresters. The drivers and passengers area of vehicles fuelled by ethanol–diesel blends should also be provided with proper ventilation, capable of minimum three air changes per minute.

2. Vehicles delivering ethanol–diesel blends should have a vapour recovery system and an electrical connection with the ground to avoid static discharge.
3. The fuel tanks in vehicles should be equipped with a flapper cap and fuel level detectors to prevent operators from opening the fuel tank cap to check the fuel level.
4. Galvanized steel should be used to make fuel tanks, and the distribution pipes should be made of plastic to avoid leakage.

13.3.2 Fuel Exposed to Atmosphere

Ethanol–diesel blends may spill and get exposed to the open atmosphere in case of any accidents while transportation and in filling stations. The density of fuel vapour, diffusivity, and source of ignition are the main factors influencing fire risk in such situations. However, ethanol–diesel blends do not have the potential to form flammable vapours when exposed to the atmosphere. A weathering effect prevents the formation of a vapour cloud over the spill. Moreover, low vapour pressure and density, high diffusivity, and high self-ignition temperature of ethanol–diesel blends help to mitigate the risk of fire.

However, when spilt and brought in contact with any sources of ignition, ethanol spills can still catch fire. Some important points should be kept in mind while dealing with such situations. They are as follows: (Safety and Procedural Requirements for type approval... 2020).

1. Ethanol fire is not always visible under normal lighting conditions. Some additives might increase the visibility of ethanol–diesel flames.
2. All ignition sources should be removed from the vicinity of the spill to prevent fire.
3. Water should never be used to extinguish ethanol fire. CO_2 or dry chemical extinguishers can be used. A special Alcohol-Resistant Foam, commonly known as *AR-AFFF*, is most effective in extinguishing ethanol fires. Vehicles using ethanol-blended fuel should be equipped with AR Foam based fire canisters.

13.4 Conclusion

A detailed review of the use of ethanol-blended diesel fuel in CI engines has been presented in this chapter. Various ways of producing renewable bioethanol have been discussed along with descriptions of several strategies implemented to use ethanol–diesel blends. The main technical barriers to the use of ethanol–diesel blends include poor miscibility of ethanol with diesel, high ignition delay, low heating value, poor lubricity, and corrosive nature. The use of emulsifiers and cosolvents solves the issue of poor miscibility. High ignition delay and low lubricity of ethanol can be dealt with by using various ignition improvers and lubricity-enhancing additives respectively. The low heating value of the fuel is compensated for by increasing the amount of fuel injected per cycle. The corrosive nature of ethanol calls for the use of ethanol-compatible materials for design purposes. Upon implementing the solutions recommended, ethanol–diesel blends can be used to fuel CI engines. The table below summarizes various aspects related to the use of ethanol as an alternate fuel in diesel engines, which have been discussed in the chapter:

Alternative fuel options for CI engines	• Amidst growing environmental concerns and dwindling reserves of non-renewable energy sources, ethanol has emerged out to be a potential candidate for a sustainable alternative energy source • Many countries worldwide, including India, have already undertaken ethanol-blended gasoline programmes. So, the infrastructure required to use ethanol as an energy source is already in place. Extensive research has shown that it is also possible to use ethanol as an alternative to diesel fuel in CI engines • Using ethanol as an energy source can help the economy of oil importing countries like India, by reducing the heavy dependency on imported fuel • As ethanol is produced by using agricultural products and crop residues as feedstocks, it can also be a boost to the agrarian economy
Production of ethanol	• Ethanol produced from renewable sources is called bioethanol. Bioethanol can be produced from four types of feedstocks: sugar, starch, lignocellulosic biomass, and algae. Depending upon the type of feedstock used, bioethanol can be classified into four categories • The majority of the bioethanol produced worldwide is of first generation, which is obtained from sugar and starch-based feedstocks. Second-generation bioethanol is produced from lignocellulosic biomass, which reduces the dependency on food crops to produce 1G bioethanol • Third and fourth-generation bioethanol uses algal biomass as feedstocks. The algae can also be genetically modified to increase its yield • The production of bioethanol from feedstocks involves three main steps: pretreatment, hydrolysis, and fermentation
Strategies for using ethanol in CI engines	• Fumigation: It involves injection of ethanol in the intake manifold of the engine using an injector, vaporizer, or carburettor • 'Splash blending' by adding cosolvents: Ethanol has limited miscibility in diesel and blends tend to undergo phase separation with time. Cosolvents like higher alcohols and THF can be added to the blends to ensure stability. The prepared fuel blend can be injected into the combustion chamber by modifying injection strategies • Emulsification: Surfactants can be used as emulsifiers to produce stable microemulsions of ethanol-in-diesel (water-in-oil). Emulsification requires gradual heating and mechanical stirring. Biodiesel, *SPAN80*, *AEP-102* etc. are some commercially available emulsifiers that can be used • Dual-fuel mode: Two approaches can be followed to use ethanol in dual-fuel mode: PFI of ethanol with DI of diesel and DI of both ethanol and diesel. Both these strategies establish a reactivity gradient in the combustion chamber, which gives more control over the combustion phasing • Ethanol shows higher ignition delay and has more tendency to cause knocking than conventional diesel. This limits the substitution of diesel energy shared by ethanol. However, the compression ratio of engines can be increased by altering the geometry of the piston bowl to power CI engines completely with ethanol. Concepts like surface ignition can also be used to initiate the start of combustion in ethanol-fuelled engines

(continued)

(continued)

Performance and emission benefits of substituting diesel fuel by ethanol	• Ethanol is an oxygenated fuel, which promotes more complete combustion and reduces the emission of PM • Ethanol has a higher latent heat of vaporization than conventional diesel fuel. Hence, when ethanol is injected in the port or is fumigated, it creates a cooling effect in the combustion chamber, which reduces the emission of NO_x in the exhaust • So, it is evident that the use of ethanol can reduce soot and NO_x simultaneously in diesel engines • The thermal efficiency of the CI engines can be increased by using ethanol-blended diesel fuel
Safety aspects of ethanol	• Ethanol is a flammable and highly volatile liquid. Hence, it possesses significant fire hazard. Ethanol-fuelled vehicles should be equipped with fire extinguishers containing Alcohol-Resistant Foam. All ignition sources should be kept at a distance. Sufficient ventilation should be provided to fuel storage areas to prevent the accumulation of fuel vapours

Upon realizing the benefits of using ethanol–diesel blends, recently, the Government of India has started to explore an initiative of blending 5% ethanol with diesel (Source: https://www.business-standard.com/industry/news/research-on-15-ethanol-blending-in-diesel-in-advanced-stages-gadkari-124101400879_1.html). Thus, the concept of using ethanol as a CI engine fuel can be regarded as a significant step towards a sustainable future and robust economy.

References

Aatmanirbhar F, Dream B (nd) Ethanol growth story. https://mopng.gov.in/files/Whatsnew/ethenol-groth-story.pdf

Abdullah B, Syed Muhammad SAF ad, Shokravi Z, Ismail S, Kassim KA, Mahmood AN, Aziz MMA (2019) Fourth generation biofuel: a review on risks and mitigation strategies. Renew Sust Energy Rev 107:37–50. https://doi.org/10.1016/j.rser.2019.02.018

Abel J, Virtanen S (2015) Corrosion of martensitic stainless steel in ethanol-containing gasoline: Influence of contamination by chloride, H_2O and acetic acid. Corros Sci 98:318–326. https://doi.org/10.1016/j.corsci.2015.05.027

Abu-Qudais M, Haddad O, Qudaisat M (nd) The effect of alcohol fumigation on diesel engine performance and emissions. www.elsevier.com/locate/enconman

Alvira P, Tomás-Pejó E, Ballesteros M, Negro MJ (2010) Pretreatment technologies for an efficient bioethanol production process based on enzymatic hydrolysis: A review. Biores Technol 101(13):4851–4861. https://doi.org/10.1016/j.biortech.2009.11.093

Ashokkumar V, Venkatkarthick R, Jayashree S, Chuetor S, Dharmaraj S, Kumar G, Chen WH, Ngamcharussrivichai C (2022) Recent advances in lignocellulosic biomass for biofuels and value-added bioproducts—a critical review. Bioresource Tech 344. https://doi.org/10.1016/j.biortech.2021.126195

Boggavarapu P, Ravikrishna R, v. (2019) Evaporating spray characteristics of a diesel-ethanol micro-emulsion. Fuel 246:104–107. https://doi.org/10.1016/j.fuel.2019.02.035

Boretti A (2012) Advantages of converting diesel engines to run as dual fuel ethanol-diesel. Appl Therm Eng 47:1–9. https://doi.org/10.1016/j.applthermaleng.2012.04.037

Britto RF, Martins CA (2014) Experimental analysis of a diesel engine operating in diesel-ethanol dual-fuel mode. Fuel 134:140–150. https://doi.org/10.1016/j.fuel.2014.05.010
Britto RF, Martins CA (2015) Emission analysis of a diesel engine operating in diesel-ethanol dual-fuel mode. Fuel 148:191–201. https://doi.org/10.1016/j.fuel.2015.01.008
Broukhiyan EMH, Lestz SS (1981) Fumigation of alcohol in a light duty automotive diesel engine
Bušić A, Mardetko N, Kundas S, Morzak G, Belskaya H, Šantek MI, Komes D, Novak S, Šantek B (2018) Bioethanol production from renewable raw materials and its separation and purification: a review. In: Food technology and biotechnology, vol 56, Issue 3, pp 289–311. University of Zagreb. https://doi.org/10.17113/ftb.56.03.18.5546
Ceballos R (2017) Bioethanol and natural resources: substrates, chemistry and engineered systems. Bioethanol Nat Res: Substrates, Chem Eng Syst https://doi.org/10.1201/9781315154299
Daroch M, Geng S, Wang G (2013) Recent advances in liquid biofuel production from algal feedstocks. Appl Energy 102:1371–1381. https://doi.org/10.1016/j.apenergy.2012.07.031
Devi A, Bajar S, Sihag P, Sheikh ZUD, Singh A, Kaur J, Bishnoi NR, Pant D (2023) A panoramic view of the technological landscape for bioethanol production from various generations of feedstocks. Bioengineered 14(1):81–112. https://doi.org/10.1080/21655979.2022.2095702
Dr. Murry Tamers (2006) Distinguishing Between 'Bio-Ethanol' and Petroleum Ethanol. Ethanol Producer Magazine
Edeh I (2020) Bioethanol production: an overview. In: Inambao F (ed) Bioethanol technologies. IntechOpen. https://doi.org/10.5772/intechopen.94895
Fırat M, Altun Ş, Okcu M, Varol Y (2022) Comparison of ethanol/diesel fuel dual direct injection (DI2) strategy with reactivity controlled compression ignition (RCCI) in a diesel research engine. Energy, 255. https://doi.org/10.1016/j.energy.2022.124556
Ghadikolaei MA, Cheung CS, Yung KF (2019) Study of combustion, performance and emissions of a diesel engine fueled with ternary fuel in blended and fumigation modes. Fuel 235:288–300. https://doi.org/10.1016/j.fuel.2018.07.089
Gnanamoorthi V, Devaradjane G (2015) Effect of compression ratio on the performance, combustion and emission of di diesel engine fueled with ethanol e Diesel blend. J Energy Inst 88(1):19–26. https://doi.org/10.1016/j.joei.2014.06.001
Goldemberg J, Guardabassi P (2010) The potential for first-generation ethanol production from sugarcane. Biofuels Bioprod Bioref 4:17–24
Hajba L, Nagy E, Hancsók J, Hajba L, Eller Z, Nagy E, Hancsok J (2011) Properties of diesel-alcohol blends. Hungarian J Ind Chem Veszprém 39:3. https://www.researchgate.net/publication/281304569
Handbook of Biofuels (nd)
Hardenberg H, Ehnert E (1981) Ignition quality determination problems with alternative fuels for compression ignition engines. SAE Tech Paper 811212. https://doi.org/10.4271/811212
He BQ, Shuai SJ, Wang JX, He H (2003) The effect of ethanol blended diesel fuels on emissions from a diesel engine. Atmos Environ 37(35):4965–4971. https://doi.org/10.1016/j.atmosenv.2003.08.029
https://energytracker.asia/pros-and-cons-of-ethanol/ Article: Eric Koons (2023, February 16) The pros and cons of ethanol: does it have a place in the future?
Huang J, Wang Y, Li S, Roskilly AP, Yu H, Li H (2009) Experimental investigation on the performance and emissions of a diesel engine fuelled with ethanol-diesel blends. Appl Therm Eng 29(11–12):2484–2490. https://doi.org/10.1016/j.applthermaleng.2008.12.016
Jambo SA, Abdulla R, Mohd Azhar SH, Marbawi H, Gansau JA, Ravindra P (2016) A review on third generation bioethanol feedstock. Renew Sustainable Energy Rev 65:756–769. https://doi.org/10.1016/j.rser.2016.07.064
Karunanithy C, Muthukumarappan K (2010) Effect of extruder parameters and moisture content of switchgrass, prairie cord grass on sugar recovery from enzymatic hydrolysis. Appl Biochem Biotechnol 162(6):1785–1803. https://doi.org/10.1007/s12010-010-8959-3. Epub 2010 Apr 27 PMID: 20419510

Kwanchareon P, Luengnaruemitchai A, Jai-In S (2007) Solubility of a diesel-biodiesel-ethanol blend, its fuel properties, and its emission characteristics from diesel engine. Fuel 86(7–8):1053–1061. https://doi.org/10.1016/j.fuel.2006.09.034

Larsen U, Johansen T, Schramm J (2009) EFP06 IEA implementing agreement on advanced motor fuels ethanol as a fuel for road transportation main report

India's Ethanol Push: A Path to Energy Security (2024) https://pib.gov.in/PressReleaseIframePage.aspx?PRID=2050907

Lee J, Lee S, Lee S (2018) Experimental investigation on the performance and emissions characteristics of ethanol/diesel dual-fuel combustion. Fuel 220:72–79. https://doi.org/10.1016/j.fuel.2018.02.002

Lee TH, Hansen AC, Li G, Lee T (2019) Effects of isopropanol-butanol-ethanol and diesel fuel blends on combustion characteristics in a constant volume chamber. Fuel 254. https://doi.org/10.1016/j.fuel.2019.06.021

Lei J, Shen L, Bi Y, Chen H (2012) A novel emulsifier for ethanol-diesel blends and its effect on performance and emissions of diesel engine. Fuel 93:305–311. https://doi.org/10.1016/j.fuel.2011.06.013

Liang J, Zhang Q, Chen Z, Zheng Z, Yang C, Ma Q (2021) The combustion and emission characteristics of diesel-ethanol blend with THF as cosolvents in a diesel engine operating with EGR. Fuel, 298. https://doi.org/10.1016/j.fuel.2021.120843

Linder J (nd) Alcoholate corrosion of aluminium in ethanol blends effects of water content, surface treatments, temperature, time and pressure

Liu H, Hu B, Jin C (2016) Effects of different alcohol additives on the solubility of hydrous ethanol/diesel fuel blends. Fuel 184:440–448. https://doi.org/10.1016/j.fuel.2016.07.037

Liu H, Wang X, Wu Y, Zhang X, Jin C, Zheng Z (2019) Effect of diesel/PODE/ethanol blends on combustion and emissions of a heavy-duty diesel engine. Fuel, 257. https://doi.org/10.1016/j.fuel.2019.116064

Morsy MH (2015) Assessment of a direct injection diesel engine fumigated with ethanol/water mixtures. Energy Convers Manage 94:406–414. https://doi.org/10.1016/j.enconman.2015.01.086

Muthuraman VS, Patel A, Shreya V, Vaidyanathan A, Reshwanth KNGL, Karthick C, Jacqueline PJ, Jan Gęca M, Ashok B, Sivagami K, Nanthagopal K (2022) Progress on compatibility issues of alcohols on automotive materials: Kinetics, challenges and future prospects- a comprehensive review. In: Process safety and environmental protection, vol 16. Institution of Chemical Engineers, pp 463–493. https://doi.org/10.1016/j.psep.2022.04.022

Nageswara Rao D, Sreenivasulu Reddy A (2023) Effects on the performance and emission characteristics of CRDI diesel engine fueled with ethanol, acid oil methyl ester biodiesel and diesel blends. Mater Today: Proc https://doi.org/10.1016/j.matpr.2023.06.065

Nour M, Attia AMA, Nada SA (2019) Improvement of CI engine combustion and performance running on ternary blends of higher alcohol (Pentanol and Octanol)/hydrous ethanol/diesel. Fuel 251:10–22. https://doi.org/10.1016/j.fuel.2019.04.026

Park SH, Youn IM, Lee CS (2011) Influence of ethanol blends on the combustion performance and exhaust emission characteristics of a four-cylinder diesel engine at various engine loads and injection timings. Fuel 90(2):748–755. https://doi.org/10.1016/j.fuel.2010.08.029

Patil AR, Desai AD (2019) Optimization of application of 2-ethyl-hexyl-nitrate on partial substitution of ethanol in CI engine for fuel economy and emission control using MADM method. SN Appl Sci 1(2). https://doi.org/10.1007/s42452-019-0163-7

Pedrozo VB, May I, Dalla Nora M, Cairns A, Zhao H (2016) Experimental analysis of ethanol dual-fuel combustion in a heavy-duty diesel engine: an optimisation at low load. Appl Energy 165:166–182. https://doi.org/10.1016/j.apenergy.2015.12.052

Pedrozo VB, May I, Guan W, Zhao H (2018) High efficiency ethanol-diesel dual-fuel combustion: a comparison against conventional diesel combustion from low to full engine load. Fuel 230:440–451. https://doi.org/10.1016/j.fuel.2018.05.034

Rakopoulos DC, Rakopoulos CD, Kakaras EC, Giakoumis EG (2008) Effects of ethanol-diesel fuel blends on the performance and exhaust emissions of heavy-duty DI diesel engine. Energy Convers Manage 49(11):3155–3162. https://doi.org/10.1016/j.enconman.2008.05.023

Reyes Y, Aranda DAG, Santander LAM, Cavado A, Belchior CRP (2009) Action principles of cosolvent additives in ethanol—diesel blends: stability studies. Energy Fuels 23(5):2731–2735. https://doi.org/10.1021/ef8010492

Ribeiro N, Pinto AC, Quintella CM, da Rocha GO, Teixeira LSG, Guarieiro LLN, Rangel M do C, Veloso MCC, Rezende MJC, da Cruz RS, de Oliveira AM, Torres EA, de Andrade JB (2007) The role of additives for diesel and diesel blended (ethanol or biodiesel) fuels: a review. Energy and Fuels 21(4):2433–2445. https://doi.org/10.1021/ef070060r

Safety and Procedural Requirements for type approval of Pure Ethanol, Flex-fuel & Ethanol-Gasoline blend vehicles, Automotive Industry Standard, 22nd December 2020

Şahin Z, Durgun O, Kurt M (2015) Experimental investigation of improving diesel combustion and engine performance by ethanol fumigation-heat release and flammability analysis. Energy Convers Manage 89:175–187. https://doi.org/10.1016/j.enconman.2014.09.053

Sarjovaara T, Alantie J, Larmi M (2013) Ethanol dual-fuel combustion concept on heavy duty engine. Energy 63:76–85. https://doi.org/10.1016/j.energy.2013.10.053

Savage LD, White RA, Cole S, Pritchett G (1986) Extended performance of alcohol fumigation in diesel engines through different multipoint alcohol injection timing cycles. Fuels Lubricants 95. https://www.jstor.org/stable/44718175

Shokravi H, Shokravi Z, Heidarrezaei M, Ong HC, Rahimian Koloor SS, Petrů M, Lau WJ, Ismail AF (2021) Fourth generation biofuel from genetically modified algal biomass: Challenges and future directions. Chemosphere 285. https://doi.org/10.1016/j.chemosphere.2021.131535

Sindhu R, Binod P, Pandey A, Ankaram SR, Duan Y, Awasthi MK (2019) Biofuel production from biomass. Current Dev Biotech Bioeng

Singh IR, Pulikkal AK (2022) Preparation, stability and biological activity of essential oil-based nano emulsions: a comprehensive review. OpenNano 8. https://doi.org/10.1016/j.onano.2022.100066

Srikanth HV, GS, Manne B, Kumar SB (2021) Niger seed oil biodiesel as an emulsifier in diesel-ethanol blends for compression ignition engines. Renew Energy 163:1467–1478. https://doi.org/10.1016/j.renene.2020.07.010

Tomás-Pejó E, Alvira P, Ballesteros M, Negro MJ (2011) Chapter 7—Pretreatment technologies for lignocellulose-to-bioethanol conversion. In: Pandey A, Larroche C, Ricke SC, Dussap C-G, Gnansounou E (eds) Biofuels. Academic Press, pp 149–176. https://doi.org/10.1016/B978-0-12-385099-7.00007-3

Tsang KS, Zhang ZH, Cheung CS, Chan TL (2010) Reducing emissions of a diesel engine using fumigation ethanol and a diesel oxidation catalyst. Energy Fuels 24(11):6156–6165. https://doi.org/10.1021/ef100899z

Van Gerpen JH, van Meter DB (1990) Emission control in diesel engine by alcohol fumigation midwest transportation center

Valera H, Agarwal AK (2019) Methanol as an alternative fuel for diesel engines. In: Energy, environment, and sustainability. Springer Nature, pp 9–33. https://doi.org/10.1007/978-981-13-328 7-6_2

Waterland LR, Venkatesh S, Unnasch S (2003) Safety and performance assessment of ethanol/diesel blends (E-Diesel). http://www.osti.gov/bridge

Wu Y, Zhang X, Zhang Z, Wang X, Geng Z, Jin C, Liu H, Yao M (2020). Effects of diesel-ethanol-THF blend fuel on the performance and exhaust emissions on a heavy-duty diesel engine. Fuel, 271. https://doi.org/10.1016/j.fuel.2020.117633

Yahuza I, Dandakouta H (2015) A performance review of ethanol-diesel blended fuel samples in compression-ignition engine. J Chem Eng Proc Tech 06(05). https://doi.org/10.4172/2157-7048.1000256

Zabed H, Sahu JN, Suely A, Boyce AN, Faruq G (2017) Bioethanol production from renewable sources: Current perspectives and technological progress. Renew Sustainable Energy Rev 71:475–501. https://doi.org/10.1016/j.rser.2016.12.076

Zhang ZH, Tsang KS, Cheung CS, Chan TL, Yao CD (2011) Effect of fumigation methanol and ethanol on the gaseous and particulate emissions of a direct-injection diesel engine. Atmos Environ 45(11):2001–2008. https://doi.org/10.1016/j.atmosenv.2010.12.019

MIX
Papier aus verantwortungsvollen Quellen
Paper from responsible sources
FSC® C105338

If you have any concerns about our products,
you can contact us on
ProductSafety@springernature.com

In case Publisher is established outside the EU,
the EU authorized representative is:
Springer Nature Customer Service Center GmbH
Europaplatz 3, 69115 Heidelberg, Germany

Printed by Libri Plureos GmbH
in Hamburg, Germany